HANDBOOK OF
GRANULAR MATERIALS

HANDBOOK OF GRANULAR MATERIALS

edited by

Scott V. Franklin
Mark D. Shattuck

CRC Press
Taylor & Francis Group
Boca Raton London New York

CRC Press is an imprint of the
Taylor & Francis Group, an **informa** business

CRC Press
Taylor & Francis Group
6000 Broken Sound Parkway NW, Suite 300
Boca Raton, FL 33487-2742

First issued in paperback 2018

ISBN 13: 978-1-138-89420-4 (pbk)
ISBN 13: 978-1-4665-0996-2 (hbk)

Library of Congress Cataloging-in-Publication Data

Handbook of granular materials / edited by Scott V. Franklin and Mark D. Shattuck.
    pages cm
  Includes bibliographical references and index.
  ISBN 978-1-4665-0996-2 (alk. paper)
  1. Granular materials--Handbooks, manuals, etc. I. Franklin, Scott V., editor. II. Shattuck, Mark D., 1965- editor.

TA418.78.H36 2015
620'.43--dc23
                                     2015016370

Visit the Taylor & Francis Web site at
http://www.taylorandfrancis.com

and the CRC Press Web site at
http://www.crcpress.com

SVF dedicates this book to his wife, Merrie, and sons Maxwell, Gabriel, and Harry, whose tolerance and support were instrumental to completing this work.

MDS dedicates this book to his children Sarah and Eli and his wife Dianna, and thanks them for their patience and support during the entire process.

Contents

Preface

This book was assembled for many reasons. The explosion of research over the last two decades has greatly advanced our understanding of granular materials, to the point where some broad themes can now be stated with some confidence. Additionally, the experimental, theoretical, and computational methods of inquiry have all seen significant development, both individually and in relation to one another. This book attempts to tie these developments together, providing guidance on how to conduct research in granular materials and also promising directions for new research.

The book is organized into three sections. Chapters 2 through 5 cover the various methods that contemporary researchers use to investigate granular materials. Chapters 6 through 10 delve into broader themes of investigation, focusing on results, not methodology. Finally, Chapters 11 through 13 describe three systems: suspensions, emulsions and foams, and colloids that can be considered as extensions of granular media. Many of the same approaches are used in these systems, although the microscopic nature often requires innovative experimental techniques.

The methods section begins with chapters on computational and experimental methods and techniques. While no review can be comprehensive, these chapters aim to provide the reader with an understanding sufficient to serve as the foundation for future study. Theoretical approaches are varied, and so we have chosen two kinetic theory and statistical approaches that are both broad enough to be applicable in a number of different situations and make specific predictions and advances so as to be widely relevant.

Recent research has been rather arbitrarily divided into static/quasi-static and dynamic effects. Chapter 6 describes computational research on static packings, while Chapter 7 describes the mechanical response of experimental packings to very small disturbances. The use of photoelastic material to directly visualize chains of force throughout the granular material has contributed much to our understanding of the response, and so Chapter 7 includes a description of this technique. The remaining chapters deal with dynamic features. Chapter 8 describes granular shear in simple systems, followed by a description of an elegant set of experiments and theory on granular avalanches, a system of critical relevance to natural phenomena such as avalanches, landslides, and earthquakes. Finally, the tendency of mixtures of two (or more) different materials to segregate under shear flows, covering both shaken and rotating systems, is discussed in Chapter 9.

The final section contains chapters that are extensions of granular systems: suspensions, emulsions and foams, and colloids. Each of these deals with various limiting cases of granular systems, for example, the absence of friction or the importance of Brownian motion. While the length and time scales of the phenomena may vary widely, the picture that emerges is complementary to that of canonical granular systems.

A beginning researcher might start, therefore, by choosing three appropriate chapters: a method, a theme, and an extension. In this way, the student learns the how, what, and why of their project. This should not be seen, of course, as implying that the remaining chapters would not in themselves be valuable, and both the editors of this book have learned a great deal in the process of reading the chapters.

Acknowledgments

The editors personally thank each of the authors who have contributed to this book. Their dedication and care come through clearly in every chapter, and the responsibility for any errors, typos, or other confusions rests solely on us. We also thank Taylor & Francis Group for suggesting this project and for their patience in its assembly.

I (Franklin) acknowledge support from the American Chemical Society's Petroleum Research Fund (PRF #51438-UR10) and the National Science Foundation (CBET #1133722). I also thank my coeditor, Mark Shattuck, whose knowledge and vision were invaluable.

I (Shattuck) acknowledge support from the National Science Foundation (DMR PREM Grant No. DMR0934206 and CBET-0968013) and from the Kavli Institute for Theoretical Physics through the National Science Foundation under Grant No. NSF PHY11-25915. I thank Scott Franklin for his enthusiasm and guidance.

Editors

Scott V. Franklin earned his bachelor's degree from the University of Chicago, Chicago, Illinois, in 1991 and his PhD from The University of Texas in Austin, Austin, Texas, in 1997. In 2000, after a two-year National Science Foundation postdoctoral fellowship in physics education research, he joined the faculty at Rochester Institute of Technology. He currently maintains a lab that focuses on experimental and computational investigations of granular and other complex materials. Recent interests of the lab include rodlike and other materials that, due solely to particle shape, can maintain a solid-like rigidity.

Mark D. Shattuck earned his bachelor's degree and MS in 1987 and 1989 from Wake Forest University, Winston-Salem, North Carolina, and his PhD from Duke University, Durham, North Carolina, in 1995. He held postdoctoral fellowships in medical physics at Duke University and in granular physics at the University of Texas before joining the faculty of The City College of New York in the Benjamin Levich Institute in 2000, where he performs experimental and computational research in soft condensed matter and granular materials. He has 15 years of experience and more than 25 publications in studies of particulate systems, and has developed a number of novel imaging and production techniques for particulate systems. He is an editor for the research journal *Granular Matter* and served on the Publication Oversight Committee of the American Physical Society for five years (with two as chair). He is a founding organizer of the annual regional meeting Northeastern Granular Materials Workshop.

Contributors

Robert P. Behringer
Department of Physics
Center for Nonlinear and Complex
  Systems
Duke University
Durham, North Carolina

Bulbul Chakraborty
Martin A. Fisher School of Physics
Brandeis University
Waltham, Massachusetts

Karin A. Dahmen
Department of Physics
University of Illinois at Urbana
  Champaign
Urbana, Illinois

Scott V. Franklin
Department of Physics
Rochester Institute of Technology
Rochester, New York

Piotr Habdas
Department of Physics
Saint Joseph's University
Philadelphia, Pennsylvania

Kimberly M. Hill
Department of Civil, Environmental
  and Geological Engineering
University of Minnesota Twin Cities
Minneapolis, Minnesota

James T. Jenkins
School of Civil and Environmental
  Engineering
Cornell University
Ithaca, New York

Arshad Kudrolli
Department of Physics
Clark University
Worcester, Massachusetts

Jeffrey F. Morris
Benjamin Levich Institute
and
Department of Chemical Engineering
The City College of New York
New York, New York

Corey S. O'Hern
Department of Mechanical
  Engineering and Materials Science
and
Department of Applied Physics
and
Department of Physics
Yale University
New Haven, Connecticut

Ashish V. Orpe
Chemical Engineering Division
National Chemical Laboratory
Pune, India

Andreea Panaitescu
Department of Physics
Clark University
Worcester, Massachusetts

Mark D. Shattuck
Benjamin Levich Institute
and
Department of Physics
The City College of New York
New York, New York

Brian P. Tighe
Process and Energy Laboratory
Delft University of Technology
Delft, the Netherlands

CHAPTER 1

Introduction

Scott V. Franklin and Mark D. Shattuck

CONTENTS

This introduction discusses historical topics of personal interest. The topics chosen are not in any way comprehensive, nor motivated by anything other than models, theories, and experiments that I found particularly enlightening. The reader is enthusiastically referred to the many existing articles and books that also give introductions to granular materials, particularly in areas with which I am less familiar (e.g., engineering, for which readers are referred to the extensive literature, e.g., [15,25]). The personal retrospective gives an emphasis to simple statistical explanations, an aesthetic property that has proven exceptionally useful for my own study of granular materials. The section concludes with a review of my personal area of research: geometrically cohesive materials.

1.1   Statistical Models

1.1.1   Avalanches, Angles of Repose, and Self-Organized Criticality

In 1987, Bak et al. [3] proposed the idea of self-organized criticality (SOC) to explain the near ubiquity of noise with a $1/f$ power spectrum, particularly in systems where such noise might originate from mechanisms with spatial extents. The power-law scaling implies the absence of a fundamental length scale, and so the challenge was to devise a system that evolved to a critical state without such a length. For their system, they chose a simplified model of granular avalanching, one whose analogs have been the focus of subsequent research to this day [28,36,42]. One attractive feature of the model is its computational simplicity; a reader with basic programming skills can likely reproduce the fundamental result in only a few hours.

The model begins with a surface profile on a discretized lattice $z(i,j)$, with z representing the height of the granular pile. A site is chosen at random and the height increased by 1, representing the addition of a single grain to the pile. Rules then redistribute the material when the height exceeds a critical value. In the original SOC paper, if the absolute height exceeded a critical value (K), then an avalanche delivered 1 grain to each of four neighbors (assuming a square lattice indexed by i and j):

$$z(i,j) > K \implies z(i,j) \rightarrow z(i,j) - 4 \tag{1.1}$$

$$z(i \pm 1, j) \rightarrow z(i \pm 1, j) + 1 \tag{1.2}$$

$$z(i, j \pm 1) \rightarrow z(i, j + 1) + 1 \tag{1.3}$$

Subsequent models [8] considered the difference in height between a point and its neighbors, with material flowing "downhill" when the height at any point exceeds that of its neighbors by a critical amount. While the specific nature of the redistribution is not important, the key point is that the transfer of material from one point to another can induce subsequent transfers, representing a broad, spatially extended event. Events can be quantified in a number of ways: by the total number of grains that moved from one position to another, by the number of sites involved in a transfer, or by the spatial geometry of the involved sites. Bak et al. found that, for a variety of system sizes in two and three dimensions, the distribution of cluster sizes is a power-law with exponent −1, indicating the absence of a characteristic event size. Because this occurred for a variety of initial conditions and system dimensions and sizes, Bak et al. asserted it was a fundamental characteristic of the system that was "self-organized" to the critical state.

Following almost immediately were two experiments of Jaeger et al. [22] on the angle of repose of a granular pile. The first reproduced the model of Bak et al., adding particles at random to a box of sand with one open side. The second consisted of a thin drum, partially filled with material, that rotated slowly. Jaeger et al. measured the capacitance across the system, which depends on the amount of (dielectric) material in that volume and is sensitive enough to capture changes involving even just a few grains (glass beads of diameter .07–.5 mm). This method was subsequently used in a number of experiments that measured the evolution of packing fraction of both spheres [24,33] and rods [48] in response to tapping.

Contrary to the prediction of self-organized criticality, Jaeger et al. observed the distribution of event size to be independent of frequency, which could be explained as the superposition of spatially extended events that occupy a very narrow range of time scales. Jaeger et al. then considered whether the discrepancy could be explained by the non-unique nature of the dynamic angle of repose. Bak et al. had assumed that avalanches begin as soon as the local angle exceeded a single value θ_r. Jaeger et al. found instead that the angle at which flow started typically exceeded θ_r by a small amount, which was related to the need for the particles to dilate slightly in order to flow. By adding energy to the pile through small vibrations (which dilated the pile and facilitated flow), Jaeger removed the excess angle from the experiment but still did not capture $1/f$ power-law scaling.

Despite the apparent quantitative failure of self-organized criticality, the model spurred a lasting interest in granular avalanching and the close study of event size distribution. Indeed, an entire chapter of this book—Chapter 9—deals with contemporary theories of avalanches.

1.1.2   Force Chains and the q Model

When forces are applied to a granular system, the discrete nature of particles, coupled with the random distribution in space, causes the force to be concentrated into "force chains," with significant load being borne by a relatively small number of particles, while many particles feel comparably small forces. A common technique to visualize this is the use of photoelasticity, where elastic deformations of a material rotate the polarization vector of light passing through the particle in both two (e.g., [17,21]) and three [19] dimensions. Viewed between two polarizers oriented 90° to one another, an unstressed particle will appear dark (since the second polarizer completely blocks the light) while the stressed particles appear bright. An example is shown in Figure 1.1, and more details are given in Chapters 3 and 7.

The mechanism by which forces can grow large is illustrated by the simple picture in Figure 1.2. Two grains support the weight of a single grain immediately above. Because the grains are offset by an angle θ, the force on each grain must be $F = W/(2\sin\theta)$ in order to balance the forces. As θ becomes small, this force becomes quite large. This magnification has significant practical consequences, as containers must be constructed to support significant lateral forces that arise.

An early attempt to explore the inhomogeneity of the force distribution was proposed by Liu et al. [19] and Coppersmith et al. [10], and termed the "q" model. While no longer the dominant paradigm for understanding force chains, its status as an early appeal to random walk statistics left a lasting mark on subsequent theories. In particular, it bears striking similarity to a later work on jamming of flow through conical hoppers in two (c.f. To et al. [45]) and three [53,54] dimensions and anisometric wedge hoppers [41].

An important result was to identify the power-law distribution of Bak's SOC model as a rare special case, with more physically realistic situations resulting in

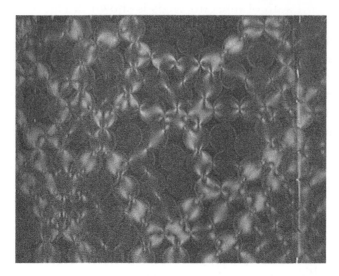

Figure 1.1   1 cm diameter photoelastic disks in a thin channel, compressed from the top. The disks are between crossed linear polarizers, and stress-induced birefringence results in the picture shown: the white streaks are force chains.

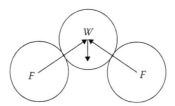

Figure 1.2   Simple illustration of two lower grains supporting a third, with weight W. Because the contact forces are at an angle with respect to gravity, they must be larger in order for their vertical components to support the upper grain's weight.

qualitatively different scalings. The model assumes a regular lattice of beads, with each row supporting the weight of the row immediately above it. The randomness of a real pile is captured by the rule that a particle distributes a random fraction of its load to each of its neighbors below. The fraction from the ith particle in a given layer to the jth particle in the layer below is labeled q_{ij}, and the total weight on the jth particle is therefore a sum over the i contacting particles from above. This leads to a recursive relation for the load on the jth particle in layer D:

$$w(D+1,j) = 1 + \sum_i q_{ij}(D)w(D,i).$$

The factor of 1 represents the weight of the particle itself, whereas the sum is over all the contacting neighbors from above, each supporting their own load $w(D,i)$ and distributing the fraction $q_{ij}(D)$ to the particle in question. The critical feature of the model is that the distributing fractions q_{ij} are *completely uncorrelated* from one another; in this way, the problem takes on the form of a random walk.

Coppersmith et al. identified two distinctly different results from this model. The simplest case, which they termed the critical $q_{0,1}$ limit, involved particles distributing either all or none of their load to a particular neighbor below. That is, the random variables q_{ij} were either 0 (no load transmitted) or 1 (all load transmitted). This is formally equivalent to a discrete random walk, with paths from particles above coalescing as they move downward. In this limit they did indeed find a power-law scaling of the weight distribution function with depth, with

$$Q_D(w) \propto D^{-a}.$$

When the random values q_{ij} were allowed to assume values other than 0 and 1, however, thus resembling a random walk with a distribution of step size, the results were qualitatively different. Allowing the distribution parameters to vary between 0 and 1 (while summing to 1 to conserve mass) was a much more physically realistic situation, more closely modeling the random packing of actual materials. In these cases, the distributions did not show a power-law scaling, decaying much more rapidly (usually exponentially). Coppersmith et al. presented both numerical and mean-field theory results to support this finding, and the agreement with direct simulations of sphere packings and experiments cast further doubt as to the applicability of self-organized criticality as the mechanism by which sandpiles form and collapse.

1.2   Flow through Hoppers

The flow of granular material through hoppers is also is well-captured by random walk-like models that, in their assertion of uncorrelated events, bear a striking similarity to that proposed by the q-model. Hopper and funnel flow have obvious industrial importance, and physicists and engineers have focused on both the flow rate through a hopper and the probability with which particles clog at the exit aperture. The latter question, intimately connected with the theory of jammed systems, presents an outstanding problem that involves two- and three-dimensional experiments and particulate and continuum theories. Clogging of particles at a hopper outlet affects a wide range of industrial processes that transport granular media through pipes, silos, and hoppers [7,12]. While early research focused on the steady-state flow rate [5], more recent attention has turned to the probability that flow stops, that is, the transition to the jammed state. Here I will describe several studies that each take a slightly different approach to understanding clogging in hoppers. An ultimate unifying theory is tantalizingly close.

1.2.1   Two-Dimensional Hoppers and Mean Field Statistical Theories

In 2001 and 2002, To et al. published a series of articles [44–46] that described a simple two-dimensional experiment and an elegant statistical theory that captured the clogging of disks flowing through a 2D hopper. The experiments allowed disks of uniform radius to flow through the hopper until a clog occurred and, when flow had stopped, took a picture of the arch that formed at the exit. They varied the size of the aperture (measured in particle diameters) and angle of the hopper side-walls. The critical question was a statistical description of the clogging arch and, if possible, an explanatory theory.

Experimentally To et al. measured an expected monotonic decrease in probability to clog as a function of aperture diameter and corresponding increase in the average number of particles that took part in a clogging arch. The clogging probability was simply the fraction of experiments that experienced a clog; given the finite system size (200 particles), it was possible for all the particles to flow through the aperture. Data from To's 2002 paper showing the probability approaching zero as the aperture width approaches five particle diameters are shown in Figure 1.3. The probability for the particles to clog drops sharply as the aperture grows larger than three particle diameters, and is practically zero (meaning the hopper empties completely) when the aperture is five diameters in size.

The two-dimensionality of the system allows an explicit enumeration of all the possible configurations of a jamming arch. In some ways, this presaged a similar undertaking of O'Hern et al. [16,51] to count all of the different configurations of N particles that occupy the same packing fraction and Franklin et al. [41] to investigate jamming in anisometric (wedge) hoppers. The experimental observations

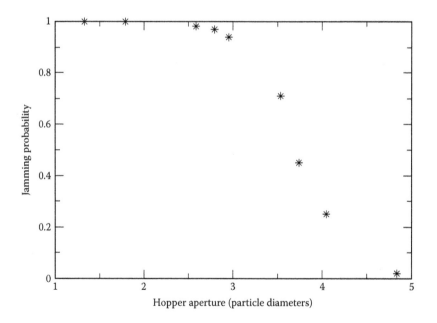

Figure 1.3   The experimental probability for a two-dimensional hopper containing 200 particles to clog as a function of hopper aperture (measured in particle diameters). Hoppers with apertures less than three particle diameters in size almost always clog; hoppers with apertures greater than four particle diameters almost never do. (Data adapted from To, K. et al., *Phys. Rev. Lett.*, 86, 71, 2001.)

can be reproduced by applying a *restricted random walk model* that assumes that a disk's location is uncorrelated with respect to those of its neighbors. For an arch to form, three constraints/conditions must be met:

- The angle θ_i between a point connecting the centers of mass of neighboring disks must be between $\pm\pi/2$. This requires the arch to continually progress across the space, i.e., disk "2" is always to the right of disk 1.

- Arches must be convex (down). That is, the angle between two disks must be smaller than the angle between the two previous disks.

- The horizontal span of the arch must exceed the aperture width (obvious).

Additionally, to enable the theoretical treatment the disks are assumed hard, so the distance between two disks in contact is the particle diameter. Given these constraints, the probability of n disks to span an arch of width x is given by

$$P_n(x) = \int_{-\pi/2}^{\pi/2} d\theta_2 \int_{-\pi/2}^{\theta_2} d\theta_3 \ldots \int_{-\pi/2}^{\theta_{n-1}} d\theta_n \, \delta\left(d - 1 - \sum_{i=2}^{n} \cos\theta_i \right) \tag{1.4}$$

From Equation 1.4, To calculated the probability that n particles would randomly configure to span a horizontal distance x, reproduced for arches of up to nine particles in length in Figure 1.4a. For a fixed aperture width D, the probability of jamming with an arch of n particles is just the cumulative sum of the individual probabilities of forming an arch with length greater than D, that is,

$$J_n(D) = \int_D^{\infty} P_n(x) \, dx. \tag{1.5}$$

In determining the total probability of a jam, regardless of arch geometry, To et al. again assumed that all possible arches were equally probable and considered the normalized sum over all arch sizes. To properly normalize the contribution from a particular geometry, they calculated the fraction of jamming $(x > D)$ arches with n particles compared to the total number of jamming arches:

$$g_d(n) \equiv \frac{J_n(D)}{\sum_n J_n(D)}.$$

The total jamming percentage is then

$$J(D) = \sum_n g_d(n) J_n(d).$$

This integrated sum is shown in Figure 1.4b. The close agreement between experiment and model gives strong support to the idea that particles in flow sample with equal probability the entire configuration space and that, at least in two dimensions, it might be possible to write down closed forms for the mechanically stable configurations.

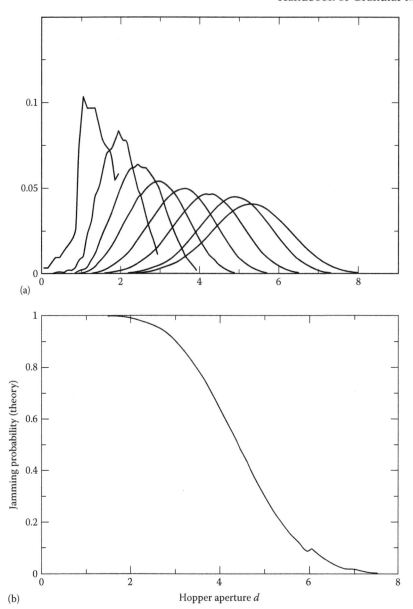

Figure 1.4   (a) Probability for an arch of N particles to span an aperture of a distance d for $N = 2,3,...9$. (b) Theoretical jamming probability as a function of aperture width calculated as in Equation 1.5. (Figures recreated after To, K. et al. *Phys. Rev. Lett.*, 86, 71, 2001.)

1.2.2   Three-Dimensional Hoppers: Cones and Wedges

Where To et al. studied the likelihood of a specific jamming configuration, Zuriguel et al. [53,54] investigated the amount of material that exited a hopper before the jam occurred. A conical hopper filled with spherical particles emptied onto a scale, which recorded the flow of material. When the drainage stopped, a puff of air broke apart the arch at the entrance and flow resumed. The event-size distribution was observed to have a broad exponential tail, which can be explained with the following simple uncorrelated statistics model. The probability of a particle (or group of particles) to flow through without jamming is modeled as p, and is independent of whether neighboring (or prior) particles have flowed. Thus the probability of exactly N particles (or groups) to exit before the system jammed is

$$P(N) = N^p(1-p)$$

the factor of $(1-p)$ indicating that the $N+1$th particle does, in fact jam. This can be rewritten as

$$P(N) = (1-p)\exp(N\log p)$$

and, since $p < 0 \Longrightarrow \log p < 0$, the large N limit goes as

$$P(N \gg 1) \sim \exp\left[-|\log p|N\right].$$

The argument and scaling are essentially unchanged if, instead of a single particle, p refers to the probability for a correlated group of particles to exit.

More recently, Janda et al. [23] connected the probability of forming an arch of specific size with the exit mass probability distribution function and also investigated how the mean flow $\langle N \rangle$ scaled with aperture diameter. A key finding was the absence of a critical aperture size above which the mean flow diverges. Rather, $\langle N \rangle$ grows with aperture diameter D as $\langle N \rangle \propto \exp[D^2]$.

Zuriguel's work focused on conical hoppers, with a circular aperture. More recently, Saraf et al. investigated the flow of round particles (acrylic spheres) through a wedge hopper with a rectangular aperture. In marked contrast with the data from conical hoppers, Saraf et al. found the probability for N particles to exit the hopper before jamming to have a broad power-law decay with exponent $\alpha = -2$, shown in Figure 1.5. After an initial plateau, the distribution falls off as a power-law with $P(N) \sim N^{-2}$.

Both the exponential tail seen in conical hoppers and the power-law distribution from wedge hoppers can be captured in a single model that postulates correlated string-like dynamics over a length scale comparable to the smallest length scale of the hopper. The probability for a correlated string of particles to exit p is now a function of the string orientation θ, as strings aligned along the length of the aperture have a high probability of passing through while those aligned across the width have a smaller probability. The experimentally observed distribution function is an average over the range of individual string exit probabilities.

Similar to models discussed previously, the strings are assumed to evenly sample all orientations and be uncorrelated with the orientation of previous (or subsequent) strings passing through the aperture. Calculating these averages,

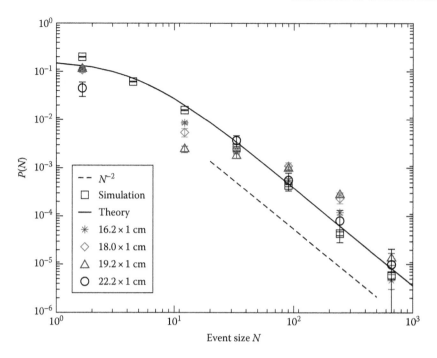

Figure 1.5 The probability for N particles to exit a wedge hopper has a broad power-law tail with $P(N) \sim N^{-2}$ (dashed line), seen in experiments, Monte Carlo simulation, and analytic theory. This is in distinct contrast to the distribution function found in conical hoppers, which decays exponentially. Shown are distribution functions for experimental hoppers of lengths $L = 16.2$–22.2 cm. Simulation and theory assume $n_c = 3$ adjacent, statistically independent cells. (From Saraf, S. and Franklin, S.V., *Physical Review E*, 83(3), 030301, March 2011.)

therefore, bears a striking resemblance to To's mean field model. The limits of the exit probability for strings parallel and perpendicular to the aperture are p_z and p_x and the uniform distribution of exit probability between $p_x < p < p_z$ takes the form

$$O(p) = \frac{\theta(p - p_x)\theta(p_z - p)}{p_z - p_x}. \tag{1.6}$$

$\theta(p-p_x)$ and $\theta(p_x-p)$ are Heaviside step functions, and $p_z-p_x \equiv \Delta p$ is a normalizing factor. The probability for exactly N strings to exit now requires the averaging over all allowable exit probabilities with the integral

$$\langle P(N) \rangle = \int_{p_x}^{p_z} p^N(1-p)O(p)dp. \tag{1.7}$$

For an isometric (round) aperture, the exit probability is independent of orientation: $p_z = p_x$. The orientational probability $O(p)$ in Equation 1.7 is functionally equivalent to a Dirac delta function at p_x, and the event size probability

$$\langle P(N) \rangle \approx \int p^N (1-p) \delta(p - p_x) \, dp = p_x^N (1 - p_x) \tag{1.8}$$

captures the exponential decay seen in conical hoppers.

The assumption that strings assume all orientations with equal likelihood allows one to calculate the indefinite integral of Equation 1.7

$$\int p^N (1-p) O(p) \, dp = \frac{1}{\Delta p} \left[\frac{p^{N+1}}{N+1} - \frac{p^{N+2}}{N+2} \right], \tag{1.9}$$

For $p < 1$, the result is a power-law, scaling as $1/N$, with an exponential cutoff (since the term in the numerator is $p^N = \exp[N \ln p]$). As $p \to 1$ the exponential cutoff occurs at larger and larger N, however, eventually disappearing when $p = 1$.

Physically, $p \to 1$ means that a string of particles has no probability of forming an arch and jamming at the hopper aperture, which implies that the aperture is longer than the string length. In this limit of $p = 1$, Equation 1.7 integrates to

$$P(N) = \frac{1}{1 - p_x} \left[\frac{1}{(N+1)(N+2)} + \frac{p_x^{N+1}}{N+1} - \frac{p_x^{N+2}}{N+2} \right] \tag{1.10}$$

and $P(N \to \infty) \propto 1/N^2$. This power-law scaling explains the experimental findings of Figure 1.5.

The probability distribution for several values of p_z is shown in Figure 1.6. The exponential cutoff is seen to occur for larger and larger values of N as p_z approaches 1. It is natural to assume that transition from exponential to power-law distribution occurs as the aperture's long length approaches the length scale over which correlated motion occurs [9].

Saraf et al. also showed how the model could be extended to still longer apertures to perhaps capture the spatially and temporally inhomogeneous flow through anisometric hoppers. The aperture is modeled as a series of n_c adjacent "cells," each of length equal to the granular string length. The probability for n_i particles to exit the ith cell is given by Equation 1.10 and, following this chapter's theme, the probabilities for different cells were assumed to be statistically independent. The average exit mass probability distribution is calculated by summing over all the different ways that N particles can exit n_c cells:

$$\langle P(N) \rangle = \sum_{n_1, n_2, \ldots = 0}^{N} \prod_{i=1}^{n_c} P(n_i) \delta \left(N - \sum_{i=1}^{n_c} n_i \right), \tag{1.11}$$

where the delta function forces the sum of the particles exiting the individual cells to total N: $n_1 + n_2 + \cdots = N$. Interested readers are referred to [41] for additional details and results.

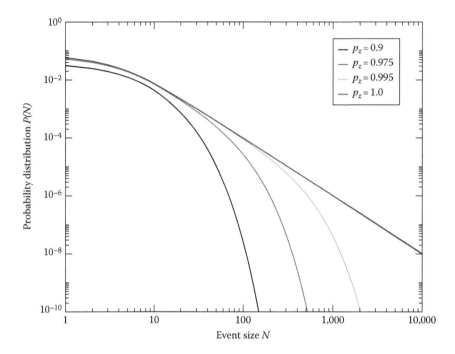

Figure 1.6 Evaluation of Equation 1.7 for values of p_z approaching 1. When $p_z < 1$ the curve has a power-law tail with an exponential cutoff. At $p_z = 1$ the exponential cutoff disappears leaving behind only the power-law tail. (From Saraf, S. and Franklin, S.V., *Physical Review E*, 83(3), 030301, March 2011.)

1.3   Irregularly Shaped Granular Materials

I conclude the section with a review of work on radically aspherical materials, primarily long, thin rods, and U-shaped staples. I will focus on statistics of the packings, which can demonstrate correlations that violate the mean-field assumptions of the previous sections, and the peculiar rigidity that particle entanglement can cause.

1.3.1   Packing of Long, Thin Rods

The packing of long, thin rods of large aspect ratio ($\alpha \equiv L/D > 10$) can result in piles that are both remarkably stable and yet form very low packing fractions. Philipse [37,38] was the first to study the solid "plug like" behavior that rods with aspect ratio above $\approx$35 displayed when poured from a container and gave a simple geometric argument, the random contact model (RCM), to explain the very low packing fractions. The model rests on a mean field assumption that particle orientations are uncorrelated and sample equally all possible orientations and is thus thematically consistent with the models discussed earlier in this chapter. Coupled with this is the idea of *excluded volume* the volume in space forbidden to another particle due to the existence of another. Onsager [35] had in fact calculated this excluded volume for spherocylinders (cylinders of length L and diameter D with two hemispherical end caps) in his study of phase transitions in colloidal suspensions, finding it to be

$$V_{ex} = \frac{\pi}{2}L^2 D + 2\pi D^2 L + \frac{4}{3}\pi D^3.$$

If orientations are uncorrelated then each particle can be assumed to occupy a volume equal to its excluded volume and the pile number density is inversely proportional to the excluded volume and proportional to the average number of contacts per particle $\langle c \rangle$:

$$N = \frac{2\langle c \rangle}{V_{ex}},$$

where the factor of 2 occurs because each contact involves two particles. The packing fraction is the product of the number density and particle volume ($V_{part} = (\pi/6)D^3 + (\pi/4)D^2 L$) and, for large aspect ratios, does indeed scale inversely with aspect ratio, confirming the model's assumptions. Subsequent experiments by Desmond and Franklin [11] found it necessary to keep the higher order terms in the excluded volume fraction for aspect ratios from 5 to 50 in a variety of cylinder packings, but the model otherwise explained 3D packings quite well.

The mean-field assumption of random contacts breaks down, however, for 2D piles formed under gravitational collapse, as most interactions between particles bring about alignment and there is not sufficient entanglement to prevent this from occurring. This was demonstrated in experiments by Stokely et al. [43], who formed 2D piles under gravity and compared both the packing fraction and orientational correlations with the mean-field predictions.

The 2D excluded area is, to first order, $A_{excl} = (2/\pi)L^2$ [4] and the predicted number density as a function of aspect ratio α therefore

$$N(L) = \frac{2\langle c \rangle}{(2L^2/\pi)} = CL^{-2}.$$

Figure 1.7 shows that the number density does indeed fall off as L^{-2} for large L, but the prefactor is significantly larger than that predicted by the RCM. Piles are $\approx 33\%$ more dense than predicted, with particles occupying $\approx 33\%$ less area than in an isotropic distribution.

The increased packing results from particle alignment, which can be seen in correlations between particle orientations. Figure 1.8 shows the angular orientation correlation function

$$\tilde{Q}(r) = \langle \cos(2\Delta\theta_{ij}) \rangle, \tag{1.12}$$

with $\Delta\theta_{ij}$ the difference in angle between particles i and j and the average over all particles whose centers of mass separation is between r and $r + \delta r$. Two particles whose centers of mass are quite close must be aligned and so $\tilde{Q}(r \to 0) \to 1$. Once particle centers of mass are separated by more than one particle length L they can in principle assume any relative orientation and so $\tilde{Q}(r \to \infty) \to 0$.

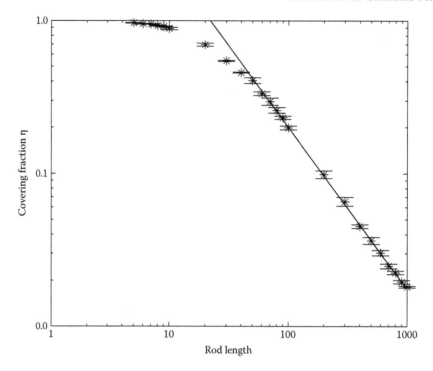

Figure 1.7   Log-log plot of packing fraction ϕ as a function of particle aspect ratio α. Data are grouped by constant dimensionless container diameter $\tilde{D} \equiv D/L$. With the exception of a few points at low $\tilde{D}$ (where significant particle alignment occurs), the data are well fit by a mean-field theory with only one free parameter.

Figure 1.8 shows the $\tilde{Q}(r)$ distribution that results from averaging Equation 1.12 over all allowable angles, assuming all are sampled equally, as well as that obtained from Monte Carlo simulations and experimental piles. The theoretical $Q(r)$ begins at 1 and rapidly decays to zero as $r \to 1$, as it must. Both the experimental and simulated piles, however, show significantly greater correlation between neighboring particles, seen in the divergence from the analytic line for $r/L > 0.5$ (between the dashed lines), and reach a nonzero asymptote once particles are separated by more than two particle lengths. This nonzero asymptote indicates a long-range ordering, with particles preferring to align with the container walls. This demonstrates the deviation from uncorrelated behavior assumed in most of the models discussed in this section and explains the enhanced packing fraction observed in experiment.

1.3.2   Geometric Cohesion in Granular Materials

Radically aspherical particles may display a cohesion brought about by entanglement. Figure 1.9 shows a pile of rods (aspect ratio 48) suspended by a single point force applied to the bottom. In this case, the force is applied by a small sphere connected to a string passing through the pile (which can be seen exiting the top of the pile) and suspended from above. The pile was formed in a large cylinder and then

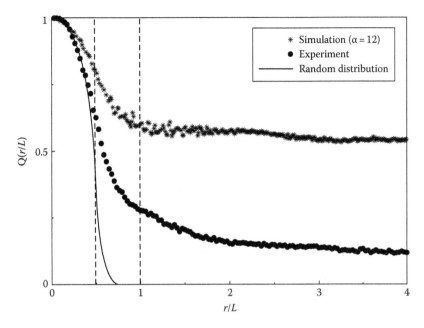

Figure 1.8 Orientational correlation function $\tilde{Q}(r/L)$ as a function of center-of-mass separation scaled by particle length. The solid line is data from an isotropic distribution, with particles assuming all allowable angles with equal probability. Both experiment (•) and simulation (∗) show an increased probability of particle alignment and (different) non-zero asymptotes for large r/L. (From Stokely, K. et al., *Phys. Rev. E.*, 67, 051302, 2003.)

pulled out of the container. Despite the absence of any other means of support, the pile maintains its original cylindrical shape and is robust to perturbations.

Because this cohesivity is caused by the particle geometry, such materials have been termed "geometrically cohesive granular materials" [18]. In these materials, the individual particle geometry enables a bulk cohesivity (as contrasted with, say, van der Waals forces in powders or capillary forces in wet materials). This behavior is most pronounced in three dimensions, as demonstrated in experiments on both long, thin rods, and radically aspherical particles (U-shaped staples). This rigidity has since been studied in the context of response to localized intruders and the collapse of columns, extending similar studies previously conducted on ordinary, round materials [1,26,27,29,30]. More recently, two systematic studies of U-shaped particles (staples) have characterized the susceptibility of piles to oscillatory disturbances and direct extensional forces.

1.3.3 Rigidity of Rodpiles

The stick-slip motion of an intruder through ordinary sand was studied by Albert et al. [1], who found force fluctuations with a characteristic $1/f^2$ scaling. Desmond and Franklin [11] repeated this experiment for rod-like materials and found three qualitatively different types of behavior. When the particle aspect ratio is low, the pile responds with local rearrangements and the drag force on the intruding

Figure 1.9   Pile of rods (aspect ratio 48) suspended by a single point force applied to the bottom. The force is applied by a single particle at the bottom which is tied to a string (exiting the pile at the top right of the image) suspended from above. The pile exhibits significant stability, resisting even large perturbing forces.

object has a random sawtooth appearance (Figure 1.10a). The linear increase in force indicates the intruder is at rest; the rapid decrease accompanies a sudden burst of motion. Importantly, the majority of the grains are at rest throughout the experiment; while individual particles near the intruder rearrange, there is no visible collective pile motion. A Fourier transform of the force vs. time data, shown in Figure 1.11, has a power-law tail that decays as f^{-2}, consistent with Albert et al.'s earlier experiments on stick-slip rearrangements in round particles.

As discovered by Philipse, piles of particles with large aspect ratios exhibit a distinct solid-like behavior. When an upward force is applied by the intruder, the pile as a whole moves upward. There is little observed relative motion between particles, and the pile acts as a solid. This behavior is reflected in the force vs. time diagram (Figure 1.10c). Two features are important to note. First, the force needed to move the pile (normalized by the total pile weight in Figure 1.10) can be many times the actual pile weight. This is a manifestation of discrete force chains [10,25] which, when deflected laterally, are amplified (see discussion earlier in this Chapter) before terminating on the container walls. The unusually large normal forces on the wall result in correspondingly large frictional forces that the force from the intruder must be overcome. The second feature to note is the spectrum of the fluctuations, shown in Figure 1.11, which has a different scaling exponent than that from granular rearrangements. The spectrum of these fluctuations decays as

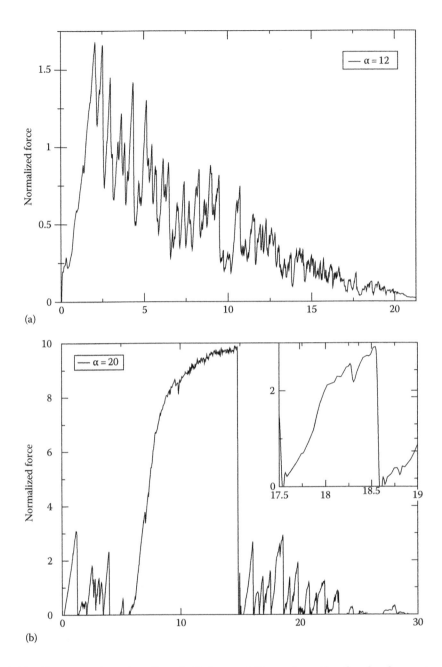

Figure 1.10 Force vs. time data for a ball dragged through a pile of rods of aspect ratio 12 (a), 20 (b), and 40 (c), all in a 5 cm tube. The low aspect ratio particles show the stick-slip behavior common in ordinary granular materials. The large aspect ratio particles, however, act as a single solid body, with small fluctuations characteristic of dry friction. Intermediate aspect ratios show both behaviors in a single experiment on both large and small (inset) time scales.   (*Continued*)

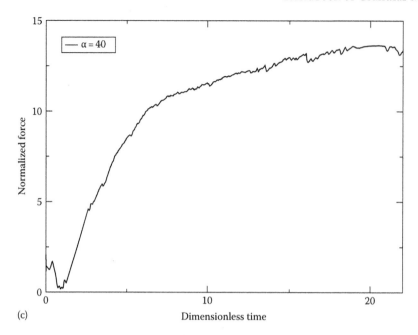

(c)

Figure 1.10 (*Continued*)   Force vs. time data for a ball dragged through a pile of rods of aspect ratio 12 (a), 20 (b), and 40 (c), all in a 5 cm tube. The low aspect ratio particles show the stick-slip behavior common in ordinary granular materials. The large aspect ratio particles, however, act as a single solid body, with small fluctuations characteristic of dry friction. Intermediate aspect ratios show both behaviors in a single experiment on both large and small (inset) time scales. (From Desmond, K. and Franklin, S.V. *Physical Review E*, 73(3), 031306, March 2006.)

f^{-1}, a behavior similar to that found in ordinary dry friction (e.g., [13]), further supporting the interpretation that the resistance to motion is primarily frictional.

Desmond et al. discovered an interesting combination of granular and solid-like behaviors displayed by particles of intermediate aspect ratios. As seen in Figure 1.10b, the pile at first responds to the intruder with the characteristic stick-slip motion associated with small aspect ratio particles. A close examination of the force (inset in Figure 1.10b), however, shows that these fluctuations contain small plateaus. As plateaus correspond to more steady upward motions, these imply small periods of time in which the pile moves as a solid. The data in Figure 1.10b show a long period of time, during the middle of the experiment, when the entire pile jams and all particles above the intruder move together. During this time, the pile is visually observed to move as a solid. Unlike the large aspect ratio particles, however, this pile is not as stable to perturbations, and eventually (at around $t = 15$ s) the pile collapses around the intruder and the stick-slip behavior resumes. The length of time spent in the solid state can be interpreted as an indication of the pile's stability. A related experiment on column collapse [47] found a similar combination of granular and solid behaviors in intermediate particle aspect ratios.

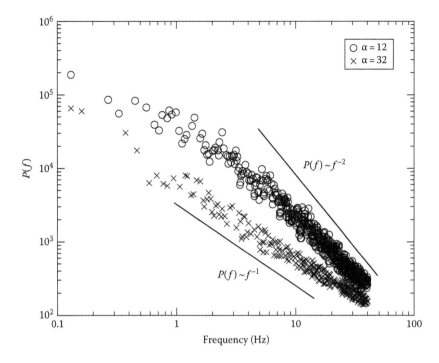

Figure 1.11   Power spectra for force fluctuations from piles exhibiting granular (○) and solid (x) behavior. Both show power-law tails, with the different exponents indicating a different mechanism for the fluctuations. The $1/f$ and and $1/f^2$ decay are consistent with previous work on, respectively, dry friction and localized granular rearrangements. Both data sets taken in a D = 2″ diameter tube. (From Desmond, K. and Franklin, S.V. *Physical Review E*, 73(3), 031306, March 2006.)

The behavior is dependent not only on the particle aspect ratio L/d but also the container diameter $\tilde{D}$. (The normalized inverse container diameter $\delta \equiv /D$ is actually used as the control parameter.) A "phase diagram" of pile behavior is shown in Figure 1.12. Figure 1.12 shows that when the aspect ratio is very small the pile behaves in a canonically granular manner. Nevertheless, the signature characteristics of the transition region plateaus in the force data and a visual observation of collective motion are seen in particles with aspect ratios as low as 8 when confined to cylinders whose diameter is twice the particle length.

1.3.4   Melting of U-Shaped Staples

U-shaped particles entangle significantly, and are considerably more solid-like than long thin rods. Gravish et al. [18] investigated the response of these particles in response to oscillatory disturbances, in the process deriving a novel characterization of rigidity. The height of a shaken pile decreased with time as a stretched exponential, introducing a time scale for the "melting." This time scale was inversely proportional to the oscillation strength (measured as the peak acceleration), and Gravish characterized rigidity by the proportionality constant of this

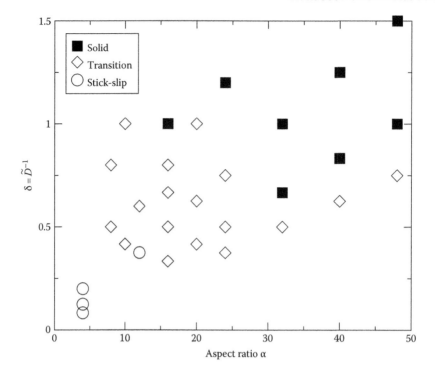

Figure 1.12   Phases exhibited by granular piles as a function of two control parameters—the aspect ratio α and the inverse container diameter $\delta \equiv \tilde{D}^{-1}$. Smaller aspect ratio particles show the stick-slip behavior of granular materials, while larger aspect ratio particles act as a solid body when the container is small enough. (From Desmond, K. and Franklin, S.V. *Physical Review E*, 73(3), 031306, March 2006.)

relationship. Figure 1.13a shows this measure as a function of the particle arm to spine "barb" ratio; first increasing, and then decreasing.

This behavior can be understood by considering the density of entanglements. An entanglement for these particles occurred when the arm of one particle passed between the arms of another. The probability of entanglement is related to the area spanned by the particle arms, and so increases with barb ratio. The particle density, however, decreases with barb ratio, similar to the behavior of large aspect ratio rods discussed earlier. The entanglement density is (roughly) proportional to the product of these two quantities. Figure 1.13b shows the entanglement density measured from Monte Carlo simulations, and reveals a peak density at approximately the same barb ratio as that maximally resistant to oscillations. Gravish et al. thus naturally concluded that rigidity is due to entanglement density, a hypothesis currently being explored in simulations.

There is a second interpretation of nonmonotonic dependence with barb ratio. As the barb ratio increases the "open" end of the particle represents a smaller fraction of the circumference. There would therefore seem to be a smaller probability of another particle finding this opening and becoming entangled. An intriguing system in which to study this would seem to be arcs of varying subtended angle.

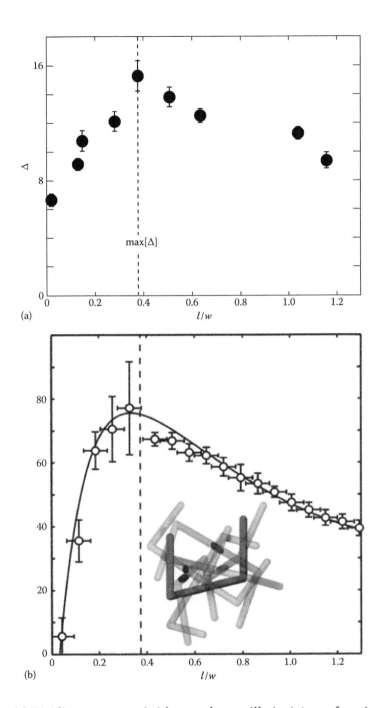

Figure 1.13   (a) Rigidity parameter (with regard to oscillation) Δ as a function of particle barb ratio l/w. Rigidity peaks when the arms are ≈40% of the spine length. (b) Entanglement density as a function of l/w from simulations and theoretical fit. The density peaks at a barb ratio comparable to that corresponding to peak rigidity. (From Gravish, N. et al. *Phys. Rev. Lett.*, 105, 128301, 2010.)

Clearly circles (subtending a full 2π radians) cannot entangle, nor are very short arcs likely to entangle in a meaningful way. This gedanken experiment would seem to recapture the maximally rigid shape without the dramatic decrease in packing fraction that the large barb ratio particles have.

1.3.5   Rheology of U-Shaped Staples

The previous sections involved experimental geometries intruder motion, column collapse, and oscillatory disturbances—originally applied to ordinary, round granular materials. We conclude with an experiment that cannot be applied to dry granular materials: extensional rheology. In an extensional test, one end of a granular sample is held fixed and the opposing end pulled with either a constant force or (attempted) constant velocity. As there are no cohesive properties in ordinary, dry granular materials, the very question of effective Young's modulus does not apply. The question only makes sense for materials that do demonstrate extensional cohesivity: granular materials that are wet, charged, or, in the case we now consider, geometrically cohesive.

The experiment consists of a cylindrical pile of U-shaped staples. Particles have a spine length of 1.3 cm and an arm length of 0.64 cm; the resulting barb ratio of $1.3/6.4 = 0.5$ is very near to that found by Gravish et al. to result in maximum entanglement. The staples, ordinary office staples, are made of wire thin enough to penetrate most voids in the pile, thus explaining the maximal entanglement. One end of the pile is held fixed; the other attached to a spring, which is then pulled with a constant speed. The instantaneous applied force and sample length can be independently measured to obtain a force-elongation curve such as Figure 1.14.

Fluctuations in Figure 1.14 are similar to those observed in response to an intruder through granular materials (e.g., [1,11]), but in this case arise due to the unsteady growth of the pile length. A yield force can be defined as the peaks in these graphs; the inset in Figure 1.14 shows peaks identified by computer analysis. Although there is a slight memory effect, with small differences in the distribution of forces that occur before and after large events, yield events can be treated as effectively independent (statistically). As a result, a single experiment effectively samples yielding at a range of pile lengths.

The most striking observation from this experiment is that longer piles are significantly weaker than shorter piles. Shown in Figure 1.15, the mean yield force decays with instantaneous sample length as a power-law with exponent $(0.86 \pm 0.12)^{-1}$. This gives a powerful clue toward the underlying mechanism, suggesting that yielding is due to a "weakest link" within the sample. Longer piles have statistically greater chances of containing a weak link and thus are more susceptible to yielding at lower applied forces.

The theory of weakest link statistics was worked out by Weibull [49] in 1939, and rests on the sole assumption that a long sample is comprised of multiple smaller elements, each with a statistically independent yielding probability. For a sample to not fail or yield, each subelement must similarly not fail. If the

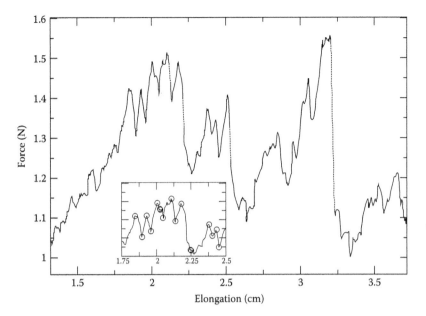

Figure 1.14 The extension of a geometrically cohesive granular material is accompanied by fluctuations in the applied force. A decrease in force implies a rapid sample growth; peaks therefore indicate yield events before a significant rearrangement. Inset shows critical points where rapid growth starts and stops identified by computer analysis. (From Franklin, S.V., *EPL*, 58004+, 2014.)

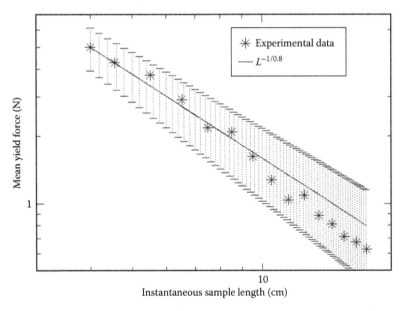

Figure 1.15 Longer samples are noticeably weaker than shorter samples. The mean yield force as a function of instantaneous sample length, with data from all 7500 identified events, shows the mean force to decay as a power-law with $L^{-1/m}$. The value of m is determined from the Weibullian analysis to be $m = 0.86 \pm 0.12$; therefore this fit has no free parameters. (Error bars represent the bounds of the power-law fit.) (From Franklin, S.V., *EPL*, 58004+, 2014.)

probability for a differential length δL to fail is $\delta Y = p(F)\delta L$, then the probability for it not to yield is $1 - \delta Y$ and the probability that the entire sample holds is

$$1 - P(F,L) = \prod_i [1 - p(F)\delta L].$$

where the product is over the N subelements.

If the probability of each subunit failing is small, this can be simplified by taking the logarithm of both sides and then Taylor expanding the resulting sum about 1:

$$\ln[1 - P(F,L)] = \sum_i \ln[1 - p(F)\delta L] \approx \sum_i \alpha p(F)\delta L$$

or, since $\sum_i \delta L = L$

$$1 - P(F,L) \approx e^{-p(F)L}.$$

$P(F,L) = 1 - e^{-p(F)L}$ is the probability that a sample of length L fails at instantaneous applied force F. In order for an experimental sample to realize that failure, however, it must first not fail at lower applied forces. Thus, the probability of actually observing failure at force F is the product of the sample not failing at lower forces and failing at that force:

$$Y(F,L) = C(L)e^{-p(F)L}\left(1 - e^{-p(F)L}\right)$$

where $C(L)$ is a normalization constant such that $\int Y(F,L)dF = 1$.

The power-law decay of mean yield force with length suggests we consider that the subunit yield probability scale similarly as a power-law with force, that is, $p(F) = F^m$. The mean force is then found by averaging over all yield forces

$$\overline{F(L)}\rangle = \int_0^\infty FP(F,L)\,dF$$

$$= CL^{1/m}\int_0^\infty F\left[e^{-\alpha ALF^m} - e^{-2\alpha ALF^m}\right]$$

$$\propto L^{-1/m}, \tag{1.13}$$

the very power-law shown in the data in Figure 1.13.

1.4   Additional Studies

The study of irregularly shaped particles is still quite new, certainly compared to that of round particles. Still, our coverage is not meant to be a comprehensive summary of cylindrical, ellipsoidal or similarly shaped particles, and interested readers are referred to a number of interesting resources. Villarruel et al. [48] found that rods of aspect ratio ≈4 constrained to a cylinder would, when tapped, align with the cylinder walls and compact. Lumay and Vandewalle [31] extended this to still larger aspect ratios, and developed a 2D lattice model that reproduced the salient features of the compaction curves. A subsequent two-dimensional experiment highlighted the importance of rotation in rod rearrangements [32]. Pournin et al. [29,40] derived an analytic expression for detecting contact between spherocylinders (a spherocylinder is composed of a cylinder and two hemispherical endcaps), an integral component of efficient computation. A spherocylinder can be defined as the locus of points equidistant from a line segment, so the problem reduces to finding the shortest distance between two segments. Pournin et al. used these results to develop discrete element method simulations that show crystallization and ordering under various excitations.

More recently, Hidalgo et al. [20] studied the role particle shape plays on stress propagation through a granular packing, while Azema and Radjï [2] examined the stress–strain relations of rod packings under shear. Attention has also been paid to the nature of random packings. Wouterse et al. [50] characterized the microstructure of random packings of spherocylinders and Zeravcic et al. [52] looked at the excitations of random ellipsoid packings at the onset of jamming.

For review, the reader is referred to Borzsonyi et al. [6], which nicely summarizes the research on shape anisotropic or *anisometric* [35] granular materials. Here one will find results pertaining to orientational ordering, the distribution of stresses throughout the sample, shear studies and fluidization, and/or the gaseous state that accompanies vigorous input of energy. We agree with and call attention to the authors, note of the lack of quantitative experiments in 3D bulk piles. It is quite likely that, as the authors state, we may expect "surprising details" to result from more complex geometries.

1.5   Conclusion

This chapter began by reviewing several investigations—theoretical, computational, and experimental—that revealed the power of uncorrelated mean-field statistics. Many of these were seen to be analogous to some form of a random walk, and the naive assumptions in the models justified by qualitative (and sometimes quantitative) agreement with observations. No attempt was made to be comprehensive, and it should be emphasized that the understanding has evolved considerably from some of the earlier interpretations. In particular, the theory of force chain generation is not nearly as simple as originally posited in the q-model, and the reader is referred to Chapter 7 for more recent theoretical innovations. Nevertheless, each can be considered to have advanced considerably the field of granular research, and it is quite likely that similarly simple models will arise again in the future to trigger new understandings.

Moving to radically aspherical shapes such as long, thin rods, and U-shaped samples, we saw that instances where mean-field approximations began to break down. These geometries introduce experimental complexities, such as choosing an appropriate system size and defining relevant dimensionless quantities, that are only now beginning to be overcome. While computational work has contributed greatly to our understanding of these particles, experiments continue to provide new phenomena that must be understood. As the field progresses, one can imagine new work that begins to consider mixtures of particles at the same particulate level of understanding that is currently only available to more narrow distributions of particle size and shape.

References

1. I. Albert, P. Tegzes, R. Albert, J. G. Sample, A. L. Barabasi, T. Vicsek, B. Hang, and P. Schiffer. Stick-slip fluctuations in granular drag. *Physical Review E*, 64:031307, 2001.

2. E. Azéma and F. Radja"i. Stress-strain behavior and geometrical properties of packings of elongated particles. *Physical Review E*, 81(5):051304, May 2010.

3. P. Bak, C. Tang, and K. Wiesenfeld. Self-organized criticality: An explanation of the 1/f Noise. *Physical Review Letters*, 59:381, 1987.

4. I. Balberg, C. H. Anderson, S. Alexander, and N. Wagner. Excluded volume and its relation to the onset of percolation. *Physical Review B*, 30:3933, 1984.

5. W. A. Beverloo, H. A. Leniger, and J. van de Velde. The flow of granular solids through orifices. *Chemical Engineering Science*, 15(3–4):260–269, 1961.

6. T. Borzsonyi and R. Stannarius. Granular materials composed of shape-anisotropic grains. *Soft Matter*, 9:7401–7418, 2013.

7. R. L. Brown and J. C. Richards. Essays on the packing and flow of powders and bulk solids. *Principles of Powder Mechanics*. Pergamon Press, Oxford, London, 1970.

8. J. Carlson and J. Langer. Properties of earthquakes generated by fault dynamics. *Physical Review Letters*, 62(22):2632–2635, May 1989.

9. J. Choi, A. Kudrolli, and M. Z. Bazant. Velocity profile of granular flows inside silos and hoppers. *Journal of Physics: Condensed Matter*, 17(24), S2533, 2005.

10. S. N. Coppersmith, C. Heng Liu, S. Majumdar, O. Narayan, and T. A. Witten. Model for force fluctuations in bead packs. *Physical Review E*, 53:4673, 1996.

11. K. Desmond and S. V. Franklin. Jamming of three-dimensional prolate granular materials. *Physical Review E*, 73(3):031306, March 2006.

12. J. Duran. *Sands, Powders, and Grains: An Introduction to the Physics of Granular Materials*. Springer-Verlag, Duran, New York, 2000.

13. H. J. Feder and J. Feder. Self-organized criticality in a stick-slip process. *Physical Review Letters*, 66:2669, 1991.

14. S. V. Franklin. Extensional rheology of entangled granular materials. *EPL (Europhysics Letters)*, 106(5), 58004+, June 2014.

15. E. Frossard. Recent advances in the physics of their mechanical behavior and applications to engineering works, *Granular Materials in Civil Engineering*. John Wiley & Sons, pp. 35–82, 2013.

16. G. J. Gao, J. Blawzdziewicz, and C. S. O'Hern. Enumeration of distinct mechanically stable disk packings in small systems. *Philosophical Magazine*, 87(3–5):425–431, January 2007.

17. J. Geng, E. Longhi, R. P. Behringer, and D. W. Howell. Memory in two-dimensional heap experiments. *Physical Review E*, 64:060301, 2001.

18. N. Gravish, P. B. Umbanhowar, and D. I. Goldman. Force and flow transition in plowed granular media. *Physical Review Letters,*, 105(12):128301, September 2010.

19. C. Liu, S. R. Nagel, D. A. Schecter, S. N. Coppersmith, S. Majumdar, O. Narayan, and T. A. Witten. Force fluctuations in bead packs. *Science*, 269:513, 1995.

20. R. C. Hidalgo, I. Zuriguel, D. Maza, and I. Pagonabarraga. Role of particle shape on the stress propagation in granular packings. *Physical Review Letters*, 103(11):118001, September 2009.

21. D. W. Howell and R. P. Behringer. Stress fluctuations in a 2D granular couette experiment: A continuous transition. *Physical Review Letters*, 82:5241, 1999.

22. H. M. Jaeger, C. Heng Liu, and S. R. Nagel. Relaxation at the angle of repose. *Physical Review Letters*, 62:40, 1989.

23. A. Janda, I. Zuriguel, A. GarcimartÃn, L. A. Pugnaloni, and D. Maza. Jamming and critical outlet size in the discharge of a two-dimensional silo. *EPL (Europhysics Letters)*, 84(4):44002, 2008.

24. J. B. Knight, C. G. Fandrich, C. N. Lau, H. M. Jaeger, and S. R. Nagel. Density relaxation in a vibrated granular material. *Physical Review E*, 51:3957, 1995.

25. D. Kolymbas (ed). *Constitutive Modelling of Granular Materials*. Springer, Berlin, Heidelberg, 2000.

26. E. Lajeunesse, A. Mangeney-Castelnau, and J. P. Vilotte. Spreading of a granular mass on a horizontal plane. *Physics of Fluids*, 16(7):2371–2381, 2004.

27. E. Lajeunesse, J. B. Monnier, and G. M. Homsy. Granular slumping on a horizontal surface. *Physics of Fluids*, 17(10):103302, 2005.

28. P. Le Doussal, A. Middleton, and K. Wiese. Statistics of static avalanches in a random pinning landscape. *Physical Review E*, 79(5), May 2009.

29. G. L. Ad and M. A. Hallworth. Axisymmetric collapses of granular columns. *Journal of Fluid Mechanics*, 508:175, 2004.

30. G. Lube, H. E. Huppert, J. Sparks, and A. Freundt. Collapses of two-dimensional granular columns. *Physical Review E*, 72:041301, 2005.

31. G. Lumay and N. Vandewalle. Compaction of anisotropic granular materials: Experiments and simulations. *Physical Review E*, 70:051314, 2004.

32. G. Lumay and N. Vandewalle. Experimental study of the compaction dynamics for two-dimensional anisotropic granular materials. *Physical Review E*, 74:021301, 2006.

33. E. R. Nowak, J. B. Knight, E. Ben-Naim, H. M. Jaeger, and S. R. Nagel. Density fluctuations in vibrated granular materials. *Physical Review E*, 57:1971, 1998.

34. C. S. O'Hern, S. A. Langner, A. J. Liu, and S. R. Nagel. Force distributions near jamming and glass transitions. *Physical Review Letters*, 86:111, 2001.

35. L. Onsager. The effects of shape on the interaction of colloidal particles. *Annals of the New York Academy of Science*, 51:627, 1949.

36. O. Perković, K. Dahmen, and J. Sethna. Avalanches, Barkhausen noise, and plain old criticality. *Physical Review Letters*, 75(24):4528–4531, December 1995.

37. A. P. Philipse. The random contact equation and its implications for (Colloidal) rods in packings, suspensions, and anisotropic powders. *Langmuir*, 12:1127, 1996.

38. A. P. Philipse and A. Verberkmoes. Statistical geometry of caging effects in random thin-rod structures. *Physica A*, 235:186, 1997.

39. L. Pournin, M. WEber, M. Tsukahara, J. A. Ferrez, M. Ramaioli, and Th Libling. Three-dimensional distinct element simulation of spherocylinder crystallization. *Granular Matter*, 7:119, 2005.

40. M. Ramaioli, L. Pournin, and Th Liebling. Vertical ordering of rods under vertical vibration. *Physical Review E (Statistical, Nonlinear, and Soft Matter Physics)*, 76(2):021304, 2007.

41. S. Saraf and S. V. Franklin. Power-law flow statistics in anisometric (wedge) hoppers. *Physical Review E*, 83(3):030301, March 2011.

42. J. P. Sethna, K. A. Dahmen, and C. R. Myers. Crackling noise. *Nature*, 410(6825):242–250, March 2001.

43. K. Stokely, A. Diacou, and S. V. Franklin. Two-dimensional packing in prolate granular materials. *Physical Review E*, 67(5):051302, 2003.

44. K. To and P.-Y. Lai. Jamming pattern in a two-dimensional hopper. *Physical Review E*, 66:011308, 2002.

45. K. To, P.-Y. Lai, and H. K. Pak. Jamming of granular flow in a two-dimensional hopper. *Physical Review Letters*, 86:71, 2001.

46. K. To, P.-Y. Lai, and H. K. Pak. Flow and jam of granular particles in a two-dimensional hopper. *Physica A*, 315:174, 2002.

47. M. Trepanier and Scott V. Franklin. Column collapse of granular rods. *Physical Review E*, 82:011308, 2010.

48. F. X. Villarruel, B. E. Lauderdale, D. M. Mueth, and H. M. Jaeger. Compaction of rods: Relaxation and ordering in vibrated, anisotropic granular material. *Physical Review E*, 61:6914, 2000.

49. W. Weibull. A statistical theory of the strength of materials. Technical Report 151, Royal Swedish Academy of the Engineering Society, Stockholm, 1939.

50. A. Wouterse, S. R. Williams, and A. P. Philipse. Effect of particle shape on the density and microstructure of random packings. *Journal of Physics: Condensed Matter*, 19(40):406215, 2007.

51. N. Xu, J. Blawzdziewicz, and C. O'Hern. Random close packing revisited: Ways to pack frictionless disks. *Physical Review E*, 71(6):061306+, June 2005.

52. Z. Zeravcic, N. Xu, A. J. Liu, S. R. Nagel, and W. van Saarloos. Excitations of ellipsoid packings near jamming. *EPL (Europhysics Letters)*, 87(2):26001, 2009.

53. I. Zuriguel, A. Garcimartin, D. Maza, L. A. Pugnaloni, and J. M. Pastor. Jamming during the discharge of granular matter from a silo. *Physical Review E*, 71:051303, 2005.

54. I. Zuriguel, L. A. Pugnaloni, A. Garcimartin, and D. Maza. Jamming during the discharge of grains from a silo described as a perocolating transition. *Physical Review E*, 68:030301, 2003.

Section I

Interpretive Frameworks

CHAPTER 2

Experimental Techniques

Mark D. Shattuck

CONTENTS

2.1 Introduction

From the beach to the sandbox, granular experiments begin at an early age. Even to a child, it is clear that sand behaves differently than ordinary fluids and solids. When poured, sand flows like a fluid, but it can form a pile. When stationary, it supports the weight of a person like a solid, but patterns can form on the surface like frozen ripples on water (Figure 2.1). Dissipation and purely repulsive

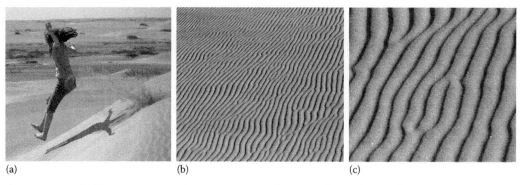

Figure 2.1   (left) Young granular experimentalist. Sand on the surface of a steep dune (near the angle of repose) behaves more like a fluid due to its proximity to failure, allowing a soft-landing for flying children. (center) Sand ripples on a dune in Baja California. This surface waveform moves slowly, on the scale of hours. (right) Close-up of ripples. The wavelength is approximately 8 cm.

interactions between grains cause these effects. In an ordinary fluid or solid, the molecules interact *elastically* through both long-range attraction and short-range repulsion. When two water molecules collide, the total shared kinetic and potential energy is conserved. When two macroscopic grains interact, energy is lost from the shared kinetic and potential energy. The lost energy flows to internal degrees of freedom as heat or radiates away as sound waves. Either way the sum of the translational kinetic energy and potential energy of the particles decreases during every interaction. Water molecules attract, leading to surface tension, solidification, and elastic behavior. Non-cohesive grains like dry sand only repel, leading to diffuse non-self-leveling surfaces, weak solids, and plastic behavior, Figure 2.1a. A strong confining pressure can mitigate some of these differences.

Excitation: One of the key elements in granular experimental design is energy input. Without energy input, dissipation causes strong damping of grain motion and quickly leads to a state with no particle motion. The molecules in water are never stationary, unless the temperature is absolute zero. By analogy, granular temperature (translational kinetic energy per particle) is constantly decreasing without energy input. The experimental method for energy input plays a critical role in the behavior of a granular system. Section 2.2 describes a number of common techniques to excite granular materials including vibration, gravity, and shear.

Visualization: The particulate nature of granular materials can be observed directly, unlike ordinary fluids and solids in which the molecules are microscopic. This provides a powerful way to observe the connection between the molecular particle scale and the continuum collective scale. However, most granular systems strongly scatter light making them opaque, even if the grains are transparent. Section 2.3 covers a range of methods to visualize the surface and the interior of granular systems, including photography, photoelasticity, index matching, x-rays, and magnetic resonance imaging.

Image analysis: Modern visualization techniques produce amazing images. Section 2.4 explains several important techniques to extract quantitative information from images including particle tracking, photoelastic force measurement, and particle imaging velocimetry.

Non-imaging techniques: Although researchers have applied high-speed visualization to the study of granular materials since the 1960s [1–3], they have also developed a host of other methods to measure particle, local, or average properties of granular materials. Section 2.5 outlines many of these techniques to measure density, position, velocity, stress, and geometrical quantities like angle of repose.

2.2   Excitation

Excitation is a critical part of granular experiments. In ordinary fluids and solids, energy conservation maintains molecular motion in the form of temperature. In granular materials, energy dissipation constantly forces "granular temperature" toward zero. Granular temperature is the mean kinetic energy of the particle's center-of-mass motion. Without excitation, the granular temperature eventually becomes zero. Continuous excitation balances dissipation, creating a sustained nonzero granular temperature. However, the sustained granular temperature may be far from equilibruim even in steady state. For example, constant shearing of an ordinary fluid will produce a nonequilibrium steady state, but the system will still be close to equilibrium. This occurs because the energy input from shearing is only a small fraction of the total temperature energy the fluid would have in equilibrium without shearing. Since it is small, any coherent energy from the shearing is quickly converted to random temperature energy. In a granular system, the equilibrium temperature energy is zero so any shearing energy is large in comparison. As a result, any steady state created will be far from equilibrum, and coherent structures tend to be more pronounced leading to inhomogeneities.

2.2.1   Vibration

Vibrations can transform a granular pile into a granular fluid. Vibro-fluidization is used in many experiments and industrial applications to enhance granular flow ability and mixing and to study the properties of granular materials [4–93]. Vibrated granular materials tend to act like fluids. For example, fluid-like complex patterns develop on the surface of a vertically vibrated granular layer [4–10] as shown in Figure 2.2. Many other interesting phenomena occur in vibrated granular systems.

A typical vibration system uses an electro-magnetic actuator to create motion. For small systems (~1 kg), a standard electro-magnetic speaker can be used, but for larger systems a purpose-built electro-magnetic shaker is required. Piezoelectric speakers can be used for very small loads and for high frequencies. Mechanical drive using a cam is also possible [49,56,94] but is limited to sinusoidal motion,

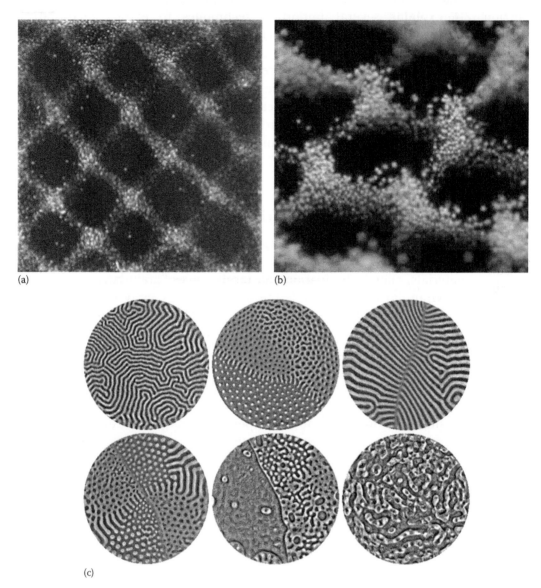

(a)

(b)

(c)

Figure 2.2   Patterns in vertically vibrated granular materials. (a) Overhead view of a square pattern produced by vibrating 60,000 0.55 mm lead spheres in a 55 mm^2 cell at $f = 22$ Hz and maximum acceleration amplitude of 3.000 times gravity. The pattern is illuminated with low angle lighting to accentuate the peaks and valleys. (b) Oblique view of the square pattern from (a). The pattern is $f/2$ subharmonic, repeating every other drive cycle. On the other cycle, the peaks become valleys and the valleys become peaks. (c) Collection of patterns at various driving frequencies and amplitudes. (top row) $f/2$ Stripes, $f/2$ Hexagons $f/4$ Stripes. (bottom row) $f/4$ Hexagons, $f/4$ Phase bubbles, $f/4$ Phase domain pattern. (Adapted from Bizon, C. et al., *Phys. Rev. Lett.*, 80, 57, 1998; Moon, S.J. et al., *Phys. Rev. E*, 65, 011301, 2002.)

and changing the amplitude is difficult. In an electro-magnetic shaker, a current flows through a coil in a static magnetic field. The force on the coil depends on the current, so arbitrary waveform generation is possible.

In an electro-magnetic shaker, the coil's motion is restricted to one direction using a flexure bearing. A second bearing is usually necessary on the drive shaft. A square cross-sectional air bearing is a good choice for the drive shaft due to low friction and high linear and rotational stiffness. The drive shaft should be connected to the shaker with a thin connecting rod [34,78]. The thin rod provides a mechanical fuse in the event of failure, protecting the shaker, air bearing, and the experiment. A rod is also stiff in only one direction and can accommodate an angular mismatch between the shaker and the air bearing and provide significant vibration isolation.

For vertical vibration, leveling presents a significant challenge. Typically, any horizontal motion will strongly affect the system, causing heaping and convection [62,72,88,95,96]. A good leveling table is needed with good vibration isolation. Feedback from resonances in the mechanical system and the shaker limits the frequency range for which the system can be leveled without adjustment.

In 1831, Faraday found spontaneous heaping and convection of grains on a thin plate vibrated by a violin bow [11]. Many researchers have investigated heaping and convection since Faraday [11,13,19,26,34–37,39,44,48–50,56,57,61,62,67, 70,71,77,84,89–91]. In the absence of horizontal vibration, air is the main cause of convection and heaping in vertically vibrated granular materials [34,56]. When a granular layer is thrown up in the air, it is more dilute and air can pass through easily. As the layer falls it compacts trapping the air between the granular layer and the bottom of the container. The air drags the grains upward causing convection and heaping. If all of the air is removed, heaping is almost completely suppressed [37,56]. The main effect of the air is through its viscosity, which is largely independent of pressure until the pressure is below ~ 10 torr. To eliminate heaping, the pressure needs to be below $\sim 10^{-2}$ torr. Sidewall friction also causes heaping and convection [19,50,57,62], which was demonstrated clearly using magnetic resonance imaging [57,62]. See (Section 2.3.5).

In vacuum with heaping suppressed, vertically vibrated granular materials form subharmonic surface wave patterns analogous to gravity-driven Faraday waves [4–10,68,97]. In three dimensions, many different patterns are possible depending on the amplitude and frequency of the shaking. Examples of the patterns are shown in Figure 2.2. In addition to space-filling patterns, localized structures called ocsillons occur in a small range of parameter space [8]. Quantitative agreement between experimental patterns and simulations of frictional hard spheres shows that only simple dissipation and friction are needed to produce the patterns [9]. The stripe patterns follow stability boundaries like fluids, in which the pattern wavelength is limited by skew-varicose and cross-roll instabilities in analogy with thermal convection patterns [66]. There are a rich variety of patterns including phase bubble patterns shown in the bottom right of the right panel of Figure 2.2 [10]. Kinks also form and their positions are controlled by subharmonic forcing and are self-centering due to subharmonic feedback from the layer hitting the shaker [68]. Recently, particle trajectories in the patterns have been measured using positron emission particle tracking (PEPT) [75]. See (Section 2.5.3). The subharmonic response is suppressed in micro-gravity [93].

The inelastic ball or plastic layer model [7,10,12,13,18,19,22,23,28] can explain the basic temporal features of vertical vibration pattern formation [7,10]. The plastic model treats the layer as a completely inelastic 1D object. It was originally used to model the air gap between the shaker and the layer including Darcy drag [12,13]. It has been modified to include Carmen–Kozney drag to predict voidage dependence on acceleration [18], slanted walls [19], and compressible gases [28]. It can predict most of the pattern transition as a function of acceleration including the transition to subharmonic $f/2$ patterns to $f/2$ hexagonal patterns to $f/4$ kinks to $f/4$ patterns to $f/4$ hexagonal patterns [7], but eventually requires corrections due to the onset of phase bubbles [10].

Two types of 2D vertical vibration are possible: one with the thin direction perpendicular to the vertical and one with the thin direction parallel. For perpendicular drive, patterns similar to the 3D patterns are observed [34,38,41, 44,48,59,60,63–65,73,81,92,98], although due to the limited dimension not as many pattern types are possible. In systems with air, bubbling occurs for strong excitation [94]. A crystallization transition occurs for systems with monodisperse particles [82,99].

For 2D vertical vibration with the thin direction parallel to the vertical, a 2D fluid is produced [78–80,100–102]. The radial distribution function and bond-order parameter are identical to those in an equilibrium hard-sphere gas [78], even during crystallization. As the granular fluid approaches crystallization it exhibits glassy behavior including caging and a Vogel-Fulcher-like divergence of relaxation times [79]. Further, the velocity distribution is nearly Maxwell–Boltzmann with a small correction, which depends only on density [80].

Low-amplitude vibration induces compaction [18,22,23,83,103,104]. Pouring frictional particles into a container usually results in a dilute packing. Vibration can reduce or eliminate friction temporarily [83] causing compaction. Sound propagation can be thought of as very low-amplitude vibration [45,46,53]. At high amplitude, vibration causes fluidization [20,42,43,52,55,58,71,105–107]. Vibration reduces drag on objects moving through granular materials [43] and enhances heat transport [29]. It is commonly used for drying [16,34].

Mixing and segregation are like two sides of a coin [25,27,40,69,76,85,87,108]. In ordinary materials, mixing is generally enhanced by vibration, but in granular materials vibration often causes segregation [84,86,109,110]. Many studies have focused on vibro-mixing [14,15,17,21,24,32], vibro-diffusion [32,42,111], and vibro-segregation [27,30,31,42,50,51,74,109].

2.2.2 Gravity Driven

For over 2000 years, the hourglass has used gravity-driven granular flow to keep time. The flow rate in an hourglass is independent of depth [112,113] making it easy to estimate subintervals. Many granular experiments use gravity to excite granular materials, including rotating drums [31,64,85–96], hoppers [1–3,20,71, 127–145], chute flows [3,33,146–155], and vertical tubes [156–159].

Rotating drums are used in many industries for mixing, coating, granulation, and drying [69]. A rotating drum consists of a container that is rotated about one or more axes. The prototypical arrangement is a cylinder rotating about its axis of symmetry. For mixing, more elaborate container shapes are used (e.g., V-shaped, paddle wheel, or double cone). Rotating drums exhibit three basic behaviors: avalanching [118,119,123], continuous [119], and centrifugal [69] depending on the rotation rate. For slow rotation, intermittent avalanches occur. The material is slowly tilted until it loses stability and flows in an avalanche. For intermediate rotation rates, the grains continuously cascade down the angled slope formed by rotation. For higher rates, the centrifugal force is so great that the grains are stuck to the confining walls and move in solid body rotation. An example of a 2D rotating drum in the continuous flow regime is shown in Figure 2.3. The cylinder of the drum is only 1% thicker than the monodisperse 3.175 mm particles. The whole drum is shown in Figure 2.3b with enlargement for two sections shown on the Figure 2.3a and c. The right region is continuously flowing down the slope setup by the rotation. On the left, the particles move in solid body rotation with the drum. The velocity profile across the centerline is shown in Figure 2.22.

Mixing occurs at different rates in the radial and axial direction in 3D drums [25,27,31,40,69,76,85,87,114,121,122,124–126]. In fact, generally, segregation is difficult to prevent when multiple particle types are present [31,121,122,126]. In long rotating tubes, axial bands quickly appear [31,121,122,126] and then coarsen [160,161]. In the radial direction, size and density segregation mechanisms can be made to cancel, allowing good mixing of dissimilar particles [124].

Gravity-driven granular discharge containers (e.g., silos and hoppers) have been used for at least 7000 years to store and deliver granular products like grain and coal [162]. Hoppers are also used in many experiments [1–3,20,71,115–117,127–145,163,164]. Hopper discharge rate depends on the 5/2 power of the orifice diameter through the Beverloo correlation [127]. Flow fields and density fluctuations have been measured using x-rays [130,131,136–138,140,141] and high-speed photography [1–3,71,129,132,134,135,139]. Of particular importance is the study of container pressure that is responsible for many industrial accidents due to silo failure during discharge [164]. Jamming of silos and hoppers is also an active area of research [145,165–169], and vibrations are often added to aid in smooth discharge [20,71,144].

Two related gravity-driven flows are chute flow [3,33,146–155] and vertical tube flow [156–159]. Chute flow generally differs from hopper discharge in that the chute is opened so that the flow experiences a free surface. Vertical tube flow is similar to hopper flow, but there is no narrowing of the flow chamber at the outlet. A similar flow-rate scaling holds for vertical tubes [156] but the velocity profile tends to be plug like [157] for slow flows. For faster flows, density waves have been observed [159].

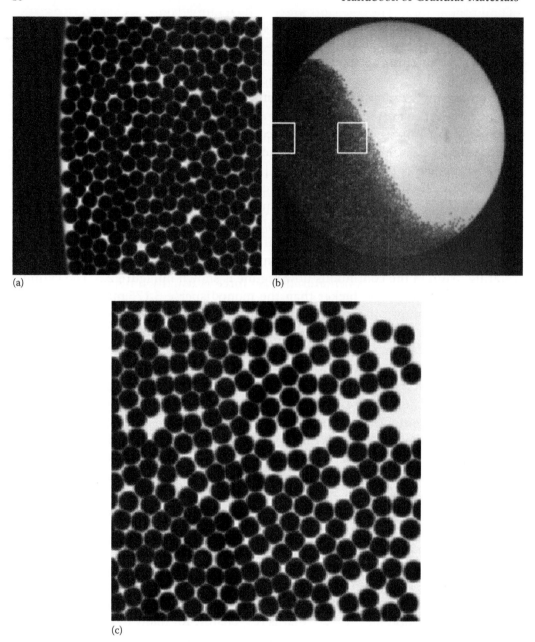

Figure 2.3 Quasi-2D rotating drum. Spherical stainless steel ball bearings are in between two glass plates separated by 1.01 particle diameters $D = 3.175$ mm. The diameter of the drum is $128D = 152.4$ mm. (b) Snapshot of full cell while rotating at 30 rpm. (a,c) Close-up image from the (a,c) squares shown in the central image.

2.2.3 Shear

In 1954, Bagnold performed one of the earliest quantitative studies of granular materials [170]. He measured the stress versus strain-rate relation for a neutrally buoyant suspension of spheres in a Couette shear device and found two regimes based on the dominance of viscosity, where the stress is linear in strain rate or grain inertia, where the stress is quadratic in strain rate. He used the scaling in the inertial regime to analyze three problems: chute flow, size segregation, and flowing gravel. Later work suggests that the liquid–solid mixture can be modeled as a Newtonian fluid with a corrected viscosity that depends on the solid's concentration [171].

In continuum theories of fluids and solids (see the chapter of Jenkins), the relation between stress and strain represents an important closure for the equations of motion. Measurements of the stress response to shear provided a unified clear picture of ordinary Newtonian fluids. In granular materials, many experiments use shear to try to build on the success in ordinary fluids and solids.

Several different geometries can be used. In simple shear, a material is placed between two parallel plates and one or both are moved along the parallel direction. Shear can induce crystallization in monodisperse particles [172–175]. Stress versus strain-rate measurements have been made in 2D [176–180] and 3D [181–187]. Couette shear is an approximation to simple shear using material between two concentric cylinders with one or both of the cylinders rotating [170,171,178,188–192]. In pure shear, a material is placed between two parallel plates and one or both are moved along the perpendicular direction in such a way that the volume of the system is conserved [193]. In pure compression, the plates are moved perpendicularly, but the volume is decreasing [194,195].

2.3   Visualization

One exciting aspect of granular experimentation is the ability to visualize systems on both the grain scale and bulk scale. However, even if particles are transparent, strong scattering from the index-of-refraction mismatch with air causes most granular materials to be opaque. This section will present a number of solutions to this problem including direct imaging of surface particles in 3D or all particles in 2D systems, index matching, x-ray imaging, and magnetic resonance imaging.

2.3.1   Photography

Direct photography is a simple and effective technique for the study of flowing and static granular systems. Because of the opaque nature of dry granular materials, direct photography can only be used for imaging 2D systems or the surface of 3D systems.

Dyed particles or naturally occurring different colors of sand are commonly used to study the flow in granular systems, for example, in tubes [157], hoppers [129,132,134], vibrating systems [71,196], and shear [197]. Often colored particles are initially set up in layers. Then flow deforms the pattern to reveal the velocity field. Using fluorescently dyed particles, stripe or grid patterns can be periodically illuminated to achieve the same effect but with less setup time and the ability to reset the pattern at any point [134]. Photographs of the surface and sides of rotating drums show axial segregation [31,121,122,126] in mixtures of granular materials. A granular analogy of the hydraulic jump in chute flows was captured using photographic visualization [150], and in a rotating drum imaging can be used to collect avalanche statistics [123]. Stereo-imaging can be used to measure strain [140]. Another interesting technique uses photo paper placed under 2D granular systems to create a shadow mask of the system. This technique was used to study 2D isotropic compression of objects using stretched rubber sheet for disks [198], polygons [199], and binary mixtures [200].

For periodic flow especially in vibrated systems, stroboscopic images are useful [6–10,12,13,34,66,146,147,201–204]. Stroboscopic imaging is especially useful when the images can be phase synchronized as well [6–10,34,66,201–204]. With phase synchronization, the frequency of the strobe is locked to a phase reference, which can be adjusted to view the motion at any phase relative to the reference.

High-speed movies have been used since the 1960s to image granular flow, first using film [1–3,135,139,151,152,154] and now using digital photography [78–80,195,205,206]. Early film systems used special movie cameras capable of up to 2000 fps [1–3,135,139]. Images were played back at slower speeds for direct observation or frame-by-frame onto a screen to trace particle position and flow fields. Other researchers determined the local density by counting the number of particles in circles of known diameter [1,2]. Later, 16 mm film exposed at 240–2880 fps was projected onto a digitizer frame-by-frame. With this technique, the centers of particles could be found to an accuracy of 2% of a particle diameter [151,152,154].

2.3.2 Photo Elasticity

Stress-induced birefringence or photoelasticity (Figure 2.4) has been an important technique for studying granular materials since 1950s [176–178,193, 194,207–214]. It can be used to visualize and measure forces in 2D packings.

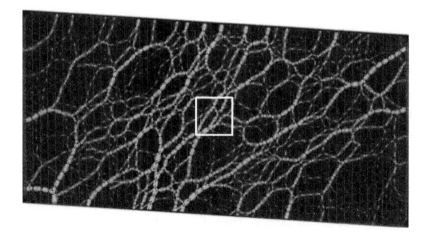

Figure 2.4   2D photoelastic disks under simple shear viewed through a dark-field circular polariscope. The force is roughly proportional to the brightness of the particles (see text). An anisotropic force network is evident. (Courtesy of J. Ren and R. P. Behringer, Duke University, Durham, NC.)

In 1969, De Jong and Verruijt [176] applied the technique to packings of disks under isobaric simple shear. They extracted the forces on each contact and constructed Maxwell force diagrams [215,216] to interpret and validate the method. They found that the forces in the Maxwell diagram were closed connected polygons verifying that their force measurements accurately represented mechanical equilibrium. Later, they combined local-averaged strain measurements with force measurements to test plastic flow rules for granular packings [177]. More recently photo-elasticity has been used for acoustic measurements [209], high-speed impact [211,214], shear [178,180,183,193,210,212,213,217–221], and compression [194,222,223].

Photoelastic force visualization and measurement are based on stress-induced birefringence. A material is birefringent if its refractive index depends on the polarization or propagation direction of light. In birefringent materials, the velocity of light v is faster for light polarized in a particular direction (fast axis) and slower for light polarized along the perpendicular slow axis. This leads to a fast and slow index of refraction, $n_{f,s} = c/v_{f,s}$, where c is the speed of light in vacuum. A birefringent material is characterized by the direction of fast axis and the index difference, $n_f - n_s$. The light in the slow direction, moving through a birefringent material of thickness t, will acquire a relative lag or retardation of $l = ct/(v_f - v_s) = t(n_f - n_s)$. Many materials, especially crystals, are inherently birefringent regardless of their stress state. However, most disordered transparent solids like glass and plastic only exhibit birefringence under internal or external stress.

In 1815, David Brewster discovered birefringence in glass and many ordinary materials like ice, human hair, and wax [224]. In 1816, he discovered stress-induced birefringence and suggested its use as a dynamometer to measure force, as it is commonly used today [225]. He outlined the basic relationships between stress and birefringence, and in 1853, James Clerk Maxwell [226] consolidated his observations into the stress-optic law for 2D stress states,

$$l = t(n_f - n_s) = tK(\sigma_1 - \sigma_2), \tag{2.1}$$

which relates the difference in the fast and slow refractive indices in a photoelastic material, $n_f - n_s$, to the principle stress difference, $\sigma_1 - \sigma_2$. K is the material-dependent stress-optic coefficient and t is the thickness of the material. The direction of the principle stresses determines the direction of the fast and slow axes.

A polariscope is used to measure stress or force in a photoelastic material. In a polariscope, polarized light travels through the birefringent sample and then through a crossed polarizer (e.g., if the incident light is horizontally polarized, then the crossed polarizer is vertical). In this arrangement, if the sample were not there, then no light would pass through the system, since the polarizer is crossed compared with the incident light. With the sample in place, the polarized light is rotated due to the lag l and some will now pass through the final polarizer. From the pattern of transmitted light, the lag can be measured and related to the stress through the stress-optic law (2.1).

To simplify the mathematical description of the polariscope, I will introduce Jones calculus. In the 1940s and 1950s, R. Clark Jones developed a simple technique (Jones Calculus) to calculate the effect of a series of optical elements on polarized light [227–234]. The calculus uses a matrix representation of polarized light traveling along the z-axis. The electric field of polarized light propagating in the z-direction can be described as the real part of the sum of two orthogonal linearly polarized complex waves,

$$
\begin{aligned}
\mathbf{E}(z,t) &= E_x(z,t)\hat{x} + E_y(z,t)\hat{y} \\
&= A_x e^{2\pi i(z/\lambda + vt + \epsilon_x)}\hat{x} + A_y e^{2\pi i(z/\lambda + vt + \epsilon_y)}\hat{y} \\
&= e^{2\pi i(z/\lambda + vt)}[A_x e^{i\epsilon_x}\hat{x} + A_y e^{i\epsilon_y}\hat{y}] \\
&= e^{2\pi i(z/\lambda + vt) + i(\epsilon_x + \epsilon_y)/2}[A_x e^{i\Delta/2}\hat{x} + A_y e^{-i\Delta/2}\hat{y}]
\end{aligned}
\tag{2.2}
$$

where

λ and v are the wavelength and frequency of the light, $\Delta = (\epsilon_x - \epsilon_y)/2$

ϵ_x and ϵ_y are the initial phases

A_x and A_y are the initial amplitudes of the two orthogonal linearly polarized waves in the x and y directions

If Δ is an integer multiple of π, then the total wave is linearly polarized in the $A_x\hat{x} + A_y\hat{y}$ direction. Otherwise, the wave is elliptically polarized with the special case of circular polarization when $A_x = A_y$. Although Jones calculus cannot represent *unpolarized* light in a simple way, Jones realized that all of the information in *polarized* light can be represented by a complex Jones vector, $\mathbf{J} = e^{i\phi}[A_x e^{i\Delta/2} \quad A_y e^{-i\Delta/2}]^T$ or 2×1 matrix, where $\phi = 2\pi(z/\lambda + vt) + (\epsilon_x + \epsilon_y)/2$. Optical elements can be represented by 2×2 matrices, which act from the left by matrix multiplication on $\mathbf{J}$ to produce a new vector representing the light leaving the element. The leading phase term $e^{i\phi}$ is not needed to describe relative phase shifts and is usually dropped for simplicity. From this description, the intensity of light is calculated by taking the modulus of the complex inner product of the Jones vector $I = \mathbf{J}^*\mathbf{J}$, where $*$ is the complex-conjugate transpose.

The power of the technique is that a complex chain of optical elements can be described using a few basic elements

$$
\mathbf{J}_x = \begin{bmatrix} 1 \\ 0 \end{bmatrix} \quad \mathbf{S}(\theta) = \begin{bmatrix} \cos\theta & -\sin\theta \\ \sin\theta & \cos\theta \end{bmatrix} \quad \mathbf{G}_x(\delta) = \begin{bmatrix} e^{-i\delta/2} & 0 \\ 0 & e^{i\delta/2} \end{bmatrix} \quad \mathbf{P}_x = \begin{bmatrix} 1 & 0 \\ 0 & 0 \end{bmatrix}.
\tag{2.3}
$$

where

The Jones vector $\mathbf{J}_x$ represents light polarized in the x-direction

$\mathbf{S}(\theta)$ rotates polarization by angle θ

$\mathbf{G}_x(\delta)$ represents a birefringent plate of thickness t with slow axis in the x-direction and nondimensional retardation of $\delta = 2\pi t(n_f - n_s)/\lambda$ where λ is the wavelength of light entering the plate and n_f and n_s are the real parts of the index of refraction in the fast and slow directions

$\mathbf{P}_x$ is a linear polarizer in the x-direction

We describe all of the components of a polariscope by combining these basic elements. The Jones vector representing light polarized at an angle θ from the x-axis, $J(\theta)$, is created by multiplying J_x by the rotation matrix.

$$J(\theta) = S(\beta)J_x = \begin{bmatrix} \cos\theta & -\sin\theta \\ \sin\theta & \cos\theta \end{bmatrix}\begin{bmatrix} 1 \\ 0 \end{bmatrix} = \begin{bmatrix} \cos\theta \\ \sin\theta \end{bmatrix}. \tag{2.4}$$

For example, $J(\pi/2) = J_y = [0\ 1]^T$ is light polarized in the y-direction. The same idea is used to create the matrix for a birefringent plate with retardation δ and slow axis at an angle β from the x-axis. In this case, two rotation matrices are needed. The first brings the light into the β oriented frame. The second rotates back to the lab frame.

$$\begin{aligned} G(\beta,\delta) &= S(\beta)G(\delta)S(-\beta) \\ &= \begin{bmatrix} \cos\beta & -\sin\beta \\ \sin\beta & \cos\beta \end{bmatrix}\begin{bmatrix} e^{-i\delta/2} & 0 \\ 0 & e^{i\delta/2} \end{bmatrix}\begin{bmatrix} \cos\beta & \sin\beta \\ -\sin\beta & \cos\beta \end{bmatrix} \\ &= \begin{bmatrix} \cos\frac{\delta}{2} - i\cos2\beta\sin\frac{\delta}{2} & -i\sin2\beta\sin\frac{\delta}{2} \\ -i\sin2\beta\sin\frac{\delta}{2} & \cos\frac{\delta}{2} + i\cos2\beta\sin\frac{\delta}{2} \end{bmatrix}. \end{aligned} \tag{2.5}$$

Then a linear polarizer oriented at angle θ is

$$\begin{aligned} P(\theta) &= S(\theta)P_xS(-\theta) \\ &= \begin{bmatrix} \cos\theta & -\sin\theta \\ \sin\theta & \cos\theta \end{bmatrix}\begin{bmatrix} 1 & 0 \\ 0 & 0 \end{bmatrix}\begin{bmatrix} \cos\theta & \sin\theta \\ -\sin\theta & \cos\theta \end{bmatrix} \\ &= \begin{bmatrix} \cos^2\theta & \cos\theta\sin\theta \\ \cos\theta\sin\theta & \sin^2\theta \end{bmatrix}. \end{aligned} \tag{2.6}$$

There are two basic types of polariscopes: linear and circular. The linear polariscope is not as common in granular experiments, but is simpler to describe. It consists of a source of linearly polarized light at angle α. This is accomplished by using a linear polarizer, but since there is no Jones vector for unpolarized light, we start with the light as it emerges from the first polarizer $J_{in}(\alpha)$. The light then enters the test birefringent material $G(\beta,\delta)$. The direction of the slow axis $\beta(x,y)$ and the lag of $\delta(x,y)$ both depend on space because the stress difference depends on space. However, since the sample is assumed thin and/or the force is uniform along the thickness in z, there is no dependence on z. This is the plane strain or plane stress assumption. Finally, the light goes through another linear polarizer, but rotated 90° from the first (i.e., at $\alpha + \pi/2$), and emerges as

$$J_{out} = P\left[\frac{\pi}{2} + \alpha\right]G(\beta,\delta)J_{in}(\alpha). \tag{2.7}$$

The intensity of the light collected by the camera is

$$I_{out} = I_0^2 J_{out}^* J_{out} = I_0^2 \sin^2[2(\alpha - \beta)]\sin^2\frac{\delta}{2}. \tag{2.8}$$

where I_0^2 is proportional to the intensity of the initial light entering the polariscope. Equation 2.1 converts δ to principle stress differences

$$I_{out} = I_0^2 \sin^2[2(\alpha - \beta)]\sin^2\left[\frac{\pi t K}{\lambda}(\sigma_1 - \sigma_2)\right]. \tag{2.9}$$

Thus, the intensity has zero contour lines if $\alpha - \beta = n\pi/2$ (isoclines) or $tK(\sigma_1 - \sigma_2) = N\lambda$ (isochromats), where n and N are integers. The isoclines determine the *direction* of the principle stresses. There is a zero contour wherever α is in the direction of either the slow or fast axis. By rotating α, the isoclines change, but the isochromats are constant and β, the direction of the slow axis, can be determined throughout the sample. The isochromats determine the magnitude of the principle stress difference. Unlike the isoclines, the isochromats depend on wavelength. Thus, if white light is used, each wavelength has zero contours at different places, leading to different colors (i.e., combinations of wavelengths) for every value of principle stress difference. Measurements of the wavelength content can be used to determine the principle stress difference. However, using the linear polariscope, the zeros of the isoclines and the isochromats overlap making it difficult to extract each. The circular polariscope allows direct measurement of the isochromats without isocline contamination.

The circular polariscope uses circularly polarized light to illuminate the test sample and the exiting light is analyzed using a circular polarizer. Linearly polarized light sent through a quarter-wave plate with slow axis at $45°$ from the initial polarization direction produces a circularly polarized output. The quarter-wave plate is a uniform birefringent material in which the optical lag δ is $90°$ or a quarter wavelength. The Jones matrix for a quarter-wave plate with slow axis angle θ with respect to the x-axis is determined from the general birefringent plate, $Q(\theta) = G(\theta, \pi/2)$. Then for the most common dark-field circular polariscope, two quarter-wave plates are inserted, one at $+45°$ and one at $-45°$ to produce

$$J_{out} = P\left[\frac{\pi}{2} + \alpha\right]Q_2\left[\frac{\pi}{4} + \alpha\right]G(\beta, \delta)Q_1\left[-\frac{\pi}{4} + \alpha\right]J_{in}(\alpha). \tag{2.10}$$

The intensity is

$$I_{out} = I_0^2 J_{out}^* J_{out} = I_0^2 \sin^2\frac{\delta}{2}. \tag{2.11}$$

In the circular polariscope, the output is independent of both α and β, and using (2.1) to convert δ to the principle stress difference

$$I_{out} = I_0^2 \sin^2\frac{\pi t K}{\lambda}(\sigma_1 - \sigma_2). \tag{2.12}$$

With this arrangement of elements (2.10) only isochromats are formed, and the background or field is dark, since from (2.11) with $\delta = 0$ the intensity is zero. Without the sample, the first quarter-wave plate produces *left* circularly polarized light and the second quarter-wave plate and linear polarizer only passes *right* circularly polarized light.

The type of circular polariscope can be categorized by four angles from right to left in (2.10): the angle of the polarizer, the slow axis angle of the two quarter-wave plates Q_1 and Q_2, and the polarization angle of the incident light. For the dark-field polariscope described by (2.10) with $\alpha = 0$ the four angles are $(90, 45, -45, 0)$. Another dark-field circular polariscope can be made using angles $(0, 45, 45, 0)$, which produces the same intensity as (2.12). However, quarter-wave plates must be made for a particular wavelength of light, since δ depends on λ. If the wrong wavelength of light is used then the $(90, 45, -45, 0)$ setup has some extra error canceling compared to the $(0, 45, 45, 0)$ version and is therefore preferred. The other common arrangement is $(0, 45, -45, 0)$, which gives an intensity of

$$I_{out} = I_0^2 \cos^2 \frac{\delta}{2}. = I_0^2 \cos^2 \frac{\pi t K}{\lambda}(\sigma_1 - \sigma_2). \tag{2.13}$$

For this arrangement, the field is bright since the intensity is maximum if $\delta = 0$. Again only isochromats are visible, but now the zero contours occur for $tK(\sigma_1 - \sigma_2) = (N + 1/2)\lambda$. This is the preferred bright field arrangement due to a similar cancelation of wave-plate error compared to $(0, 45, 45, 90)$, which also gives (2.13). It is useful to note that the dark-field *linear* polariscope described in (2.7) can be achieved using the same setup with angles $(90 + \alpha, 90 + \alpha, \alpha, \alpha)$ just by rotating the quarter-wave plate so they are aligned. Many commercial polariscopes allow the linear polarizers and quarter-wave plates to be rotated individually to change from linear to circular and from dark to bright field without removing the quarter-wave plates and in tandem to adjust α.

Using a combination of linear and circular measurements, the direction and magnitude of the principle stress difference can be determined and the full 2D stress tensor can be reconstructed. However, in granular applications, the internal stress distribution of each particle is less important than the forces between particles. Section 2.4.3 explores techniques to extract forces directly from photoelastic images.

2.3.3 Index Matching

Index of refraction matching is a common technique to overcome the opaque nature of granular materials from strong scattering due to the index of refraction difference between transparent solids and air. In this technique, the air is replaced by a liquid, which has the same index as the transparent granular material. While index matching does allow internal optical measurement of granular materials, it is invasive. The presence of the fluid typically changes the behavior of a granular system by increasing the effects of the fluid viscosity. Liquid viscosities are at least two orders of magnitude larger than gases. Care must be taken in comparing the results from fluid-saturated granular systems to those of dry systems. The presence of liquid is most important in flowing systems, but lubrication can affect static systems as well.

An example of index matching is shown in Figure 2.5. In panel (a), borosilicate glass spheres in a borosilicate cylindrical container are shown in front of an image. The image is obscured by the scattering from the transparent glass spheres.

(a)   (b)   (c)   (d)

Figure 2.5   (a) Borosilicate glass spheres and container in air. (b) Borosilicate glass spheres in a container filled with index-matching oil. Here, there is no refraction from the particles, but the container acts like a lens. (c) Borosilicate glass spheres in a container filled with index-matching oil and with the container immersed in index-matching oil. Here, there is no refraction from the particles or the container. (d) Borosilicate glass spheres in container filled with dyed index-matching oil and with the container immersed in index-matching oil. Here the image is not refracted, but the intensity depends on the average amount of fluid in the path of the light. The particles are visible, but the image behind the cell is undistorted.

In panel (b), corn oil replaces the air, and the particles no longer scatter light. However, the cylindrical container now acts as a lens, and the image behind is distored. In panel (c), the cylindrical container is placed in a second container with flat walls and filled with corn oil. Now the particles and container are index matched and the background image is clear and undistorted.

Borosilicate glass and vegetable oil are only one of the common index-matching systems. Many specially blended fluids have been designed to index match different types of glass and other transparent materials [235]. Glass and oil have an advantage for studying granular systems because the density contrast with the fluid is still high enough that particle inertia can be dominant. On the other hand, if index matching and density matching are desired, PMMA particles in Triton X-100, anhydrous zinc chloride, and water [236–243] are often used to study suspensions.

Index matching alone simply makes the particle invisible. To make the particles visible, a dye is added to the fluid and shown in panel (d) of Figure 2.5. Now the intensity of transmitted light is proportional to the summed density of particles in the container, since only the fluid attenuates the light. This technique is analogous to an x-ray radiograph, in which the absorption of light is caused by the presence of the dye in the fluid. Another method replaces a small fraction of the particles with colored particles as tracers. The position and motion of the tracers can be tracked using the techniques of (Section 2.4).

A very powerful method to visualize index-matched particles uses fluorescent dyes and a laser sheet to illuminate a 2D cross section of a granular sample [238–243]. A schematic of the technique is shown on the left of Figure 2.6. Since the particles are index matched the light passes through without refraction and the laser sheet illuminates only a thin section of the sample. In the approximately

(a)

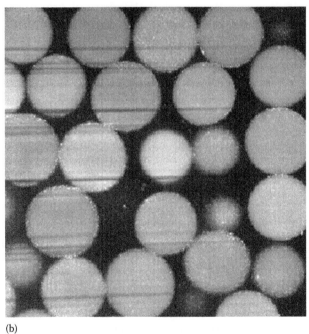

(b)

Figure 2.6 (a) Schematic of the laser sheet technique to visualize particle positions in 3D. Particles are index matched so that the laser sheet can be transmitted through the sample. The particles are labeled with a fluorescent dye that is activated by the laser light, but fluoresces at a different wavelength, causing the particles to glow. Once an image is captured, the laser sheet is moved to a new position (slice). Typical image sets may contain 1024 slices, each with 1024×1024 pixels. (b) A typical experimental image of a single slice through a 3D granular packing of spheres.

2D illuminated section, the dyed particles fluoresce and an image is taken perpendicular to the laser sheet, giving a clear slice of the 3D structure. This image is used directly to follow particles that are currently in the slice or the laser is then scanned to produce sufficient slices to reconstruct a full 3D image of the sample.

On the right of Figure 2.6, an example slice is shown from a system of index-matched spheres in a cubical container. The particles are made of polyacrylamide and are swollen in a 0.05 mg mL^{-1} Rhodamine 6G solution. Rhodamine 6G is a fluorescent dye that is excited by 530 nm light and emits at 560 nm. Because swollen hydrogels like polyacrylamide are mostly water, their index of refraction is matched to water. This system has been used to measure contact areas and relate them to contact forces [244]. The laser sheet in this image is formed by using a cylindrical lens, but it can also be created with acousto-optic modulators. In the image, the laser light enters from the right. The dark streaks emanating from the edges of some of the particles are shadows caused by surface effects. This artifact increases as the number of particles increases. By collecting many 2D images at different positions, a full 3D image can be obtained.

2.3.4 X-Ray

While x-ray imaging is predominately used in medical applications, researchers have used x-rays to measure the internal structure of granular systems since the 1960s [130,131,136–138,140,141,245–255]. Much of the early research focused on direct 2D density measurements in hoppers. X-rays are high-energy photons, which are not refracted by most materials, but can be absorbed. X-rays sent through a sample are differentially absorbed depending on the thickness and radio-opacity of the material. For a granular system made of a single material only the cumulative thickness of grains in the x-ray path determines the absorption. Thus, the intensity of x-rays reaching the detector is inversely proportional to the density of the granular material along the x-rays path. Lead [136,137] or iodine tracers [141,245] are used to enhance the contrast due to high x-ray radio-opacity of those materials.

Early x-ray images called radiographs were captured using photographic plates or film, which limited the number and speed at which images could be taken. Modern fluoroscopes use digital detectors to allow the acquisition of many frames per second. Fluoroscopes have been used to measure density waves in flowing granular hoppers [141,245].

X-rays can also be used to reconstruct 3D density information from multiple 2D projections taken at different angles. The technique called x-ray computed tomography (CT) is based on the Radon transform [256,257] and was developed for medical imaging in the 1960s and 1970s. Recently, smaller x-ray CT machines (μCT) have been developed for the nonmedical market and used to study granular materials [247–255,258]. An example scan is shown in Figure 2.7. The left image is a single reconstructed slice through a collection of long rods. Any slice can be created retrospectively from the 3D data set. The middle image shows a 3D surface reconstruction of the entire pile of iodine-soaked wooden rods. The right image is an enlargement.

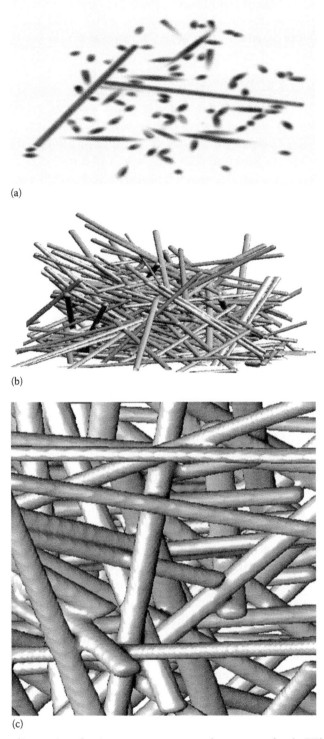

(a)

(b)

(c)

Figure 2.7   Three-dimensional micro x-ray computed tomography (μCT) scan of a pile of iodine soaked and dried 10 cm bamboo rods with aspect ratio $\alpha = 30$. The iodine enhances the x-ray absorption. (a) Single slice from the reconstructed image. (b) Surface reconstruction of the entire pile. (c) Close-up of a section of the pile.

μCT is a powerful technique to image the 3D structure of granular materials. The main drawback is the time to acquire all of the views needed for the reconstuctions, which is typically on the order of minutes.

2.3.5 Magnetic Resonance Imaging

Magnetic resonance imaging (MRI) is another, primarily medical technique, which has been successfully applied to granular materials [57,61,62,65,106,107,120, 124–126,259–267]. MRI was developed in the 1970s [268–270] and is an extension of nuclear magnetic resonance (NMR) developed in the 1940s [271,272]. MRI has been used to study granular materials since the 1990s. In particular, MRI has been used to measure the local velocity and density fields in a rotating drum [120,261,262,265,266], to observe convection due to shaking [57,61,62,260], to observe mixing in rotating drums [124–126,261,273,274], and visualize vibrating systems [65,106,107,263,267].

There are a number of important MRI techniques that are applicable to the study of granular materials. All are based on combining NMR with linear field gradients to encode spatial information into the frequency of precession.

When a nucleus with a magnetic moment, or spin, is placed in a magnetic field, the magnetic moment precesses about the magnetic field direction with a frequency ω_L proportional to the field strength. For example, for protons in $H_0 = 2$ T magnetic field,

$$\omega_L = 2\pi\gamma H_0 = 2\pi(85.5 \text{ MHz}) = 537 \times 10^6 \text{ rad s}^{-1}, \tag{2.14}$$

where $\gamma = 4.258 \times 10^7$ Hz T^{-1} is called the gyromagnetic ratio. In a large system in equilibrium, there is no net signal from this oscillation since all of the spins are at different phases. The average magnetization is constant and parallel to the applied magnetic field. If the spin system is taken out of equilibrium, as discussed later, it will eventually exponentially relax back to equilibrium with a time constant T_1 through interactions between the spins and the environment.

If a small oscillating magnetic field is also applied, there is a net absorption of energy from the oscillating field when its frequency is equal to the frequency of precession. This resonant excitation field causes a synchronous rotation of the magnetization toward the transverse plane, perpendicular to the static field. If the excitation field is turned off at the point when the magnetization reaches the transverse plane, a coherent magnetic signal oscillating at ω_L is produced. This type of excitation is called a 90° pulse since it rotates the magnetization 90° into the transverse plane. The signal can be measured by induction [272] with a resonant pickup coil and is proportional to the sum of the signals from all spins in the pickup coil's volume of sensitivity V_c. The signal will decay exponentially with a decay constant T_2^* due to two effects: (1) coupling with other spins and (2) small difference in the local magnetic field. This signal is called a free induction decay. NMR is the process of resonant absorption of energy by nuclear spins causing a coherent magnetic signal and can be used to determine many details of atomic interactions and molecular structure.

NMR is also the basis of MRI. Because ω_L is proportional to the magnetic field strength, a field, which varies linearly with position, can be applied after the excitation and used to encode spatial information. This can be done during the free induction decay or later during a spin echo. A spin echo can be produced by applying a second excitation pulse at a time τ after an initial 90° pulse, which rotates the spins by 180°. Before the 180° pulse, spins will rotate at slightly different rates due to small differences in the magnetic field, which is the cause of the second type of decay discussed earlier. Faster spins will get ahead of slower ones. However, this type of decay is reversed by the 180° pulse. The 180° pulse flips the phase of the spins causing the faster spins to be behind the slower spins. As the spins continue to evolve, the fast spins catch back up to the slower spins and an echo is formed at a time 2τ [275–277]. The magnitude of the echo signal still decays exponentially as a function of τ with decay constant T_2 due to the first mechanism of spin–spin interaction, which is irreversible.

If a magnetic field gradient in the x-direction, G_x, is applied during a free induction decay or a spin echo, then for times short compared to T_2^*, the signal

$$S'(t) \propto \exp(i\omega_L t) \int_{V_c} n(\mathbf{x}) \exp(2\pi i \gamma G_x x t)\,d\mathbf{x}, \tag{2.15}$$

where

x is the x-position of a spin
$n(\mathbf{x})$ is the number density of spins at position $\mathbf{x}$

NMR signals are usually measured using quadrature demodulation at ω_L so $S(t) = S'(t)\exp(-i\omega_L t)$ contains both the real and imaginary parts of (2.15). $S(t)$ can be recast as the k_x component of the Fourier transform of the spin density $n(\mathbf{x})$,

$$S(k_x) \propto \int_{V_c} n(\mathbf{x}) \exp(i k_x x)\,d\mathbf{x}, \tag{2.16}$$

by identifying $2\pi\gamma G_x t$ as k_x. Then, one line in "k-space" will be determined in time from $S(k_x) = S(t)$. Typically, multiple experiments are needed to completely sample the full 3D k-space

$$S(\mathbf{k}) \propto \int_{V_c} n(\mathbf{x}) \exp(i\mathbf{k} \cdot \mathbf{x})\,d\mathbf{x}. \tag{2.17}$$

Many different methods exist to sample k-space, including radial sampling, spiral sampling, and raster sampling. For more details, see, for example, [277]. Once sufficient sampling is complete, the spin density $n(\mathbf{x})$ is recovered by Fourier transform.

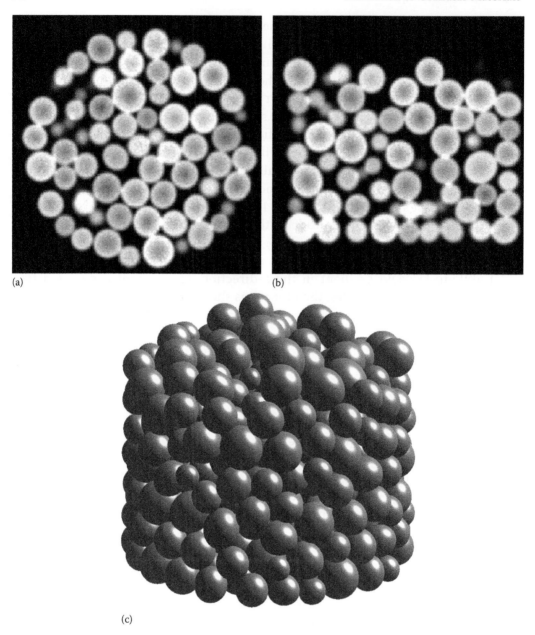

Figure 2.8 (a) Axial slice from a 3D magnetic resonance scan of 362 mustard seeds with mean diameter $D = 2.0 \pm 0.3$ mm in a 17 mm diameter tube. (b) Longitudinal slice. (c) 3D reconstruction of the particles.

To apply MRI to granular materials, particles, which have nuclear spins, are needed. The most common is hydrogen, and most medical scanners are designed to image hydrogen. One significant problem is that the relaxation time T_2 is usually very short for solids, and imaging them is difficult. To overcome this difficulty, fluid-saturated solids are often used. Seeds are a cheap source of oil-saturated solids. Common examples of high oil content seeds are poppy and mustard. A 3D MRI scan of mustard seeds is shown in Figure 2.8. On the left is an axial image of a cylindrical container of 362 mustard seeds. The center shows a longitudinal slice. From the full 3D data set any 2D slice may be retrospectively reconstructed. Using the techniques of (Section 2.4) the center positions were extracted. The mustard seeds are slightly prolate ellipsoids on average, and three principle diameters, D_1, D_2, D_3, sorted from small to large, and three orientation angles were also extracted. A 3D surface reconstruction of the packing using the measured positions, diameters, and angles is shown on the right.

Statistical properties of the mustard seed packing are shown in Figures 2.9 and 2.10. From D_1, D_2, D_3 a mean diameter $D = \sum_n D_n / 3$ and three scaled principle diameters $\sigma_n = D_n / D$ were defined. The histogram of mean diameters is shown on the left of Figure 2.9. The average diameter is 2.05 mm, and the standard deviation is 0.25 mm. To categorize the shape of the seeds, the ratio of the small-to-middle and middle-to-large is examined by plotting the value of $1 - \sigma_1 / \sigma_2$ versus $1 - \sigma_2 / \sigma_3$ for each particle. Both of these are zero for spheres. If the first is large, then the shape is oblate; if the second is large, then the shape is prolate; if both are large, then the shape is triaxial. The data for each seed are shown in the center panel of Figure 2.9. The categories are defined based on 10% deviations from spherical. The majority of points are less than 10% away from spherical but more on the prolate side. To further investigate the shape, a sphericity parameter

$$\Psi = \frac{\pi^{\frac{1}{3}}(6V_p)^{\frac{2}{3}}}{S_p} \tag{2.18}$$

is defined as the ratio of the surface area of a sphere with the same volume V_p as the particle to the actual surface area of the particle S_p. $\Psi = 1$ for a sphere and $\Psi < 1$ for all other shapes. The histogram of Ψ in the right panel of Figure 2.9 shows that the particles are nearly spherical.

Figure 2.10 gives information about the particle interactions. On the left, the distribution of contacts per particle shows that particles have on average six contacts. This is below the isostatic number of 10 for frictionless ellipsoids, but is consistent with frictional particles. The radial distribution function is shown in the center panel normalized by the mean particle diameter. The broadness of the first peak is consistent with the standard deviation of the mean particle diameters. From the contacts, an estimate of the overlap or deformation can be made by assuming that the particles are perfect ellipsoids. The histogram of these deformations is shown in the right panel of Figure 2.10. It follows an exponential distribution with scale $0.004D$.

MRI is also sensitive to motion and can be used to measure velocity and deformations. Three main techniques are used: (1) spin tagging, (2) phase contrast, and (3) Fourier velocity encoding.

Spin tagging consists of preferentially exciting a region of the sample, allowing the spins to move, then imaging the spins normally by sampling k-space. Selective excitation is accomplished by using a gradient field during the excitation pulse.

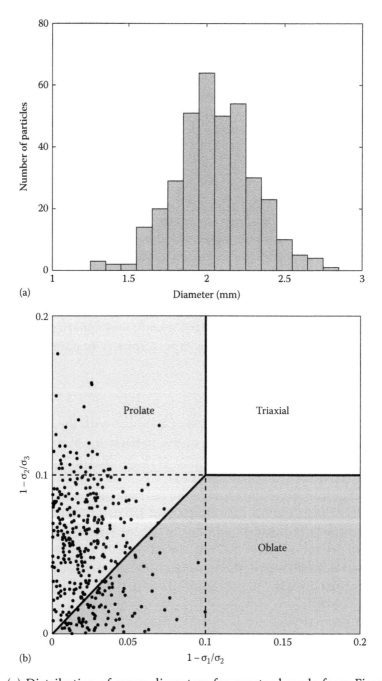

Figure 2.9   (a) Distribution of mean diameters for mustard seeds from Figure 2.8. The seeds are modeled as ellipsoids. (b) Categorization of the shape (slightly prolate) of the seeds. $\sigma_1, \sigma_2, \sigma_3$ are the sorted (small-to-large) principle diameters scaled by the mean diameter.   (*Continued*)

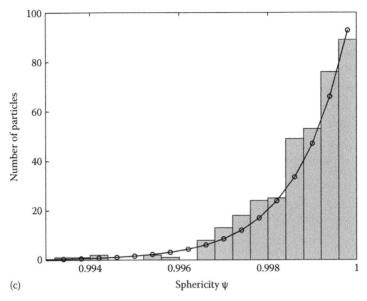

(c)

Figure 2.9 (*Continued*)   (c) Distribution of sphericity (see text).

The gradient field will change the precession frequency so that only the area near the zero of the gradient will be in resonance and be excited. For a linear gradient and a single excitation frequency the excited portion will be a plane or slice. The thickness of the slice will be determined by the bandwidth of the excitation pulse. A schematic is shown in Figure 2.11. The initial excitation is shown in stripes for a tube of fluid flowing to the right. At a later time τ_f the excited fluid has moved to the position shown by the hash marks. A standard imaging sequence is used to measure the new positions after the delay. This technique is called "time-of-flight" [278]. For this to work, the excitation time and measurement time must be short compared to the time it takes a spin to move across a voxel. The flight time τ_f needs to be long enough that the spins have moved sufficiently to be measured and $\tau_f < T_2$ to ensure enough signal is available. Finally, most imaging techniques require multiple excitations to sufficiently sample k-space. Therefore, the flow must be either steady or periodic during the full acquisition of k-space. This technique has been used to measure velocity fields in suspensions [259] and granular rotating drums [120].

More complicated modulated excitation pulses can be used to create excitation regions with other geometries like stripes and grids for motion measurements [279,280]. This technique was used to measure convection in vertically vibrated granular materials [57].

The phase contrast method was developed in the 1980s [281,282]. This method encodes velocity in the phase angle of the standard imaging signal (2.16). Usually, after Fourier transform, the phase angle of the image is discarded and only the magnitude that is proportional to the spin density $n(\mathbf{x})$, is retained. However, using a bipolar gradient, velocity can be encoded into the phase signal. A bipolar gradient consists of two gradient pulses of equal area, one that has the gradient direction reversed. In this situation, a stationary spin processes faster during the first pulse and slower during the second pulse. The net result is that all stationary spins have the same phase relation before and after the application of a bipolar gradient. Spins that move in the gradient direction between the application of the first and second pulses will not return to their initial phase. Under the approximation that the pulse width is small, the phase change of a moving spin is proportional to the distance traveled in the direction of the field gradient and therefore proportional to the average velocity. By retaining

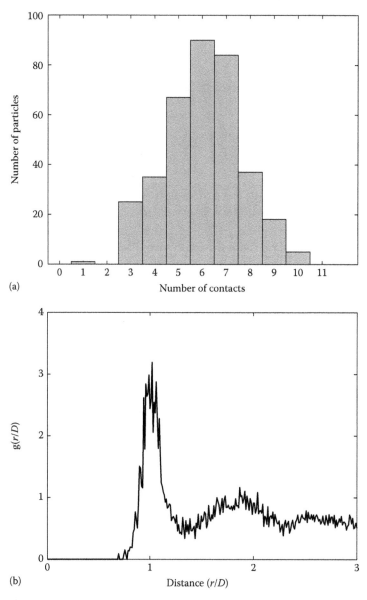

Figure 2.10   (a) Histogram of the number of contact for each particle from Figure 2.8. (b) Radial distribution function for particles. The distance is scaled by the average mean diameter. Spreading of the peak at 1 is due to diameter variability.   (*Continued*)

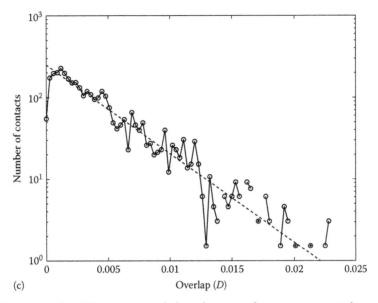

(c)   Overlap (D)

Figure 2.10 (*Continued*)   (c) Exponential distribution of apparent particle overlaps with exponential scale of $0.004D$.

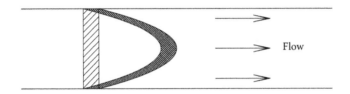

Figure 2.11   Diagram showing the time-of-flight flow measurement technique. Spins are excited in a thin slice, shown with stripes. They are allowed to flow to the position shown with hash marks, where their position is measured.

the phase portion of the signal after Fourier transform the velocity can be computed. Phase contrast methods have been applied to suspensions [259] and granular materials [120].

The dynamic range of phase contrast is limited by signal to noise and is not well suited for flows with a wide range of velocities. It is able to measure velocity in all three directions. As with time-of-flight techniques, the flow field must be either stationary or periodic. In practice, phase images are very sensitive to inhomogeneities in the static field, the gradients, and the excitation field. As a result, the background phase must be removed before this technique can be used successfully.

Fourier velocity encoding was developed theoretically by [283] and verified experimentally by [284]. For details see [277]. Fourier velocity encoding takes phase contrast one step further by directly encoding the velocity into the velocity Fourier space or "q-space" in the same way position is encoded into k-space.

Using multiple bipolar gradients of different strengths a six-dimensional k and q signal space can be defined

$$S(\mathbf{k},\mathbf{q}) \propto \int_{V_c} n(\mathbf{x},\mathbf{v}) \exp(i\mathbf{k}\cdot\mathbf{x} + i\mathbf{q}\cdot\mathbf{v}) \, d\mathbf{x}\, d\mathbf{v}, \tag{2.19}$$

where
  $\mathbf{v}$ is the velocity of a spin
  $n(\mathbf{x},\mathbf{v})$ is the number of spins per unit 6D space velocity volume
  In practice, the full six dimensions are usually not all measured because of time
    constraints.

As an example, the flow of water through two long pipes with inner diameters of 1.765 cm placed side by side was measured using velocity encoding. For this case, $n(\mathbf{x},\mathbf{v})$ was limited to $n(x,y,v_z)$, where x and y are in the plane transverse to the flow and v_z is the direction of the flow. At one end, the pipes are connected together, and at the other end one is used as the inlet and the other as the outlet. Since both the inflow and the outflow are imaged, the net flow rate must be zero. Without any adjustable parameter, we find that the net flow rate is 3.2×10^{-3} cm^3 s^{-1} which is only 2.7% of the flow rate through one of the pipes. The flow profile is shown in Figure 2.12a and the velocity along a line through the center of both pipes is shown in Figure 2.12b. The dotted line is the theoretical parabolic flow with no adjustable parameters based on the separately measured total flow rate. One of the strengths of the velocity encoding technique is the ability to determine the velocity distribution. In Figure 2.12c, the measured and the theoretical distributions are shown for both pipes. The agreement is good with no fitting parameters.

Another example using mustard seeds as a granular material is shown in Figures 2.13 and 2.14. In this case [106,107], 55 mustard seeds with mean diameter $D = 2.04$ mm are placed in a cylindrical container of diameter 9 mm $= 4.5D$ and shaken vertically along the cylinder axis. Velocity encoding is used to measure the spin density $n(z,v_z)$, where z is the position along the cylinder axis and v_z is the velocity. $n(z,v_z)$ is measured at 16 phases of the shaking cycle and 8 are shown in Figure 2.13. The position of the bottom of the container is shown as a white horizontal line with a vertical tick to mark the zero of velocity. Figure 2.14 shows $n(1.7D,v_z)$ (circles) and $n(6D,v_z)$ (squares) from phase $2\pi/8$. Even though the vertically averaged velocity distribution $\int n(z,v_z)\,dz$ (thick black line) is not Gaussian, the individual distributions at specific heights fit well by a Gaussian (gray lines).

In summary, MRI is a powerful technique to see inside a 3D sample and measure velocity and deformation. Materials are generally restricted to liquids or liquid-impregnated solids like seeds, or to special designed particles. Some MRI systems can also image waxes or soft rubber. Almost no metal or magnetic

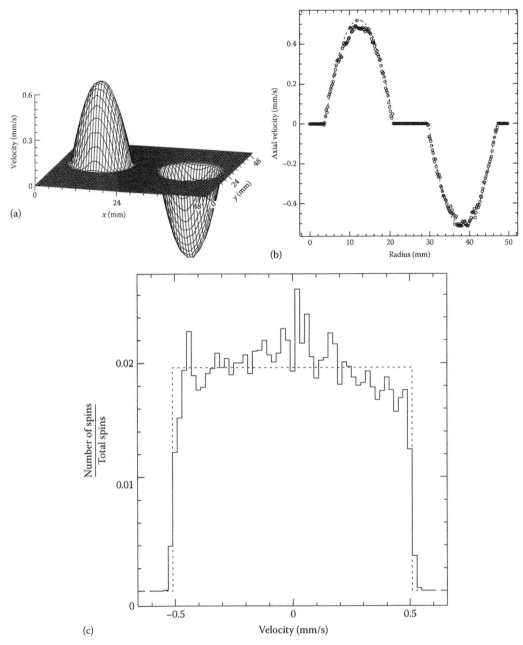

Figure 2.12   Fluid flow in twin counter flowing 1.765 cm diameter pipes. The velocity is averaged over a 1 mm thick slice. The in-plane resolution is 195 μm, and the velocity resolution is 40 μm s^{-1}. The maximum velocity is 520 μm s^{-1}. (a) Measured velocity profile in both pipes. (b) Measured velocity (circles) along a line going through the centers of both pipes. The theoretical velocity profile based on the flow rate and the diameter of the tube is shown as a dotted line. There are no fitted parameters. (c) Measured distribution of velocity through both pipes (solid) and theoretical velocity distribution (dotted) with no fitting parameters.

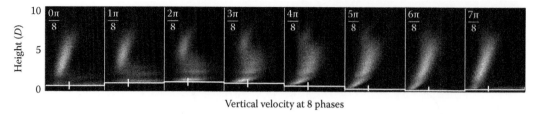

Vertical velocity at 8 phases

Figure 2.13   Position and velocity distributions for a vertically vibrated cylindrical container of mustard seeds. The average probability of a particle having a vertical position z (vertical axis) and vertical velocity v_z (horizontal axis) at 8 equally spaced phases, ϕ of the cycle, $n(z, v_z, \phi)$, in a system of 55 vertically vibrated mustard seeds as measured by magnetic resonance. The horizontal white line shows the position of the bottom of the container at each phase and the length is equal to a velocity of 1 m s^{-1}. The short vertical line is the zero of v_z. (From Huntley, J.M. et al., *Proc. R. Soc. A*, 463, 2519, 2007; Mantle, M.D. et al. *Powder Technology*, 197, 164, 2008.)

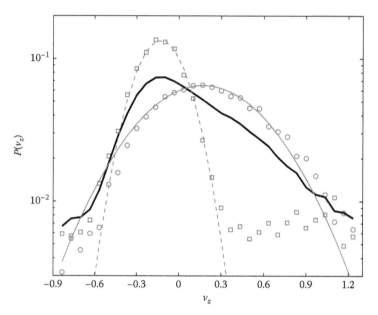

Figure 2.14   Distribution of vertical velocities $n(v_z)$ for system described in Figure 2.13 at phase $\phi = 2\pi/8$ and at two heights: one near the bottom at $z = 1.7D$ (circles) and one near the top at $z = 6D$ (squares). Gaussian fits are shown as light lines. The dark line is the height-averaged velocity distribution, which is not Gaussian. (From Huntley, J.M. et al., *Proc. R. Soc. A*, 463, 2007; Mantle, M.D. et al.)

materials may be used. Typically, scans take a long time, minutes to hours. It is good for quasi-static flows like rotate, scan, repeat [124], for stationary flow, or for periodic flows where stroboscopic phase averaging can be used [106,107]. There are some fast techniques like echo-planar imaging and spiral sampling. The fastest techniques may allow 2D scans at 30 fps.

2.4   Image Analysis

The visual nature of granular materials makes image analysis a key part of many experimental techniques. Although image analysis has been used since the 1940s [1–3,6–10,12,13,31,34,66,71,78–80,114,121–123,126,129,132,134, 135,139,140,146,147,150–152,154,157,198,200–204,285] the advent of inexpensive digial cameras has revolutionized many experimental imaging techniques. Gone are the days of projecting images onto a screen to extract quantitative data [3,114,135]. Now inexpensive "webcams" can deliver digital images directly to the computer at high resolution (1920×1080) and high speed (60 fps). Fourth generation programming languages like MATLAB®, IDL, and Python make the development of image analysis software tractable. For example, all of the techniques discussed in the following Sections 2.4.1 through 2.4.3 have been implemented by the author in MATLAB.

2.4.1   Particle Tracking

Many of the visualization techniques discussed in (Section 2.3) produce images of granular particles. Researchers use particle tracking techniques to extract more quantitative information from images [3,41,48,52,55,78–80,83,105,151,152,154, 239,241,242,286,287]. Early tracking techniques used movie film at up to 2000 fps with tracking done directly by projecting individual image frames [3]. Later, movie frames were projected onto digitizers to bring them into computers for analysis producing particle tracks at 2880 fps with accuracy of 1/50 of a particle diameter [151,152,154]. Currently, fast high-resolution images are taken directly by digital cameras either connected to computers or connected to high-capacity digital storage media.

The task of particle tracking breaks easily into two parts: locating the particles in a particular image frame (location) and determining where a particle has moved from one frame to the next (connection).

Connection: The second task is generally straightforward and will not be discussed in depth. The general outline consists of finding the permutation of particle indices from frame n to frame $n+1$, which minimizes the maximum distance traveled by a particle. If the particles move less than half a particle diameter D from one frame to the next then the process is relatively easy. Find the permutation of m, $P(m)$, such that $\delta_{m,P(m)} = \left| \mathbf{x}_m(n\Delta) - \mathbf{x}_{P(m)}((n+1)\Delta) \right| < D/2$ for all m, where Δ is the time between frames. If the particles move less than $D/2$ in a time Δ then only one particle can be less than $D/2$ away in the next frame. If particles move more than $D/2$ in one frame then the problem is ill-posed and there is no way to identify a particle from one frame in the next definitively. However, in practice, there are methods to find the best guess for the proper connection.

The typical approach finds all (m,k) pairs with $\delta_{m,k} < L$, where L is the maximum distance that a particle might travel. Then test all possible $k = P_q(m)$ for each q combination to find the minimum $\sum_m \delta_{m,P_q(m)}$. As earlier, if $L < D/2$, then there will only be one (m,k) for each of the N particles and the permutation $P(m)$, which minimizes the total distance, difference is unique. But as L gets larger there will be more than N pairs. For example, particles 1 and 2 in frame one may be within L of particles 3 and 4 in frame two and all other particles only have one pair within L. Then the problem pairs are $(1,2)$, $(1,4)$, $(2,3)$, and $(2,4)$, but only two permutations $[P(1) = 3, P(2) = 4]$ and $[P(1) = 4, P(2) = 3]$ need to be tested.

Another case that can occur when $L > D/2$ is that all or some of the particles have a mean motion which when corrected for would yield smaller frame-to-frame distances. In this case, modify $\delta_{m,P(m)} = |\mathbf{x}_m(n\delta) + \mathbf{V}(\mathbf{x}) - \mathbf{x}_{P(m)}((n+1)\delta)|$, where $\mathbf{V}(\mathbf{x})$ is the possible space-dependent mean motion of particles at position $\mathbf{x}$. For example, $\mathbf{V}$ would be a constant in a sedimenting flow. It is possible to measure $\mathbf{V}(\mathbf{x})$ from the data by iteration in steady flow. Assume $\mathbf{V}(\mathbf{x}) = 0$ and calculate $P_0(m)$. Using $P_0(m)$, calculate $\mathbf{V}_1(\mathbf{x})$ by averaging the velocity field. Then use $\mathbf{V}_1(\mathbf{x})$ to calculate $P_1(m)$ and repeat until $P_n(m)$ does not change.

Location: Many techniques exist to locate particles and other objects in images. For a review, see a modern text on machine vision (e.g., [228]). Several special techniques exist specifically for circular objects. For example, the centroid method [286] and the Hough transform [52,55,105,289,290] have been applied to circular granular materials. Since digital images represent light intensity near discrete grid points or pixels, the spatial accuracy of the information obtained from the image depends on the pixel size or resolution. Most techniques can determine an object's position to pixel resolution (i.e., the center of the object is within a certain pixel); however, to measure many important quantities sub-pixel accuracy is needed. For example, velocity distributions require both high-spatial and high-temporal resolution since velocities depend on the difference in position and time. Another example is the radial distribution function (the probability that a particle is a given distance from another particle). This function also depends on position differences. With the proper tracking technique both the velocity distribution [80] and radial distribution function [79] can be measured with high precision, which compares well with simulation and theory. This section will focus on a general sub-pixel accurate technique based on least squares fitting (LSQ). LSQ can not only locate particles, but also extract more general information from images, for example, shape (Section 2.3.5), forces (Section 2.4.3), and velocity (Section 2.4.2). Further, many tracking techniques, including centroid and Hough transforms, can be expressed in the LSQ framework.

To be concrete, assume that the image to be tracked is of the form

$$I_{exp}(\mathbf{x}, t) = \sum_{n=1}^{N} I_p^*(\mathbf{x} - \mathbf{x}_n(t); D, \dots), \qquad (2.20)$$

where

    N is the number of particles

    $\mathbf{x}_n(t)$ is the position of particle n at time t

    $I_p^*(\mathbf{x}; D, \dots)$ is a function describing the shape of a particle centered at the origin

The particle shape function may depend on the diameter of the particle D or any other properties of the particles or the imaging system. To track particles with actual shape, I_p^*, in an image described by (2.20), an ideal particle function I_p is defined. For spherical particles imaged in backlighting and normalized so that the particles are bright,

$$I_p(\mathbf{x}; D, w) = \left[1 - \tanh\left(\frac{|\mathbf{x}| - D/2}{w}\right)\right] / 2 \tag{2.21}$$

is a good choice. This function smoothly varies between 1 inside the particle, 1/2 at D, and 0 outside of the particle. The parameter w determines how sharply the function changes ($\tanh(1) \simeq 76\%$ change over a range of $\pm w$). In principle, w can be related to the focus of the imaging system through the point spread function. Smaller values of w correspond to sharper focus. A value of $w = 0.5$ pixels is typical of a good digital imaging system with a sharp focus. Figure 2.15 shows an

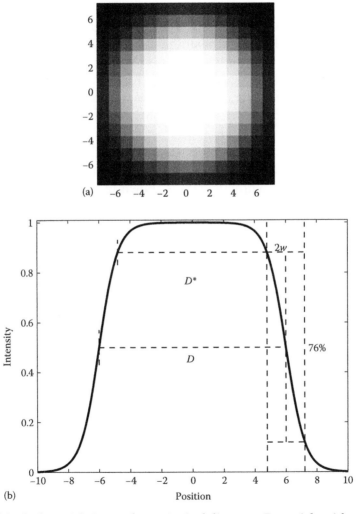

Figure 2.15 (a) Ideal particle image for a 12-pixel diameter D particle with $w = 1.2$ pixels $\simeq D/10$. (b) Central cross section of ideal particle image. D is the point where the intensity equals 1/2 of its maximum value. $D^* \simeq D - 2w$ is the real diameter of the particle. Calibration is required to obtain an accurate measure of D^* in terms of D.

image of a $D = 12$, $w = 1.2$ pixel ideal particle and a cross section through it. The actual diameter of the particle D^* is usually less than D and may be as small as $D - 2w$. However, to measure D^* accurately a separate calibration is needed. D should be thought of as a fitting parameter related to the diameter of the particle. The basic idea of LSQ is to compare the actual image to a calculated image based on (2.20) and find the positions and parameters that minimize the difference. In that way, the LSQ determines the most likely position of a particle (i.e., the position $\mathbf{x}_0$ that minimizes the weighted squared difference between the actual image $I(\mathbf{x})$ and the calculated ideal particle image $I_p(\mathbf{x} - \mathbf{x}_0)$). Specifically, the minimum of the weighted squared difference χ^2 is given by

$$\chi^2(\mathbf{x}_0; D, w) = \int W(\mathbf{x} - \mathbf{x}_0)[I(\mathbf{x}) - I_p(\mathbf{x} - \mathbf{x}_0; D, w)]^2 \, d\mathbf{x}, \qquad (2.22)$$

where W is the weight function. The domain of integration is over the area of the experimental image I. However, the domain of $\mathbf{x}_0$ is larger. In general, if the size of the image is Lx by Ly and the size of I_p is sx by sy then the range of integration is [0 Lx] and [0 Ly] but the range of $\mathbf{x}_0$ is [−sx Lx+sx] and [−sy Ly+sy]. When $\mathbf{x}_0$ is the position of a particle center, then χ^2 will be minimum. Therefore, minimizing χ^2 over $\mathbf{x}_0$ will produce the particle's position. In fact, if all of the particles are the same then there will be a minimum in χ^2 at the position of each particle. If the image was given by (2.20) with $I_p^* = I_p$ then χ^2 would be zero for each $\mathbf{x}_0 = \mathbf{x}_n$. The process of finding the particles is equivalent to finding all of the minima of χ^2. An example of χ^2 for synthetic data is shown in Figure 2.16. In the figure, χ^2 is inverted and normalized to the interval [0,1] for display; therefore, bright spots represent the positions of the minima of (2.22).

One strength of this method is that χ^2 can be expressed in terms of convolutions. Convolution and cross-correlation can be evaluated quickly using Fourier

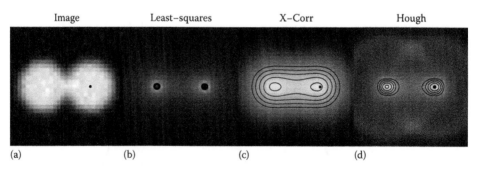

| Image | Least–squares | X–Corr | Hough |

(a)   (b)   (c)   (d)

Figure 2.16 Comparison of circle center finding techniques. (a) Synthetic 29×29 pixel image of two particles, generated using (2.20) with $D = 12 + (\pi - 3)$ pixels, $w = 1.3$ pixels, $\mathbf{x}_{1,2} = (\pm D/2, 0)$ and signal to noise of 40. An irrational position is used to insure that the exact position is not obtained due to truncation. (b–d) Likelihood map scaled to the interval [0,1] for finding a particle at a given pixel for least square, cross-correlation, and Hough transform. Interpolated contour lines are at $(0.975, 0.875, 0.775, 0.675, 0.575)$. The black dot in each image is the position of the right particle.

transforms and are implemented in hardware on the graphics processing unit of most computers with a video card. Expanding (2.22):

$$\chi^2(\mathbf{x}_0; D, w) = \int W(\mathbf{x} - \mathbf{x}_0)[I(\mathbf{x})^2 - 2I(\mathbf{x})I_p(\mathbf{x} - \mathbf{x}_0; D, w) + I_p(\mathbf{x} - \mathbf{x}_0; D, w)^2]\, d\mathbf{x},$$

$$(2.23)$$

$$\chi^2(\mathbf{x}_0; D, w) = I^2 \otimes W - 2I \otimes (WI_p) + \langle WI_p^2 \rangle,$$
$$(2.24)$$

where

$$f \otimes g = [f \otimes g](\mathbf{x}_0) = \int f(\mathbf{x})g(\mathbf{x} - \mathbf{x}_0)\, d\mathbf{x} \qquad (2.25)$$

is a modified cross-correlation (the sign of $\mathbf{x}_0$ is usually +) and $\langle f \rangle = 1 \otimes f$, which is not simply a constant, but a function of $\mathbf{x}_0$ since the domain of integration is over the image only. To evaluate (2.25) using convolution defined as

$$f * g = [f * g](\mathbf{x}_0) = \int f(\mathbf{x})g(\mathbf{x}_0 - \mathbf{x})\, d\mathbf{x}, \qquad (2.26)$$

care must be taken if W or I_p are not symmetric since

$$f \otimes g = [f(\mathbf{x}) \otimes g(\mathbf{x})](\mathbf{x}_0) = [f(\mathbf{x}) * g(-\mathbf{x})](-\mathbf{x}_0). \qquad (2.27)$$

Several choices are possible for the weight function. If

$$W = 1, \; \chi^2(\mathbf{x}_0; D, w) = \int I^2\, d\mathbf{x} - 2I \otimes I_p + \langle I_p^2 \rangle. \qquad (2.28)$$

In this case, the first term does not depend on $\mathbf{x}_0$ and the last term only depends on the $\mathbf{x}_0$ near the edge of the image. Thus, $I \otimes I_p$ will be maximum wherever $\mathbf{x}_0$ is the particle's position. The fact that the cross-correlation is maximum near particle centers is the starting point for many particle tracking techniques. It is clear that the basis for this ansatz is the minimization of χ^2. Cross-correlation alone works well for images in which the particles are well separated and have good signal to noise. This is because the weight function is very broad making the fit sensitive to the fact that the ideal particle image has zeros all around it, in contrast with the real image, in which other particles are nearby. Figure 2.16c shows an image normalized to the interval $[0, 1]$ of $-\chi^2$ from Equation 2.28 with $D = 12.14$ for synthetic data. The maximum shown by the contours is shifted slightly from the actual particle position. However, the position will likely be accurate to a pixel. This choice of W also produces a broad peak to maximize making it somewhat more sensitive to noise.

A more compact weight function often produces better results. A simple compact function is the ideal particle itself $W(\mathbf{x}) = I_W = I_p(\mathbf{x}, D_W, w)$. D_W sets the size of the region of interest with the most compact weight that is usefully given by $D_W = D$. Then only information near the particle will be used. For

$$W = I_W; \; \chi^2(\mathbf{x}_0; D, w) = I^2 \otimes I_W - 2I \otimes (I_W I_p) + \langle I_W I_p^2 \rangle. \qquad (2.29)$$

The first term of χ^2 shows that only the area of size I_W around each point $\mathbf{x}_0$ is important. The last term is a constant except near the edges of the image. Near the edge, all of the terms get smaller due to the fact that there is less overlap between the experimental image I and the region of interest I_W. Dividing through by the last term normalizes this effect at the boundary and defines a new χ^2 for an arbitrary weight function:

$$\chi^2(\mathbf{x}_0; D, w) = \frac{I^2 \otimes W - 2I \otimes W I_p}{< W I_p^2 >} + 1. \tag{2.30}$$

This normalized χ^2 is still minimized at the positions of particles and can be used to find particles that have centers outside of the image. An image normalized to the interval $[0, 1]$ of $1/\chi$ from (2.30) is shown in Figure 2.16b. The inverse is used to make the minimum bright for display.

For comparison, the Hough transform is also shown in panel (d) of Figure 2.16. The Hough transform [289,290] uses the edges or contour of an object to find the center and has been applied to granular materials [52,55,105]. For an object contour with center at the origin defined by the parametric function of length l and parameters q_n, $\mathbf{c}(l; q_n)$, the Hough transform of image $I_e(\mathbf{x})$ is

$$H(\mathbf{x}_0; q_n) = \int_{\mathbf{c}} I_e(\mathbf{x}_0 + \mathbf{c}(l; q_n)) \, dl. \tag{2.31}$$

For a circle, the only parameter is the particle diameter $q_1 = D$. The image I_e is an image of the edges or contours of the object to be tracked and is often determined from the gradient of the original image I, for example, $I_e(\mathbf{x}) = |\nabla I(\mathbf{x})|$. The Hough transform converts from image space to parameter space. It sums the intensity of all points that are on a contour $\mathbf{c}(l; q_n)$ centered at $\mathbf{x}_0$ for all values of $\mathbf{x}_0$ and the parameters q_n. If the image contains a contour like $\mathbf{c}(l; q_n)$ then $H(\mathbf{x}_0; q_n)$ will be large when $\mathbf{x}_0$ is at the center. The maxima of $H(\mathbf{x}_0; q_n)$ determine the positions $\mathbf{x}_0$ and parameters q_n of the contours in the image $I(\mathbf{x})$. It turns out that (2.31) can also be expressed as an LSQ fit of the edge image I_e to an image of the contour,

$$\chi_H^2(\mathbf{x}_0; q_n) = \int [I_e(\mathbf{x}) - C(\mathbf{x} - \mathbf{x}_0; q_n)]^2 \, d\mathbf{x}, \tag{2.32}$$

where

$$C(\mathbf{x}; q_n) = \int_{\mathbf{c}} \delta(\mathbf{x} - \mathbf{c}(l; q_n)) \, dl \tag{2.33}$$

is an image of the contour. Expanding (2.32) and noting that

$$H(\mathbf{x}_0; q_n) = \int I_e(\mathbf{x}) C(\mathbf{x} - \mathbf{x}_0; q_n) \, d\mathbf{x}$$

$$= \int_{\mathbf{c}} \int I_e(\mathbf{x}) \delta(\mathbf{x} - \mathbf{x}_0 - \mathbf{c}(l; q_n)) \, d\mathbf{x} \, dl, \tag{2.34}$$

$$\chi_H^2(\mathbf{x}_0;q_n) = -2H(\mathbf{x};q_n) + \int I_e(\mathbf{x})^2 + C(\mathbf{x}-\mathbf{x}_0;q_n)^2\,d\mathbf{x}. \tag{2.35}$$

The second term is a constant and the third term is constant except near the boundaries of the image. For the Hough transform to work I must be an image of the edges or contours of the objects, and a contour function must be available. Many techniques are available for edge detection (e.g., see [291]), but generally an approximation to the gradient squared of the image is used to create an image of the edges. An example of $H(\mathbf{x}_0;D = 12.14, \sigma = 1/2)$ scaled to the interval $[0,1]$ for a synthetic image is shown in Figure 2.16d using (2.34) with

$$C(\mathbf{x};D,\sigma) = \exp\left(-\frac{(|\mathbf{x}|-D/2)^2}{2\sigma^2}\right) \tag{2.36}$$

as an approximation to (2.33). The peaks are nearly as sharp as for Figure 2.16b, but there are other small false peaks as well. The false peaks are at the two points that are a distance $D/2$ from the edges of both particles. If a third particle is near one of these points, the position of the peak maybe shifted. As with the convolution method this usually will not affect the determination of the center to within one pixel.

Next, a test of the LSQ method on experimental data is presented. The left panel of Figure 2.17 shows an image of monodisperse particles. The raw image, from the rotating drum shown in Figure 2.3c, is taken using backlighting so that the particles appears dark. The displayed image has been normalized using a background image taken without particles and has been scaled using the bright and dark peaks from the histogram of the normalized image so that intensities are approximately 1 inside and 0 outside of the particles. The image resolution is approximately 16 pixels per particle and the signal to noise is approximately 50. The focus is relatively sharp ($w = 0.64$). This image represents a typical image to

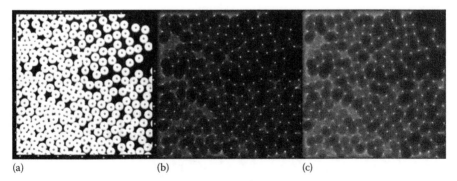

(a)   (b)   (c)

Figure 2.17   Center locations for the rotating drum image from Figure 2.3c found using (2.30). (a) Original image corrected for background lighting variations, normalized, and inverted. Two hundred and fifty nine full or partial particles are identified by gray dots. Dots off of the image are at the centers of particles, which are in partial view. (b) Image of the reciprocal of (2.30), $1/\chi^2$, for $D = 16$ and $w = 1$. The reciprocal is used to visualize the minima. (c) Same as (b) with gray scale adjusted to enhance the smaller peaks.

be tracked. In the center χ^2 from (2.30) using $I_p(\mathbf{x}; 16, 1)$ and $I_W = I_p$ is shown. For visualization, the inverse of χ^2 is shown so that particle centers appear as bright peaks. On the right $1/\chi^2$ is redisplayed with the gray scale adjusted to enhance smaller peaks. The gray dots on the left image show the 259 particles found. Using this technique on the full data sets from the rotating drum, 1×10^7 particles were identified at 1000 time points without any missed particles, provided the particle centers were not more than $D/2$ pixels outside of the image frame.

The LSQ, as well as the convolution and Hough techniques, gives particle centers to approximately one pixel accuracy. One method to get sub-pixel accuracy is to interpolate the peaks found by these techniques. This method will typically give about $1/10$ of a pixel accuracy for dilute images and somewhat less for dense images. However, using LSQ on the whole image allows significantly higher sub-pixel accuracy. To do this, we will look for a modified χ^2 fitting function

$$\chi^2(\mathbf{x}_n; D, w) = \int [I(\mathbf{x}) - I_c(\mathbf{x}, \mathbf{x}_n)]^2 \, d\mathbf{x}, \tag{2.37}$$

where

$$I_c(\mathbf{x}, \mathbf{x}_n) = \sum_n W_n(\mathbf{x}) I_p(\mathbf{x} - \mathbf{x}_n; D, w) \tag{2.38}$$

is an entire calculated image and W_n is a function which is one inside the Voronoi volume of particle n and zero outside. W_n takes care of particles that are overlapping. An example calculated I_c is shown in the center of Figure 2.18 with the experimental image on the left and the squared difference between I and I_c on the right. I_c is calculated using $D = 16$, $w = 1$, and the pixel accurate particle centers $\mathbf{x}_n$ found previously.

To find sub-pixel accurate particle positions minimize (2.37) over $\mathbf{x}_n$. This is equivalent to solving,

$$\frac{\partial \chi^2(\mathbf{x}_n^*; D, w)}{\partial \mathbf{x}_n^*} = 0, \tag{2.39}$$

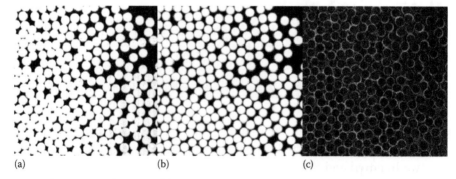

(a)   (b)   (c)

Figure 2.18   (a) Original image. (b) Calculated image. (c) Squared difference between the measured image and the calculated image. The particle positions that minimize this function are typically accurate to a small fraction of pixel.

$\chi_0^2 = 937.4$ $\chi_1^2 = 615.2$ $\chi_2^2 = 180.0$

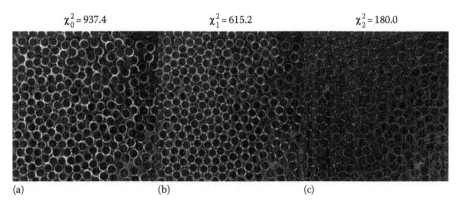

(a) (b) (c)

Figure 2.19 Squared differences between the experimental image and the calculated image at three steps of minimization. (a) For the pixel accurate initial guess, repeated from Figure 2.18c. (b) After minimization of the particle position. (c) After minimization over D, w, and particle position. The centers found using this image are accurate to approximately $1/80$ of a pixel. Using calibration measurements in the same apparatus (see text) 13774 ± 8 pixels equals one meter giving an accuracy of $0.91\,\mu m$ or $1/1320$ of a particle diameter.

for the $\mathbf{x}_n^*$. Because of the good initial pixel accurate guess, Newton's method [292] can be used. The result is shown in Figure 2.19. The left panel shows $(I - I_c^0)^2$ for the initial guess with a $\chi^2 = \chi_0^2 = 937.4$ and the center panel shows the result after minimization. The value of χ^2 has decreased to $\chi_1^2 = 615.2$. Further, the difference image shows that the error is now symmetric with respect to each particle and further improvements will need to come from adjusting D and w. That is, finding D^* and w^* from the simultaneous solution of

$$\frac{\partial \chi^2(\mathbf{x}_0; D^*, w^*)}{\partial D^*} = 0 \quad \text{and} \quad \frac{\partial \chi^2(\mathbf{x}_0; D^*, w^*)}{\partial w^*} = 0. \tag{2.40}$$

Using Newton's method to solve these equations brings the χ^2 down to 321.4. With the new improved values of D and w, (2.37) is reminimized with respect to $\mathbf{x}_n$ to produce the result shown in the right panel of Figure 2.19 with a final $\chi^2 = \chi_2^2 = 180.0$. The process could be repeated, but in this case χ^2 is only reduced to 178.8 after reminimizing with respect to D, w, and $\mathbf{x}_n$ again.

The final positions found have an accuracy of about $1/80$ of a pixel with particles that are 16.5 pixels for an accuracy of $1/1320$ of a particle diameter. This value is determined using the identical apparatus, but with particles glued to the glass confining walls. Images are then taken for various rotation angles. Assuming the relative distance between the particles cannot change and the absolute positions only depend on a single rotation angle (and two translations if the axle is not true), an estimate of the positional error can be made. We find an absolute error of $0.91 \pm 0.02\,\mu m$ in position. The error in determining the actual diameter D^* of the particle from the fitted D alone is a little larger at $1.42\pm0.09\,\mu m$, but if a nonlinear fit of $D^*(D, w; a, b, c) = aD + bw + cDw$ is used the accuracy can be regained. This technique can be used with nominal resolutions down to three pixels per particle, but with reduced sub-pixel accuracy.

If the nominal resolution is increased to 60 pixels per particle, then $1/1000$ of a pixel accuracy can be achieved. Combining this high resolution with two camera synchronized stereoscopic measurements 3D tracking can be achieved in thin containers or in dilute samples. An example is shown in Figure 2.20, with an in-plane position resolution of 55 ± 5 nm for 3.175 mm particle or $1/58,000$ of a particle diameter at 60,000 fps. The accuracy in the third dimension is $470 + / - 50$ nm or about a factor of 10 less accurate. With this technique, it is possible to measure positions with an accuracy of a fraction of the wavelength of light used to illuminate the image. This is possible because at 60 pixels per particle more than 2500 measurements (one for each pixel in the particle) go into each position calculation.

2.4.2   Particle Image Velocimetry

Section 2.4.1 describes techniques to track individual particles from high-resolution images like the one in Figure 2.3c. For these techniques at least three pixels per particle are needed. However, for images like Figure 2.3b with about two pixels per particle a different technique is needed. Particle imaging velocimetry PIV [293] is a useful method for extracting motion information from images in which individual particles are not distinct. PIV has been applied to avalanches [294] and 2D vertical vibration [73] . PIV can also be understood using the LSQ framework described in (Section 2.4.1).

The idea of PIV is to find the most likely location of a small area from an image at time t in a later image at time $t + \Delta t$. This location will be an approximation of the displacement $\Delta \mathbf{x}$ for the center of the small area. In the LSQ framework, the "ideal particle" becomes a small patch from the frame at t and the "image" becomes the frame at $t + \Delta t$. A simple adaptation of (2.22) gives

$$\chi^2(\mathbf{x}_0, \Delta \mathbf{x}_0; t, \Delta t) = \frac{1}{lh} \int_{x_0-l/2}^{x_0+l/2} \int_{y_0-h/2}^{y_0+h/2} [I(\mathbf{x}, t+\Delta t) - I(\mathbf{x} - \Delta \mathbf{x}_0, t)]^2 \, dx \, dy, \quad (2.41)$$

where

   $I(\mathbf{x}, t)$ is the image at position $\mathbf{x}$ and time t

   $\chi^2(\mathbf{x}_0, \Delta \mathbf{x}_0; t, \Delta t)$ is the average squared difference between the image at time $t + \Delta t$ and a small rectangular patch of size $l \times h$ centered at $\mathbf{x}_0$ at time t, shifted by $\Delta \mathbf{x}_0$. Then, for each position $\mathbf{x}_0$ of interest the displacement $\Delta \mathbf{x}$ or velocity $\Delta \mathbf{x}/\Delta t$ can be determined by finding the minimum of χ^2 over $\Delta \mathbf{x}_0$

   The parameters l and h should be chosen such that they are small enough that the velocity is approximately constant over the region, but large enough that there are distinct features in the region.

An example showing $1/\chi^2$ for the image in the center of Figure 2.3 is shown in Figure 2.21. The small 9×9 patch from a point $\mathbf{x}_0$ near the surface flow at time t is shown on the left as a white square. $1/\chi^2$ is shown in the center. The bright spot is the minimum of χ^2 or maximum of $1/\chi^2$. The arrow represents the approximate displacement of the patch at time $t + \Delta t$. The displacement can be

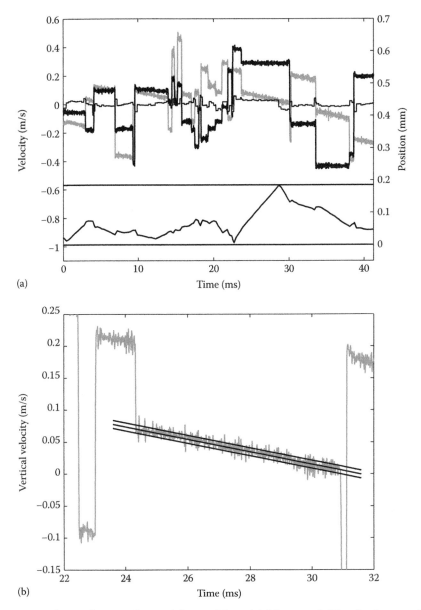

(a)

(b)

Figure 2.20 High-resolution 3D particle tracking. (a) Measured 3D velocity as a function of time for one of the particles in a quasi-2D vertically vibrated cell. (thick black line) Horizontal x-velocity, (thick gray line) vertical z-velocity, (thin black line) depth y-velocity in the thin direction. The velocity in the thin direction is significantly reduced due to inelasticity with the walls and the small impact parameter in that direction. In the thin direction, the cell is 3.360 mm or 1.058 times the particle diameter of 3.175 mm. That leaves 185 μm in which the particle can move. The lower portion shows the particle trajectory in the thin y-direction. The horizontal lines are separated by 185 μm. (b) Enlargement of measured vertical velocity as a function of time (gray line). The three black lines show a free fall trajectory with a slope of −9.81 m s^{-2}. The distance between the outer lines represents the 68% confidence region based on a measured position error of ±55 nm.

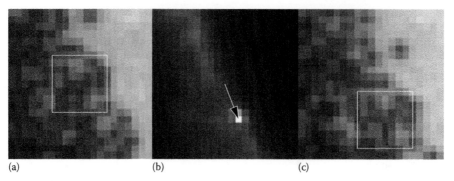

(a)   (b)   (c)

Figure 2.21   Demonstration of PIV technique in rotating drum. (a,c) Small section of image from Figure 2.3b at two times. The small box in the left image most closely matches the small box in the right image. The offset represents the average displacement of material between the two frames. (b) Map of $1/\chi^2$ (see text). The peak represents the most likely position of the material from the box in the left image in the right image and the arrow is the displacement vector.

refined using interpolation to produce sub-pixel accuracy of about $1/10$ to $1/20$ of a pixel. The image at $t + \Delta t$ is shown on the right with a square representing the most likely position of the patch in the new frame. By adjusting the value of Δt, the sensitivity can be retrospectively adjusted. Large Δt are used for slow moving regions and small Δt for fast moving regions.

Using PIV, the entire flow field can be mapped for each point in the rotating drum. Further, since the flow is steady a temporal average can also be used. The result is shown in Figure 2.22 and compared to the direct particle tracking technique from (Section 2.4.1). Although the entire flow field was mapped, for clarity only a single cross section along the horizontal midline is shown. The left plot shows the average particle velocities during one revolution calculated using direct particle tracking from images like those in Figure 2.3a and c. The black dots are local velocities, and the gray dashed line shown on both the left and the right for comparison is spatially averaged over one particle diameter. The black dashed line is the velocity of the drum. The particle velocity approaches the drum velocity away from the flowing surface. On the right the vertical velocity calculated from PIV is shown. It matches the velocity from particle tracking with no adjustable parameters. The main difference is that the PIV technique is sensitive to any motion, including the rotation of the glass plates of the drum. Small imperfections on the surface of the glass are detected in the absence of particles, and the measured velocity follows the velocity of the drum. While this may be undesirable in some situations, in this case it provides an in situ measurement of the rotation rate.

2.4.3   Photoelastic Force

In (Section 2.3.2), photoelasticity was introduced as a technique to visualize stress in granular materials. In this section, I will describe three techniques to directly

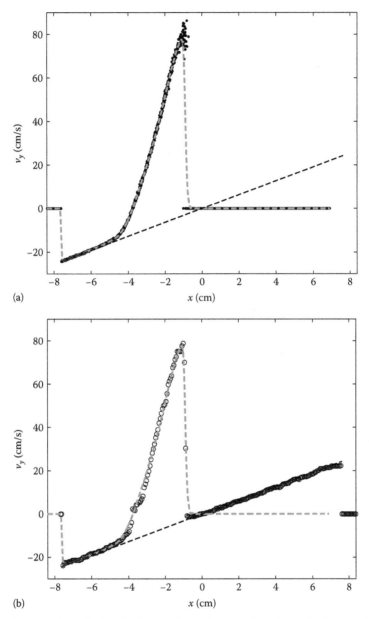

Figure 2.22 Measured vertical velocity v_y from the rotating drum in Figure 2.3 as a function of horizontal position x across the midline. (a) Velocity measured using direct particle tracking both with (black dots) and without (gray dashed line) spatial averaging over one particle diameter. (b) Velocity measured using PIV (circles) with spatially averaged velocity from particle tracking (gray dashed line) for comparison. (both) Black dashed line for solid body rotation velocity.

extract the forces in 2D granular materials. (1) *Fringe-Order Method*: The fringe method was introduced to granular materials in the 1960s by de Jong et al. [176,177] to visualize force chains, extract forces, produce Maxwell constructions, and in order to test a plastic flow rule based on a decomposition into free-rotating sliding units and a non-coaxial stress-strain-rate relation. (2) *Squared Gradient Method*: Later [178,180,211,214,295,296] forces were measured by calibrating the square of the gradient of photoelastic images to known forces. (3) *Least Squares Method*: The least squares method [210,219,220,297] compares the actual photoelastic image to a calculated image and finds the forces that minimize the difference.

2.4.3.1 Fringe-Order Method

In principle, the forces from a photoelastic image (Figure 2.4) could be calculated by integrating the stresses, but in practice this is difficult. The forces are so concentrated that the stress practically diverges, and from (2.12) the isochromats are very close together. One early method to extract forces [176,177] used a dark-field circular polariscope to measure the intensity from (2.12) with monochromatic light. By using monochromatic light λ is a constant and the isochromats appear as dark and light bands or fringes, Figure 2.4. The dark fringes occur when

$$tK(\sigma_1 - \sigma_2) = N\lambda, \tag{2.42}$$

and N is called the fringe order. Notice that for the 0th fringe, $N = 0$, $\sigma_1 = \sigma_2$ and there is no principle stress difference. If the 0th fringe can be determined then the other fringe orders can be counted from there. The starting point to extract the force is the 2D (x, y) plane stress or plane strain solution for a concentrated surface force on a half-space $(y < 0)$ of thickness t [298]. The stress in cylindrical coordinates is

$$\sigma_r = -\frac{2}{\pi r}[P\sin\theta + Q\cos\theta], \sigma_\theta = 0, \tau_{r\theta} = 0, \tag{2.43}$$

where
  r is the distance from the application point of the concentrated force
  P and Q are, respectively, the forces per unit length in the directions normal and tangent to the half-space
  θ is the angle measured from the x-axis

The line forces are independent of the coordinate along the z-direction. The principle stress difference is twice the maximum shear stress,

$$\tau_{max} = (\sigma_1 - \sigma_2)/2 = \sqrt{(\sigma_r - \sigma_\theta)^2/4 + \tau_{r\theta}^2}. \tag{2.44}$$

Considering only normal forces in (2.43) the isobars of

$$\tau_{max} = \frac{P}{\pi D} \tag{2.45}$$

are on circles tangent to the x-axis with diameter D. From (2.42) $2KF = \pi DN\lambda$, where $F = Pt$ is the total force. If F_0, N_0, and D_0 are known for a force on a calibration particle, then measurements of N and D for a contact on a test particle yield the force from

$$F = F_0 \frac{DN}{D_0 N_0}. \tag{2.46}$$

An image of a single particle under diametrical load is shown in Figure 2.23a. The compressive forces and displacements are measured concurrently using a texture analyzer. The center image of Figure 2.23 shows a close-up with three circles overlaid on each of the first, second, and third dark fringes with diameter D_1, D_2, D_3, and from (2.46) for constant F the ratios are $D_n/D_1 = 1/n$. From a similar calibration image, the value of $F_0/(D_0 N_0)$ is found to be 0.0159 ± 0.0015 N per pixels order. From this image $D_2 = 7.1 \pm 0.5$ pixels for the second-order fringe giving $F = 0.23 \pm 0.04$ N. This value along with two others are shown in the right frame of Figure 2.23 as well as the force–displacement measurement from the texture analyzer. The major source of error for this technique is the contamination from the other forces distorting the circular shape of the single-point force solution (2.43). This distortion is clear in the first-order fringe and makes the determination of the best diameter difficult. This also makes automation of this technique difficult, and the three measurements shown are all done by hand. Further, the error tends to grow with the force since the lower-order fringes become more distorted and higher-order fringes must be used (D_4 and D_8 for the other points), but the error in determining the diameter is fixed by the image resolution.

The direction of the force can also be determined from the angle that the tangent circles make with the boundary of the particle. This can be seen in Figure 2.24, which is a close-up of the image in Figure 2.4. This angle is a consequence of the general point force solution (2.43) when $Q \neq 0$.

2.4.3.2 Squared Gradient Method

One of the most obvious features of a photoelastic image like the one in Figure 2.4 is the connection of the forces into 1D structures called force chains. This suggests that the brightness of the image is determined by the force. This idea is strengthened by the fact that in a dark-field polariscope from (2.12) no light is transmitted if the force is zero. However, it turns out that the relation between force and transmission is more complicated. Combining (2.12) with (2.43) and (2.44), the average intensity for a particle with diameter D and a normal point force F is

$$\bar{I} = \frac{4}{\pi D^2} \int I_{out}(r,\theta)\, r\, dr\, d\theta = \frac{4}{\pi D^2} \int I_0^2 \sin^2\left[\frac{2K}{\lambda}\frac{F\sin\theta}{r}\right] r\, dr\, d\theta, \tag{2.47}$$

where

$$I_{out}(r,\theta) = I_0^2 \sin^2\left[\frac{2K}{\lambda}\frac{F\sin\theta}{r}\right]. \tag{2.48}$$

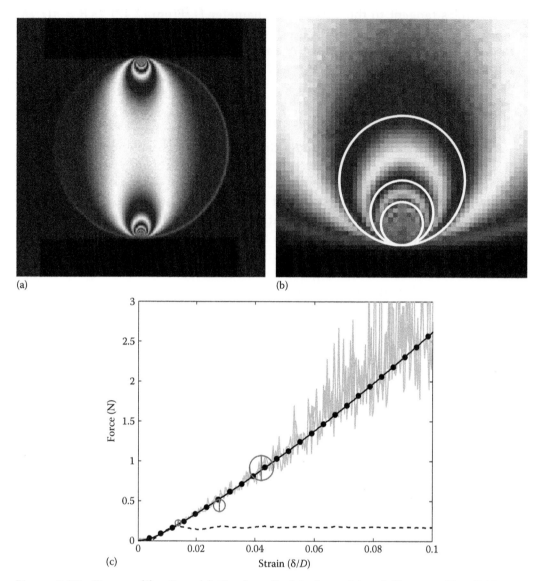

(a)

(b)

(c)

Figure 2.23　Force calibration. (a) Single cylindrical particle of diameter $D = 5.08$ mm and thickness $T = 3.05$ mm under diametrical compressive force $F = 0.206$ N and strain $\delta/D = 1.38\%$ in a texture analyzer. (b) Demonstration that circular fringes follow (2.46). The circles in descending size are for fringe order $N = 1, 2, 3$. The diameters are $D_1/1$, $D_1/2$, and $D_1/3$. The force in any image is determined from the fringe order and diameter using the known fringe order and in a calibration image. The direction of the force can be determined from the angle the fringe circle makes with the edge of the particle. (c) Several methods to determine the force as a function displacement strain δ/D. Black points are from the force gauge of the texture analyzer. The dark line uses least squares fitting. The light line is from the average squared gradient. The dashed line is from average intensity. The large circles are from the fringe-order method at $\delta/D = 1.38\%$, 2.80%, and 4.21%. The diameter of the large circles is determined by the estimated force error.

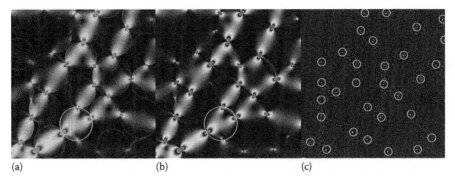

Figure 2.24 Enlargement of small area (rectangle Figure 2.4) from 2D photoelastic disks under simple shear viewed through a dark-field circular polariscope. (a) Experiment. (Courtesy of J. Ren and R. P. Behringer, Duke University, Durham, NC.) (b) Calculated reconstruction used to extract the interparticle forces. (c) Least squares image used to determine the initial guesses for the contact points and final contact positions (circles).

If the argument of the sine is small then $\bar{I}$ will be proportional to F. However, when the argument is large the dependence is complicated. In Figure 2.23, the average intensity for a single particle under diametrical compressive force as a function of compressive strain is shown on the right as a dashed line. The line is scaled so that for small strains ($\delta/D < 1\%$) the value is equal to the force. However, for large strains the value saturates and is independent of the strain. The force chains that are evident in a typical photoelastic image are due to an effective clipping of the force above a certain strain. To extend the range of this method, a common technique [178,180,211,214,295,296] is to look at the average squared gradient of the intensity. By taking the square of the gradient of the intensity in (2.49), we have

$$|\nabla I_{out}(r,\theta)|^2 = \frac{16 I_0^4 K^2 F^2}{\lambda^2 r^4} \sin^2\left[\frac{2K}{\lambda}\frac{F\sin\theta}{r}\right]. \tag{2.49}$$

Now the average will have a direct dependence on F regardless of the strength. The dependence will be nonlinear so it must be fit from known data. An example of force measurement using average squared gradient is shown in Figure 2.23c as a light gray line. The relation between the average squared gradient and F is measured in a similar data set and then used to convert the measured values from this data set to force. The results are best for intermediate forces ($0.2\,\text{N} < F < 1.2\,\text{N}$) where the absolute error is 7%. At high forces ($F > 1.2\,\text{N}$) there is significant noise as the wavelength of the spatial oscillations approaches the resolution of the image and the typical absolute error is over 20%. At low forces, the gradients are small and noise begins to contribute more to the gradient signal. The deviation at small force is not easily seen in the figure because the magnitude is small, but the typical absolute error for $F < 0.2\,\text{N}$ is 20% and reaches 100% at $F = 0.01\,\text{N}$.

2.4.3.3 Least Squares Method

The preceding methods allow quantitative measurement of interparticle forces. The gradient squared method provides a similar accuracy to the fringe method and is easily automated directly in a computer. However, the accuracy can be improved. One method [210,219,220,297] to improve accuracy is based on an extension of the least-squared method for particle location (Section 2.4.1). The idea is to calculate an image that depends on the interparticle positions *and forces* and use LSQ to determine the measured forces. A comparison of a real photoelastic image (a) and a calculated image (b) is shown in Figure 2.24 for an enlarged section of Figure 2.4, which is enclosed in a box. To calculate this image, an equation for the stress as a function of the interparticle forces is needed.

The starting point, as in the fringe method, is the 2D (x, y) plane stress or plane strain solution for a concentrated surface force on a half-space $(y < 0)$ of thickness t [298]. The stress in cylindrical coordinates centered at the concentrated force is given in (2.43), but for this development it is easier to work in coordinates with origin at the particles' center. Converting to rectangular coordinates with an applied concentrated force $\mathbf{F} = (Q\hat{x} - P\hat{y})t$ pointing downward at position $R\hat{y}$ on the edge of a particle with radius R,

$$\sigma_x = -\frac{2}{\pi A^2}[Px^2(R - y) - Qx^3],$$

$$\sigma_y = -\frac{2}{\pi A^2}[P(R - y)^3 - Qx(R - y)^2], \tag{2.50}$$

$$\tau_{xy} = \frac{2}{\pi A^2}[Px(R - y)^2 + Qx^2(R - y)^2],$$

where $A = x^2 + (R - y)^2$. Converting back to polar coordinates (r, θ), now centered on the particle, and allowing a more general force applied with normal component Pt and tangential component Qt at angle ϕ with respect to the x-axis (i.e., $\mathbf{F} = [(P\cos\phi + Q\sin\phi)\hat{x} + (P\sin\phi - Q\cos\phi)\hat{y}]t$), (2.50) becomes

$$\sigma_r = -\frac{2}{\pi RA^2}\left\{P[\rho - C]^2[1 - \rho C] + Q\rho S[\rho - C]^2\right\},$$

$$\sigma_\theta = -\frac{2}{\pi RA^2}\left\{PS^2[1 - \rho C] + Q\rho S^3\right\}, \tag{2.51}$$

$$\tau_{r\theta} = -\frac{2}{\pi RA^2}\left\{PS[\rho - C][1 - \rho C] + Q\rho S^2[\rho - C]\right\},$$

where $\rho = r/R$ is measured from the center of the particle, $C = \cos(\theta - \phi)$, $S = \sin(\theta - \phi)$, and $A = 1 - 2\rho\cos(\theta - \phi) + \rho^2$. Then by superposition a particle with N_f forces will have a total stress,

$$\sigma_x^{Tot} = \sum_{n=1}^{N_f} \sigma_x(P_n, Q_n, \phi_n), \tag{2.52}$$

where x stands for any of the components r, θ, or $r\theta$. It is easily checked that (2.51) satisfies the stress equations, and since they are linear (2.52) does as well. However, these solutions were developed for half-plane boundary conditions. To find the solution for a cylinder, the surface traction boundary conditions must be examined. The stress on the boundary ($\rho = 1$) can be calculated from (2.51).

$$
T_r = \sigma_r = -\frac{1}{2\pi R}\left[P[1 - \cos(\theta - \phi)] + Q\sin(\theta - \phi)\right],
$$

$$
\sigma_\theta = -\frac{1}{2\pi R}\left[P[\cos(\theta - \phi) + 1] - Q\sin^3(\theta - \phi)/A_1^2\right], \tag{2.53}
$$

$$
T_\theta = \tau_{r\theta} = -\frac{1}{2\pi R}\left[P\sin(\theta - \phi) + Q[1 + \cos(\theta - \phi)]\right],
$$

where $A_1 = 1 - \cos(\theta - \phi)$. Because the surface of the cylinder lies on the coordinate surface $r = R$ or $\rho = 1$, the normal surface traction $T_r = \sigma_r$ and the tangential surface traction $T_\theta = \sigma_{r\theta}$. In equilibrium, both tractions must be zero. As an example, consider the diametrical compression shown in Figure 2.23. In this case $N_f = 2$ and there are two forces of equal magnitude, so that $P_1 = P_2 = P$, the tangential forces are zero $Q_1 = Q_2 = 0$, and $\phi_{1,2} = \pm\pi/2$. For these values, (2.52) reduces to $\sigma_r = \sigma_\theta = -P/(\pi R), \sigma_{r\theta} = 0$. Therefore, there is an unbalanced radial compression due to the change in boundary conditions caused by the total pressure $\Pi = 2Pt/(2\pi Rt) = P/(\pi R)$. To remove this, a constant tension solution $\sigma_r = \sigma_\theta = +P/(2\pi R), \sigma_{r\theta} = 0$ is added to (2.51) and (2.53) giving

$$
\sigma_r = -\frac{2}{\pi RA^2}\left\{P[\rho - C]^2[1 - \rho C] + Q\rho S[\rho - C]^2\right\} + \frac{P}{2\pi R},
$$

$$
\sigma_\theta = -\frac{2}{\pi RA^2}\left\{P[S^2[1 - \rho C]] + Q\rho S^3\right\} + \frac{P}{2\pi R}, \tag{2.54}
$$

$$
\tau_{r\theta} = -\frac{2}{\pi RA^2}\left\{PS[\rho - C][1 - \rho C] + Q\rho S^2[\rho - C]\right\},
$$

and

$$
T_r = \sigma_r = -\frac{1}{2\pi R}\left[-P\cos(\theta - \phi) + Q\sin(\theta - \phi)\right],
$$

$$
\sigma_\theta = -\frac{1}{2\pi R}\left[P\cos(\theta - \phi) - Q\sin^3(\theta - \phi)/A_1^2\right], \tag{2.55}
$$

$$
T_\theta = \tau_{r\theta} = -\frac{1}{2\pi R}\left[P\sin(\theta - \phi) + Q[1 + \cos(\theta - \phi)]\right].
$$

Now combining (2.55) with (2.52) for a cylinder with N_f forces $\mathbf{F}_n = [(P_n\cos\phi_n + Q_n\sin\phi_n)\hat{x} + (P_n\sin\phi_n - Q_n\cos\phi_n)\hat{y}]t$ at angle ϕ_n and using identities, $\sin(\alpha\pm\beta) = \sin\alpha\cos\beta \pm \cos\alpha\sin\beta$, $\cos(\alpha\pm\beta) = \cos\alpha\cos\beta \mp \sin\alpha\sin\beta$ gives

$$T_r = \frac{\cos\theta}{2\pi R}\left[\sum_{n=1}^{N_f} P_n\cos\phi_n + Q_n\sin\phi_n\right] + \frac{\sin\theta}{2\pi R}\left[\sum_{n=1}^{N_f} P_n\sin\phi_n - Q_n\cos\phi_n\right],$$

$$T_\theta = \frac{\cos\theta}{2\pi R}\left[\sum_{n=1}^{N_f} P_n\sin\phi_n - Q_n\cos\phi_n\right] - \frac{\sin\theta}{2\pi R}\left[\sum_{n=1}^{N_f} P_n\cos\phi_n + Q_n\sin\phi_n\right]$$

$$-\frac{\left[\sum_{n=1}^{N_f} Q_n\right]}{2\pi R}. \tag{2.56}$$

The mechanical equilibrium conditions for each particle that the sum of forces $\sum_n \mathbf{F}_n = 0$ and the sum of torques $\sum_n \mathbf{F}_n \times R(\cos\phi_n\hat{x} + \sin\phi_n\hat{y}) = R\sum_n Q_n\hat{z} = 0$ guarantee that the expressions in square brackets in (2.56) are zero and therefore $T_r = 0$ and $T_\theta = 0$ and

$$\sigma_r = -\sum_{n=1}^{N_f}\frac{2}{\pi R A_n^2}\left\{P_n[\rho - C_n]^2[1 - \rho C_n] + Q_n\rho S_n[\rho - C_n]^2 - P_n A_n^2/4\right\},$$

$$\sigma_\theta = -\sum_{n=1}^{N_f}\frac{2}{\pi R A_n^2}\left\{P_n S_n^2[1 - \rho C_n] + Q_n\rho S_n^3 - P_n A_n^2/4\right\}, \tag{2.57}$$

$$\tau_{r\theta} = -\sum_{n=1}^{N_f}\frac{2}{\pi R A_n^2}\left\{P_n S_n[\rho - C_n][1 - \rho C_n] + Q_n\rho S_n^2[\rho - C_n]\right\},$$

where $C_n = \cos(\theta - \phi_n)$, $S_n = \sin(\theta - \phi_n)$, and $A_n = 1 - 2\rho\cos(\theta - \phi_n) + \rho^2$ is a solution to the equilibrium stress equation for a cylinder.

Combining (2.57), (2.44), and (2.12) for the dark-field polariscope yields the full equation for a photoelastic image of a single disk:

$$I_p(\mathbf{x}; D, C, I_0, N_f, P_n, Q_n, \phi_n) = I_0\sin^2\left(C\sqrt{(\sigma_r - \sigma_\theta)^2/4 + \tau_{r\theta}^2}\right)\Theta(D/2 - |\mathbf{x}|), \tag{2.58}$$

where

$C = 2\pi t K/\lambda$, $D = 2R$ is the diameter

I_0 is the overall intensity

the Heaviside step function $\Theta(x)$, which is zero for $x < 0$ and one for $x > 0$, sets the intensity to zero outside of the particle

To create an entire image $I_c(\mathbf{x})$ of N particles like the one in Figure 2.24b set

$$I_c(\mathbf{x}) = \sum_{m=1}^{N} I_p(\mathbf{x} - \mathbf{x}_m; D_m, C, I_m, N_f(m), P_{nm}, Q_{nm}, \phi_{nm}), \tag{2.59}$$

where

$\mathbf{x}_m$ is the position

I_m is the intensity

$N_f(m)$ is the number of contacts

D_m is the diameter of the mth particle

P_{nm}, Q_{nm}, and ϕ_{nm} are the P, Q, and ϕ values for the nth force on the mth particle

To determine the forces on a particle m minimize

$$\chi^2 = \int (I(\mathbf{x}) - I_p(\mathbf{x} - \mathbf{x}_m; D_m, C, I_m, N_f(m), P_{nm}, Q_{nm}, \phi_{nm}))^2 \, d\mathbf{x}, \qquad (2.60)$$

with respect to $\mathbf{x}_m, D_m, C, I_m, P_{nm}, Q_{nm}$, and ϕ_{nm}, assuming the total number of contacts $N_f(m)$ is predetermined. $I(\mathbf{x})$ is the experimental image. So for the mth particle there are $5 + 3N_f(m)$ parameters and for the whole image there are $5N + 3\sum_m N_f(m)$ total parameters. However, the solution is only good for particles in equilibrium so there are also constraints. First, the total force $\mathbf{F}_m$ and torque τ_m on each particle are zero. This leads to three constraints for each particle m:

$$F_{mx} = 0 = \sum_{n=1}^{N_f} P_{nm} \cos \phi_{nm} + Q_{nm} \sin \phi_{nm},$$

$$F_{my} = 0 = \sum_{n=1}^{N_f} P_{nm} \sin \phi_{nm} - Q_{nm} \cos \phi_{nm}, \qquad (2.61)$$

$$\tau_m = 0 = \sum_{n=1}^{N_f} Q_{nm}.$$

A fourth constraint comes from the total pressure on a particle

$$\Pi_m = \sum_{n=1}^{N_f(m)} \frac{P_{nm} t}{2\pi R_m t} = \sum_{n=1}^{N_f(m)} \frac{P_{nm} + Q_{nm}}{2\pi R_m}. \qquad (2.62)$$

The second expression is true since $\sum_n Q_{nm} = 0$. To determine Π_m, notice that every term in (2.57) contains either P_n or Q_n, and multiplying every P_n and Q_n by a constant is equivalent to multiplying stress component and the pressure by the same constant. Therefore, the pressure Π_m cannot be determined independently from C in (2.58) and must be found by calibration. Therefore, without loss of generality $\sum_n P_n$ can be set to one or equivalently define $C_m = 2\pi R_m C \Pi_m$, which must be found for each particle. To extract C and Π_m separately, a calibration with a known Π_m is required. After subtracting the constraints there are $1 + 3N_f(m)$ parameters for each particle and $N + 3\sum_m N_f(m)$ total parameters for a complete image if (2.60) is fit for each particle separately.

To fit (2.60), a good initial guess for the parameters is needed. A good starting point is the positions of all of the contacts and the particles. This can be obtained using the techniques of (Section 2.4.1). An example used to find the contacts is shown in the right panel of Figure 2.24. Equation 2.30 is used with an I_p appropriate for the image of a contact, for example, from (2.43). The image of $1/\chi^2$ is shown so that minima are bright along with the final values of the positions of the contacts as circles. The peaks of this image along with the centers of the particle are used as an initial guess for $\mathbf{x}_m$, $N_f(m)$, ϕ_{nm}, and D_m. A guess for P_{nm} can be determined from the average squared gradient near each particle, but setting $P_{nm} = 1/N_f(m)$ usually works. The Q_{nm} are initially set to zero. The experimental image is background corrected and scaled between zero and one so that I_m should be close to one. C_m is set based on the average squared gradient near each particle and a calibration parameter. To take into account the constraints (2.61) and (2.62) four parameters must be eliminated. Many choices are possible, but one that works well is to eliminate two of the Ps and two of the Qs. To eliminate P_1, P_2, Q_1, and Q_2, rearrange (2.61) and solve

$$\begin{bmatrix} \Delta C_1 & \Delta S_1 \\ \Delta S_1 & -\Delta C_1 \end{bmatrix} \begin{bmatrix} P_1 \\ Q_1 \end{bmatrix} = -\begin{bmatrix} \cos\phi_2 \\ \sin\phi_2 \end{bmatrix} - \sum_{n=3}^{N_f} \begin{bmatrix} P_n\Delta C_1 + Q_n\Delta S_1 \\ P_n\Delta S_1 - Q_n\Delta C_1 \end{bmatrix}, \tag{2.63}$$

for P_1 and Q_1, where $\Delta C_n = \cos\phi_n - \cos\phi_2$ and $\Delta S_n = \sin\phi_n - \sin\phi_2$. Then

$$P_2 = 1 - P_1 - \sum_{n=3}^{N_f} P_n \quad \text{and} \quad Q_2 = -Q_1 - \sum_{n=3}^{N_f} Q_n. \tag{2.64}$$

There are always at least two contacts. If $N_f=2$, then $P_1 = P_2 = 1/2$ and $Q_1 = -Q_2 = -\cot[(\phi_1-\phi_2)/2]/2$. For a single particle under diametrical load as shown in Figure 2.23a $(\phi_1-\phi_2) = \pi$ and $Q_1 = Q_2 = 0$. For $N_f > 2$, then (2.63) is solved numerically. An example of minimizing (2.60) subject to constraints (2.61) and (2.62) using (2.63) and (2.64) is shown in Figure 2.25. The image shows an enlargement of the circled particle from the experimental image in Figure 2.24, the calculated particle image using the fitted parameters and the integrand from (2.60) (i.e., the square difference between experiment and calculated image).

As a final step, the individual particle fits are used as an initial guess to fit the full image using

$$\chi^2 = \int \left(I(\mathbf{x}) - \sum_{m=1}^{N} I_p(\mathbf{x} - \mathbf{x}_m; D_m, C, I_m, N_f(m), P_{nm}, Q_{nm}, \phi_{nm}) \right)^2 d\mathbf{x}. \tag{2.65}$$

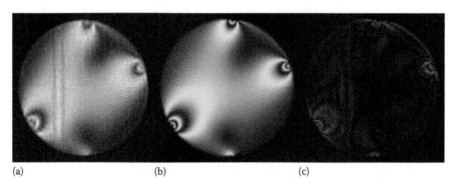

(a)   (b)   (c)

Figure 2.25 Details of single particle from Figure 2.24. Experimental image (a), calculated image (b), and square difference (c) shown.

For the whole image there are more constraints. For example, each particle–particle contact is shared by two particles and the forces must be equal in magnitude and opposite direction. From the original $\sum_m N_f(m)$ contacts assume N_w are boundary contacts, then there are $N_w + \sum_m N_f(m)/2 = N_c$ unique contacts. Numbering the unique contacts from $k = 1$ to N_c, it is useful to resubscript P_{nm} and Q_{nm} to P_k and Q_k. Each unique contact has one P, Q, and ϕ. For each particle, there is one $\mathbf{x}_m$, D_m, and I_m. There is one C for the image. In total, there are $3(N_w + \sum_m N_c(m)/2) + 4N + 1 = 3N_c/2 + 4N + 1$ parameters. Some of these are not independent. In particular, there are $3N$ constraints from equilibrium and the total pressure Π and C cannot be determined independently. As earlier, set $\sum_k P_k = 1$. Then $\Pi = 1/p$ where p is the perimeter and $C' = pC \sum_k P_k$, where C' is the value found by fitting. Subtracting the constraints, $3N_c/2 + N$ parameters remain. One useful method to determine these parameters fixes $\mathbf{x}_m$, D_m, I_m, and ϕ_{nm}. Then $(z-3)N$ parameters remain, where $z = 2N_c/N$ is the average number of contacts per particle. When z is minimum $z = 3 + 1/N$ [299] then there is only one free parameter and all Ps and Qs can be determined from force and torque balance and only C' or the total pressure is determined by fitting. Usually, $z > 3 + 1/N$ and some force parameters will need to be fit. The force and torque balance equations (2.61) can be written more compactly as a matrix equation [299,300]

$$C(\phi_k, D_m)F = 0, \tag{2.66}$$

where

C is a $3N \times zN$ matrix

$F = [P_k, Q_k]^T$ is a $zN \times 1$ matrix containing all the $2N_c = zN$ unique Ps and Qs

The null space of C, null(C) is a $zN \times (z-3)N$ matrix and defines a basis for the $(z-3)N \times 1$ matrix f of independent variables. The initial values of f = $(\text{null}(C))^T F$, where F is the initial guess from the individual particle fits. Now (2.65) can be evaluated using f through F = null(C)f. It is clear from (2.66) that xF is still a force-balanced solution for all x. Then, the values of F are normalized by $\sum_k P_k$. In addition to normalizing the forces, two other things must be checked. First, for cohesionless particles all $P_k >= 0$. Second, if the maximum friction coefficient μ is known then $|Q_k| <= \mu P_k$. These constraints can be included using a constrained minimization algorithm. After minimizing over f then it is straightforward to minimize over the remaining parameters starting with ϕ_k then followed by $\mathbf{x}_m$, D_m, and I_m. The process may be repeated as needed to obtain an optimal solution. Using this technique, all of the forces, positions, etc., were measured from the image in Figure 2.4 and a small section is shown in the center panel of Figure 2.24.

To convert the fitting parameters into SI units, calibration is needed. Length is calibrated using an object of measured size in the image, for example, the particles, the container, or an object placed in the image specifically for that purpose. To calibrate the total force and measure C, the single particle fitting technique was applied to a cylinder under diametrical load as shown in Figure 2.23a. The force was measured by fitting C, $\mathbf{x}$, D, and I under the assumption that $\phi_1 - \phi_2 = \pi$, and therefore, $P_1 = P_2 = 1/2$ and $Q_1 = Q_2 = 0$. The results are shown in Figure 2.23.

The solid line in the right panel has one additional calibration parameter α relating the fitted C to the force measured in Newtons from the texture analyzer. Then, $\alpha = 0.5347 \, \text{N/m}$ is determined from the fit such that $P'_m = \alpha C_m P_m$, where P'_m is the force per unit length in Newtons per meter, C_m is the unitless fitted value for the mth C value in (2.60), and P_m is the normalized unitless force per unit length for the mth force. The typical error in measuring the force is uniform over the range displayed and equal to $\pm 0.0016 \, \text{N}$.

From the stress–strain curve shown it is possible to extract Young's modulus E given the Poisson ratio v of the material. In principle, v could be measured as well by monitoring the volume of the particle during compression. The displacement δ as a function of force F for a D diameter, t thickness cylinder made of material with E_1 and v_1 between two plates with E_2 and v_2 [301] is

$$
\begin{aligned}
\delta &= \frac{2F(V_1 + V_2)}{t}\left[1 + \ln\left(\frac{2t^3}{(V_1 + V_2)FD}\right)\right] \\
&\approx \frac{2FV}{t}\left[1 + \ln\left(\frac{2t^3}{VFD}\right)\right],
\end{aligned}
\tag{2.67}
$$

where $V_n = (1 - v_n^2)/(\pi E_n)$. For this case, $V = V_1 \gg V_2$ and V_2 can be neglected. The stress–strain relation can be fit to (2.67), but sometimes it is useful to have the inverse. Unfortunately, the ln correction is not small and to invert the relation between force and displacement for a cylinder between two plates the Lambert W function $W_k(z)$ must be used. $W_k(z)$ is the k-branched multivalued solution of the equation $z = W(z)e^{W(z)}$. The function is described in detail in [302]. Using $W_k(z)$, the force

$$
F = \frac{2t^3}{VD}\exp\left[W_{-1}\left(\frac{-\delta D}{4t^2 e}\right) + 1\right].
\tag{2.68}
$$

Given $t = 3.048 \, \text{mm}$ and $D = 5.080 \, \text{mm}$ a fit to this equation for V yields $5.36 \times 10^{-8} \, \text{m}^2 \text{N}^{-1}$, and assuming $v = 0.5$ then $E = 4.45 \, \text{MPa}$, which is in agreement with the nominal value given by the manufacturer of 4 GPa. From this value of E, the value of the strain-optical coefficient is found to be 0.0087. The strain-optical coefficient has a nominal value given by the manufacturer of 0.009 and is the unitless strain version of the stress-optical coefficient K.

One common problem with using (2.67) and (2.68) is that it is difficult to get perfect cylinders. A small bowing of the axial surface as shown in the left panel of Figure 2.26 has significant effects on the force–displacement curve. The equations relate the *force F* and the displacement, but photoelasticity is sensitive to the *force per unit length P = F/L*. If the cylinder is bowed, then when the particle first touches the effective thickness $L < t$ and the photoelastic measurement overestimates the force. In the right panel of Figure 2.26, $F/P = L/t$ is plotted for a particle with a small curvature. The curvature shown on the left is to scale and the radius of the curvature is about eight times the cylinder radius. From geometry, the function L/t can be obtained assuming linear deformation. L/t and F/P are plotted in the right panel and agree well.

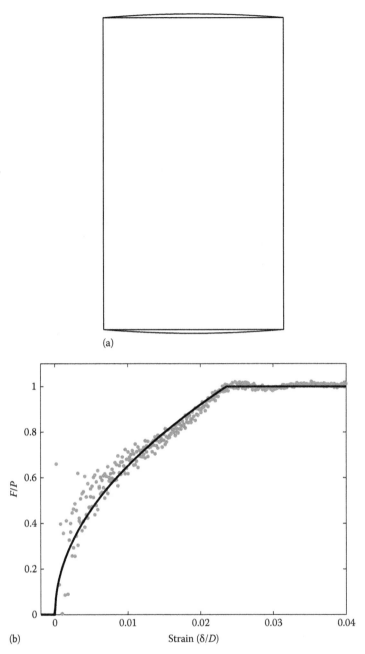

(a)

(b)   Strain (δ/D)

Figure 2.26   (a) Side view of cylindrical particle of diameter $D = 5.08$ mm and thickness $T = 3.05$ mm. The bowing of the surface causes the contact length to depend on force. (b) The photoelastic signal depends on the force per unit length $P = F/L$ and the texture analyzer measures the force F. If the radius of the cylinder is independent of thickness then F is proportional to P. In some particles, the side is bowed (a) and F/P is not a constant. The plot shows F/P (dots) for a bowed particle as a function of strain displacement δ/D. The line is a calculation of L assuming the bow is circular with diameter 38 mm. For this particle, the photoelastic measurement overestimates the true force during the first 2.2% strain.

2.5   Non-Imaging Techniques

From simple angle of repose and carbon paper force measurements to complex triaxial deformation tests and miniature transmitters, a large number of non-imaging techniques are used to study granular materials. A short survey of these techniques is covered in the following sections.

2.5.1   Geometry

Perhaps the most striking difference between sand and water is the formation of a pile when poured. The angle that a pile makes with the horizontal is known as the angle of repose θ. Although Coulomb studied the problem of determining the angle of repose from particle properties [303] and posited his friction law as a solution, research continues on the subject even today [304–313]. Both a static and a dynamic angle of repose can be measured. If a pile is slowly tilted, it will fail at the static angle of repose reducing the angle to the dynamic angle of repose. The method of creating the pile can significantly change the angle of repose, and at least seven different angles of repose have been defined [309,314]. The angle of repose is related to the particle friction coefficient μ.

Often the static friction coefficient is approximated by finding the angle of repose through $\mu_s = \tan\theta$. Direct measures of μ_s are also common from $\mu_s = F_f/F_n$, where F_f is the tangential force needed to move a particle under a given normal load F_n. A simple technique is to place a block of the material on a planar substrate and tip the plane until the material moves. Then, the tangent of the tip angle is μ_s. Sliding or kinetic friction μ_k can also be measured in a similar way from the tip angle needed to stop the material from sliding. Unfortunately, both μ_s and μ_k depend on a number of nonmaterial properties, including the shape of the surface, the roughness of the surface, the static charging, humidity, surface contaminants, and all of the properties of the substrate or other particles. Surface roughness can be determined using wavelength scanning interferometry [315].

Shape and particle size distribution affect not only friction but also other properties of granular materials. Typical measurements for nearly spherical particles include sphericity like (2.18) or ellipsoid approximations as shown in Figure 2.9. These require detailed measurements of individual particles. Simpler measures include average diameter, average volume, or average mass. A rolling test can sort particles by sphericity. The further or faster particles roll down an incline the more spherical they are. For elongated particles or flat particles, the ratio of the longest to the shortest dimension defines an aspect ratio. For diverse particle collections, the distribution of any of these measures may be important. Log-normal distribution of mean radius is common in industrial and natural granular materials.

2.5.2   Volume Fraction and Concentration

In thermodynamics, the key extensive macroscopic variables are the number N, energy E, and volume V of the system. Because the kinetic energy of granular particles is constantly dissipated by inelastic collisions and the elastic potential energy is often irrelevant (hard particle approximation), the number and volume play a more central role in granular materials. Therefore, the intensive quantities, the number density $n = N/V$ and the volume fraction $\phi = V_p n$ are important quantities to measure. The volume fraction measures the fraction of the system volume taken up by the volume of the N particles NV_p. In systems with only one type of particle ϕ is often the most important system variable. The global volume fraction can be measured from the apparent volume of the container combined with the number of particles in the container. A simple technique to measure the number of particles uses an analytic balance to measure the weight of all of the grains divided by the average weight of a single grain. If ϕ is uniform, this simple technique is usually sufficient.

A number of other techniques have been used in more complicated situations. Capacitive techniques [28,92] use a lock-in amplifier and a bridge circuit to measure the change in dielectric constant due to the number of particles in between the plates of a capacitor. With proper calibration this technique can yield accurate local results, since the capacitance is only sensitive to the material near the plates.

If there are multiple types of granular particles, then the component number density for the ith type particle $n_i = N_i/V$ is also important. Measurements of n_i can be difficult without imaging, but researchers have used direct counting to make estimates, including the use of a novel miniature vacuum to statistically sample concentration during the progression of axial mixing in a rotating drum [114]. Another technique made use of radioactive tracers [111] and a miniaturized detector used previously on industrial scale [316]. The miniature scintillation detector scanned the container to measure the concentration of radioactive tracers in a vertically vibrated sand tube [32]. The tracers were initially placed in a small region in the middle of the tube and then observed as they diffused during shaking to estimate enhanced self-diffusion.

2.5.3   Position

Measuring the positions of particles in 3D samples is a difficult task and imaging techniques all have limitations. One non-imaging technique is based on PEPT [317–319]. PEPT uses triangulation of positron emission to determine the 3D position of a tracer particle within a granular system. The tracer particle is mechanically indistinguishable from the other particles and is created by irradiating a glass particle with a beam of ^{3}He to create a radioisotope of F that decays by the emission of positrons. PEPT has been used in a number of granular applications [75,317–325] including measurements in vertically vibrated pattern-forming systems [75]. The technique is limited in the number of tracers that can be used, typically one, but has high time resolution of <2 ms and spatial resolution of about 1 mm.

2.5.4   Velocity

A number of granular researchers [33,149,153,155,326] have made use of fiber-optic velocimetry. The technique uses temporal cross-correlation of fiber-optic sensor signals placed a fixed distance apart. Particles passing both sensors will be correlated at a time determined by their velocity component in the direction of the separation. Multiple fibers can be used to increase accuracy or to measure other velocity components.

Diffusing-wave spectroscopy (DWS) [327,328] is a technique, which can determine the velocity distribution in granular materials. It has been applied to gravity-driven flows [143,329]. The technique is an extension of single scattering spectroscopy techniques to the case of strongly multiply-scattered light. In the multiply scattered case, light is assumed to diffuse through a sample in a random walk. Imagine a very short-time coherent pulse of light entering a strongly scattering medium. Each photon will follow a different scattered path with a different path length. Therefore, each photon will emerge at a different time based on the path length traveled and the short pulse will be broadened when it leaves the sample. Encoded in the broadened pulse is the path-length distribution. From the assumption of diffusion and the path-length distribution, information about the sample can be extracted, including the particle size distribution [335] and the average speed of the particles [332]. The typical DWS experiment is not carried out using a short pulse, but continuous illumination and the distribution of path lengths are extracted using photon–photon correlations.

2.5.5 Stress and Force

Many ingenuous methods have been used to measure stress and forces in granular materials. Here, only a cursory examination is made. A classic test, which has likely been used since before recorded history is the slump test or column collapse. When using mortar or mud for construction, the builder needs to know if there is enough water in the mixture. In a slump test, the material is placed in a tall open container and then the open end is flipped over onto a planar substrate. Then the container is removed rapidly so that initially the material maintains the shape of the container. Over time the material will slump (deform under the force of gravity). If the material is soft, it will deform rapidly and form a short pile with a large horizontal extent. If the material is hard, it will deform slowly and produce a tall narrow pile. The ancient builder could determine the correct amount of water by observing the slumping characteristics. Similar tests are used today in many industries including concrete and cooking. To create accurate standards several quantities may be measured, including the final height or width, height or width after a fixed time, time to reach final height or width, and time to reach preset height or width.

Another old test method still in use today is the penetration or indention test [222,223,336–338]. In this test, an object is pressed into a material with a known force and the penetration depth is recorded, or the object is pushed to a fixed depth and the force is measured. For granular material, the test is sensitive to the preparation of the system as well as the shape and surface properties of the indenter. Commercial testing units are common, and the test is used extensively in soil analysis. Granular impact is related but less controlled as the impacting object is dropped or shot into the granular material [211,214,339–352].

Commercial sensors are used to measure pressure, strain, acceleration, and sound propagation. These sensors are usually mounted on walls of granular experiments, for example, on hoppers [142] and shear cells [183]. However, there are sensors developed in the late 1950s, which can be placed into a granular material and report measurements wirelessly. The "rf-pill" was developed for medical applications [353–357] to measure the pressure in the digestive tract. The rf-pill is a small transmitter and sensor, which has been used to measure stress in vertical tube flow [158] and pressure in hopper discharge [128,133].

A simple technique to measure the maximum contact force on a wall uses carbon paper [358]. A sheet of ordinary paper is placed between the wall and a sheet of carbon paper, and the granular material is added. The pressure on the carbon paper produces a circular mark with a diameter that is proportional to the maximum force during the lifetime of the contact. Using a calibration from known forces, a scale can be made to convert the diameter of the mark to the force.

Drag on an object is a common fluid test. In fluids, the drag force is typically linear in the velocity for small velocities with a cross-over to quadratic at higher speeds. For granular materials, the force is usually independent of speed [359–366], but analyzing the dependence on other properties like cross-sectional area can be used as a characterization.

Much of the non-imaging work on granular material is in the form of stress–strain analysis [181,183,185,309,367–370]. A granular material is placed in a container with either constant pressure or constant volume walls and then strained using constant stress or constant strain. During the application of a constant strain, the bulk stress needed to produce the strain is measured or vice versa for constant stress. The relation between stress and strain is the foundation of continuum theories of granular materials.

References

1. R.L. Brown and J.C. Richards. Profile of flow granules through apertures. *Transactions Instrumentation and Chemical Engineers*, 38:243, 1960.

2. R.L. Brown and J.C. Richards. Kinematics of the flow of dry powders and bulk solids. *Rheologica Acta*, 4(3):153–165, 1965.

3. A. W. Roberts. An investigation of the gravity flow of noncohesive granular materials through discharge chutes. *Journal of Manufacturing Science Engineering*, 91:373–381, 1969.

4. F. Dinkelacker, A. Hübler, and E. Lüscher. Pattern formation of powder on a vibrating disc. *Biological Cybernetics*, 56(1):51–56, 1987.

5. B. Thomas, M.O. Mason, Y.A. Liu, and A.M. Squires. Identifying states in shallow vibrated beds. *Powder Technology*, 57(4):267–280, 1989.

6. F. Melo, P. Umbanhowar, and H.L. Swinney. Transition to parametric wave patterns in a vertically oscillated granular layer. *Physical Review Letters*, 72:172–175, 1994.

7. F. Melo, P.B. Umbanhowar, and H.L. Swinney. Hexagons, kinks, and disorder in oscillated granular layers. *Physical Review Letters*, 75(21):3838–3841, 1995.

8. P. Umbanhowar, F. Melo, and H.L. Swinney. Localized excitations in a vertically vibrated granular layer. *Nature*, 382:793–796, 1996.

9. C. Bizon, M.D. Shattuck, J.B. Swift, W.D. McCormick, and H.L. Swinney. Patterns in 3d vertically oscillated granular layers: Simulation and experiment. *Physical Review Letters*, 80(1):57–60, 1998.

10. S.J. Moon, M.D. Shattuck, C. Bizon, D.I. Goldman, J.B. Swift, and H.L. Swinney. Phase bubbles and spatiotemporal chaos in granular patterns. *Physical Review E*, 65(1):011301/1–10, 2002.

11. M. Faraday. On a peculiar class of acoustical figures; and on certain forms assumed by groups of particles upon vibrating elastic surfaces. *Philosophical Transactions of Royal Society London*, 121:299–340, 1831.

12. W. Kroll. Über das verhalten von schüttgut in lotrechi schwingenden gefäßen. *Forschung auf dem Gebiet des Ingenieurwesens A*, 20(1):2–15, 1954.

13. W. Kroll. Fließerscheinungen an haufwerken in schwingenden gefäßen. *Chemie Ingenieur Technik*, 27(1):33–38, 1955.

14. J.L. Olsen and E.G. Rippie. Segregation kinetics of particulate solids systems i. influence of particle size and particle size distribution. *Journal of Pharmaceutical Sciences*, 53(2):147–150, 1964.

15. E.G. Rippie, J.L. Olsen, and M.D. Faiman. Segregation kinetics of particulate solids systems ii. particle densitysize interactions and wall effects. *Journal of Pharmaceutical Sciences*, 53(11):1360–1363, 1964.

16. V.A. Chlenov and N.V. Mikhailov. Some properties of a vibrating fluidized bed. *Journal of Engineering Physics*, 9(2):137–139, 1965.

17. M.D. Faiman and E.G. Rippie. Segregation kinetics of particulate solids systems iii. dependence on agitation intensity. *Journal of Pharmaceutical Sciences*, 54(5):719–722, 1965.

18. T. Yoshida and Y. Kousaka. Mechanism of vibratory packing of granular solids. *Chemical Engineering*, 30(11):1019–1025,a1, 1966.

19. H. Takahashi, A. Suzuki, and T. Tanaka. Behaviour of a particle bed in the field of vibration i. analysis of particle motion in a vibrating vessel. *Powder Technology*, 2(2):65–71, 1968.

20. A. Suzuki, H. Takahashi, and T. Tanaka. Behaviour of a particle bed in the field of vibration ii. flow of particles through slits in the bottom of a vibrating vessel. *Powder Technology*, 2(2):72–77, 1968.

21. A. Suzuki, H. Takahashi, and T. Tanaka. Behaviour of a particle bed in the field of vibration ili. mixing of particles in a vibrating vessel. *Powder Technology*, 2(2):78–81, 1968.

22. T.G. Owe Berg, R.L. McDonald, and R.J. Trainor Jr. The packing of spheres. *Powder Technology*, 3(1):183–188, 1970.

23. W.A. Gray and G.T. Rhodes. Energy transfer during vibratory compaction of powders. *Powder Technology*, 6(5):271–281, 1972.

24. K. Ahmad and I. J. Smalley. Observation of particle segregation in vibrated granular systems. *Powder Technology*, 8(12):69–75, 1973.

25. M.H. Cooke, D.J. Stephens, and J. Bridgwater. Powder mixing a literature survey. *Powder Technology*, 15(1):1–20, 1976.

26. G. Rátkai. Particle flow and mixing in vertically vibrated beds. *Powder Technology*, 15(2):187–192, 1976.

27. J.C. Williams. The segregation of particulate materials. a review. *Powder Technology*, 15(2):245–251, 1976.

28. R.G. Gutman. Vibrated beds of powders Part I: A theoretical model for the vibrated bed. *Transactions Instrumentation of Chemical Engineers*, 54:174–184, 1976.

29. R.G. Gutman. Vibrated beds of powders Part II: Heat transfer in and energy dissipation of a vibrated bed. *Transactions Instrumentation of Chemical Engineers*, 54:251–257, 1976.

30. D.S. Parsons. Particle segregation in fine powders by tapping as simulation of jostling during transportation. *Powder Technology*, 13(2):269–277, 1976.

31. J. Bridgwater. Fundamental powder mixing mechanisms. *Powder Technology*, 15(2):215–236, 1976.

32. C.F. Harwood. Powder segregation due to vibration. *Powder Technology*, 16(1):51–57, 1977.

33. S.B. Savage. Gravity flow of cohesionless granular materials in chutes and channels. *Journal of Fluid Mechanics*, 92:53–96, 5 1979.

34. B. Thomas, Y.A. Liu, R. Chan, and A.M. Squires. A method for observing phase-dependent phenomena in cyclic systems: Application to study of dynamics of vibrated beds of granular solids. *Powder Technology*, 52(1):77–92, 1987.

35. S.B. Savage. Streaming motions in a bed of vibrationally fluidized dry granular material. *Journal of Fluid Mechanics*, 194:457–478, 1988.

36. P. Evesque and J. Rajchenbach. Instability in a sand heap. *Physical Review Letters*, 62:44–46, Jan 1989.

37. C. Laroche, S. Douady, and S. Fauve. Convective flow of granular masses under vertical vibrations. *Journal de Physique*, 50(7):699–706, 1989.

38. S. Douady, S. Fauve, and C. Laroche. Subharmonic instabilities and defects in a granular layer under vertical vibrations. *EPL (Europhysics Letters)*, 8(7):621, 1989.

39. P. Evesque, E. Szmatula, and J.-P. Denis. Surface fluidization of a sand pile. *EPL (Europhysics Letters)*, 12(7):623, 1990.

40. L.T. Fan, Y. Chen, and F.S. Lai. Recent developments in solids mixing. *Powder Technology*, 61(3):255–287, 1990.

41. E. Clément and J. Rajchenbach. Fluidization of a bidimensional powder. *Europhysics Letters*, 16(2):133–138, 1991.

42. O. Zik and J. Stavans. Self-diffusion in granular flows. *EPL (Europhysics Letters)*, 16(3):255, 1991.

43. O. Zik, J. Stavans, and Y. Rabin. Mobility of a sphere in vibrated granular media. *EPL (Europhysics Letters)*, 17(4):315–319, 1992.

44. E. Clément, J. Duran, and J. Rajchenbach. Experimental study of heaping in a two-dimensional "sand pile". *Physical Review Letters*, 69:1189–1192, August 1992.

45. C.H. Liu and S.R. Nagel. Sound in sand. *Physical Review Letters*, 68:2301–2304, April 1992.

46. C.H. Liu and S.R. Nagel. Sound in a granular material: Disorder and nonlinearity. *Physical Review B*, 48:15646–15650, December 1993.

47. E. Clément, S. Luding, A. Blumen, J. Rajchenbach, and J. Duran. Fluidization, condensation and clusterization of a vibrating column of beads. *International Journal of Modern Physics B*, 07(09n10):1807–1827, 1993.

48. J. Rajchenbach, E. Clément, and J. Duran. Experimental study of bidimensional models of sand. *International Journal of Modern Physics B*, 07(09n10):1789–1798, 1993.

49. H.K. Pak and R.P. Behringer. Surface waves in vertically vibrated granular materials. *Physical Review Letters*, 71:1832–1835, September 1993.

50. J.B. Knight, H. M. Jaeger, and S.R. Nagel. Vibration-induced size separation in granular media: The convection connection. *Physical Review Letters*, 70:3728–3731, June 1993.

51. J. Duran, T. Mazozi, E. Clément, and J. Rajchenbach. Size segregation in a two-dimensional sandpile: Convection and arching effects. *Physical Review E*, 50:5138–5141, December 1994.

52. S. Warr, G.T.H. Jacques, and J.M. Huntley. Tracking the translational and rotational motion of granular particles: Use of high-speed photography and image processing. *Powder Technology*, 81(1):41–56, 1994.

53. C.H. Liu and S.R. Nagel. Sound and vibration in granular materials. *Journal of Physics: Condensed Matter*, 6(23A):A433, 1994.

54. S. Luding, E. Clément, A. Blumen, J. Rajchenbach, and J. Duran. Anomalous energy dissipation in molecular-dynamics simulations of grains: The "detachment effect". *Physical Review E*, 50(5):4113–4120, 1994.

55. S. Warr, J.M. Huntley, and G.T. H. Jacques. Fluidization of a two-dimensional granular system: Experimental study and scaling behavior. *Physical Review E*, 52:5583–5595, November 1995.

56. H.K. Pak, E. Van Doorn, and R.P. Behringer. Effects of gases on granular materials under vertical vibration. *Physical Review Letters*, 74:4643–4646, 1995.

57. E.E. Ehrichs, H. M. Jaeger, G.S. Karczmar, J.B. Knight, V.Y. Kuperman, and S.R. Nagel. Granular convection observed by magnetic resonance imaging. *Science*, 267(5204):1632–1634, 1995.

58. C.E. Brennen, S. Ghosh, and C.R. Wassgren. Vertical oscillation of a bed of granular material. *Journal of Applied Mechanics*, 63:156–161, 1996.

59. C.R. Wassgren, C.E. Brennen, and M.L. Hunt. Vertical vibration of a deep bed of granular material in a container. *Transactions of the ASME*, 63:712, 1996.

60. E. Clément, L. Vanel, J. Rajchenbach, and J. Duran. Pattern formation in a vibrated two-dimensional granular layer. *Physical Review E*, 53:2972–2975, March 1996.

61. J.B. Knight, E.E. Ehrichs, V.Yu. Kuperman, J.K. Flint, H.M. Jaeger, and S.R. Nagel. Experimental study of granular convection. *Physical Review E*, 54:5726–5738, November 1996.

62. J.B. Knight. External boundaries and internal shear bands in granular convection. *Physical Review E*, 55:6016–6023, May 1997.

63. E. van Doorn and R.P. Behringer. Dilation of a vibrated granular layer. *EPL (Europhysics Letters)*, 40(4):387, 1997.

64. E. van Doorn and R.P. Behringer. Onset and evolution of a wavy instability in shaken sand. *Physics Letters A*, 235(5):469–474, 1997.

65. A. Caprihan, E. Fukushima, A.D. Rosato, and M. Kos. Magnetic resonance imaging of vibrating granular beds by spatial scanning. *Review of Scientific Instruments*, 68(11):4217–4220, 1997.

66. J.R. De Bruyn, C. Bizon, M.D. Shattuck, D. Goldman, J.B. Swift, and H.L. Swinney. Continuum-type stability balloon in oscillated granular layers. *Physical Review Letters*, 81(7):1421–1424, 1998.

67. M. Medved, D. Dawson, H.M. Jaeger, and S.R. Nagel. Convection in horizontally vibrated granular material. *Chaos: An Interdisciplinary Journal of Nonlinear Science*, 9(3):691–696, 1999.

68. I.S. Aranson, D. Blair, W.K. Kwok, G. Karapetrov, U. Welp, G.W. Crabtree, V.M. Vinokur, and L.S. Tsimring. Controlled dynamics of interfaces in a vibrated granular layer. *Physical Review Letters*, 82:731–734, January 1999.

69. J.M. Ottino and D.V. Khakhar. Mixing and segregation of granular materials. *Annual Review of Fluid Mechanics*, 32(1):55–91, 2000.

70. S.S. Hsiau and C.H Chen. Granular convection cells in a vertical shaker. *Powder Technology*, 111(3):210–217, 2000.

71. C.R. Wassgren, M.L. Hunt, P.J. Freese, J. Palamara, and C.E. Brennen. Effects of vertical vibration on hopper flows of granular material. *Physics of Fluids (1994-present)*, 14(10):3439–3448, 2002.

72. M. Medved. Connections between response modes in a horizontally driven granular material. *Physical Review E*, 65:021305, January 2002.

73. R. Deng and C.-H. Wang. Particle image velocimetry study on the pattern formation in a vertically vibrated granular bed. *Physics of Fluids (1994-present)*, 15(12):3718–3729, 2003.

74. D.N. Fernando and C.R. Wassgren. Effects of vibration method and wall boundaries on size segregation in granular beds. *Physics of Fluids (1994-present)*, 15(11):3458–3467, 2003.

75. Y.S. Wong, C.H. Gan, C.H. Wang, X. Fan, D.J. Parker, A. Ingram, and J.P.K. Seville. Instabilities in vertically vibrated granular beds at the single particle scale. *Physics of Fluids (1994-present)*, 18(4), 2006.

76. L. Pernenkil and C.L. Cooney. A review on the continuous blending of powders. *Chemical Engineering Science*, 61(2):720–742, 2006.

77. G.M. Rodríguez-Liñán and Y. Nahmad-Molinari. Granular convection driven by shearing inertial forces. *Physical Review E*, 73:011302, January 2006.

78. P.M. Reis, R.A. Ingale, and M.D. Shattuck. Crystallization of a quasi-twodimensional granular fluid. *Physical Review Letters*, 96(25):258001/1–4, 2006.

79. P.M. Reis, R.A. Ingale, and M.D. Shattuck. Caging dynamics in a granular fluid. *Physical Review Letters*, 98(18):188301/1–4, 2007.

80. P.M. Reis, R.A. Ingale, and M.D. Shattuck. Forcing independent velocity distributions in an experimental granular fluid. *Physical Review E*, 75(5):051311/1–14, 2007.

81. P. Eshuis, K. van der Weele, D. van der Meer, R. Bos, and D. Lohse. Phase diagram of vertically shaken granular matter. *Physics of Fluids (1994-present)*, 19(12), 2007.

82. M.D. Shattuck, R.A. Ingale, and P.M. Reis. Granular thermodynamics. *AIP Conference Proceedings*, 1145(1):43–50, 2009.

83. G.-J. Gao, J. Blawzdziewicz, C.S. O'Hern, and M. Shattuck. Experimental demonstration of nonuniform frequency distributions of granular packings. *Physical Review E*, 80(6):061304, December 2009.

84. M. Majid and P. Walzel. Convection and segregation in vertically vibrated granular beds. *Powder Technology*, 192(3):311–317, 2009.

85. J. Bridgwater. Mixing of particles and powders: Where next? *Particuology*, 8(6):563–567, 2010. Special Issue In Honor of Professor Mooson Kwauk on his 90th Birthday.

86. C.H. Tai, S.S. Hsiau, and C.A. Kruelle. Density segregation in a vertically vibrated granular bed. *Powder Technology*, 204(23):255–262, 2010.

87. J. Bridgwater. Mixing of powders and granular materials by mechanical means a perspective. *Particuology*, 10(4):397–427, 2012.

88. T. Pöschel, D.E. Rosenkranz, and J.A.C. Gallas. Recurrent inflation and collapse in horizontally shaken granular materials. *Physical Review E*, 85:031307, March 2012.

89. P. Eshuis, K. van der Weele, M. Alam, H. van Gerner, M. van der Hoef, H. Kuipers, S. Luding, D. van der Meer, and D. Lohse. Buoyancy driven convection in vertically shaken granular matter: Experiment, numerics, and theory. *Granular Matter*, 15(6):893–911, 2013.

90. F. Zhang, L. Wang, C. Liu, P. Wu, and S. Zhan. Patterns of convective flow in a vertically vibrated granular bed. *Physics Letters A*, 378(1819):1303–1308, 2014.

91. T.M. Yamada and H. Katsuragi. Scaling of convective velocity in a vertically vibrated granular bed. *Planetary and Space Science*, 100(0):79–86, 2014.

92. K. Asencio, W. Bramer-Escamilla, G. Gutirrez, and I. Snchez. Electrical capacitance sensor array to measure density profiles of a vibrated granular bed. *Powder Technology*, 270, Part A(0):10–19, 2015.

93. J.E. Kollmer, M. Tupy, M. Heckel, A. Sack, and T. Pöschel. Absence of subharmonic response in vibrated granular systems under microgravity conditions. *Physical Review Applied*, 3:024007, February 2015.

94. H.K. Pak and R.P. Behringer. Bubbling in vertically vibrated granular materials. *Nature*, 371:231–233, 1994.

95. K. Liffman, G. Metcalfe, and P. Cleary. Granular convection and transport due to horizontal shaking. *Physical Review Letters*, 79:4574–4576, December 1997.

96. G. Metcalfe, S.G.K. Tennakoon, L. Kondic, D.G. Schaeffer, and R.P. Behringer. Granular friction, coulomb failure, and the fluid-solid transition for horizontally shaken granular materials. *Physical Review E*, 65:031302, February 2002.

97. C. Bizon, M.D. Shattuck, and J.B. Swift. Linear stability analysis of a vertically oscillated granular layer. *Physical Review E*, 60(6):7210–7216, 1999.

98. J. Duran, T. Mazozi, E. Clement, and J. Rajchenbach. Decompaction modes of a two-dimensional "sandpile" under vibration: Model and experiments. *Physical Review E*, 50:3092–3099, October 1994.

99. M. Hanifpour, N. Francois, V. Robins, A. Kingston, S.M. VaezAllaei, and M. Saadatfar. Structural and mechanical features of the order-disorder transition in experimental hard-spherepackings. *Physical Review E*, 91:062202, June 2015.

100. J.S. Olafsen and J.S. Urbach. Clustering, order, and collapse in a driven granular monolayer. *Physical Review Letters*, 81(20):4369–4372, 1998.

101. K. Feitosa and N. Menon. Breakdown of energy equipartition in a 2d binary vibrated granular gas. *Physical Review Letters*, 88:198301, April 2002.

102. J.S. Olafsen and J.S. Urbach. Two-dimensional melting far from equilibrium in a granular monolayer. *Physical Review Letters*, 95:098002, August 2005.

103. J.B. Knight, C.H. Fandrich, C.N. Lau, H.M. Jaeger, and S.R. Nagel. Density relaxation in a vibrated granular material. *Physical Review E*, 51(5):3957, 1995.

104. E.R. Nowak, J.B. Knight, E. Ben-Naim, H.M. Jaeger, and S.R. Nagel. Density fluctuations in vibrated granular materials. *Physical Review E*, 57:1971–1982, February 1998.

105. R.D. Wildman and J.M. Huntley. Novel method for measurement of granular temperature distributions in two-dimensional vibro-fluidised beds. *Powder Technology*, 113(12):14–22, 2000.

106. J.M. Huntley, T.W. Martin, M.D. Mantle, M.D. Shattuck, A.J. Sederman, R.D. Wildman, L.F. Gladden, and N.A. Halliwell. NMR measurements and hydrodynamic simulations of phase-resolved velocity distributions within a three-dimensional vibrofluidized granular bed. *Proceedings of the Royal Society A*, 463(2086):2519–2542, 2007.

107. M.D. Mantle, A.J. Sederman, L.F. Gladden, J.M. Huntley, T.W. Martin, R.D. Wildman, and M.D. Shattuck. MRI investigations of particle motion within a three-dimensional vibro-fluidized granular bed. *Powder Technology*, 197:164–179, 2008.

108. A. Kudrolli. Size separation in vibrated granular matter. *Reports on Progress in Physics*, 67(3):209, 2004.

109. A. Rosato, K.J. Strandburg, F. Prinz, and R.H. Swendsen. Why the brazil nuts are on top: Size segregation of particulate matter by shaking. *Physical Review Letters*, 58:1038–1040, March 1987.

110. P.M. Reis, T. Sykes, and T. Mullin. Phases of granular segregation in a binary mixture. *Physical Review E*, 74:051306, November 2006.

111. C.F. Harwood, K. Walanski, and R. Semmler. A miniature scintillation probe for use in powder tracer studies. *Instrumentation Science & Technology*, 6(3):227–237, 1975.

112. R.T. Balmer. The operation of sand clocks and their medieval development. *Technology and Culture*, 19(4):615–632, 1978.

113. A.A. Mills, S. Day, and S. Parkes. Mechanics of the sandglass. *European Journal of Physics*, 17(3):97, 1996.

114. R. Hogg, D.S. Cahn, T.W. Healy, and D.W. Fuerstenau. Diffusional mixing in an ideal system. *Chemical Engineering Science*, 21(11):1025–1038, 1966.

115. R.M. Nedderman, U. Tüzün, S.B. Savage, and G.T. Houlsby. The flow of granular materials-i: Discharge rates from hoppers. *Chemical Engineering Science*, 37(11):1597–1609, 1982.

116. U. Tüuzün, G.T. Houlsby, R.M. Nedderman, and S.B. Savage. The flow of granular materials-ii velocity distributions in slow flow. *Chemical Engineering Science*, 37(12):1691–1709, 1982.

117. S.B. Savage, R.M. Nedderman, U. Tüzün, and G.T. Houlsby. The flow of granular materials-iii rapid shear flows. *Chemical Engineering Science*, 38(2):189–195, 1983.

118. P. Evesque and J. Rajchenbach. Characterization of glass bead avalanches by using the technique of a rotating cylinder. *Competes Rendus Academia Science Series* II (Paris), 307:223–226, 1988.

119. J. Rajchenbach. Flow in powders: From discrete avalanches to continuous regime. *Physical Review Letters*, 65:2221–2224, October 1990.

120. M. Nakagawa, S.A. Altobelli, A. Caprihan, E. Fukushima, and E.-K. Jeong. Noninvasive measurements of granular flows by magnetic resonance imaging. *Experiments in Fluids*, 16(1):54–60, 1993.

121. K.M. Hill and J. Kakalios. Reversible axial segregation of binary mixtures of granular materials. *Physical Review E*, 49:R3610–R3613, May 1994.

122. O. Zik, Dov Levine, S.G. Lipson, S. Shtrikman, and J. Stavans. Rotationally induced segregation of granular materials. *Physical Review Letters*, 73:644–647, August 1994.

123. F. Cantelaube, Y. Limon-Duparcmeur, D. Bideau, and G.H Ristow. Geometrical analysis of avalanches in a 2d drum. *Journal of Physics I France*, 5(5):581–596, 1995.

124. G. Metcalfe and M. Shattuck. Pattern formation during mixing and segregation of flowing granular materials. *Physica A: Statistical Mechanics and its Applications*, 233(3–4):709–717, 1996.

125. M. Nakagawa, S.A. Altobelli, A. Caprihan, and E. Fukushima. NMRI study: Asial migration of radially segregated core of granular mixtures in a horizontal rotating cylinder. *Chemical Engineering Science*, 52(23):4423–4428, 1997.

126. K.M. Hill, A. Caprihan, and J. Kakalios. Axial segregation of granular media rotated in a drum mixer: Pattern evolution. *Physical Review E*, 56:4386–4393, October 1997.

127. W.A. Beverloo, H.A. Leniger, and J. van de Velde. The flow of granular solids through orifices. *Chemical Engineering Science*, 15(3–4):260–269, 1961.

128. M.F. Handley and M.G. Perry. Measurements of stresses in flowing granular materials. *Rheologica Acta*, 4(3):225–235, 1965.

129. G.C. Gardner. The region of flow when discharging granular materials from bin-hopper systems. *Chemical Engineering Science*, 21(3):261–273, 1966.

130. J.D. Athey, J.O. Cutress, and R.F. Pulfer. X-ray investigations of flowing powders. *Chemical Engineering Science*, 21(9):835–836, 1966.

131. J.O. Cutress and R.F. Pulfer. X-ray investigations of flowing powders. *Powder Technology*, 1(4):213–220, 1967.

132. W. Reisner. The behaviour of granular materials in flow out of hoppers. *Powder Technology*, 1(5):257–264, 1968.

133. M.F. Handley and M.G. Perry. Stresses in granular materials flowing in converging hopper sections. *Powder Technology*, 1(5):245–251, 1968.

134. W.G. Pariseau. Discontinuous velocity, fields in gravity flows of granular materials through slots. *Powder Technology*, 3(1):218–226, 1970.

135. T.G. Owe Berg, R.L. McDonald, and R.J. Trainor Jr. Two-dimensional flow of spheres. *Powder Technology*, 3(1):56, 1970.

136. P.L. Bransby, P.M. Blair-Fish, and R.G. James. An investigation of the flow of granular materials. *Powder Technology*, 8(56):197–206, 1973.

137. J. Lee, S.C. Cowin, and J.S. Templeton. An experimental study of the kinematics of flow through hoppers. *Transactions of The Society of Rheology (1957–1977)*, 18(2):247–269, 1974.

138. P.L. Bransby and P.M. Blair-Fish. Initial deformations during mass flow from a bunker: Observation and idealizations. *Powder Technology*, 11(3):273–288, 1975.

139. O.R. Walton. Particle dynamics modeling of geological materials, Rep. UCRL-52915. PhD thesis, University of California, Oakland, CA, Lawrence Livermore National Laboratory, 1980.

140. R.L. Michalowski. Flow of granular material through a plane hopper. *Powder Technology*, 39(1):29–40, 1984.

141. G. William Baxter, R.P. Behringer, T. Fagert, and G. Allan Johnson. Pattern formation in flowing sand. *Physical Review Letters*, 62:2825–2828, June 1989.

142. G.W. Baxter, R. Leone, and R.P. Behringer. Experimental test of time scales in flowing sand. *EPL (Europhysics Letters)*, 21(5):569, 1993.

143. N. Menon and D.J. Durian. Diffusing-wave spectroscopy of dynamics in a three-dimensional granular flow. *Science*, 275(5308):1920–1922, 1997.

144. M.L. Hunt, R.C. Weathers, A.T. Lee, C.E. Brennen, and C.R. Wassgren. Effects of horizontal vibration on hopper flows of granular materials. *Physics of Fluids (1994-present)*, 11(1):68–75, 1999.

145. K. Desmond and S.V. Franklin. Jamming of three-dimensional prolate granular materials. *Physical Review E*, 73:031306, March 2006.

146. E. Ono. Powder flow down a chute. *Oyobuturi*, 36(5):347–351, 1967.

147. R. Utsumi and K. Ueda. The distribution of velocity among the granular solids flowing in an inclined open channel. *Journal of the Society of Materials Science, Japan*, 20(213):773–775, 1971.

148. D.A. Augenstein and R. Hogg. An experimental study of the flow of dry powders over inclined surfaces. *Powder Technology*, 19(2):205–215, 1978.

149. M. Ishida and T. Shirai. Velocity distributions in the flow of solid particles in an inclined open channel. *Journal of Chemical Engineering of Japan*, 12(1): 46–50, 1979.

150. C.E. Brennen, K. Sieck, and J. Paslaski. Hydraulic jumps in granular material flow. *Powder Technology*, 35(1):31–37, 1983.

151. T.G. Drake and R.L. Shreve. High-speed motion pictures of nearly steady, uniform, two-dimensional, inertial flows of granular material. *Journal of Rheology (1978-present)*, 30(5):981–993, 1986.

152. T.G. Drake. Structural features in granular flows. *Journal of Geophysical Research: Solid Earth*, 95(B6):8681–8696, 1990.

153. P.C. Johnson, P. Nott, and R. Jackson. Frictional-collisional equations of motion for participate flows and their application to chutes. *Journal of Fluid Mechanics*, 210:501–535, 1 1990.

154. T.G. Drake. Granular flow: Physical experiments and their implications for microstructural theories. *Journal of Fluid Mechanics*, 225:121–152, 4 1991.

155. S.S. Hsiau and M.L. Hunt. Shear-induced particle diffusion and longitudinal velocity fluctuations in a granular-flow mixing layer. *Journal of Fluid Mechanics*, 251:299–313, 6 1993.

156. E.C. Bingham and R.W. Wikoff. The flow of dry sand through capillary tubes. *Journal of Rheology (1929–1932)*, 2(4):395–400, 1931.

157. M.S. Brinn, S.J. Friedmen, F.A. Gluckert, and R.L. Pigford. Heat transfer to granularmatgerials. *Industrial & Engineering Chemistry*, 40(6):1050–1061, 1948.

158. J.W. Delaplaine. Forces acting in flowing beds of solids. *AIChE Journal*, 2(1):127–138, 1956.

159. J. Duran, T. Mazozi, S. Luding, E. Clément, and J. Rajchenbach. Discontinuous decompaction of a falling sandpile. *Physical Review E*, 53:1923–1930, February 1996.

160. K. Choo, T.C.A. Molteno, and S.W. Morris. Traveling granular segregation patterns in a long drum mixer. *Physical Review Letters*, 79:2975–2978, October 1997.

161. K. Choo, M.W. Baker, T.C.A. Molteno, and S.W. Morris. Dynamics of granular segregation patterns in a long drum mixer. *Physical Review E*, 58(5):6115–6123, November 1998.

162. Y. Garfnkel, D. Ben-Shlomo, and T. Kuperman. Large-scale storage of grain surplus in the sixth millennium BC: The silos of teltsaf. *Antiquity*, 83:309–325, 6 2009.

163. T.V. Nguyen, C.E. Brennen, and R.H. Sabersky. Funnel flow in hoppers. *Journal of Applied Mechanics*, 47(4):729–735, 1980.

164. G. Gutiérrez, C. Colonnello, P. Boltenhagen, R. Darias, J.R. Peralta-Fabi, F. Brau, and E. Clément. Silo collapse under granular discharge. *Physical Review Letters*, 114:018001, January 2015.

165. K. To, P.-Y. Lai, and H.K. Pak. Jamming of granular flow in a two-dimensional hopper. *Physical Review Letters*, 86:71–74, January 2001.

166. I. Zuriguel, A. Garcimartín, D. Maza, L.A. Pugnaloni, and J.M. Pastor. Jamming during the discharge of granular matter from a silo. *Physical Review E*, 71:051303, May 2005.

167. A. Garcimartín, I. Zuriguel, L.A. Pugnaloni, and A. Janda. Shape of jamming arches in two-dimensional deposits of granular materials. *Physical Review E*, 82:031306, September 2010.

168. C.C. Thomas and D.J. Durian. Geometry dependence of the clogging transition in tilted hoppers. *Physical Review E*, 87:052201, May 2013.

169. C.C. Thomas and D.J. Durian. Fraction of clogging configurations sampled by granular hopper flow. *Physical Review Letters*, 114:178001, April 2015.

170. R.A. Bagnold. Experiments on a gravity-free dispersion of large solid spheres in a newtonian fluid under shear. *Proceedings of the Royal Society of London Series A. Mathematical and Physical Sciences*, 225(1160):49–63, 1954.

171. M.L. Hunt, R. Zenit, C.S. Campbell, and C.E. Brennen. Revisiting the 1954 suspension experiments of r. a. bagnold. *Journal of Fluid Mechanics*, 452:1–24, 2 2002.

172. J.-C. Tsai and J.P. Gollub. Slowly sheared dense granular flows: Crystallization and nonunique final states. *Physical Review E*, 70:031303, September 2004.

173. K.E. Daniels and R.P. Behringer. Hysteresis and competition between disorder and crystallization in sheared and vibrated granular flow. *Physical Review Letters*, 94:168001, April 2005.

174. K.E. Daniels and R.P Behringer. Characterization of a freezing/melting transition in a vibrated and sheared granular medium. *Journal of Statistical Mechanics: Theory and Experiment*, 2006(07):P07018, 2006.

175. A. Panaitescu, K.A. Reddy, and A. Kudrolli. Nucleation and crystal growth in sheared granular sphere packings. *Physical Review Letters*, 108:108001, March 2012.

176. G. De Josselin De Jong, and A. Verruijt. Etude photo-elastique d'un empile-ment de disques. *Cahiers du groupe Francais de Rheologie*, 2(1):73–86, 1969.

177. A. Drescher and G. de Josselin de Jong. Photoelastic verification of a mechan-ical model for the flow of a granular material. *Journal of the Mechanics and Physics of Solids*, 20(5):337–340, 1972.

178. D. Howell, R.P. Behringer, and C. Veje. Stress fluctuations in a 2d gran-ular couette experiment: A continuous transition. *Physical Review Letters*, 82:5241–5244, June 1999.

179. R.P. Behringer, D. Bi, B. Chakraborty, S. Henkes, and R.R. Hartley. Why do granular materials stiffen with shear rate? Test of novel stress-based statis-tics. *Physical Review Letters*, 101:268301, December 2008.

180. S. Farhadi and R.P. Behringer. Dynamics of sheared ellipses and circular disks: Effects of particle shape. *Physical Review Letters*, 112:148301, April 2014.

181. S.B. Savage and M. Sayed. Stresses developed by dry cohesionless granu-lar materials sheared in an annular shear cell. *Journal of Fluid Mechanics*, 142:391–430, 5 1984.

182. D.M. Hanes and D.L. Inman. Observations of rapidly flowing granular-fluid materials. *Journal of Fluid Mechanics*, 150:357–380, 1 1985.

183. B. Miller, C. O'Hern, and R.P. Behringer. Stress fluctuations for continuously sheared granular materials. *Physical Review Letters*, 77:3110–3113, October 1996.

184. G.I. Tardos, M. Irfan Khan, and D.G. Schaeffer. Forces on a slowly rotating, rough cylinder in a couette device containing a dry, frictional powder. *Physics of Fluids*, 10(2):335–341, 1998.

185. G.I. Tardos, S. McNamara, and I. Talu. Slow and intermediate flow of a fric-tional bulk powder in the couette geometry. *Powder Technology*, 131(1):23–39, 2003.

186. C. Coste. Shearing of a confined granular layer: Tangential stress and dila-tancy. *Physical Review E*, 70:051302, November 2004.

187. K. Lu, E.E. Brodsky, and H.P. Kavehpour. Shear-weakening of the transitional regime for granular flow. *Journal of Fluid Mechanics*, 587:347–372, 9 2007.

188. R. Khosropour, J. Zirinsky, H.K. Pak, and R.P. Behringer. Convection and size segregation in a couette flow of granular material. *Physical Review E*, 56:4467–4473, October 1997.

189. W. Losert, L. Bocquet, T.C. Lubensky, and J.P. Gollub. Particle dynamics in sheared granular matter. *Physical Review Letters*, 85:1428–1431, August 2000.

190. M. Toiya, J. Stambaugh, and W. Losert. Transient and oscillatory granular shear flow. *Physical Review Letters*, 93:088001, August 2004.

191. V. Mehandia, K.J. Gutam, and P.R. Nott. Anomalous stress profile in a sheared granular column. *Physical Review Letters*, 109:128002, September 2012.

192. N. Murdoch, B. Rozitis, K. Nordstrom, S. F. Green, P. Michel, T.-L.de Lophem, and W. Losert. Granular convection in microgravity. *Physical Review Letters*, 110:018307, January 2013.

193. D. Bi, J. Zhang, B. Chakraborty, and R.P. Behringer. Jamming by shear. *Nature*, 480:355–358, December 2011.

194. L. Oger, J.-C. Charmet, D. Bideau, and J.-P. Troadec. Mechanical properties of pulverulentmatter 2d. *Competes Rendus Academia Science Series II (Paris)*, 302(6):277–280, 1986.

195. X. Cheng. Experimental study of the jamming transition at zero temperature. *Physical Review E*, 81:031301, March 2010.

196. M.E. Möbius, X. Cheng, P. Eshuis, G.S. Karczmar, S.R. Nagel, and H.M. Jaeger. Effect of air on granular size separation in a vibrated granular bed. *Physical Review E*, 72:011304, July 2005.

197. J. Adam, J.L. Urai, B. Wieneke, O. Oncken, K. Pfeiffer, N. Kukowski, J. Lohrmann, S. Hoth, W. van der Zee, and J. Schmatz. Shear localisation and strain distribution during tectonicfaultingnew insights from granular-flow experiments and high-resolution optical image correlation techniques. *Journal of Structural Geology*, 27(2):283–301, 2005.

198. T. Ivan Quickenden and G.K. Tan. Random packing in two dimensions and the structure of monolayers. *Journal of Colloid and Interface Science*, 48(3):382–393, 1974.

199. M. Ammi, D. Bideau, and J.P. Troadec. Geometrical structure of disordered packings of regular polygons; comparison with disc packings structures. *Journal of Physics D: Applied Physics*, 20(4):424, 1987.

200. D. Bideau, A. Gervois, L. Oger, and J.P. Troadec. Geometrical properties of disordered packings of hard disks. *Journal of Physics France*, 47(10):1697–1707, 1986.

201. P.B. Umbanhowar, F. Melo, and H.L. Swinney. Periodic, aperiodic, and transient patterns in vibrated granular layers. *Physica A*, 249(1-4):1–9, 1998.

202. J.R. de Bruyn, B.C. Lewis, M.D. Shattuck, and H.L. Swinney. Spiral patterns in oscillated granular layers. *Physical Review E*, 63(4):041305/1–12, 2001.

203. D.I. Goldman, M.D. Shattuck, H.L. Swinney, and G.H. Gunaratne. Emergence of order in an oscillated granular layer. *Physica A*, 306(1-4):180–8, 2002.

204. D.I. Goldman, M.D. Shattuck, S.J. Moon, J.B. Swift, and H.L. Swinney. Lattice dynamics and melting of a nonequilibrium pattern. *Physical Review Letters*, 90(10):104302/1–4, 2003.

205. S. Dorbolo, L. Maquet, M. Brandenbourger, F. Ludewig, G. Lumay, H. Caps, N. Vandewalle, S. Rondia, M. Mlard, J. van Loon, A. Dowson, and S. Vincent-Bonnieu. Influence of the gravity on the discharge of a silo. *Granular Matter*, 15(3):263–273, 2013.

206. T. Miller, P. Rognon, B. Metzger, and I. Einav. Eddy viscosity in dense granular flows. *Physical Review Letters*, 111:058002, August 2013.

207. T. Wakabayashi. Photo-elastic method for determination of stress in powdered mass. *Journal of the Physical Society of Japan*, 5(5):383–385, 1950.

208. H.G.B. Allersma. Determination of the stress distribution in assemblies of photoelastic particles. *Experimental Mechanics*, 22(9):336–341, 1982.

209. E.T. Owens, S. Couvreur, and K.E. Daniels. Spatiotemporally resolved acoustics in a photoelastic granular material. *AIP Conference Proceedings*, 1145(1):447–450, 2009.

210. J. Zhang, T.S. Majmudar, A. Tordesillas, and R.P. Behringer. Statistical properties of a 2d granular material subjected to cyclic shear. *Granular Matter*, 12(2):159–172, 2010.

211. A.H. Clark, L. Kondic, and R.P. Behringer. Particle scale dynamics in granular impact. *Physical Review Letters*, 109:238302, December 2012.

212. J. Ren, J.A. Dijksman, and R.P. Behringer. Reynolds pressure and relaxation in a sheared granular system. *Physical Review Letters*, 110:018302, January 2013.

213. R.P. Behringer, D. Bi, B. Chakraborty, A. Clark, J. Dijksman, J. Ren, and J. Zhang. Statistical properties of granular materials near jamming. *Journal of Statistical Mechanics: Theory and Experiment*, 2014(6):P06004, 2014.

214. A.H. Clark, A.J. Petersen, and R.P. Behringer. Collisional model for granular impact dynamics. *Physical Review E*, 89:012201, January 2014.

215. J.C. Maxwell F.R.S. XLV. On reciprocal figures and diagrams of forces. *Philosophical Magazine Series 4*, 27(182):250–261, 1864.

216. J. Clerk Maxwell F.R.S. On bow's method of drawing diagrams in graphical statics, with illustrations from peaucellier's linkage. *Proceedings of the Cambridge Philosophical Society*, 2:407–421, 1876.

217. C.T. Veje, D.W. Howell, and R.P. Behringer. Kinematics of a two-dimensional granularcouette experiment at the transition to shearing. *Physical Review E*, 59:739–745, January 1999.

218. R.P. Behringer, D. Howell, L. Kondic, S. Tennakoon, and C. Veje. Predictability and granular materials. *Physica D: Nonlinear Phenomena*, 133(14):1–17, 1999.

219. T.S. Majmudar and R.P. Behringer. Contact force measurements and stress induced anisotropy in granular materials. *Nature*, 435:1079–1082, June 2005.

220. T.S. Majmudar, M. Sperl, S. Luding, and R.P. Behringer. Jamming transition in granular systems. *Physical Review Letters*, 98:058001, January 2007.

221. B. Utter and R.P. Behringer. Experimental measures of affine and nonaffine deformation in granular shear. *Physical Review Letters*, 100:208302, May 2008.

222. J. Geng, D. Howell, E. Longhi, R.P. Behringer, G. Reydellet, L. Vanel, E. Clément, and S. Luding. Footprints in sand: The response of a granular material to local perturbations. *Physical Review Letters*, 87:035506, July 2001.

223. J. Geng, G. Reydellet, E. Clment, and R.P. Behringer. Greens function measurements of force transmission in 2d granular materials. *Physica D: Nonlinear Phenomena*, 182(34):274–303, 2003.

224. D. Brewster. Experiments on the depolarisation of light as exhibited by various mineral, animal, and vegetable bodies, with a reference of the phenomena to the general principles of polarisation. *Philosophical Transactions of the Royal Society of London*, 105:29–53, 1815.

225. D. Brewster. On the communication of the structure of doubly refracting crystals to glass, muriate of soda, fluor spar, and other substances, by mechanical compression and dilatation. *Philosophical Transactions of the Royal Society of London*, 106:156–178, 1816.

226. J.C. Maxwell. IV. On the equilibrium of elastic solids. *Transactions of the Royal Society of Edinburgh*, 20:87–120, 1 1853.

227. R.C. Jones. A new calculus for the treatment of optical systems. *Journal of Optical Society America*, 31(7):488–493, July 1941.

228. H. Hurwitz Jr. and R.C. Jones. A new calculus for the treatment of optical systems. *Journal of Optical Society America*, 31(7):493–495, July 1941.

229. R.C. Jones. A new calculus for the treatment of optical systems. *Journal of Optical Society America*, 31(7):500–503, July 1941.

230. R.C. Jones. A new calculus for the treatment of optical systems. IV. *Journal of Optical Society America*, 32(8):486–493, August 1942.

231. R.C. Jones. A new calculus for the treatment of optical systems V. A more general formulation, and description of another calculus. *Journal of Optical Society America*, 37(2):107, February 1947.

232. R.C. Jones. A new calculus for the treatment of optical systems VI. Experimental determination of the matrix. *Journal of Optical Society America*, 37(2):110, February 1947.

233. R.C. Jones. A new calculus for the treatment of optical systems. VII. Properties of the n-matrices. *Journal of Optical Society America*, 38(8):671–683, August 1948.

234. R.C. Jones. New calculus for the treatment of optical systems. VIII. Electromagnetic theory. *Journal of Optical Society America*, 46(2):126–131, February 1956.

235. S. Wiederseiner, N. Andreini, G. Epely-Chauvin, and C. Ancey. Refractive-index and density matching in concentrated particle suspensions: A review. *Experiments in Fluids*, 50(5):1183–1206, 2011.

236. S.C. Jana, B. Kapoor, and A. Acrivos. Apparent wall slip velocity coefficients in concentrated suspensions of noncolloidal particles. *Journal of Rheology*, 39(6), 1995.

237. G.P. Krishnan, S. Beimfohr, and D.T. Leighton. Shear-induced radial segregation in bidisperse suspensions. *Journal of Fluid Mechanics*, 321:371–393, 8 1996.

238. J.-C. Tsai, G.A. Voth, and J.P. Gollub. Internal granular dynamics, shear-induced crystallization, and compaction steps. *Physical Review Letters*, 91:064301, August 2003.

239. O. Pouliquen, M. Belzons, and M. Nicolas. Fluctuating particle motion during shear induced granular compaction. *Physical Review Letters*, 91:014301, July 2003.

240. S. Siavoshi, A.V. Orpe, and A. Kudrolli. Friction of a slider on a granular layer: Nonmonotonic thickness dependence and effect of boundary conditions. *Physical Review E*, 73:010301, January 2006.

241. S. Slotterback, M. Toiya, L. Goff, J.F. Douglas, and W. Losert. Correlation between particle motion and voronoi-cell-shape fluctuations during the compaction of granular matter. *Physical Review Letters*, 101:258001, December 2008.

242. K.A. Lörincz and P. Schall. Visualization of displacement fields in a sheared granular system. *Soft Matter*, 6:3044–3049, 2010.

243. J.A. Dijksman, F. Rietz, K.A. Lörincz, M. van Hecke, and W. Losert. Invited article: Refractive index matched scanning of dense granular materials. *Review of Scientific Instruments*, 83(1), 2012.

244. S. Mukhopadhyay and J. Peixinho. Packings of deformable spheres. *Physical Review E*, 84:011302, July 2011.

245. G.W. Baxter, R. Leone, G.A. Jonhson, and R.P. Behringer. Time-dependence, scaling and pattern formation for flowing sand. *European Journal of Mechanics B-Fluids*, 10(2,S):181–186, 1991.

246. D.M. Mueth, G.F. Debregeas, G.S. Karczmar, P.J. Eng, S.R. Nagel, and H.M. Jaeger. Signatures of granular microstructure in dense shear flows. *Nature*, 406(6794):385–389, July 27 2000.

247. G.T. Seidler, G. Martinez, L.H. Seeley, K.H. Kim, E.A. Behne, S. Zaranek, B.D. Chapman, S.M. Heald, and D.L. Brewe. Granule-by-granule reconstruction of a sandpile from x-ray microtomography data. *Physical Review E*, 62:8175–8181, December 2000.

248. P. Richard, P. Philippe, F. Barbe, S. Bourlès, X. Thibault, and D. Bideau. Analysis by x-ray microtomography of a granular packing undergoing compaction. *Physical Review E*, 68:020301, August 2003.

249. T. Aste, M. Saadatfar, and T.J. Senden. Geometrical structure of disordered sphere packings. *Physical Review E*, 71:061302, June 2005.

250. M. Jerkins, M. Schröter, H.L. Swinney, T.J. Senden, M. Saadatfar, and T. Aste. Onset of mechanical stability in random packings of frictional spheres. *Physical Review Letters*, 101:018301, July 2008.

251. H. Lee, M. Brandyberry, A. Tudor, and K. Matous. Three-dimensional reconstruction of statistically optimal unit cells of polydisperse particulate composites from microtomography. *Physical Review E*, 80:061301, December 2009.

252. Y. Fu, Y. Xi, Y. Cao, and Y. Wang. X-ray microtomography study of the compaction process of rods under tapping. *Physical Review E*, 85:051311, May 2012.

253. V. Yadav, J.-Y. Chastaing, and A. Kudrolli. Effect of aspect ratio on the development of order in vibrated granular rods. *Physical Review E*, 88:052203, November 2013.

254. A. Gillman, K. Matouš, and S. Atkinson. Microstructure-statistics-property relations of anisotropic polydisperse particulate composites using tomography. *Physical Review E*, 87:022208, February 2013.

255. A.G. Athanassiadis, P.J. La Rivire, E. Sidky, C. Pelizzari, X. Pan, and H.M. Jaeger. X-ray tomography system to investigate granular materials during mechanical loading. *Review of Scientific Instruments*, 85(8), 2014.

256. J. Radon. Uber die bestimmung von funktionen durch ihre integralwerte langs gewisser mannigfaltigkeiten [on the determination of functions from their integrals along certain manifolds]. *Ber Saechsische Akad Wiss*, 29:262, 1917.

257. J. Radon. On the determination of functions from their integral values along certain manifolds. *Medical Imaging, IEEE Transactions on*, 5(4):170–176, December 1986.

258. A. Kudrolli and J.P. Gollub. Localized spatiotemporal chaos in surface waves. *Physical Review E*, 54:R1052–R1055, 1996.

259. S.A. Altobelli, R.C. Givler, and E. Fukushima. Velocity and concentration measurements of suspensions by nuclear magnetic resonance imaging. *Journal of Rheology (1978-present)*, 35(5):721–734, 1991.

260. V.Yu. Kuperman. Nuclear magnetic resonance measurements of diffusion in granular media. *Physical Review Letters*, 77:1178–1181, August 1996.

261. G.H. Ristow and M. Nakagawa. Shape dynamics of interfacial front in rotating cylinders. *Physical Review E*, 59:2044–2048, February 1999.

262. J.D. Seymour, A. Caprihan, S.A. Altobelli, and E. Fukushima. Pulsed gradient spin echo nuclear magnetic resonance imaging of diffusion in granular flow. *Physical Review Letters*, 84:266–269, January 2000.

263. X. Yang, C. Huan, D. Candela, R.W. Mair, and R.L. Walsworth. Measurements of grain motion in a dense, three-dimensional granular fluid. *Physical Review Letters*, 88:044301, January 2002.

264. L.F. Gladden. Magnetic resonance: Ongoing and future role in chemical engineering research. *AIChE Journal*, 49(1):2–9, 2003.

265. L. Sanfratello, A. Caprihan, and E. Fukushima. Velocity depth profile of granular matter in a horizontal rotating drum. *Granular Matter*, 9(1-2):1–6, 2007.

266. P. Moucheront, F. Bertrand, G. Koval, L. Tocquer, S. Rodts, J.-N. Roux, A. Corfdir, and F. Chevoirw. MRI investigation of granular interface rheology using a new cylinder shear apparatus. *Magnetic Resonance Imaging*, 28(6):910–918, 2010.

267. J.M. Huntley, T. Tarvaz, M.D. Mantle, A.J. Sederman, L.F. Gladden, N.A. Sheikh, and R.D. Wildman. Nuclear magnetic resonance measurements of velocity distributions in an ultrasonically vibrated granular bed. *Philosophical Transactions of the Royal Society of London A: Mathematical, Physical and Engineering Sciences*, 372(2015), 2014.

268. P.C. Lauterbur. Imaging formation by induced local interactions: Examples employing nuclear magnetic resonance. *Nature*, 242:190–191, 1973.

269. P. Mansfield and P.K. Grannell. Diffraction in solids? *Journal of Physics C*, 6:190, 1973.

270. A. Kumar, D. Welti, and R.R. Ernst. {NMR} fourierzeugmatography. *Journal of Magnetic Resonance (1969)*, 18(1):69–83, 1975.

271. E.M. Purcell, H.C. Torrey, and R.V. Pound. Resonance absorption by nuclear magnetic moments in a solid. *Physical Review*, 69:37, 1946.

272. F. Bloch, W.W. Hansen, and M. Packard. Nuclear induction. *Physical Review*, 69:127, 1946.

273. D. Fischer, T. Finger, F. Angenstein, and R. Stannarius. Diffusive and subdiffusive axial transport of granular material in rotating mixers. *Physical Review E*, 80:061302, December 2009.

274. T.T.M. Nguyễn, A.J. Sederman, M.D. Mantle, and L.F. Gladden. Segregation in horizontal rotating cylinders using magnetic resonance imaging. *Physical Review E*, 84:011304, July 2011.

275. E.L. Hahn. Spin echoes. *Physical Review*, 74:580, 1950.

276. C.P. Slichter. *Principles of Magnetic Resonance*. Springer-Verlag, New York, 1978.

277. P.T. Callaghan. *Principles of Nuclear Magnetic Resonance Microscopy*. Clarendon Press, Oxford, London, 1991.

278. J.R. Singer. Blood flow measurements by NMR of the intact body. *IEEE Transactions on Nuclear Science*, 27(3):1245–1249, 1980.

279. L. Axel and L. Dougherty. MR imaging of motion with spatial modulation of magnetization. *Radiology*, 171(3):841–845, 1989. PMID: 2717762.

280. L. Axel and L. Dougherty. Heart wall motion: Improved method of spatial modulation of magnetization for MR imaging. *Radiology*, 172(2):349–350, 1989. PMID: 2748813.

281. M. O'Donnell. NMR blood flow imaging using multiecho, phase contrast sequences. *Medical Physics*, 12(1):59–64, 1985.

282. C.L. Dumoulin. Magnetic resonance angiography. *Radiology*, 161(3):717–720, 1986.

283. P.R. Moran. A flow velocity zeugmatographic interlace for NMR imaging in humans. *Magnetic Resonance Imaging*, 1:197–203, 1982.

284. T.W. Redpath, D.G. Norris, R.A. Jones, and J.M.S. Hutchison. A new method of NMR flow imaging. *Physical Medical Biology*, 29(7):891–898, 1984.

285. A. Panaitescu and A. Kudrolli. Spatial distribution functions of random packed granular spheres obtained by direct particle imaging. *Physical Review E*, 81:060301, June 2010.

286. J.C. Crocker and D.G. Grier. Methods of digital video microscopy for colloidal studies. *Journal of Colloid and Interface Science*, 179(1):298–310, 1996.

287. E.C. Rericha, C. Bizon, M.D. Shattuck, and H.L. Swinney. Shocks in supersonic sand. *Physical Review Letters*, 88(1):014302/1–4, 2002.

288. E.R. Davies. *Computer and Machine Vision: Theory, Algorithms, Practicalities*, 4th edn. Academic Press. Elsevier, Boston, MA, 2012.

289. R.O. Duda and P.E. Hart. Use of the hough transformation to detect lines and curves in pictures. *Communication of the ACM*, 15(1):11–15, January 1972.

290. E.R. Davies. Chapter 12—circle and ellipse detection. In E.R. Davies, ed., *Computer and Machine Vision*, 4th edn, pp. 303–332. Academic Press, Boston, MA, 2012.

291. E.R. Davies. A skimming technique for fast accurate edge detection. *Signal Processing*, 26(1):1–16, 1992.

292. W.H. Press, S.A. Teukolsky, W.T. Vetterling, and B.P. Flannery. *Numerical Recipes in C*, 2nd edn. *The Art of Scientific Computing*. Cambridge University Press, New York, NY, USA, 1992.

293. R.J. Adrian. Twenty years of particle image velocimetry. *Experiments in Fluids*, 39(2):159–169, 2005.

294. S.P. Pudasaini, S.-S. Hsiau, Y. Wang, and K. Hutter. Velocity measurements in dry granular avalanches using particle image velocimetry technique and comparison with theoretical predictions. *Physics of Fluids (1994-present)*, 17(9), 2005.

295. E.T. Owens and K.E. Daniels. Sound propagation and force chains in granular materials. *EPL (Europhysics Letters)*, 94(5):54005, 2011.

296. P. Mort, J.N. Michaels, R.P. Behringer, C.S. Campbell, L. Kondic, M. Kheiripour Langroudi, M. Shattuck, J. Tang, G.I. Tardos, and C. Wassgren. Dense granular flow a collaborative study. *Powder Technology*, 2015.

297. J. Zhang, T.S. Majmudar, M. Sperl, and R.P. Behringer. Jamming for a 2d granular material. *Soft Matter*, 6:2982–2991, 2010.

298. A. Flamant. Sur la répartition des pressions dans un solide rectangulaire chargé transversalement. *Competes Rendus Academie Sciences Paris*, 114:1465–1468, 1892.

299. S.F. Edwards. The equations of stress in a granular material. *Physica A: Statistical Mechanics and its Applications*, 249(14):226–231, 1998.

300. T. Unger, J. Kertész, and D.E. Wolf. Force indeterminacy in the jammed state of hard disks. *Physical Review Letters*, 94:178001, May 2005.

301. J. Puttock and E.G. Thwaite. Elastic compression of spheres and cylinders at point and line contact. Commonwealth Scientific and Industrial Research Organisation, Australia; National standards Laboratory; technical paper. Commonwealth Scientific and Industrial Research Organization, 1969.

302. R.M. Corless, G.H. Gonnet, D.E.G. Hare, D.J. Jeffrey, and D.E. Knuth. On the lambertw function. *Advances in Computational Mathematics*, 5(1):329–359, 1996.

303. C.A. Coulomb. Memoires de mathematiques et de physique presentes a l'academie royale des sciences par divers savans et lusdans les assemblees. *Académie royale des Sciences par divers Savants*, 7:343, 1773.

304. P. Bak, C. Tang, and K. Wiesenfeld. Self-organized criticality. *Physical Review A*, 38:364–374, July 1988.

305. H.M. Jaeger, C.-h. Liu, and S.R. Nagel. Relaxation at the angle of repose. *Physical Review Letters*, 62:40–43, January 1989.

306. G.A. Held, D.H. Solina, H. Solina, D.T. Keane, W.J. Haag, P.M. Horn, and G. Grinstein. Experimental study of critical-mass fluctuations in an evolving sandpile. *Physical Review Letters*, 65:1120–1123, August 1990.

307. J. Lee and H.J. Herrmann. Angle of repose and angle of marginal stability: Molecular dynamics of granular particles. *Journal of Physics A: Mathematical and General*, 26:373–383, 1993.

308. J.J. Alonso, J.-P. Hovi, and H.J. Herrmann. Lattice model for the calculation of the angle of repose from microscopic grain properties. *Physical Review E*, 58:672–680, July 1998.

309. H.J. Herrmann. On the shape of a sandpile. In H.J. Herrmann, J.-P. Hovi, and S. Luding, eds, *Physics of Dry Granular Media*, volume 350 of NATO ASI Series E: Applied Sciences, pp. 319–338. Kluwer Academic publishers, 1998.

310. A. Samadani and A. Kudrolli. Angle of repose and segregation in cohesive granular matter. *Physical Review E*, 64:051301, October 2001.

311. M. de Sousa Vieira. Breakdown of self-organized criticality in sandpiles. *Physical Review E*, 66:051306, November 2002.

312. M.G. Kleinhans, H. Markies, S.J. de Vet, A.C. in 't Veld, and F.N. Postema. Static and dynamic angles of repose in loose granular materials under reduced gravity. *Journal of Geophysical Research: Planets*, 116(E11), 2011.

313. H. Yang, R. Li, P. Kong, Q.C. Sun, M.J. Biggs, and V. Zivkovic. Avalanche dynamics of granular materials under the slumping regime in a rotating drum as revealed byspeckle visibility spectroscopy. *Physical Review E*, 91:042206, April 2015.

314. K. Wieghardt. Experiments in granular flow. *Annual Review of Fluid Mechanics*, 7(1):89–114, 1975.

315. P.D. Ruiz, Y. Zhou, J.M. Huntley, and R.D. Wildman. Depth resolved whole-field displacement measurement using wavelength scanning interferometry. *Journal of Optics A: Pure and Applied Optics*, 6(7):679, 2004.

316. A.M. Hoffman. Radioactive tracer techniques in solid propellant mixing. *Industrial & Engineering Chemistry*, 52(9):781–782, 1960.

317. D.J. Parker, C.J. Broadbent, P. Fowles, M.R. Hawkesworth, and P. McNeil. Positron emission particle tracking—a technique for studying flow within engineering equipment. *Nuclear Instruments and Methods in Physics Research Section A: Accelerators, Spectrometers, Detectors and Associated Equipment*, 326(3):592–607, 1993.

318. D.J. Parker, M.R. Hawkesworth, C.J. Broadbent, P. Fowles, T.D. Fryer, and P.A. McNeil. Industrial positron-based imaging: Principles and applications. *Nuclear Instruments and Methods in Physics Research Section A: Accelerators, Spectrometers, Detectors and Associated Equipment*, 348(23):583–592, 1994.

319. D.J. Parker, D.A. Allen, D.M. Benton, P. Fowles, P.A. McNeil, M. Tan, and T.D. Beynon. Developments in particle tracking using the birmingham positron camera. *Nuclear Instruments and Methods in Physics Research Section A: Accelerators, Spectrometers, Detectors and Associated Equipment*, 392(13):421–426, 1997. Position-Sensitive Detectors Conference 1996.

320. R.D. Wildman, J.M. Huntley, and D.J. Parker. Convection in highly flu-idized three-dimensional granular beds. *Physical Review Letters*, 86:3304–3307, April 2001.

321. R.D. Wildman and D.J. Parker. Coexistence of two granular temperatures in binary vibrofluidized beds. *Physical Review Letters*, 88:064301, January 2002.

322. R.D. Wildman, T.W. Martin, J.M. Huntley, J.T. Jenkins, H. Viswanathan, X. Fen, and D.J. Parker. Experimental investigation and kinetic-theory-based model of a rapid granular shear flow. *Journal of Fluid Mechanics*, 602:63–79, 5 2008.

323. R.D. Wildman, C.M. Hrenya, J.M. Huntley, T. Leadbeater, and D.J. Parker. Experimental determination of temperature profiles in a sheared granular bed containing two and three sizes of particles. *Granular Matter*, 14(2):215–220, 2012.

324. B.F.C. Laurent and P.W. Cleary. Comparative study by {PEPT} and {DEM} for flow and mixing in a ploughshare mixer. *Powder Technology*, 228(0):171–186, 2012.

325. C.R.K. Windows-Yule, G.J.M. Douglas, and D.J. Parker. Competition between geometrically induced and density-driven segregation mechanisms in vibrofluidized granular systems. *Physical Review E*, 91:032205, March 2015.

326. T. Shirai, M. Ishida, Y. Ito, N. Inoue, and S. Kobayashi. Rotation of a horizontal disk within an aerated particle bed. *Journal of Chemical Engineering of Japan*, 10(1):40–45, 1977.

327. G. Maret and P.E. Wolf. Multiple light scattering from disordered media. The effect of brownian motion of scatterers. *ZeitschriftfrPhysik B Condensed Matter*, 65(4):409–413, 1987.

328. D.J. Pine, D.A. Weitz, P.M. Chaikin, and E. Herbolzheimer. Diffusing wave spectroscopy. *Physical Review Letters*, 60:1134–1137, March 1988.

329. N. Menon and D.J. Durian. Particle motions in a gas-fluidized bed of sand. *Physical Review Letters*, 79:3407–3410, November 1997.

330. A.R. Abate, H. Katsuragi, and D.J. Durian. Avalanche statistics and time-resolved grain dynamics for a driven heap. *Physical Review E*, 76:061301, December 2007.

331. H.G.M. Ruis, P. Venema, and E. van der Linden. Diffusing wave spectroscopy used to study the influence of shear on aggregation. *Langmuir*, 24(14):7117–7123, 2008.

332. V. Zivkovic, M.J. Biggs, D.H. Glass, and L. Xie. Particle dynamics and granular temperatures in dense fluidized beds as revealed by diffusing wave spectroscopy. *Advanced Powder Technology*, 20(3):227–233, 2009.

333. K. Kim and H. Pak. Diffusing-wave spectroscopy study of microscopic dynamics of three-dimensional granular systems. *Soft Matter*, 6:2894–2900, 2010.

334. V. Zivkovic, M.J. Biggs, and D.H. Glass. Scaling of granular temperature in a vibrated granular bed. *Physical Review E*, 83:031308, March 2011.

335. F. Scheffold. Particle sizing with diffusing wave spectroscopy. *Journal of Dispersion Science & Technology*, 23(5):591, 2002.

336. M. Yamashiro and Y. Yuasa. A penetration test for granular materials using various tip angles and penetrometer shapes. *Powder Technology*, 34(1):99–103, 1983.

337. A.-B. Huang, M.Y. Ma, and J.S. Lee. A micromechanical study of penetration tests in granular material. *Mechanics of Materials*, 16(12):133–139, 1993. Special Issue on Mechanics of Granular Materials.

338. G. Hill, S. Yeung, and S.A. Koehler. Scaling vertical drag forces in granular media. *EPL (Europhysics Letters)*, 72(1):137, 2005.

339. W.A. Allen, E.B. Mayfield, and H.L. Morrison. Dynamics of a projectile penetrating sand. *Journal of Applied Physics*, 28(3):370–376, 1957.

340. M.J. Forrestal and V.K. Luk. Penetration into soil targets. *International Journal of Impact Engineering*, 12(3):427–444, 1992.

341. A.M. Walsh, K.E. Holloway, P. Habdas, and J.R. de Bruyn. Morphology and scaling of impact craters in granular media. *Physical Review Letters*, 91:104301, September 2003.

342. K.A. Newhall and D.J. Durian. Projectile-shape dependence of impact craters in loose granular media. *Physical Review E*, 68:060301, December 2003.

343. J.R. de Bruyn and A.M. Walsh. Penetration of spheres into loose granular media. *Canadian Journal of Physics*, 82(6):439–446, 2004.

344. M.P. Ciamarra, A.H. Lara, A.T. Lee, D.I. Goldman, I. Vishik, and H.L. Swinney. Dynamics of drag and force distributions for projectile impact in a granular medium. *Physical Review Letters*, 92(19):194301/1–4, 2004.

345. M.A. Ambroso, C.R. Santore, A.R. Abate, and D.J. Durian. Penetration depth for shallow impact cratering. *Physical Review E*, 71:051305, May 2005.

346. E.L. Nelson, H. Katsuragi, P. Mayor, and D.J. Durian. Projectile interactions in granular impact cratering. *Physical Review Letters*, 101:068001, August 2008.

347. D.I. Goldman and P. Umbanhowar. Scaling and dynamics of sphere and disk impact into granular media. *Physical Review E*, 77:021308, February 2008.

348. A. Seguin, Y. Bertho, P. Gondret, and J. Crassous. Sphere penetration by impact in a granular medium: A collisional process. *EPL (Europhysics Letters)*, 88(4):44002, 2009.

349. A.H. Clark and R.P. Behringer. Granular impact model as an energy-depth relation. *EPL (Europhysics Letters)*, 101(6):64001, 2013.

350. T.A. Brzinski, P. Mayor, and D.J. Durian. Depth-dependent resistance of granular media to vertical penetration. *Physical Review Letters*, 111:168002, October 2013.

351. N. Nordstrom, K.E. Lim, M. Harrington, and W. Losert. Granular dynamics during impact. *Physical Review Letters*, 112:228002, June 2014.

352. M. Omidvar, M. Iskander, and S. Bless. Response of granular media to rapid penetration. *International Journal of Impact Engineering*, 66(0):60–82, 2014.

353. J.T. Farrar, V.K. Zworykin, and J. Baum. Pressure-sensitive telemetering capsule for study of gastrointestinal motility. *Science*, 126(3280):975–976, 1957.

354. R.S. Mackay and B. Jacobson. Endoradiosonde. *Nature*, 179:1239–1240, 1957.

355. J.T. Farrar, C. Berkley, and V.K. Zworykin. Telemetering of intraenteric pressure in man by an externally energized wireless capsule. *Science*, 131(3416):1814, 1960.

356. A.M. Connell and E.N. Rowlands. Wireless telemetering from the digestive tract. *Gut*, 1(3):266–272, 1960.

357. B.W. Watson, B. Ross, and A.W. Kay. Telemetering from within the body using a pressure-sensitive radio pill. *Gut*, 3(2):181–186, 1962.

358. D.M. Mueth, H.M. Jaeger, and S.R. Nagel. Force distribution in a granular medium. *Physical Review E*, 57:3164–3169, March 1998.

359. K. Wieghardt. Forces in granular flow. *Mechanics Research Communications*, 1(1):3–7, 1974.

360. R. Albert, M.A. Pfeifer, A.-L. Barabási, and P. Schiffer. Slow drag in a granular medium. *Physical Review Letters*, 82:205–208, January 1999.

361. I. Albert, P. Tegzes, B. Kahng, R. Albert, J.G. Sample, M. Pfeifer, A.-L. Barabási, T. Vicsek, and P. Schiffer. Jamming and fluctuations in granular drag. *Physical Review Letters*, 84:5122–5125, May 2000.

362. I. Albert, J.G. Sample, A.J. Morss, S. Rajagopalan, A.-L. Barabási, and P. Schiffer. Granular drag on a discrete object: Shape effects on jamming. *Physical Review E*, 64:061303, November 2001.

363. J. Geng and R.P. Behringer. Slow drag in two-dimensional granular media. *Physical Review E*, 71:011302, January 2005.

364. R. Soller and S.A. Koehler. Drag and lift on rotating vanes in granular beds. *Physical Review E*, 74:021305, August 2006.

365. J. Goldsmith, H. Guo, S.N. Hunt, Mingjiang Tao, and S. Koehler. Drag on intruders in granular beds: A boundary layer approach. *Physical Review E*, 88:030201, September 2013.

366. N. Gravish, P.B. Umbanhowar, and D.I. Goldman. Force and flow at the onset of drag in plowed granular media. *Physical Review E*, 89:042202, April 2014.

367. A.L. Demirel and S. Granick. Friction fluctuations and friction memory in stick-slip motion. *Physical Review Letters*, 77:4330–4333, November 1996.

368. J.L. Anthony and C. Marone. Influence of particle characteristics on granular friction. *Journal of Geophysical Research: Solid Earth*, 110(B8), 2005.

369. Y.R. Li and A. Aydin. Behavior of rounded granular materials in direct shear: Mechanisms and quantification of fluctuations. *Engineering Geology*, 115(12):96–104, 2010.

370. A.F. Cabalar. Stress fluctuations in granular material response during cyclic direct shear test. *Granular Matter*, 1–8, 2015.

CHAPTER 3

Computational Methods

Corey S. O'Hern

CONTENTS

This chapter describes numerical and simulation techniques that are used in computational studies of dense granular materials. In particular, we will focus on computational methods to generate static packings of model granular materials and to measure their response to compression and simple shear deformation modes.

There are two main classes of computational methods for generating static packings of granular materials: hard- and soft-particle methods. For the first, particles interact via a hard-core potential for which the pair interaction energy is infinite if the particles overlap, whereas it is zero otherwise. The packing-generation protocol can involve a number of different types of boundary conditions and driving mechanisms for obtaining a broad ensemble of static packings including isotropic compression, simple, and pure shear. The hard-particle methods include various geometrical [7,15,40] and Monte Carlo [36] techniques and molecular dynamics (MD) of hard particles [16] coupled with a series of compressions of the system (or increases in particle sizes), for example, the Lubachevsky–Stillinger algorithm [37]. Soft-particle methods employ energy minimization techniques or dissipative MD simulations [10] to relax the forces between over-lapping particles after each successive compression of the system for dilute initial configurations (or each successive decompression of the system for dense, overcompressed initial configurations). Many of the geometrical and Monte Carlo packing-generation methods employ "single-particle" moves, displacing one particle at a time independently, while energy minimization and MD techniques employ "collective" moves involving the cooperative motion of many particles. As particular examples, we describe packing-generation algorithms for (1) frictionless disks and spheres, (2) frictional spherical particles, (3) frictionless non-spherical composite particles such as dimers, and elongated particles including (4) ellipsoids, (5) spherocylinders, and (6) superquadrics.

3.1   Monte Carlo Methods

The basic Monte Carlo method for generating hard-particle packings combines single particle movements with system compressions, as illustrated in Figure 3.1 for a bidisperse collection of frictionless disks. In Figure 3.1 half of the disks are large and half are small, with diameter ratio $\sigma_l/\sigma_s = 1.4$ ($\sigma_{l,s}$ are the diameters of the large/small disks). The system is initialized with N randomly placed points and the disks grow uniformly until the closest pair is at contact (panel a). At this point, each disk is given a random displacement in the x- and y-directions (with the root-mean-square displacement for a given particle set by the average distance between its three closest neighbors). The move is accepted if it does not give rise to particle overlaps as shown in panel (b). After N attempts, the particle sizes

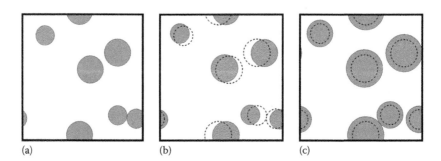

(a)   (b)   (c)

Figure 3.1   Illustration of the Monte Carlo packing-generation method for a system of $N = 6$ bidisperse frictionless disks (half small and half large with diameter ratio 1.4) in 2D. (a) The x- and y-coordinates for $N = 6$ random points are first generated in a square cell with periodic boundary conditions and the particles are grown uniformly until the closest pair of disks is in contact. (b) An attempt is made to move each particle randomly from the original (dashed outline) to the new position (shaded disk), and the move is accepted if it does not give rise to particle overlap. (c) The disks are expanded uniformly from the original (dashed outline) to the new size (shaded disk) until the two closest disks touch. This process is repeated for N_I iterations to obtain a single static packing.

are again increased uniformly until the nearest pair is at contact (panel c). This process of moving and growing the particles is repeated for N_I iterations to yield a single static packing. Further, N_c configurations are generated by seeding the Monte Carlo method with N_c independent sets of N initial points.

The results for $N = 6$ bidisperse hard frictionless disks in a unit square with periodic boundary conditions are shown in Figure 3.2. For this system, the packing fraction converges to its final value within $N_I \approx 10^3$ iterations. Figure 3.2b compares the distribution of final packing fractions from the Monte Carlo packing-generation method (for $N_c = 10^5$ static packings) to that for the soft-particle MD method. (See Section 3.2) The Monte Carlo method generates each of the mechanically stable (MS) packings obtained from the soft-particle MD method, but with

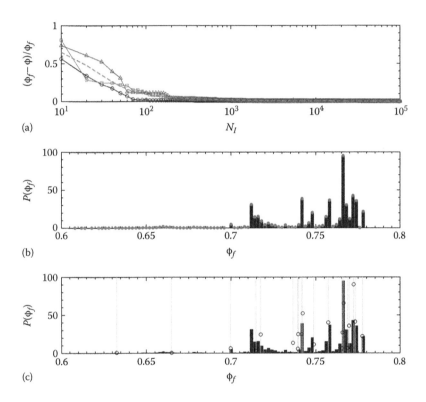

Figure 3.2   (a) The deviation in the packing fraction ϕ from its final value ϕ_f as a function of the number of iterations N_I of the Monte Carlo packing-generation method for three independent trials (different symbols) using bidisperse frictionless disks as in Figure 3.1. The dashed line indicates an average over $N_c = 10^3$ independent trials. (b) The probability distribution of final packing fractions after $N_I = 10^5$ iterations obtained from $N_c = 10^4$ (filled circles) and 10^5 (rectangles) independent trials. (c) The probability distribution of final packing fractions (rectangles) after $N_I = 10^5$ iterations and $N_c = 10^5$ trials. The open circles give the probabilities for obtaining the 20 distinct mechanically stable packings of frictionless bidisperse disks for $N = 6$ using soft-particle molecular dynamics with the packing fractions indicated by the solid vertical lines.

different probabilities, as well as many additional MS configurations (Figure 3.2c) that are not obtained from the soft-particle MD method.

To characterize the structural properties of the packings generated via the Monte Carlo method, we calculated the number of interparticle contacts (defined as $r_{ij} \leq \delta_c$, where δ_c is a small cutoff, typically such that $r_{ij}/\sigma_{ij} - 1 \leq 10^{-3}$, where $\sigma_{ij} = (\sigma_i + \sigma_j)/2$ and σ_i is the diameter of disk i), the $N \times N$ distance matrix for each configuration

$$D_{ij} = |\vec{r}_i - \vec{r}_j|, \tag{3.1}$$

where $\vec{r}_i$ is the location of particle i, and its second invariant

$$q_2 = \frac{1}{2}((TrD)^2 - TrD^2). \tag{3.2}$$

The cumulative distribution of interparticle separations $\overline{C}(r_{ij}/\sigma_{ij} - 1)$ scaled by N for packings of bidisperse frictionless hard disks generated using the "basic" Monte Carlo method is shown in Figure 3.3a and b for several values of the total number of iterations N_I and two system sizes $N = 6$ and 12. As N_I increases, a plateau develops in the cumulative distribution, which indicates a bimodal distribution of separations r_{ij}. The probability distribution of contact number differences $N_c^{iso} - N_c$ over all packings obtained from the Monte Carlo method is shown in Figure 3.3c. $N_c^{iso} = 2(N-N_r)-1$ is the isostatic number of contacts for frictionless disks, where the number of contacts matches the number of degrees of freedom and N_r is the number of rattler particles with two or fewer interparticle contacts. For $N = 6$, the Monte Carlo method generates similar probabilities for MS packings ($N_c = N_c^{iso}$) and first-order saddle points ($N_c^{iso} - N_c = 1$), whereas higher-order saddle points ($N_c^{iso} - N_c > 1$) have frequencies of 1% or less. As N increases, the probability for obtaining first- and second-order saddles increases above that for MS packings. Thus, at large N, the basic Monte Carlo method becomes inefficient at generating MS packings.

A scatter plot of the second invariant of the distance matrix q_2 for the $N_c = 10^5$ packings obtained from the "basic" Monte Carlo packing-generation method (after $N_I = 10^5$ iterations) versus the packing fraction is shown in Figure 3.4a. The first-order saddle packings form 1D lines (or geometrical families [12,13]) in the space of q_2 versus ϕ, whereas MS packings occur as discrete points. Two first-order saddle packings that belong to the same geometrical family with the same network of interparticle contacts (but different q_2) are shown in Figure 3.4b and c. As q_2 varies from -3.1 to -3 along this geometrical family (cf. Figure 3.4a), particle 1 moves toward 2. This geometrical family terminates in a MS packing when particle 1 comes into contact with 2. The packings that are missing two contacts (second-order saddle packings) form more diffuse lines in the q_2-ϕ plane. Improved sampling of the higher-order saddle packings is necessary to determine to what extent they form lines or fill in 2D areas.

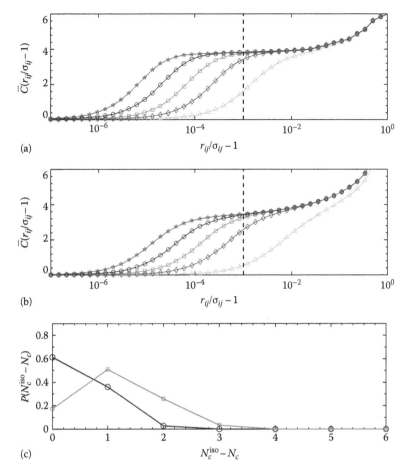

(a)

(b)

(c)

Figure 3.3   Cumulative distribution $\overline{C}(r_{ij}/\sigma_{ij} - 1)$ of interparticle separations $r_{ij}/\sigma_{ij} - 1$ normalized by N for packings generated using the "basic" Monte Carlo method for $N_I = 10^4$ (triangles), 10^5 (diamonds), 10^6 (squares), 10^7 (circles), and 10^8 (stars) and two system sizes (a) $N = 6$ and (b) 12. The dashed vertical line indicates the interparticle separation δ_c below which we define two particles to be in contact. (c) The distribution of contact numbers $N_c^{iso} - N_c$ using the criterion that contacting particles satisfy $r_{ij}/\sigma_{ij} - 1 \leq \delta_c = 10^{-3}$ for $N = 6$ (circles) and 12 (squares).

3.2   Molecular Dynamics Methods

Over the past 10 years, the O'Hern group has developed a suite of discrete element simulations to generate static granular packings using several packing-generation protocols including slow and fast isotropic compression [10,11,33], gravitational deposition [13], simple shear [12], and vibration [39]. O'Hern and coworkers have studied packings of frictionless [10–13] and frictional [26] disks, particles with anisotropic shapes including dimers [34], ellipses [20,31,35], ellipsoids [31], and bumpy particles [26], and particles with purely repulsive and adhesive contact interactions [17].

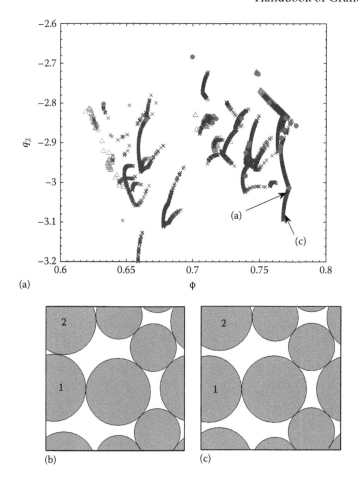

Figure 3.4   (a) The second invariant $q_2 = ((TrD)^2 - (TrD)^2)/2$ of the distance matrix D plotted versus the packing fraction ϕ for $N_c = 10^5$ bidisperse frictionless hard disk packings with $N = 6$ generated via the "basic" Monte Carlo method (after $N_I = 10^5$ iterations) with one (exes) or two (triangles) missing contacts $N_c^{iso} - N_c = 1$ or 2 relative to the isostatic value for frictionless disks. The MS packings with $N_c = N_c^{iso}$ are indicated by the filled circles. Two configurations shown in panels (b) and (c) with $N_c^{iso} - N_c = 1$ that belong to a single geometrical family are labeled (b) and (c) in panel (a).

3.2.1   Frictionless Spherical Particles

We describe two numerical procedures for creating isotropically compressed MS packings of frictionless particles at jamming onset: (1) a Lubachevsky–Stillinger (LS) [18,19] method for hard particles and (2) a dissipative MD method for soft particles that interact via purely repulsive contact forces [13].

For the LS method, N particles with mass m are placed with random positions in a square box with unit length ($L = 1$) in 2D (or cube with unit length in 3D) with no particle overlaps and a Gaussian velocity distribution at temperature T. The particle positions are updated using energy- and momentum-conserving event-driven dynamics for hard spherical particles for a time period t, after which all particle sizes are increased by $r_{ij}^{\min}/2$, where $r_{ij}^{\min}$ is the smallest separation between all pairs of particles. The particles are moved forward in time at constant velocity until the next interparticle collision. The velocities of the two colliding particles are updated and all of the particles are again moved forward in time at constant velocity. The process is repeated for a given number of collisions [18,19].

Figure 3.5a shows the collision frequency f (in units of $\sqrt{T/(m\sigma_s^2)}$) versus packing fraction ϕ for the LS method for a 2D bidisperse system with $N = 6$ disks. The collision frequency diverges as $f \sim (\phi_J - \phi)^{-1}$ near a MS packing with ϕ_J [30]; the compressions are stopped when the collision frequency exceeds a large value, $f_{\max} = 10^6$.

For the MD soft-particle packing-generation method, N particles are again given random, nonoverlapping positions in the simulation cell, this time with zero velocities. Collisions are now not assumed instantaneous, but rather the particles interact via the purely repulsive linear spring potential (with $\alpha = 2$):

$$V(r_{ij}) = \frac{\epsilon}{\alpha}\left(1 - \frac{r_{ij}}{\sigma_{ij}}\right)^{\alpha}\Theta\left(1 - \frac{r_{ij}}{\sigma_{ij}}\right), \tag{3.3}$$

where

    r_{ij} is the center-to-center separation between particles i and j
    ϵ is the strength of the repulsive interaction
    $\sigma_{ij} = (\sigma_i + \sigma_j)/2$ is the average diameter of particles i and j
    Θ is the Heaviside step function that sets the interaction to zero beyond $r_{ij} > \sigma_{ij}$

Particle sizes are again increased uniformly by $\Delta\phi$, and the total potential energy $V = \sum_{ij} V(r_{ij})$ is minimized using overdamped MD simulations. If the system is below jamming onset, the potential energy can be minimized to zero. In this case, we terminate the energy minimization when $V/N\epsilon < V_{\text{tol}} = 10^{-16}$. In contrast, when the system is above jamming onset, the minimized potential energy is nonzero, and we terminate the minimization when $(V_t - V_{t+1})/(V_t + V_{t+1}) < V_{\text{tol}} = 10^{-16}$, where V_t is the total potential energy at time t, as shown in Figure 3.5b. Each time the system switches from below to above jamming onset or vice versa, the packing fraction increment is decreased by a factor of 2. We terminate the MD packing-generation procedure when the minimized total potential energy obeys $V_{\text{tol}} < V/N\epsilon < 2V_{\text{tol}}$. The MS packings obtained from this method are accurate to 10^{-8} in packing fraction and particle positions.

Using the LS and MD packing-generation methods, we performed more than 10^6 independent trials for each system size to enumerate nearly all MS static packings in small 2D bidisperse and 3D monodisperse systems. Figure 3.6 shows a 2D projection of the paths in configuration space for the LS and MD packing-generation methods with two different initial conditions for 2D bidisperse systems with $N = 6$. We plot $\vec{r}_p = \frac{1}{5}\sum_{i=2}^{6}[(x_i - x_1)\hat{x} + (y_i - y_1)\hat{y}]$, where particle 1 is fixed at the origin. For the calculation of $\vec{r}_p$, the configurations from the LS and MD methods were aligned so that the particle positions and labels were identical

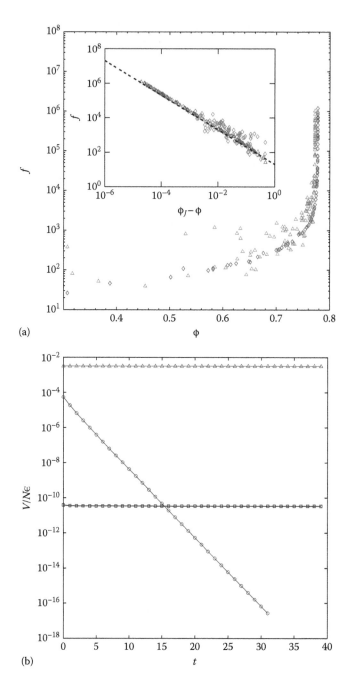

(a)

(b)

Figure 3.5   (a) Collision frequency f for the LS packing-generation procedure as a function of packing fraction ϕ for two compression rates (triangles: higher, diamonds: lower) for a 2D bidisperse system with $N = 6$. The inset shows f versus $\phi_J - \phi$ on a $\log_{10}-\log_{10}$ scale for the data in the main panel. The dashed line has slope -1. (b) The total potential energy $V/N\epsilon$ as a function of time during energy minimization for the MD packing-generation procedure at three packing fractions, two above jamming onset (squares and triangles) and one below (circles) for 2D bidisperse systems with $N = 6$.

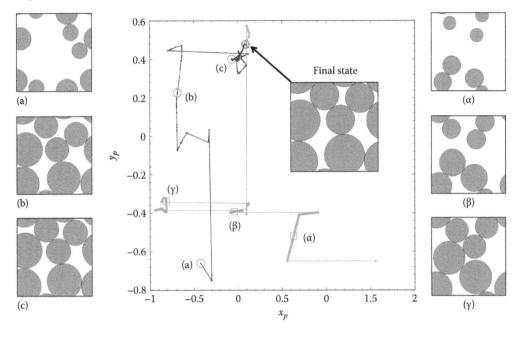

Figure 3.6 Two-dimensional projection of the paths in configuration space for the LS (circles) and MD (squares) packing-generation methods with two different initial conditions for 2D bidisperse systems with $N = 6$. For these initial conditions, both methods generate the same MS packing. For the calculation of $\vec{r}_p = \frac{1}{5}\sum_{i=2}^{6}[(x_i - x_1)\hat{x} + (y_i - y_1)\hat{y}]$, where particle 1 was fixed at the origin (in the lower left corner), the configurations from the LS and MD methods were aligned so that the particle positions and labels were identical in the final MS packing.

in the final MS packings. For the MD and LS methods, the packing-generation process was initialized with the configurations in shapshots (a) and (α), respectively. For the MD method, the system populates configurations (a), (b), and (c), and terminates at a particular MS packing (center of Figure 3.6). In contrast, for the LS method, the system visits different configurations (α), (β), and (γ), but ends at the same MS packing obtained from the MD method.

Even though the LS and MD packing-generation methods yield the same ensemble of MS packings at jamming onset, the probabilities with which the MS packings occur can vary strongly with the method. Figure 3.7 shows the number of times that each of the 20 distinct MS packings for $N = 6$ bidisperse disks was obtained over 10^6 trials. The MS packing probabilities increase with packing fraction, although there are fluctuations. The LS and MD methods give similar results for the MS packing probabilities for highly probable packings, but not the low-probability packings. Thermal fluctuations cause the low-probability packings to become more rare for the LS method compared to the MD method with overdamped dynamics. The MS packing probabilities for the LS and MD methods will approach each other in the fast compression rate limit for LS. In contrast, only

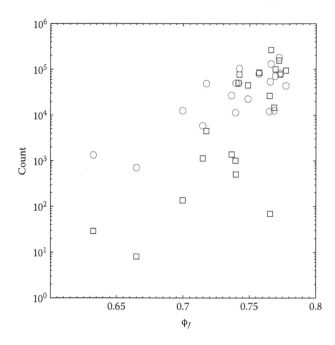

Figure 3.7   Number of times out of $N_t = 10^6$ trials that each of the 20 distinct MS packings at ϕ_J is obtained using the LS (squares) and MD (circles) packing-generation methods for 2D bidisperse systems with $N = 6$.

the densest packing will be obtained using the LS method in the slow compression rate limit.

3.2.2   Frictional Spherical Particles

To provide a specific example of soft-particle packing-generation methods, we describe discrete element simulations of bidisperse mixtures of frictional disks subjected to isotropic compression in 2D, implementing the Cundall–Strack model for static friction [8]. The position of the center of mass of each disk i (with mass m) obeys the equation of motion

$$m\frac{d^2\vec{r_i}}{dt^2} = \vec{F_i^r} + \vec{F_i^d} + \vec{F_i^s}, \tag{3.4}$$

which is the sum of an elastic contact force, a viscous damping (relative to a stationary background) force, and a contact friction force. ($\vec{r}_{ij}$ is the separation vector between the centers of disks i and j, and the sum over j indicates a sum over particles in contact with disk i). The elastic force $\vec{F_i^r} = \sum_j \vec{F_{ij}^r} = -\sum_j \frac{dV}{dr_{ij}} \hat{r}_{ij}$, and V is the purely repulsive elastic energy (Equation 3.3, here with $\alpha = 2$) between overlapping disks i and j. The viscous damping force is $\vec{F_i^d} = -b\vec{v_i}$, with a damping coefficient $b\sigma_s/\sqrt{m\epsilon} = 2$ chosen to be in the overdamped limit. The force on disk i arising from static friction with contacting disks j is $\vec{F}_i^s = \sum_j F_{ij}^t \hat{t}_{ij}$ with pair tangential forces at each contact, given by

$$F_{ij}^t = -k_s u_{t_{ij}},$$ (3.5)

where
    $\hat{t}_{ij}$ indicates the tangential direction (with $\hat{t}_{ij} \cdot \hat{r}_{ij} = 0$ and $\hat{r}_{ij} \times \hat{t}_{ij} = \hat{z}$)
    $k_s = \epsilon/(3\sigma_s)$ is the stiffness of the tangential spring

The relative tangential displacement $u_{t_{ij}}$ between disks i and j is obtained by solving

$$\frac{d\vec{u}_{t_{ij}}}{dt} = \vec{v}_{t_{ij}} - \frac{\left(\vec{u}_{t_{ij}} \cdot \vec{v}_{ij}\right)\vec{r}_{ij}}{r_{ij}^2},$$ (3.6)

with $\vec{v}_{t_{ij}} = \hat{t}_{ij}(\vec{v}_{ij} \cdot \hat{t}_{ij})$, and $\vec{v}_{ij} = \vec{v}_i - \vec{v}_j$. During the lifetime of each contact, the Coulomb sliding condition $|\vec{F}_{ij}^t| \leq \mu|\vec{F}_{ij}^r|$ is enforced, where μ is the static friction coefficient. When a contact between disks i and j breaks, $u_{t_{ij}}$ is reset to zero. This model for static friction has been implemented in a number of discrete element simulations of static and flowing granular media [8,38].

The rotational degree of freedom for each disk θ_i obeys

$$I_i \frac{d^2\theta_i}{dt^2} = \tau_i - b_t \frac{d\theta_i}{dt},$$ (3.7)

where the torque on disk i

$$\tau_i = \frac{1}{2} \sum_j \left[\vec{r}_{ij} \times (F_{ij}^t \hat{t}_{ij})\right] \cdot \hat{z},$$ (3.8)

the moment of inertia $I_i = m\sigma_i^2/8$, and the rotational damping coefficient $b_t/(\sigma_s\sqrt{m\epsilon}) = 2$ is in the overdamped limit.

The soft-particle MD packing-generation algorithm for frictional particles proceeds as follows. First, disks are placed at random positions in a square cell of length L at low packing fraction $\phi = 0.01$ with zero translational and rotational velocities. The disks are enlarged by an initial packing fraction increment $\Delta\phi = 10^{-3}$ and, after each compression, Equations 3.4 and 3.7 are integrated until the maximum particle kinetic energy, net force, and net torque are sufficiently small (here $K_{max} = \max_i K_i/\epsilon < 10^{-15}$, $F_{max} = \max_i|\vec{F}_i|\sigma_s/\epsilon < 10^{-11}$, and $\tau_{max} = \max_i|\tau_i|/\epsilon < 10^{-11}$, where $K_i = m\vec{v}_i^2/2 + I_i\omega_i^2/2$). These conditions ensure that all particles in the system are in force and torque balance after each compression.

A single MD relaxation during the packing-generation procedure is shown in Figure 3.8 for $N = 6$ bidisperse systems of frictional disks with $\mu = 0.05$ at packing fractions (a) below and (b) above the onset of jamming. For both (a) and (b), the maximum kinetic energy and torque on a single disk reach their terminal

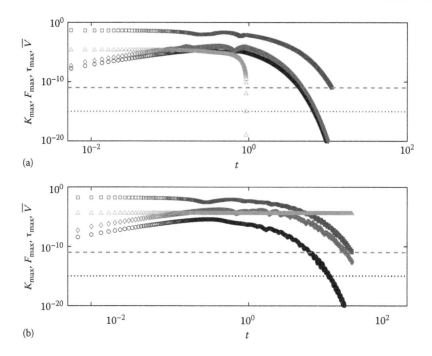

(a)

(b)

Figure 3.8   Maximum kinetic energy K_{max} (circles), force magnitude F_{max} (squares), and torque τ_{max} (diamonds) on a given disk, and total potential energy per particle $V/\epsilon N$ (triangles) during a single MD relaxation during the packing-generation procedure for a packing fraction (a) below and (b) above the onset of jamming for $N = 6$ bidisperse frictional disks with $\mu = 0.05$. The dashed (dotted) horizontal line indicates F_{max}, $\tau_{max} = 10^{-11}$ (K_{max}, $\overline{V} = 10^{-16}$).

values before the maximum force on a disk does. In panel (a), the final total potential energy per particle $V/(N\epsilon) = \overline{V} < 10^{-16}$, whereas $\overline{V} > 10^{-16}$ in panel (b). To generate "just-touching" jammed packings, we successively compress relaxed "unjammed" packings, or decompress relaxed "jammed" packings, with successively smaller packing fraction increments $\Delta\phi$ each time the process switches from compression to decompression or vice versa. The packing-generation algorithm is stopped when the relaxed total potential energy per particle satisfies $10^{-16} < \overline{V} < 2 \times 10^{-16}$.

3.2.3   Boundary Conditions

All of the packing-generation methods discussed to this point in the chapter employed periodic boundary conditions in cubic cells with side length L. An illustration of periodic boundary conditions for a 2D MS packing of bidisperse disks with $N = 6$ is provided in Figure 3.9. Eight image cells surround the central cell to specify the interparticle interactions near the boundaries. (In 3D, 26 image cells surround the central cell.) We see that disk 5 interacts with disks 3, 4, and 6 in the central cell. In contrast, disk 6 interacts with disk 5 in the central cell, 2' is the

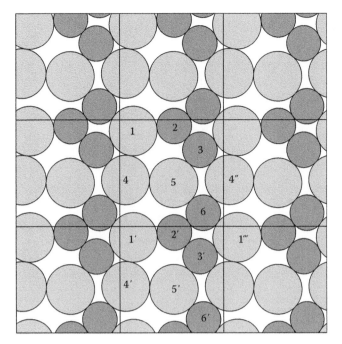

Figure 3.9   Illustration of periodic boundary conditions for a mechanically stable packing of $N = 6$ bidisperse disks in a square cell. The disks in the central cell are labeled 1 to 6. The disks are replicated in the eight adjacent image cells. The disks in the bottom image cell are labeled 1' to 6'. Note that disk 5 in the central cell is in contact with disks 3, 4, and 6 in the same cell, while disk 6 in the central cell is in contact with disk 5 in the central cell, 2' in the bottom image, 4'' in the right image cell, and 1''' in the bottom right image cell.

lower image, 4'' in the right image, and 1''' in the lower right image. (Note that if the diameter of one of the particles is larger than half of the box size, $\sigma > L/2$, it can interact with more than one image of the same particle.) To maintain continuous evolution of the particle trajectories, the particles may move out of the central cell during the packing-generation process, but this does not affect the interparticle separations and forces.

These systems can also be used to study the linear (i.e., the shear modulus) and nonlinear response (e.g., shear-induced jamming) of static packings to applied simple shear strain. To implement simple shear, an affine shear is applied to all particles i in the central cell,

$$x_i' = x_i + \Delta\gamma y_i, \tag{3.9}$$

where
   $\Delta\gamma$ is the strain increment
   x is the shear flow direction
   y is the shear gradient direction, coupled with simple shear-periodic (Lees–Edwards [1]) boundary conditions

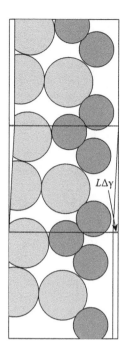

Figure 3.10   Illustration of the implementation of simple shear in Lees–Edwards boundary conditions for a packing of bidisperse disks. The particles in the main cell are given the affine deformation in Equation 3.9. The image cells above and below in the central cell in the shear gradient direction are also shifted by $\pm L\Delta\gamma$ in the shear flow direction.

Shear-periodic boundary conditions are illustrated in Figure 3.10 for 2D systems. The top/bottom row of image cells is shifted to the right/left by $\Delta\gamma L$. As in the quiescent case, we identify the particles in the main or image cells that contact each of the particles in the main cell and, following the application of shear strain, relax the system using energy minimization at fixed locations of the image cells. This is followed by the application of successive shear strain increments and energy minimization to a prescribed total shear strain γ.

Simple shear deformation can also be implemented by shifting top and bottom rigid boundaries by $\pm\Delta\gamma L$ in the shear flow direction and applying periodic boundary conditions in the directions perpendicular to the shear gradient direction with or without affine distortions in the bulk, as shown in Figure 3.11. After each movement of the boundaries, the system can be relaxed using energy minimization. Boundary-driven shear allows one to model the interactions between the sheared material and the container and set up strongly nonlinear velocity profiles [39].

3.2.4   Non Spherical Composite Particles

In this section, we focus on nonspherical composite particles in 2D formed from collections of disks that are rigidly connected together, for example, dimer- and trimer-shaped particles [34,35]. Packings of these anisotropic particles display

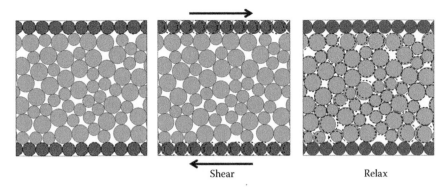

Shear   Relax

Figure 3.11   Illustration of the implementation of simple shear in 2D systems with fixed walls (dark shading) in the shear gradient direction and periodic boundary conditions in the shear flow direction. In this example, only the boundaries were moved with no affine bulk (light shading) deformation, followed by energy minimization at fixed locations of the boundaries. The particle positions before and after a shear strain step (center) and before and after the energy minimization (right) are shown.

interesting structural and mechanical properties, for example, shear strengthening and large yield strength [21]. Composite particles with circular asperities have also been employed to model friction in granular packings [2] and flows [6].

Figure 3.12a through c shows static packings composed of bidisperse (50–50 by number) dimer- and trimer-shaped particles in 2D. The schematic in Figure 3.12d defines the geometries. For dimers, the geometry is defined by the aspect ratio $\alpha_D = L_D/\sigma$. Trimers are specified by three shape parameters, $\sin(\theta/2) = \beta_T/\alpha_T$, $\beta_T = h_T/\sigma - 1$, and $\alpha_T = (L_T/\sigma - 1)/2$. To frustrate positional and orientational ordering, we consider bidisperse mixtures of composite particles that satisfy $\alpha_D^l/\alpha_D^s = 1.4$ (for dimers) and $\alpha_T^l/\alpha_T^s = \beta_T^l/\beta_T^s = 1.4$ (for trimers). Figure 3.13 shows the distribution of packing fractions at jamming onset for systems

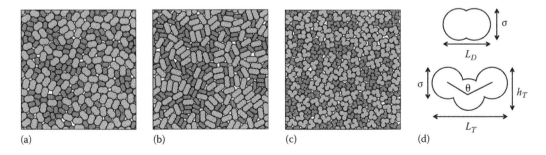

(a)   (b)   (c)   (d)

Figure 3.12   Mechanically stable static packings composed of 50–50 bidisperse mixtures (by number) of $N = 256$ frictionless (a) dimers with aspect ratio $\alpha_D = L_D/\sigma = 1.5$ and size ratio $\alpha_D^l/\alpha_D^s = 1.4$, trimers with (b) $\theta = 180°$, $\alpha_T = 0.5$, and $\beta_T = 0.0$, and (c) $\theta = 90°$, $\alpha_T = 0.5$, and $\beta_T = 0.5$, both with size ratio $\alpha_T^l/\alpha_T^s = \beta_T^l/\beta_T^s = 1.4$. (d) Schematic that defines the shape parameters for rigid dimer- and trimer-shaped particles.

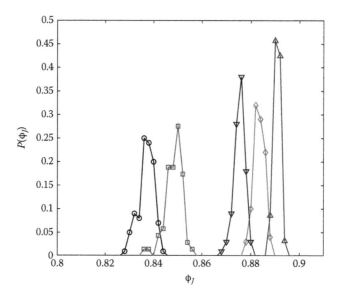

Figure 3.13 Distribution of the packing fractions $P(\phi_J)$ at jamming onset for 50–50 bidisperse mixtures by number of $N = 256$ frictionless disks with diameter ratio $\sigma_l/\sigma_s = 1.4$ (circles), trimers with $\theta = 180°$, $\alpha_T = 0.5$, and $\beta_T = 0.0$ (squares) and $\theta = 90°$, $\alpha_T = 0.5$, and $\beta_T = 0.5$ (downward triangles) both with size ratio $\alpha_T^l/\alpha_T^s = \beta_T^l/\beta_T^s = 1.4$, dimers with $\alpha_D = L_D/\sigma = 1.5$ and $\alpha_D^l/\alpha_D^s = 1.4$ (diamonds), and ellipses with ratio of the major to minor axes $\alpha = 1.5$ and $\alpha_l/\alpha_s = 1.4$ (upward triangles).

composed of $N = 256$ bidisperse, frictionless disks as well as elongated shapes, including dimers, trimers, and ellipses at roughly the same aspect ratio. The location of the most probable packing fraction at jamming onset increases in the following order for this particular set of shapes and aspect ratios: disks, linear trimers, nonlinear trimers, dimers, and ellipses. The widths of the packing fraction distributions are $\sim 1\% - 2\%$ for all shapes at $N = 256$ and decrease as a power law with increasing system size.

3.2.4.1 Contact Detection

Identifying physically reasonable interaction laws between dimer-shaped particles is not trivial. When dimer packings are generated at jamming onset with only infinitesimal overlaps (c.f. Figure 3.14a), overlaps between pairs of lobes on different dimers are distinct. (See Figure 3.15b). Computational studies have shown that, at the onset of jamming, dimer packings are typically isostatic with $z = z_{\text{iso}} = 2d_f$ contacts per particle, where $d_f = 3$ in 2D [34]. However, when dimer packings are sufficiently overcompressed (Figure 3.14b), the area of overlap between one pair of lobes merges with that between another pair of lobes (Figure 3.15c and d). Thus, two different contact numbers may be defined: a "lobe–lobe" number z_{ll} that treats all lobe–lobe interactions as distinct, and a single contact number z_s in which all merged overlaps are considered as a single contact. These need not be identical, and the difference between the two reveals

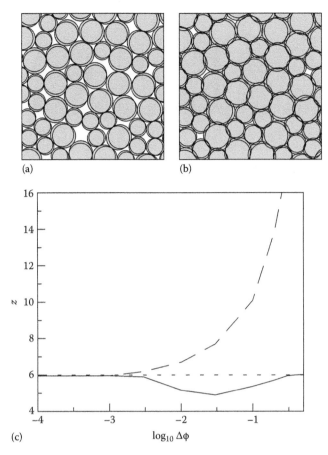

Figure 3.14 Static packing of $N = 48$ bidisperse dimers with aspect ratio $\alpha_D = 1.1$ and compression (a) $\Delta\phi = 10^{-4}$ and (b) 0.2. The contact number is $z = z_{\text{iso}} = 2(3N-1)/N \approx 5.96$ in (a) and $z_{ll} = 13.58 > z_{\text{iso}}$ (counting all lobe–lobe interactions) or $z_s = 5.67 < z_{\text{iso}}$ (merged overlap areas counted as a single contact) in (b). (c) Contact number z_{ll} (dashed line) and z_s (solid line) versus overcompression $\Delta\phi$ averaged over ≈ 50 configurations. The dotted line indicates $z = z_{\text{iso}} = 6$.

information about the packing. For the dimer packing Figure 3.14b, the lobe–lobe contact is $z_{ll} = 13.5$, much greater than the isostatic number $z_{iso} = 6$, whereas $z_s = 5.67 < z_{iso}$.

Figure 3.14c shows the two definitions of the contact number z_{ll} and z_s as a function of compression $\Delta\phi$ for $N = 48$ bidisperse dimer packings with aspect ratio $\alpha_D = 1.1$. For low compressions ($\Delta\phi < 10^{-3}$) $z_{ll} = z_s = z_{iso}$. For $\Delta\phi > 10^{-3}$, z_{ll} begins to increase, reaching ≈ 8 near $\Delta\phi = 3 \times 10^{-2}$ when all lobe–lobe contacts have formed between dimers initially in contact at $\Delta\phi = 0$. z_{ll} continues to increase to 24 when all lobe–lobe contacts have formed between Voronoi neighbors (defined at $\Delta\phi = 0$). In contrast, z_s decreases below z_{iso} for $\Delta\phi > 10^{-3}$ as some of the lobe–lobe interactions merge into extended contacts. z_s returns to the isostatic value 6 at large $\Delta\phi$ when all Voronoi neighbors (defined at $\Delta\phi = 0$) are in contact.

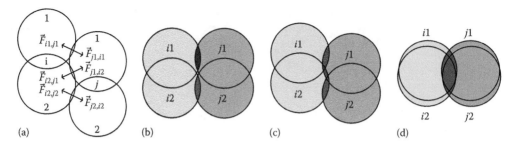

(a)   (b)   (c)   (d)

Figure 3.15 (a) Schematic of the possible repulsive interactions between dimer lobes. $\vec{F}_{ik_i,jk_j}$ is the pair force on lobe k_i on dimer i arising from lobe k_j on dimer j. (b) Two overlapping dimers (i and j) with aspect ratio $\alpha_D = 1.5$. The area of overlap between lobes $i1$ and $j1$ (dark gray) is separate from the area of overlap between lobes $i2$ and $j2$ (light gray). (c) and (d) Two overlapping dimers at $\alpha_D = 1.5$ and 1.1, respectively, for which the areas of overlap between the lobes have merged together (dark gray).

3.2.4.2 Force Law in Dimers

In previous studies of dimer packings at jamming onset [34,35], interactions between dimers were modeled as pair forces $\vec{F}_{ik_i,jk_j}$ between lobe k_i on dimer i and lobe k_j on dimer j. The total force between dimers i and j was given by the sum of forces between all pairs of lobes, $\vec{F}_{ij}^{\text{pair}} = \vec{F}_{i1,j1} + \vec{F}_{i1,j2} + \vec{F}_{i2,j1} + \vec{F}_{i2,j2}$, as illustrated in Figure 3.15a. This method yields physically reasonable results in the low compression limit when the overlap areas between lobe pairs $i1$ and $j1$ and lobe pairs $i2$ and $j2$ are separated as shown in Figure 3.15a. However, for dimer pairs where overlapped areas are merged (Figure 3.15b and c), this method multiply counts the overlapped areas and overestimates $\vec{F}_{ij}^{\text{pair}}$. For the dimers in Figure 3.15d, the force calculated without multiple counting should be $\vec{F}_{ij} \approx \vec{F}_{i1,j1} \approx \vec{F}_{i1,j2} \approx \vec{F}_{i2,j1} \approx \vec{F}_{i2,j2}$ because lobes $i1$ and $i2$ (and lobes $j1$ and $j2$) are nearly in the same positions, but the force calculated by summing over all pair forces between interacting lobes is $\vec{F}^{\text{pair}} = \vec{F}_{i1,j1} + \vec{F}_{i1,j2} + \vec{F}_{i2,j1} + \vec{F}_{i2,j2} \approx 4\vec{F}_{ij}$.

The multiple counting of lobe interactions for compressed dimer-shaped particles gives rise to anomalous behavior for the structural and mechanical properties, which do not approach those for disks in the $\alpha_D \to 1$ limit. We now describe a micromechanical model based on the area of overlap between compressed dimer-shaped particles that result in structural and vibrational properties that converge smoothly to those in the $\alpha_D \to 1$ limit at all values of the compression $\Delta\phi$.

As shown in Figure 3.16, the total overlap area between two dimers with the same sized constituent disks can be expressed in terms of the area of overlap between two (A_{2a}, A_{2b}, A_{2c}, A_{2d}), three (A_{3a}, A_{3b}, A_{3c}, A_{3d}), and four lobes (A_4)

$$A_T = A_{2a} + A_{2b} + A_{2c} + A_{2d} - A_{3a} - A_{3b} - A_{3c} - A_{3d} + A_4. \tag{3.10}$$

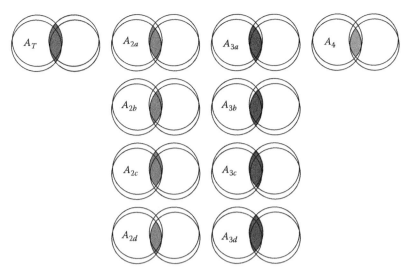

Figure 3.16 The total overlap area A_T for the pair of dimers in Figure 3.15d can be decomposed into the areas of overlap between two (A_{2a}, A_{2b}, A_{2c}, and A_{2d}), three (A_{3a}, A_{3b}, A_{3c}, and A_{3d}), and four lobes (A_4).

Each of these areas (formed from two, three, and four lobe intersections) can be further decomposed as the sum of n circular segments (1 per pair of overlapping lobes) with areas

$$A_{\text{seg}} = \frac{1}{4}\left(\sigma^2 \arctan\left(\frac{a}{\sqrt{\sigma^2 - a^2}}\right) - a\sqrt{\sigma^2 - a^2}\right), \tag{3.11}$$

where

σ is the diameter of the circular lobe that forms the segment
a is the length of the chord that terminates the segment and an interior triangle with area

$$A_{\text{tri}} = \left|\frac{x_{12}(y_{23} - y_{31}) + x_{23}(y_{31} - y_{12}) + x_{31}(y_{12} - y_{23})}{2}\right|, \tag{3.12}$$

where $\vec{r}_{12}$, $\vec{r}_{23}$, and $\vec{r}_{31}$ are the locations of the points of intersections of the lobes. (See Figure 3.17.)

For disks that are not the same size (diameters σ_i and σ_j and center-to-center separation r_{ij}), the area of overlap $A_{ij}(r_{ij})$ is

$$A_{ij}(r_{ij}) = \frac{1}{4}\left(\sigma_i^2 \arctan\left(a_{ij}\Big/\sqrt{\sigma_i^2 - a_{ij}^2}\right) - a_{ij}\sqrt{\sigma_i^2 - a_{ij}^2}\right.$$
$$\left. + \sigma_j^2 \arctan\left(a_{ij}\Big/\sqrt{\sigma_j^2 - a_{ij}^2}\right) - a_{ij}\sqrt{\sigma_j^2 - a_{ij}^2}\right), \tag{3.13}$$

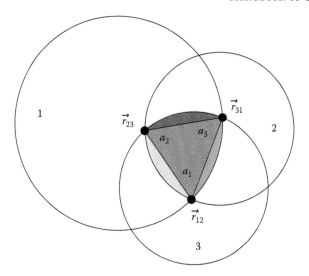

Figure 3.17 The area of intersection of three dimer lobes, 1, 2, and 3, can be decomposed into a triangle (dark gray) and three circular segments (light gray). (The area of intersection between two and four dimer lobes can be decomposed into a triangle and two and four circular segments, respectively.) The locations of the lobe intersection points are $\vec{r}_{12}$, $\vec{r}_{23}$, and $\vec{r}_{31}$, and the lengths of the lines connecting these points are a_1, a_2, and a_3.

where

$$a_{ij} = \sigma_{ij}^2 \sqrt{\left(1 - (r_{ij}/\sigma_{ij})^2\right)\left((r_{ij}/\sigma_{ij})^2 - (\sigma_i - \sigma_j)^2/(2\sigma_{ij}^2)\right)}/r_{ij} \qquad (3.14)$$

and $\sigma_{ij} = (\sigma_i + \sigma_j)/2$. To lowest order in r_{ij}/σ_{ij}, this is $\mathcal{A}_{ij}^{\text{approx}}(r_{ij}) = \frac{2\sqrt{2}}{3}\sigma_{ij}\sqrt{\sigma_i \sigma_j}$ $(1 - r_{ij}/\sigma_{ij})^{3/2}$.

Given a particular amount of overlap A between a given number of dimer lobes, for example, A_{2a} between two dimer lobes, we calculate their effective separation r_{ij}^{eff} (assuming the lobes are individual disks) by inverting Equation 3.13, that is, $r_{ij}^{\text{eff}}/\sigma_{ij} = \mathcal{A}^{-1}(A)$. We assume that this effective separation between overlapping lobes gives rise to a purely repulsive linear spring potential, $V = \frac{\epsilon}{2}$ $(1 - r_{ij}^{\text{eff}}/\sigma_{ij})^2$ (Equation 3.3), and do this for each of the nine area terms in Equation 3.10 to obtain the total potential energy between dimers i and j:

$$V_{ij} = \frac{\epsilon}{2}\Big(\left(1 - \mathcal{A}_{ij}^{-1}(A_{2a})\right)^2 + \left(1 - \mathcal{A}_{ij}^{-1}(A_{2b})\right)^2 + \left(1 - \mathcal{A}_{ij}^{-1}(A_{2c})\right)^2$$

$$+ \left(1 - \mathcal{A}_{ij}^{-1}(A_{2d})\right)^2 - \left(1 - \mathcal{A}_{ij}^{-1}(A_{3a})\right)^2 - \left(1 - \mathcal{A}_{ij}^{-1}(A_{3b})\right)^2$$

$$- \left(1 - \mathcal{A}_{ij}^{-1}(A_{3c})\right)^2 - \left(1 - \mathcal{A}_{ij}^{-1}(A_{3d})\right)^2 + \left(1 - \mathcal{A}_{ij}^{-1}(A_4)\right)^2\Big). \qquad (3.15)$$

Keeping only leading order terms in the area of overlap reduces this to

$$V_{ij}^{\text{approx}} = \frac{3^{4/3}\epsilon}{8\sigma_{ij}^{4/3}(\sigma_i\sigma_j)^{2/3}}$$
$$\times \left(A_{2a}^{4/3} + A_{2b}^{4/3} + A_{2c}^{4/3} + A_{2d}^{4/3} - A_{3a}^{4/3} - A_{3b}^{4/3} - A_{3c}^{4/3} - A_{3d}^{4/3} + A_4^{4/3} \right) \quad (3.16)$$

The potential energy in Equation 3.15 converges to the repulsive linear spring interaction potential used in Refs. [34,35] to generate static dimer packings in the $\Delta\phi \rightarrow 0$ limit for all α_D. Equation 3.15 allows the numerical calculation of the total potential energy $V = \sum_{i>j} V_{ij}$, force on dimer i, $\vec{F}_i = -dV/d\vec{r}_i$, torque on dimer i, $T_i = -dV/d\theta_i$, where θ_i is the angle that dimer i makes with the horizontal axis, and the dynamical matrix elements $M_{\xi_i\xi_j}$, where $\xi_i = x_i$, y_i, and $\sigma_i\theta_i$ give the center of mass location and orientation of dimer i.

Figure 3.18e shows the effective force law (force versus displacement) between two parallel dimers with aspect ratio $\alpha_D = 1.3$ undergoing compression for the micromechanical model in which (1) lobe interactions are multiply counted or (2) the interaction potential is given by Equation 3.15. The two force laws are the same as long as overlaps between lobes have not merged, or $\delta_m/\sigma < 0.021$ for the configuration in Figure 3.18a. Beyond δ_m, the two force laws differ. The force law based on the total area of overlap converges to linear behavior $F \sim \delta$ more quickly than the one that multiply counts lobe interactions, for example, it is not sensitive to the formation of the fourth lobe contact at $\delta/\sigma = \delta_4/\sigma = 0.075$. In future studies, these results can be compared to finite element analyses of linear elastic particles with complex shapes.

3.2.4.3 Characteristics of Dimer Packings

Figure 3.19a shows the two definitions of the contact number, z_{ll} and z_s, for static packings of $N = 48$ bidisperse dimers versus $\Delta\phi$ over a wide range of aspect ratios. Similar to Figure 3.14c, z_{ll} increases with $\Delta\phi$ and z_s initially decreases. For $\Delta\phi \lesssim 0.03$, changes in contact number are mainly due to the merging of areas of overlap, while for $\Delta\phi \gtrsim 0.03$, changes in contact number are due to additional contacts from neighbors that were not in contact at $\Delta\phi = 0$. The packings may be divided into three regimes, defined by the degree of multiple counting of lobe contacts: $z_{ll} = z_s = z_{iso}$ (no multiple counting), $z_s < z_{ll} < 2z_s$ (some lobes multiply counted, and $z_{ll} > z_s$ (all lobes multiply counted). In Figure 3.19b, we show the different regimes for z_{ll} and z_s as a function of $\Delta\phi$ and α_D. The regime $z_{ll} = z_s$ is bounded by $(\alpha_D - 1) \gtrsim \Delta\phi^{1/2}$. In contrast, the $z_{ll} > 2z_s$ regime occurs when $(\alpha_D - 1) \lesssim \Delta\phi$. The $z_s \geq z_{ll} < 2z_s$ regime occurs for $\Delta\phi \lesssim (\alpha_D - 1) \lesssim \Delta\phi^{1/2}$.

Figure 3.20 shows the vibrational density of states $D(\omega)$ (in the harmonic approximation) for static packings of dimer-shaped particles with and without multiple counting of the lobe interactions as a function of $\Delta\phi$ and α_D. With multiple counting, the maximum frequency in $D(\omega)$ can increase significantly above

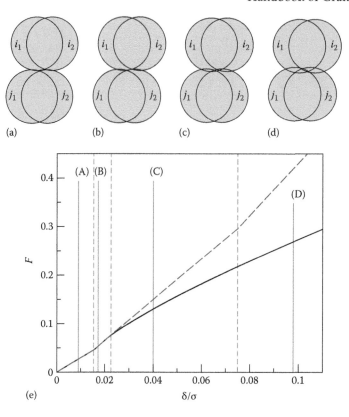

(e)

Figure 3.18 Snapshots during the compression of two parallel dimers i and j with aspect ratio $\alpha_D = 1.3$ (a) at $\delta = 0.009\sigma$, two distinct contacts between lobe pairs i_1 and j_1 and i_2 and j_2 exist and at (b) $\delta = 0.017\sigma$, three distinct contacts between lobes i_1 and j_1, i_2 and j_2, and i_1 and j_2 exist. (c) A single extended area of overlap is found at $\delta = 0.045\sigma$. (d) For $\delta = 0.098\sigma$, the extended overlap area includes interactions between two, three, and four lobes. (e) Magnitude of the inter-dimer normal force F versus displacement δ/σ when compressing dimers i and j. The vertical solid lines correspond to the configurations in (a–d). The dashed vertical lines (from small to large δ/σ) correspond to the displacement at which the third interaction between dimer lobes is formed, the contact areas between different lobe pairs merge, and the fourth contact between lobes is formed.

the disk value to $\omega_{max} \approx 4$ (even for nearly circular particles with $\alpha_D \approx 1$) as lobe–lobe interactions are added. Without multiple counting, the maximum frequency for small aspect ratio dimer packings remains near the disk value $\omega_{max} \approx 2$ with increasing $\Delta\phi$ until next-nearest neighbors are made, after which it increases to $\omega_{max} \approx 2.5$.

Figure 3.21 shows the vibrational density of states $D(\omega)$ for static packings of bidisperse dimers as a function of compression $\Delta\phi$ on a logarithmic scale to focus on the low-frequency behavior. $D(\omega)$ contains a strong peak of mainly rotational modes at low frequency. With multiple counting of the lobe interactions in the total potential energy, the location of the low-frequency peak ω_{min} does

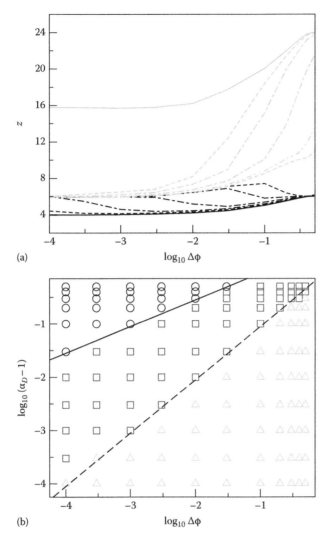

(a)

(b)

Figure 3.19 (a) Contact numbers z_{ll} (gray) and z_s (black) versus $\Delta\phi$ for static packings of bidisperse dimers with aspect ratio $\alpha_D = 1.0$ (solid line), 1.01 (dashed line), 1.03 (dash–dash–dotted line), 1.1 (dash–dotted line), 1.3 (dot–dot–dashed line), and 1.5 (dotted). (b) The $\Delta\phi$ and $\alpha_D - 1$ parameter space decomposed into regions where $z_{ll} = z_s$ (circles), $z_s < z_{ll} \leq 2z_s$ (squares), and $z_{ll} > 2z_s$ (triangles). The solid (dashed) line has slope 1/2 (1).

not depend on $\Delta\phi$. However, with the interaction potential in Equation 3.15, for the regime in $\Delta\phi$ where the overlaps between lobes have merged, the location of the low-frequency peak scales as $\omega_{min} \sim (\Delta\phi)^{1/4}$. Figure 3.22 shows that, with multiple-counting of the lobe interactions, the low-frequency peak ω_{min} scales linearly with aspect ratio $\alpha_D - 1$. In contrast, implementing the interaction potential in Equation 3.15 yields a different scaling behavior, $\omega_{min} \sim (\alpha_D - 1)^{1/2}$, after the overlap areas merge.

Together, the results in Figures 3.20 through 3.22 show that the low-frequency peak in $D(\omega)$ for the interaction model in Equation 3.15 scales as $\omega^1_{min} \sim (\alpha_D - 1)$ before the areas of overlap between lobes have merged, whereas it scales as $\omega^2_{min} \sim (\alpha_D - 1)^{1/2}(\Delta\phi)^{1/4}$ after the areas of overlap have merged. ω^1_{min} for dimers has similar scaling behavior as the peak ω_2 in the intermediate frequency band of $D(\omega)$ for ellipse packings [34]. The low-frequency peak ω^2_{min} shows similar scaling behavior as the peak in the lowest-frequency band in $D(\omega)$ for ellipse packings, $\omega_1 \sim (\alpha - 1)^{1/2}\Delta\phi^{1/2}$ [34], albeit with a different scaling exponent for the $\Delta\phi$ dependence.

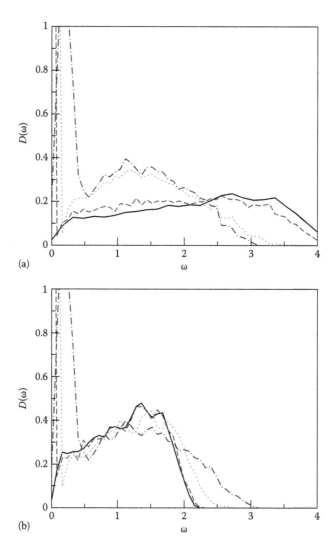

Figure 3.20 The vibrational density of states $D(\omega)$ (in the harmonic approximation) for static packings of $N = 48$ bidisperse dimers at aspect ratio $\alpha_D = 1$ (solid line), 1.01 (dashed line), 1.1 (dot–dashed line), and 1.3 (dot–dot–dashed line) at compression $\Delta\phi = 0.03$ (a) with and (b) without multiple counting of the interactions between lobes.   (*Continued*)

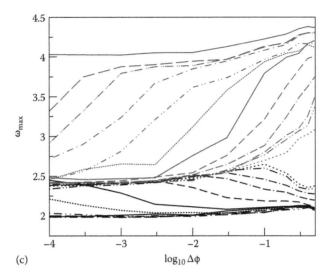

(c)

Figure 3.20 (*Continued*)   The vibrational density of states $D(\omega)$ (in the harmonic approximation) for static packings of $N = 48$ bidisperse dimers at aspect ratio $\alpha_D = 1$ (solid line), 1.01 (dashed line), 1.1 (dot–dashed line), and 1.3 (dot–dot–dashed line) at compression $\Delta\phi = 0.03$ (c) Largest frequency ω_{max} at which $D(\omega_{max}) = 0.1$ versus $\Delta\phi$ with (gray) and without (black) multiple counting of the interactions between lobes for aspect ratio $\alpha_D = 1.0$ (solid line), 1.01 (dashed line), 1.03 (dash–dash–dotted line), 1.1 (dash–dotted line), 1.3 (dot–dot–dashed line), and 1.5 (dotted line).

3.2.5   Elongated Particles

The structural and mechanical properties of static packings of elongated particles differ strongly from those for static packings of spherical particles. For example, a number of studies have shown that the packing fraction of static packings first increases from random close packing for spherical particles [4] with increasing aspect ratio α and then decreases as $1/\alpha$ in the large aspect ratio limit [28]. Also, there are data that suggest that static packings of elongated particles are hypostatic, with fewer contacts than necessary to stabilize all degrees of freedom [5,9]. However, additional studies are required to identify rattler particles, particles with multiple contacts, and contacts between particles that are nearly parallel.

In what follows, we describe methods to calculate contact distances and forces between elongated particles including ellipsoids, spherocylinders, and superquadrics.

3.2.5.1   Ellipsoidal Particles

For completeness, we summarize the techniques for generating static packings of ellipsoidal particles and studying their mechanical properties here. More detailed expositions may be found in Refs. [20,31,32].

It is assumed that, in both 2D and 3D, particles interact via the pairwise, purely repulsive linear spring potential (Equation 3.3 with $\alpha = 2$). This equation is generalized so that r_{ij} is the center-to-center separation between nonspherical particles i and j and σ_{ij} is the orientation-dependent center-to-center separation at which particles i and j come into contact as shown in Figure 3.23.

Perram and Wertheim developed an efficient method for calculating the exact contact distance between ellipsoidal particles with any aspect ratio and size distribution in 2D and 3D [3,14,27]. In their formulation, the contact distance between particles i and j is obtained from

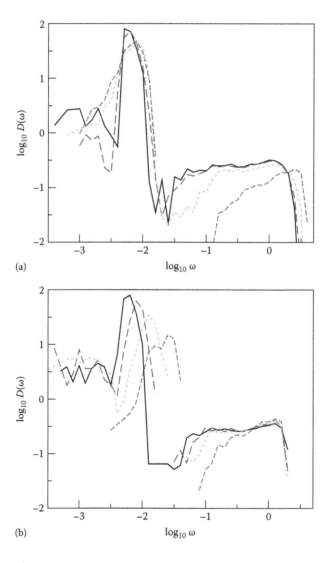

Figure 3.21   (a) The vibrational density of states $D(\omega)$ for static packings of $N = 48$ bidisperse dimers with aspect ratio $\alpha_D = 1.01$ and compression $\Delta\phi = 10^{-4}$ (solid line), 10^{-3} (dashed line), 10^{-2} (dotted line), and 10^{-1} (dot–dashed line) with (a) and without (b) multiple counting of the lobe interactions.   (Continued)

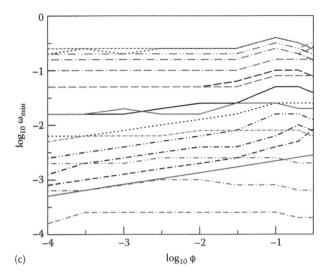

(c)

Figure 3.21 (*Continued*) (c) Smallest frequency ω_{min} at which $D(\omega_{min} = 0.1)$ versus $\Delta\phi$ with (gray) and without (black) multiple counting of the lobe interactions for aspect ratio $\alpha_D = 1.0003$ (solid line), 1.01 (dashed line), 1.1 (dash–dash–dotted line), and 1.5 (dash–dotted line). The dotted line has slope 1/4.

$$\sigma_{ij} = \min_\lambda \sigma_{ij}(\lambda),$$

$$\sigma_{ij}(\lambda) = \frac{\sigma_{ij}^0(\lambda)}{\sqrt{1 - \frac{\chi(\lambda)}{2} \sum_\pm \frac{(\beta(\lambda)\hat{r}_{ij} \cdot \hat{\mu}_i \pm \beta(\lambda)^{-1}\hat{r}_{ij} \cdot \hat{\mu}_j)^2}{1 \pm \chi(\lambda)\hat{\mu}_i \cdot \hat{\mu}_j}}},$$

$$\sigma_{ij}^0(\lambda) = \frac{1}{2}\sqrt{\frac{b_i^2}{\lambda} + \frac{b_j^2}{1-\lambda}},$$

$$\chi(\lambda) = \left(\frac{(a_i^2 - b_i^2)(a_j^2 - b_j^2)}{(a_j^2 + \frac{1-\lambda}{\lambda}b_i^2)(a_i^2 + \frac{\lambda}{1-\lambda}b_j^2)}\right)^{1/2},$$

$$\beta(\lambda) = \left(\frac{(a_i^2 - b_i^2)(a_j^2 + \frac{1-\lambda}{\lambda}b_i^2)}{(a_j^2 - b_j^2)(a_i^2 + \frac{\lambda}{1-\lambda}b_j^2)}\right)^{1/4}, \tag{3.17}$$

where a and b are the major and minor axes of the ellipsoids, respectively. The expression for $\sigma_{ij}(\lambda)$ is minimized over the parameter λ to find the minimum contact distance $\sigma_{ij} = \sigma_{ij}(\lambda_{min})$ between ellipsoidal particles i and j.

Since σ_{ij} depends explicitly on particle positions $\vec{r}_i$ and orientations, we can calculate the forces $\vec{F}_{ij} = -\nabla V_{ij}$ and torque $T_{ij} = [\vec{r}_{ij}^c \times \vec{F}_{ij}]\cdot\hat{z}$, where $\vec{r}_{ij}^c$ is the point of contact between ellipsoidal particles i and j. The forces and torques are then used in the soft-particle MD packing-generation method described in Section 3.2.2 to obtain static packings of ellipsoidal particles.

To investigate the mechanical properties of static packings of ellipsoidal particles, one can calculate the eigenvalues of the dynamical matrix and the resulting density of vibrational modes in the harmonic approximation. The dynamical matrix is defined as

$$M_{kl} = \frac{\partial^2 V}{\partial \xi_k \partial \xi_l},$$ (3.18)

where

ξ_k (with $k = 1, \ldots, d_f N$) represent the $d_f N$ degrees of freedom in the system
d_f is the number of degrees of freedom per particle

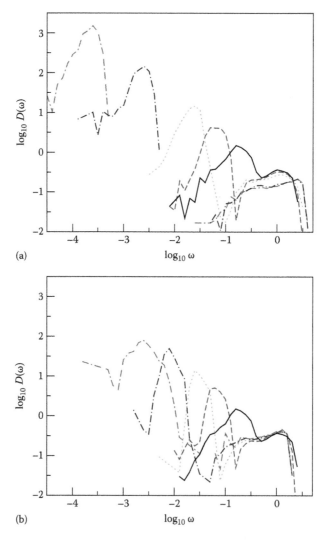

(a)

(b)

Figure 3.22 The vibrational density of states $D(\omega)$ for static packings of $N = 48$ bidisperse dimers at compression $\Delta\phi = 0.03$ with aspect ratio $\alpha_D = 1.0003$ (solid line), 1.003 (dashed line), 1.03 (dotted line), 1.1 (dot–dashed line), and 1.2 (dash–dash–dotted line) with (a) and without (b) multiple counting of the lobe interactions.   (*Continued*)

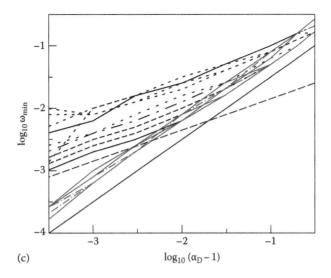

(c)

Figure 3.22 (*Continued*)   The vibrational density of states $D(\omega)$ for static packings of $N = 48$ bidisperse dimers at compression $\Delta\phi = 0.03$ with aspect ratio $\alpha_D = 1.0003$ (solid line), 1.003 (dashed line), 1.03 (dotted line), 1.1 (dot–dashed line), and 1.2 (dash–dash–dotted line) with (c) The location of the low-frequency peak in $D(\omega)$, $\omega_{\min}$, versus $\Delta\phi$ with (gray) and without (black) multiple counting of the lobe interactions for compressions $\Delta\phi = 0.0003$ (solid line), 0.003 (dashed line), 0.03 (dash–dash–dotted line), 0.1 (dash–dotted line), 0.3 (dot–dot–dashed line), and 0.5 (dotted line). The solid (dashed) line has slope 1 (1/2).

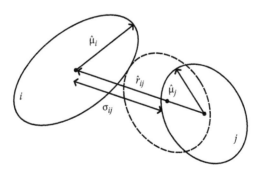

Figure 3.23   Definition of the contact distance σ_{ij} for ellipsoidal particles i and j with unit vectors $\hat{\mu}_i$ and $\hat{\mu}_j$ that characterize the orientations of their major axes. σ_{ij} is the center-to-center separation at which ellipsoidal particles first touch when they are brought together along $\hat{r}_{ij}$ at fixed orientation.

In 2D, $d_f = 3$ with

$$\vec{\xi} = \{x_1, x_2, \ldots, x_N, y_1, y_2, \ldots, y_N, l_2\theta_1, l_2\theta_2, \ldots, l_2\theta_N\} \qquad (3.19)$$

and in 3D for prolate ellipsoids $d_f = 5$ with

$$\vec{\xi} = \{x_1, x_2, \ldots, x_N, y_1, y_2, \ldots, y_N, z_1, z_2, \ldots, z_N, l_\theta^1\theta_1, l_\theta^2\theta_2, \ldots, l_\theta^N\theta_N,$$
$$l_3\phi_1, l_3\phi_2, \ldots, l_3\phi_N\}, \qquad (3.20)$$

where
  θ_i is the polar angle
  ϕ_i is the azimuthal angle in spherical polar coordinates, $l_2 = \sqrt{a^2 + b^2}/2$, $l_3 = \sqrt{(a^2 + b^2)/5}$, and $l_\theta^i = \sqrt{(2b^2 + (a^2 - b^2)\sin^2\phi_i)/5}$

See Ref. [31] for complete expressions for the forces, torques, and dynamical matrix entries in 2D.

3.2.5.2 Spherocylinders

A spherocylinder is a cylinder (length l diameter σ) with hemispherical endcaps (see Figure 3.24a). Because of this construction, all points on the surface lie a distance $\sigma/2$ from the line segment along the cylinder axis. The question of whether two spherocylinders overlap, and therefore interact through contact forces, is thus reduced to finding the shortest distance between two line segments. If this distance is less than σ (for equal-sized spherocylinders), the spherocylinders overlap and experience equal and opposite repulsive forces. Pournin [29] has shown that the distance between two points on different line segments is given by

$$d_{ij}^2 = \left|\vec{r}_{ij}\right|^2 + 2\lambda_j\hat{u}_j \cdot \vec{r}_{ij} - 2\lambda_i\hat{u}_j \cdot \vec{r}_{ij} - 2\lambda_i\lambda_j\hat{u}_i \cdot \hat{u}_j + \lambda_i^2, \qquad (3.21)$$

where
  λ_i and λ_j indicate the distance that the two points are from the ends of the spherocylinders
  $\hat{u}_i$ and $\hat{u}_j$ are unit vectors along the spherocylinder axes
  r_{ij} is the distance between the spherocylinder centers of mass

The shortest distance is found by minimizing Equation 3.21 with respect to λ_i and λ_j. This minimization can be determined analytically and is thus computationally quite efficient. Purely repulsive linear spring forces between spherocylinders can be implemented using Equation 3.3 with d_{ij} replacing r_{ij}. Packings of frictionless spherocylinders can be generated using the successive compression plus energy minimization technique, similar to the algorithms developed in Refs. [24,25] and discussed earlier in this chapter. Sample static packings of frictionless spherocylinders with aspect ratio $\alpha = l/\sigma = 5$ and 50 are shown in Figures 3.24b.

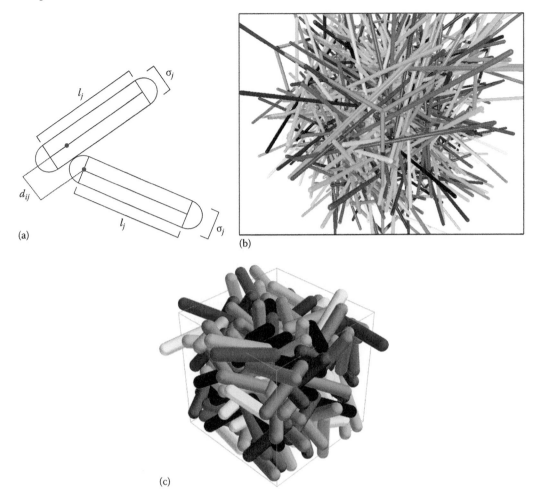

Figure 3.24   (a) Two spherocylinders with lengths l_i and l_j and diameters σ_i and σ_j. The shortest distance between the central line segments d_{ij} is indicated. When $d_{ij} < (\sigma_i + \sigma_j)/2$, the two spherocylinders interact. (b and c) Static packings of spherocylinders with aspect ratio $\alpha = l/\sigma = 5$ and 50.

3.2.5.3   Modeling Irregular Particles with Superquadrics

Superquadrics are a family of shapes whose surfaces are defined by the equation

$$\left(\frac{x}{r_x}\right)^{n_x} + \left(\frac{y}{r_y}\right)^{n_y} + \left(\frac{z}{r_z}\right)^{n_z} = 1, \tag{3.22}$$

where

   r_x, r_y, and r_z are the axes lengths in the $\hat{x}, \hat{y}$, and $\hat{z}$ directions
   n_x, n_y, and n_z are positive real numbers that determine the curvature of the shape

In particular, if $n_x = n_y = n_z = 2$ the particle is elliptical, and as n_x, n_y, and $n_z \to \infty$ the particle takes the shape of a rectangular box with sharp corners.

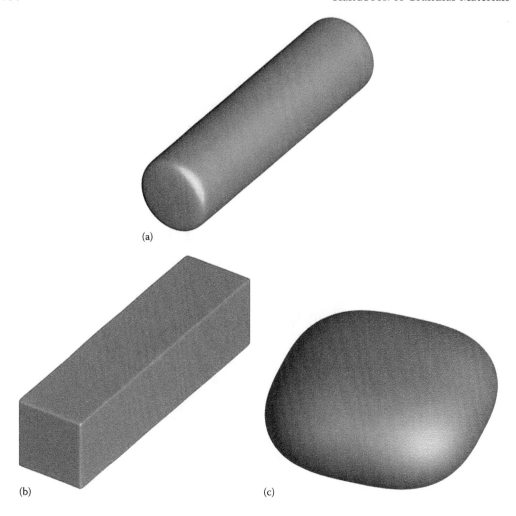

Figure 3.25   Three examples of superquadric shapes. (a) $n_x = n_y = 2, n_z = 50$ gives a cylinder; (b) $n_x = n_y = n_z = 50$ gives a rectangular box; (c) $n_x = n_z = 3, n_y = 2$ gives a briquette shape.

Superquadrics can model particles with concave or convex shapes as well as blunt or sharp edges. Several examples of superquadrics are shown in Figure 3.25.

Superquadrics allow for a far more flexible range of particle shape at the cost of computational efficiency, as the detection of contacts cannot be calculated analytically. Studies of 2D superquadrics [22,23] have established an efficient numerical method that involves minimization in one fewer dimension than the system. Essentially, the equation that defines the surface of the ith particle is rewritten as

$$f^i(x,y,z) \equiv \left(\frac{x}{r_x}\right)^{n_x} + \left(\frac{y}{r_y}\right)^{n_y} + \left(\frac{z}{r_z}\right)^{n_z} - 1.$$

$f^i(x,y,z)$ is less than 1 for all points in the particle, equal to 1 for all points on the surface, and greater than 1 for all points outside the particle. This function is then minimized *on the surface* of the jth particle, that is, on a 2D surface. The two particles are in contact if $f^i_{min} \leq 1$. The forces and torques can then be determined with the discrete element modeling (DEM) technique discussed earlier, as with spheres and spherocylinders.

3.3   Conclusions

This chapter described computational methods to generate static packings of frictionless spheres as well as nonspherical composite particles such as rigid dimers and elongated particles including ellipsoids, spherocylinders, and superquadrics. These methods generate force- and torque-balanced, MS particle packings at jamming onset with only infinitesimal interparticle overlaps. Both hard- (Monte Carlo and Lubachevsky–Stillinger) and soft-particle (dissipative MD) methods were reviewed, and we showed that hard- and soft-particle methods generate the same MS packings of spherical particles at jamming onset although with different probabilities of occurrence. We also reviewed methods to prepare static packings of frictional spherical particles using the Cundall–Strack model. Finally, we presented novel results that identify the regime of aspect ratio and compression where the composite particle model for nonspherical particle shapes yields consistent results for the contact number and density of vibrational modes (in the harmonic approximation).

References

1. M. P. Allen and D. J. Tildesley. *Computer Simulation of Liquids*. Oxford University Press, Oxford, U.K., 1987.

2. F. Alonso-Marroquin. Spheropolygons: A new method to simulate conservative and dissipative interactions between 2d complex-shaped rigid bodies. *Europhys. Lett.*, 83:14001, 2008.

3. B. J. Berne and P. Pechukas. Gaussian model potentials for molecular interactions. *J. Comp. Phys.*, 56:4213, 1972.

4. J. G. Berryman. Random close packing of hard spheres and disks. *Phys. Rev. A*, 27, 1983.

5. J. Blouwofff and S. Fraden. The coordination number of granular cylinders. *Europhys. Lett.*, 76:1095, 2006.

6. V. Buchholtz and T. Pöschel. Numerical investigations of the evolution of sandpiles. *Phys. A*, 202:390, 1994.

7. A. S. Clarke and H. Jónsson. Structural changes accompanying densification of random hard-sphere packings. *Phys. Rev. E*, 47:3976, 1991.

8. P. A. Cundall and O. D. L. Strack. A discrete numerical model for granular assemblies. *Géotechnique*, 29:47, 1979.

9. A. Donev, R. Connelly, F. H. Stillinger, and S. Torquato. Underconstrained jammed packings of non spherical hard particles: Ellipses and ellipsoids. *Phys. Rev. E*, 75:051304, 2007.

10. G.-J. Gao, J. Blawzdziewicz, and C. S. O'Hern. Frequency distribution of mechanically stable disk packings. *Phys. Rev. E*, 74:061304, 2006.

11. G.-J. Gao, J. Blawzdziewicz, and C. S. O'Hern. Enumeration of distinct mechanically stable disk packings in small systems. *Phil. Mag. B*, 87:425, 2007.

12. G.-J. Gao, J. Blawzdziewicz, and C. S. O'Hern. Geometrical families of mechanically stable granular packings. *Phys. Rev. E*, 80:061303, 2009.

13. G.-J. Gao, J. Blawzdziewicz, C. S. O'Hern, and M. D. Shattuck. Experimental demonstration of nonuniform frequency distributions of granular packings. *Phys. Rev. E*, 80:061304, 2009.

14. J. G. Gay and B. J. Berne. Modification of the overlap potential to mimic a linear site-site potential. *J. Comp. Phys.*, 74:3316, 1981.

15. W. S. Jodrey and E. M. Tory. Computer simulation of close random packing of equal spheres. *Phys. Rev. A*, 32:2347, 1985.

16. R. L. Kansal, T. M. Truskett, and S. Torquato. Nonequilibrium hard-disk packings with controlled orientational. *J. Chem. Phys.*, 113:4844, 2000.

17. G. Lois, J. Blawzdziewicz, and C. S. O'Hern. Jamming transition and new percolation universality classes in particulate systems with attraction. *Phys. Rev. Lett.*, 100:028001, 2008.

18. B. D. Lubachevsky and F. H. Stillinger. Geometric properties of random disk packings. *J. Stat. Phys.*, 60:561, 1990.

19. B. D. Lubachevsky, F. H. Stillinger, and E. N. Pinson. Disks vs. spheres: Contrasting properties of random packings. *J. Stat. Phys.*, 64:501, 1991.

20. M. Mailman, C. F. Schreck, B. Chakraborty, and C. S. O'Hern. Jamming in systems composed of frictionless ellipse-shaped particles. *Phys. Rev. Lett.*, 102:255501, 2009.

21. M. Z. Miskin and H. M. Jaeger. Adapting granular materials through artificial evolution. *Nat. Mater.*, 12:326, 2013.

22. G. G. W. Mustoe and M. Miyata. Material flow analyses of noncircular-shaped granular media using discrete element methods. *J. Eng. Mech.*, 127:1017, 2001.

23. G. G. W. Mustoe, M. Miyata, and M. Nakagawa. Discrete element methods for mechanical analysis of systems of general shaped bodies. In *Proceedings of the 5th International Conference on Computational Structures Technology*, pp. 219–224. Civil-Comp Press, Edinburgh, 2000.

24. C. S. O'Hern, S. A. Langer, A. J. Liu, and S. R. Nagel. Random packings of frictionless particles. *Phys. Rev. Lett.*, 88:075507–1–4, 2002.

25. C. S. O'Hern, S. A. Langer, A. J. Liu, and S. R. Nagel. Force distributions near jamming and glass transitions. *Phys. Rev. Lett.*, 86:111, 2001.

26. S. Papanikolaou, C. S. O'Hern, and M. D. Shattuck. Isostaticity at frictional jamming. *Phys. Rev. Lett.*, 110:198002, 2013.

27. J. W. Perram and M. S. Wertheim. Statistical mechanics of hard ellipsoids. I. Overlap algorithm and the contact function. *J. Comp. Phys.*, 58:409, 1985.

28. A. P. Phillpse. The random contact equation and its implication for (colloidal) rods in packings, suspensions, and anisotropic powders. *Langmuir*, 12:1127, 1996.

29. L. Pournin, M. WEber, M. Tsukahara, J. A. Ferrez, M. Ramaioli, and T. M. Libling. Three-dimensional distinct element simulation of spherocylinder crystallization. *Granular Matter*, 7:119, 2005.

30. M. D. Rintoul and S. Torquato. Computer simulations of dense hard-sphere systems. *J. Chem. Phys.*, 105:9258, 1996.

31. C. F. Schreck, M. Mailman, B. Chakraborty, and C. S. O'Hern. Constraints and vibrations in static packings of ellipsoidal particles. *Phys. Rev. E*, 85:061305, 2012.

32. C. F. Schreck and C. S. O'Hern. Computational methods to study jammed systems. In J. S. Olafsen, ed., *Experimental and Computational Techniques in Soft Condensed Matter Physics*. Cambridge University Press, New York, 2010.

33. C. F. Schreck, T. Shen, and C. S. O'Hern. Packings: Static. In M. D. Shattuck and S. F. Franklin, eds., *Handbook of Granular Materials*. CRC Press, New York, 2013.

34. C. F. Schreck, N. Xu, and C. S. O'Hern. A comparison of jamming behavior in systems composed of dimer- and ellipse-shaped particles. *Soft Matter*, 6:2960, 2010.

35. T. Shen, C. F. Schreck, B. Chakraborty, D. E. Freed, and C. S. O'Hern. Structural relaxation in dense liquids composed of anisotropic particles. *Phys. Rev. E*, 86:041303, 2012.

36. R. J. Speedy. Random jammed packings of hard discs and spheres. *J. Phys.: Condens. Matter*, 10:4185, 1998.

37. F. H. Stillinger and Lubachevsky B. D. Geometric properties of random disk packings. *J. Stat. Phys.*, 60:561, 1990.

38. O. R. Walton. Numerical simulation of inclined chute flows of monodisperse inelastic, frictional spheres. *Mech. Mater.*, 16, 1993.

39. N. Xu, C. S. O'Hern, and L. Kondic. Stabilization of nonlinear velocity profiles in athermal systems undergoing planar shear flow. *Phys. Rev. E*, 72:041504, 2005.

40. A. Z. Zinchenko. Algorithm for random close packing of spheres with periodic boundary conditions. *J. Comp. Phys.*, 114:298, 1994.

CHAPTER 4

Kinetic Theories for Collisional Grain Flows

James T. Jenkins

CONTENTS

4.1   Introduction

There is an obvious analogy between the colliding macroscopic grains of a granular material with high energy and the agitated molecules of a dense gas. This is supported by the experimental observation that when a granular material is rapidly sheared, the shear and normal stresses required to maintain the motion vary with the square of the shear rate (Bagnold 1954, Savage 1972, Savage and McKeown 1983, Savage and Sayed 1984, Hanes and Inman 1985). The interpretation of these observations is that, at high shear rates, the dominant mechanism of momentum transfer is collision between grains, with the interstitial liquid or gas playing a relatively minor role. In this chapter, we explore how methods from the kinetic theory of dense gases (e.g., Reif 1969, Chapman and Cowling 1970) have been adopted and extended to derive balance laws, constitutive relations, and boundary conditions for systems of macroscopic, dissipative, grains. Our intention is to provide an accessible overview of continuum theories that have their origin in the collisions between pairs of spherical particles and the simplest relevant statistical characterizations of the likelihood of such interactions. More elaborate and exhaustive treatments of the kinetic theory for granular flows exist. The book by Brilliantov and Pöschel (2004) is an example, and the review of Goldhirsch (2003) provides access to many others.

There are at least two important differences between granular and atomic gases: collisions between grains are invariably inelastic and typical granular flows involve spatial variations over far fewer grains than their molecular counterparts. The additional mechanism of dissipation makes steady granular shearing flows possible in situations where they do not exist for elastic molecules, and the limited extent of the flow over which the spatial gradients occur means that the influence of boundaries is far more pervasive.

We begin with the simplest kinetic theory, that for identical, smooth, nearly elastic, spheres (Lun et al. 1984, Jenkins and Richman 1985) and derive balance laws for the means of the mass density, velocity, and the kinetic energy of the velocity fluctuations (Section 4.2). To the order of the approximations used in determining the velocity distribution function, the expressions for the stress tensor and the flux of fluctuation energy are identical to those for a dense gas of perfectly elastic spheres. The only nonclassical term is a collisional rate of dissipation per unit volume that is present in the energy balance. This can then be extended in a straightforward manner to incorporate friction between grains.

Differences in the boundaries that drive the shearing flows are almost certainly responsible for the quantitative differences in experiments. A derivation of boundary conditions (e.g., Richman 1988) highlights the exceptional nature of steady, homogeneous, simple shearing. Analyzing this steady, simple shearing flow makes direct contact between the predictions of the kinetic theory and the experimentally observed quadratic dependence of the shear and normal stress on the shear rate.

Next, we consider a steady, inhomogeneous shearing flow between identical bumpy boundaries, in which the working of the slip velocity through the shear stress at the boundary is not balanced by the rate of collisional dissipation. The thickness of the shearing flow, the relative velocity of the boundaries, and the shear and normal stress must be related in order for a steady flow solution to exist. Two different types of solutions are found, depending upon whether the boundaries provide or remove fluctuation energy to or from the flow.

We conclude the chapter by considering the two limiting cases of high dissipation (in moderately dense flows) and, finally, very dense flows. Shearing flows of very dissipative spheres, and dilute flows in particular, involve significant anisotropy in the strength of the velocity fluctuations. As a consequence, such flows fall outside of the domain of simple kinetic theory, in which these fluctuations are assumed to be isotropic. Chou and Richman (1998) have investigated anisotropy in the strength of the velocity fluctuations and its influence on the components of the stress in steady, uniform shearing over a range of volume fractions, and anisotropy has been incorporated into extensions of the elementary kinetic theory that are quadratic, rather than linear, in the spatial gradients of the mean fields (e.g., Sela et al. 1996). Constitutive equations for very dissipative interactions that are linear in the spatial gradients and based on an isotropic measure of the strength of the velocity fluctuations were derived by Garzo and Dufty (1999). Despite the limiting assumptions, the theory does seem to provide an adequate description of moderately dense shearing flows of very dissipative spheres and forms the basis of a phenomenological extension of kinetic theory to very dense, dissipative flows.

Very dense shearing flows involve multiple and/or repeated collisions that violate the assumption of instantaneous, uncorrelated, binary collisions on which simple kinetic theory is based. Because of the practical importance of very dense shearing flows in a gravitational field, we describe an extension of the simple kinetic theory that maintains much of the structure of the simpler theory. We phrase and solve a boundary value problem for a dense, inclined flow of dissipative spheres using this extension of the kinetic theory. The numerical solution of this problem reproduces the features of such flows observed in numerical simulations. Finally, we compare the profiles that result from a numerical solution with those of a simple, analytic, approximation theory.

4.2   Balance Laws and Distribution Functions

Consider smooth spheres with a diameter σ and a mass m moving with particle velocity $\mathbf{c}$. The dynamics of a collision between two such spheres, labeled 1 and 2, is determined in terms of their relative velocity, $\mathbf{g} \equiv \mathbf{c}_1 - \mathbf{c}_2$, prior to the collision, the unit vector, $\mathbf{k}$, from the center of particle 1 to that of particle 2 at collision, and a coefficient of restitution, e, with $0 \le e \le 1$, by

$$\mathbf{g}' \cdot \mathbf{k} = -e(\mathbf{g} \cdot \mathbf{k}),$$

where the prime indicates a quantity immediately after the collision. Then, for any property Ψ that is a function of the particle velocity, its collisional change $\Delta\Psi$, defined by

$$\Delta\Psi \equiv \Psi_1' + \Psi_2' - \Psi_1 - \Psi_2,$$

may easily be calculated. For example, the total change of kinetic energy in a collision is found, by considering $\Psi = mc^2/2$, with $c^2 = \mathbf{c} \cdot \mathbf{c}$, to be

$$\Delta\left(\frac{1}{2}mc^2\right) = -\frac{1}{4}m\left(1 - e^2\right)(\mathbf{g} \cdot \mathbf{k})^2.$$

The distribution of velocities is given by a function $f^{(1)}$ of $\mathbf{c}$, position $\mathbf{x}$, and time t, defined so that the number n of particles per unit volume at $\mathbf{x}$ and t is

$$n(\mathbf{x}, t) = \int f^{(1)}(\mathbf{c}, \mathbf{x}, t)\, d\mathbf{c},$$

where the integration is over the entire volume of velocity space. The mean $\langle \Psi \rangle$ of a particle property is defined in terms of n and the velocity distribution function $f^{(1)}$ by

$$\langle \Psi \rangle = \frac{1}{n} \int \Psi f^{(1)}(\mathbf{c}) \, d\mathbf{c},$$

where the dependence on $\mathbf{x}$ and t is understood. For example, the mean mass density ρ is mn, the mean velocity $\mathbf{u}$ is $\langle \mathbf{c} \rangle$, the fluctuation velocity $\mathbf{C}$ is $\mathbf{c} - \mathbf{u}$, and the granular temperature T is $\langle C^2 \rangle / 3$.

Balance laws for the mean fields ρ, $\mathbf{u}$, T, and higher moments of the velocity distribution function may be obtained using Chapman's transport equation (e.g., Reif 1965, pp. 525–527). This equation is simply the expression of the fact that the time rate of change of the average of any function of the particle velocity in an infinitesimal volume fixed in space is the sum of its rate of change due to (1) an external force, (2) its net flux through the surface of the volume, and (3) collisions between particles.

The frequency of collisions is given in terms of the complete pair distribution function $f^{(2)}(\mathbf{c}_1, \mathbf{x}_1, \mathbf{c}_2, \mathbf{x}_2, t)$ governing the likelihood that at time t, spheres with velocities near $\mathbf{c}_1$ and $\mathbf{c}_2$ will be located near $\mathbf{x}_1$ and $\mathbf{x}_2$, respectively. For particles that are in contact, $\mathbf{x}_2 = \mathbf{x}_1 + \sigma \mathbf{k}$. When collisions involving particles whose centers are both within the volume are distinguished from collisions that involve a particle with its center inside the volume and a particle with its center outside the volume, and when the spatial gradient of the complete pair distribution function for pairs of colliding particles is small, the transport equation may be written as

$$\frac{\partial}{\partial t} \langle n\Psi \rangle = \left\langle n \frac{\partial \Psi}{\partial c_i} \frac{F_i}{m} \right\rangle - \frac{\partial}{\partial x_i} [\langle nc_i \Psi \rangle + \Theta_i(\Psi)] - \frac{\partial u_j}{\partial x_i} \Theta_i \left(\frac{\partial \Psi}{\partial C_j} \right) + \Phi(\Psi), \tag{4.1}$$

where

F is the external force acting on a particle

$\Theta(\Psi)$ is the collisional flux of Ψ

$$\Theta_i(\Psi) \equiv -\frac{1}{2} \sigma \iiint_{\mathbf{g} \cdot \mathbf{k} \geq 0} (\Psi_1' - \Psi_1) k_i f^{(2)}(\mathbf{c}_1, \mathbf{x}, \mathbf{c}_2, \mathbf{x} + \sigma \mathbf{k}, t) \sigma^2 (\mathbf{g} \cdot \mathbf{k}) \, d\Omega \, d\mathbf{c}_1 \, d\mathbf{c}_2,$$

with $d\Omega$ the element of solid angle centered at $\mathbf{k}$; and $\Phi(\Psi)$ is the rate of collisional production of Ψ

$$\Phi(\Psi) \equiv \frac{1}{2} \iiint_{\mathbf{g} \cdot \mathbf{k} \geq 0} \Delta \Psi f^{(2)}(\mathbf{c}_1, \mathbf{x} - \sigma \mathbf{k}, \mathbf{c}_2, \mathbf{x}, t) \sigma^2 (\mathbf{g} \cdot \mathbf{k}) \, d\Omega \, d\mathbf{c}_1 \, d\mathbf{c}_2.$$

The collisional flux is not present in the kinetic theory of dilute gases, and the third term on the right-hand side of the transport equation must be included when the particle property is a function of $\mathbf{C}$ rather than $\mathbf{c}$.

The balance laws for mass, momentum, and energy result from the transport equation when $\mathbf{\Psi}$ is taken to be m, $m\mathbf{c}$, and $mC^2/2$, respectively. These have the familiar forms

$$\dot{\rho} + \rho u_{i,i} = 0, \tag{4.2}$$

$$\rho \dot{u}_i = t_{ik,k} + nF_i, \tag{4.3}$$

and

$$\frac{3}{2}\rho \dot{T} = -q_{i,i} + t_{ik}D_{ik} - \gamma, \tag{4.4}$$

where
   the dot denotes a time derivative calculated with respect to the mean velocity
   $\mathbf{t}$ is the symmetric stress tensor
   $\mathbf{q}$ is the flux of fluctuation energy
   $\mathbf{D}$ is the symmetric part of the velocity gradients

Importantly, γ is the collisional rate of dissipation of fluctuation energy per unit volume, the only nonclassical term in the equations.

The stress and the flux of fluctuation energy are due to both transport between collisions and transfer in collisions. The transport parts are $-\rho\langle \mathbf{CC}\rangle$ and $\rho\langle \mathbf{C}C^2\rangle/2$, respectively. The collisional transfers, expected to dominate at relatively high number densities, depend upon the exchange of momentum and energy in a collision and the frequency of collision. As an alternative to the number density, the solid volume fraction, $v \equiv \pi n \sigma^3/6$, is often employed to characterize the density of the system. Then, $\rho = \rho_s v$, where ρ_s is the mass density of the material of the spheres.

4.2.1 Radial Distribution Function

For moderately dense systems, say $v \le 0.49$, at which a first-order phase transition from a random to order collisional gas is first possible (Adler and Wainwright 1957), the complete pair distribution function $f^{(2)}$ for a colliding pair is assumed to be the product of the single particle distribution function $f^{(1)}$ of each sphere, evaluated at its center, and a factor g_0 that incorporates the influence of the volume occupied by the spheres on their collision frequency. This factor is the equilibrium radial distribution function, evaluated at the point of contact. It is given as a function of v by Carnahan and Starling (1970) as

$$g_0(v) = \frac{(2-v)}{2(1-v)^3}. \tag{4.5}$$

In dilute systems, g_0 is unity, the positions of the centers of the spheres are not distinguished, and the presumed absence of correlation in position and velocity is called the assumption of molecular chaos. In distinguishing between the positions of the centers of a pair of colliding spheres, the possibility of collisional transfers is incorporated into the dense theory for even the simplest of velocity distribution functions (e.g., Jenkins and Savage 1983). No matter which distribution function is used, g_0 determines the density dependence of the resulting theory.

Here, we employ (4.5) and variants of it that are more appropriate for dense and very dissipative systems to describe the dependence of the transport coefficients on the volume fraction. The product $G \equiv v g_0$ will be shown to arise prominently in the expression for the mean free path or, equivalently, the distance between the edges of the spheres, quantities that rapidly go to zero as the random close packed limit is approached (volume fraction $v \to 0.64$). We note that as the limit is approached, G increases rapidly (e.g., at $v = 0.5$, $G = 3$). In what follows, we make use of this to simplify the dependence of the theory on the volume fraction. Also, for volume fractions above 0.49, a more accurate form of g_0 introduced by Torquato (1985) replaces the singularity of (4.5) at unity with a singularity at the value of 0.64 appropriate to random close packing of identical, frictionless spheres:

$$\hat{g}_0(v) = g_0(0.49) \frac{(0.64 - 0.49)}{(0.64 - v)}. \tag{4.6}$$

Finally, numerical simulations (Mitarai and Nakanishi 2007, Reddy and Kumaran 2007) indicate that for dense shearing flows of very dissipative spheres, the singularity depends on the amount of dissipation in a collision and that it is less than the value at random close packing. We employ the various forms of G associated with these collisional radial distribution functions as we consider theories that apply for increasing volume fractions and increasing amounts of collisional dissipation.

4.2.2 Velocity Distribution Function

For elastic spheres in thermal equilibrium, the velocity distribution function is Maxwellian

$$f_0^{(1)} = \frac{n}{(2\pi T)^{3/2}} \exp\left(-\frac{C^2}{2T}\right).$$

For smooth but inelastic spheres, there is no such equilibrium state. When the collisions are inelastic, the simplest state possible is that of steady, homogeneous shearing. However, if not too much energy is dissipated in collisions, it is plausible to assume that $f^{(1)}$ does not differ much from $f_0^{(1)}$. Consequently, for nearly elastic spheres, $f^{(1)}$ is taken to be a perturbation of the Maxwellian

$$f^{(1)} = \left[1 + \frac{\widehat{A}_{ik} C_i C_k}{2T^2} - \frac{b_i C_i \left(5 - C^2/T\right)}{10 T^2}\right] f_0^{(1)}, \tag{4.7}$$

where

the coefficient $\widehat{\mathbf{A}}$ is the deviatoric part of the second moment $\langle \mathbf{CC} \rangle$ of $f^{(1)}$

$\mathbf{b}$ is the contraction $\langle \mathbf{CC}^2 \rangle$ of its third moment

They are determined as functions of the mean fields ρ, $\mathbf{u}$, and T and their spatial gradients as approximate solutions of the balance laws resulting from (4.1) that govern their evolution, as follows.

Let L and U be, respectively, a characteristic length and velocity of a typical flow. Then, when it is assumed that $\widehat{\mathbf{A}}/T$, $\mathbf{b}/T^{3/2}$, σ/L, and $(1-e)^{1/2}$ are all small and that $U/T^{1/2}$ and $G \equiv \nu g_0$ are near one, these balance laws are satisfied identically at lowest order provided that

$$\frac{\widehat{A}_{ik}}{T} = -\frac{\sqrt{\pi}}{6}\left(1 + \frac{5}{8}\frac{1}{G}\right)\frac{\sigma}{T^{1/2}}\hat{D}_{ik}, \tag{4.8}$$

where the hat denotes the deviatoric part, and

$$\frac{b_i}{T^{3/2}} = -\frac{15\sqrt{\pi}}{16}\left(1 + \frac{5}{12}\frac{1}{G}\right)\frac{\sigma}{T}T_{,i}.$$

Up to an error proportional to the square of quantities assumed to be small, these expressions are identical to those for perfectly elastic spheres as given, for example, in Section 16.34 of Chapman and Cowling (1970). Consequently, so also are the expressions for the pressure tensor and the energy flux vector calculated by employing the velocity distribution function (4.7) and Enskog's extension of the assumption of molecular chaos.

4.3   Moderately Dense Flows

4.3.1   Slightly Dissipative Collisions

4.3.1.1   Constitutive Relations

Derivations for the stress tensor and energy flux vector that appear in the momentum and energy balance equations (Equations 4.3 and 4.4) can be found in Chapman and Cowling (1970, Section 16.41). The stress tensor is

$$t_{ij} = (-p + \varpi D_{kk})\delta_{ij} + 2\mu\hat{D}_{ij}, \tag{4.9}$$

expressed in terms of the pressure

$$p = \rho T(1 + 4G),$$

bulk viscosity

$$\varpi = \frac{8}{3\sqrt{\pi}}\rho\sigma T^{1/2}G,$$

and shear viscosity

$$\mu = \frac{8}{5\sqrt{\pi}} \rho \sigma G T^{1/2} J, \tag{4.10}$$

with

$$J \equiv 1 + \frac{\pi}{12} \left(1 + \frac{5}{8G}\right)^2.$$

Recall that G increases sharply with volume fraction; in this limit, J simplifies to $J = 1 + \pi/12$.

The energy flux vector is (Chapman and Cowling, 1970, Sec. 16.42)

$$q_i = -\kappa T_{,i}, \tag{4.11}$$

where the analog of the coefficient of thermal conductivity k is

$$k = \frac{4M}{\sqrt{\pi}} \rho \sigma G T^{1/2}, \tag{4.12}$$

with

$$M \equiv 1 + \frac{9\pi}{32} \left(1 + \frac{5}{12G}\right)^2.$$

The only nonclassical quantity is the rate of dissipation γ, whose form, at lowest order, is independent of the perturbations $\mathbf{A}$ and $\mathbf{b}$

$$\gamma = \frac{24}{\sqrt{\pi}} (1 - e) \frac{\rho T^{3/2}}{\sigma} G. \tag{4.13}$$

It is the presence of γ in the energy Equation (4.4) that allows steady solutions for inelastic spheres in situations where none are possible in the classical kinetic theory.

Extenstion to Frictional Particles

Jenkins and Zhang (2002) indicate how sliding friction and the resulting coupling between the translational and rotational degrees of freedom of the spheres may be taken into account in a simple, but crude way for relatively small coefficients of sliding friction by assuming that the balances of angular momentum and rotational fluctuation energy are satisfied as in a steady, homogeneous shearing flow. Then the average rotation of the spheres is half the vorticity of the mean velocity and the energy of the rotational fluctuations can be determined in terms of that of the translational fluctuations. This permits the rotational energy to be replaced in the rate of collisional dissipation of translational energy and an effective coefficient of restitution may then be defined that includes dependence on the coefficient of sliding friction. Berzi and Jenkins (2010) extend the method to the other extreme—large coefficients of sliding friction. More complete treatments of frictional spheres that involve the full equations of balance for angular momentum and rotational energy are possible (e.g., Lun 1991), but are far more complicated.

4.3.1.2 Boundary Conditions

Assuming that the velocity distribution function (4.7) applies in the neighborhood of a boundary, boundary conditions on the stress tensor and the energy flux vector may be derived by considering the mean rate of transfer of momentum and energy in collisions between the flowing spheres and the boundary, taking into account the shielding of a wall sphere from collisions with flow spheres by its neighboring wall spheres (e.g., Richman 1988).

At a point on a rigid boundary translating with velocity $\mathbf{U}$, the mean flow velocity $\mathbf{u}$ will, in general, differ from $\mathbf{U}$, and slip will occur. The slip velocity $\mathbf{v}$ is $\mathbf{U} - \mathbf{u}$. Because the boundary is impenetrable, at a point with inward unit normal $\mathbf{N}$, $\mathbf{v} \cdot \mathbf{N} = 0$.

Over a unit area of the boundary, the rate $\mathbf{M}$ at which momentum is supplied by the boundary must be balanced by the rate at which it is removed by the flow,

$$M_i = -t_{ij}N_j. \tag{4.14}$$

The corresponding rate at which energy is supplied by the boundary is $\mathbf{M} \cdot \mathbf{U} - D$, where D is the rate of dissipation in collisions. This must balance the rate at which energy is removed by the flow,

$$M_i U_i - D = -u_i t_{ij} N_j + q_i N_i,$$

or with (4.14) and the definition of the slip velocity,

$$M_i v_i - D = q_i N_i, \tag{4.15}$$

Note that the energy flux at the boundary may be positive or negative, depending upon the relative magnitudes of the slip working and the rate of dissipation.

Consider a boundary consisting of a flat plate with smooth spheres of diameter d fixed to it (see Figure 4.1). The centers of the spheres are assumed to be

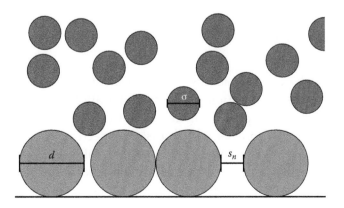

Figure 4.1   Bumpy boundary.

positioned randomly in such a way that the average distance between their edges is s; a natural measure of the roughness of the boundary is $\sin\theta \equiv (d+s)/(d+\sigma)$. The value of θ for given values of σ, d, and s may be calculated by determining the average depth of penetration of flow spheres around a typical wall sphere. For example, for a random close packing of wall spheres with the same diameter as the flow spheres, $\theta \doteq \pi/6$. Neither σ nor d is assumed to differ much from their average, $\overline{\sigma} \equiv (d+\sigma)/2$. Also, the coefficient of restitution e_w for a collision between a sphere of the flow and a sphere of the wall is, like e, assumed to be near one.

If now the velocity distribution function (4.7) is used to calculate the rate at which collisions over a unit area of the wall supply momentum to the flow, the result is (Richman 1988)

$$M_i = \rho\chi T\left[N_i + \left(\frac{2}{\pi}\right)^{1/2}\left(H\frac{v_i}{T^{1/2}} + I_{ijk}\frac{\overline{\sigma}u_{k,j}}{T^{1/2}}\right)\right], \tag{4.16}$$

where

$$H \equiv \frac{2}{3}[2\csc^2\theta(1-\cos\theta) - \cos\theta], \tag{4.17}$$

$$\begin{aligned}
I_{ijk} &\equiv (2\cos\theta N_i N_k + H\delta_{ik})N_j \\
&\quad + \frac{1}{2}\left[\left(5\sin^2\theta - 4\right)N_i N_j N_k - \sin^2\theta\left(N_i\delta_{jk} + N_j\delta_{ik} + N_k\delta_{ij}\right)\right] \\
&\quad \times \left[1 + \frac{\pi}{12\sqrt{2}}\frac{\sigma}{\overline{\sigma}}\left(\frac{5}{8}\frac{1}{G} + 1\right)\right],
\end{aligned}$$

and χ is a factor, roughly corresponding to g_0 in the flow, providing the influence of the size and spacing of the wall spheres on the frequency of collision at the wall.

The corresponding expression for the rate of dissipation per unit area is

$$D = 2\left(\frac{2}{\pi}\right)^{1/2}\rho\chi(1-e_w)T^{3/2}(1-\cos\theta)\csc^2\theta. \tag{4.18}$$

4.3.1.3 Shearing Flows

Homogeneous shearing

Consider a moderately dense flow of identical, smooth, nearly elastic spheres maintained by the relative motion of identical, parallel, bumpy walls a fixed distance $L + 2\overline{\sigma}$ apart. The upper wall moves with constant velocity U in the x direction, the lower wall moves with the same speed in the opposite direction (see Figure 4.2). The nonvanishing x components of the flow velocity and slip velocity are denoted by u and v, respectively. In general, the flow velocity, granular temperature, and solid volume fraction will be functions of the transverse coordinate y, measured from the centerline. A prime will denote a derivative with respect to y.

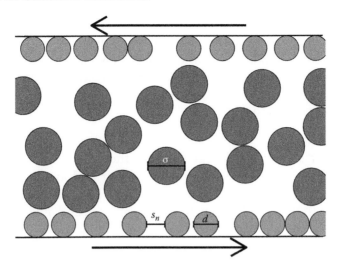

Figure 4.2   Bumppy boundary shear.

Simple shear, in which $u = u'y$, with u', T, and v constants, is exceptional. It occurs when the diameter and solid volume fraction of the flow spheres, the diameter and separation of the wall hemispheres, and the two coefficients of restitution are such that in the interior and at the boundaries, the rate at which fluctuation energy is produced by the shear stress working through the mean velocity gradient or through the slip velocity is precisely equal to the local rate at which it is dissipated in collisions. We first treat this homogeneous flow, noting that the apparent rate of shear $2U/L$ differs from u' by $2v/L$ because of slip at the boundary.

If external forces are assumed to be absent or if the weight of material over a unit area is a small fraction of the pressure, the x and y components of (4.3) require that the shear stress $S \equiv t_{xy}$ and the pressure $p \equiv (t_{xx} + t_{yy} + t_{zz})/3$ be constant, and the energy equation (4.4) reduces to

$$S \frac{du}{dy} - \gamma = 0.$$

With the definitions of the stress tensor (Equation 4.9), energy flux vector (Equation 4.10), and dissipation γ (Equation 4.13), this determines the granular temperature

$$T = \frac{1}{15} \frac{\sigma^2}{(1-e)} \left(\frac{du}{dy} \right)^2 \left[1 + \frac{\pi}{12} \left(1 + \frac{5}{8G} \right)^2 \right]. \tag{4.19}$$

Then, the tractions required to maintain the flow are

$$p = \rho T (1 + 4G) \tag{4.20}$$

and

$$S = \frac{8}{5\sqrt{\pi}} \rho \sigma G T^{1/2} \left[1 + \frac{\pi}{12} \left(1 + \frac{5}{8G} \right)^2 \right] \frac{du}{dy}. \tag{4.21}$$

With T given by (4.19), the normal stress and shear stress vary with the square of the shear rate. This is what is seen in the experiments (Bagnold 1954, Savage 1972, Savage and McKeown 1983, Savage and Sayed 1984, Hanes and Inman 1985). The results of the numerical simulations of Walton and Braun (1986) are in excellent agreement with the values of the predicted stress, provided that the coefficient of restitution is greater than 0.8.

Inhomogeneous shearing

Next, consider steady rectilinear flows in the $x - y$ plane in which the rate of production of fluctuation energy in the interior and at the boundaries is not balanced by the rates of collisional dissipation. The boundaries are again assumed to be identical, parallel, and bumpy. In this case, u', v, and T will depend upon y. Then, with $Q \equiv q_y$, the balance of fluctuation energy (4.4) reduces to

$$\frac{dQ}{dy} - S\frac{du}{dy} + \gamma = 0.$$

In flows of moderate density, we neglect the contributions from particle transport to the fluxes of momentum and energy, keeping only those terms proportional to G (terms proportional to $1/G$ are neglected compared to unity). With this assumption, the pressure p and shear stress S in the limit of high volume fraction become

$$p = 4\rho GT \tag{4.22}$$

and

$$S = \frac{8J}{5\sqrt{\pi}}\rho\sigma T^{1/2}G\frac{du}{dy}.$$

As noted earlier, in the dense limit $J \approx 1 + \pi/12$. Eliminating G between these two equations results in a relation between the velocity gradient and the temperature

$$\frac{du}{dy} = \frac{5\sqrt{\pi}}{2J}\frac{T^{1/2}}{\sigma}\frac{S}{p}. \tag{4.23}$$

Similarly, the energy flux (4.11), with (4.12), is found to be

$$Q = -\frac{4M}{\sqrt{\pi}}\rho\sigma T^{1/2}G\frac{dT}{dy}, \tag{4.24}$$

where $M \equiv 1 + 9\pi/32$. Finally, the dissipation rate per unit volume is given by (4.13). In these constitutive relations, introduced by Haff (1983) and Jenkins and Savage (1983), the dependence upon v has been replaced by a dependence upon p and T, and the dependence of p on y is often known.

Equation (4.22) may be used to write the energy flux and the rate of dissipation in terms of T and the constant pressure p and, when these are used with (4.23) in the balance of energy, the resulting equation may be written in terms of $w \equiv T^{1/2}$, called the fluctuation velocity:

$$\sigma^2 \frac{d^2 w}{dy^2} - k^2 w = 0,$$

where

$$k^2 \equiv \left[3(1-e) - \frac{5\pi}{4J} \left(\frac{S}{p} \right)^2 \right] \frac{1}{M}.$$

In the approximation employed for moderately dense flows, there is a common dependence of the transport coefficients and the rate of dissipation upon the volume fraction that may be eliminated in favor of the temperature and pressure. As a consequence, the energy equation, when expressed in terms of the fluctuation velocity w, has a particularly simple form and may be solved analytically.

There are two different types of solutions, depending on the nature of k. When k is real and positive, the rate of collisional dissipation in the flow is greater than the rate of production of fluctuation energy by the shear stress working through the gradients of mean velocity and the solution involves hyperbolic functions. This solution is

$$w(y) = w_1 \frac{\cosh(ky/\sigma)}{\cosh(kL/2\sigma)} \tag{4.25}$$

in which $w_1 \equiv w(L/2)$ and it has been insured that $w'(0) = 0$. When k is imaginary, the solution has the same structure, but involves trigonometric functions.

These steady solutions are possible only if the boundary conditions permit them. That is, if fluctuation energy is produced in the interior, then, in order for the flow to be steady, the nature of boundaries, the slip velocity at their surface, and their separation must be such that the flux of energy from the flow equals that dissipated by the boundaries. This imposes conditions on the slip velocity and gap thickness, derived as follows.

For the boundary at $y = L/2$, the components of the momentum flux (Equation 4.16) are

$$M_y = -\rho\chi w^2,$$

and

$$M_x = \left(\frac{2}{\pi} \right)^{1/2} \rho\chi w \frac{2}{3} [2\csc^2\theta (1 - \cos\theta) - \cos\theta](v - \bar{\sigma}u')$$
$$+ \left(\frac{2}{\pi} \right)^{1/2} \rho\chi w \frac{1}{2} \bar{\sigma}u' \left(1 + \frac{\sigma}{\bar{\sigma}} \frac{\pi}{12\sqrt{2}} \right) \sin^2\theta,$$

The normal component of the boundary condition (4.14) is

$$p = \rho \chi w^2;\tag{4.26}$$

while the tangential component, $M_x = S$, may be rewritten by employing Equations 4.23 and 4.26 to eliminate, respectively, χ and u' from the expression for M_x. Solving for v/w_1,

$$\frac{v}{w_1} = \left(\frac{\pi}{2}\right)^{1/2} f\frac{S}{p},\tag{4.27}$$

where

$$f \equiv \frac{3}{2}\frac{1 - \left[5(\overline{\sigma}/\sigma)/(2\sqrt{2}J)\right]\left[1 + \pi(\sigma/\overline{\sigma})/(12\sqrt{2})\right]\sin^2\theta}{2\csc^2\theta(1 - \cos\theta) - \cos\theta} + \frac{5}{\sqrt{2}J}\frac{\overline{\sigma}}{\sigma}$$

determines the slip velocity.

The gap thickness L is determined from the energy flux condition (4.15), $Sv - D = Q$. Substituting for the slip velocity (4.27), dissipation (4.18), and using the constitutive relation (4.24) for the energy flux and (4.22) for G, the energy boundary condition at $L/2$ may be written as

$$\sigma\frac{dw}{dy} = bw,\tag{4.28}$$

where

$$b \equiv \left[\left(\frac{\pi}{2}\right)f\left(\frac{S}{p}\right)^2 - 2(1 - e_w)(1 - \cos\theta)\csc^2\theta\right]\frac{1}{M\sqrt{2}}.$$

When the explicit expression (4.25) for w is employed in (4.28), the energy flux boundary condition yields the relation

$$\tanh\left(\frac{kL}{2\sigma}\right) = \frac{b}{k},\tag{4.29}$$

determining L/σ, the gap thickness in sphere diameters, provided that $0 \leq b \leq k$ and $k^2 > 0$. If $k^2 < 0$, then the solution for $w(y)$ involves trigonometric functions and the relation corresponding to (4.29) is

$$\tan\left(\frac{KL}{2\sigma}\right) = -\frac{b}{K},\tag{4.30}$$

where

$$K^2 \equiv \left[\frac{5\pi}{4J}\left(\frac{S}{p}\right)^2 - 3(1 - e)\right]\frac{1}{M}.$$

In (4.30), a nonnegative thickness requires that $b \leq 0$. Solutions for other than the principal value of the argument lead to negative temperatures in the flow and must be discarded. The gap width at which steady, uniform flows are possible is determined by the stress ratio and the properties of the flow spheres and the boundaries.

The velocity profile across the gap can be found by integrating the relation between velocity gradient and temperature (4.23) assuming the solution (4.25) and obvious symmetry consequences (i.e., the velocity must vanish on the centerline)

$$u(y) = \frac{5\sqrt{\pi}\,w_1}{2J}\,\frac{\sinh(ky/\sigma)}{k}\,\frac{S}{\cosh(kL/2\sigma)}\,\frac{S}{N}.$$

Upon evaluating this at $L/2$, using the definition of slip velocity (4.27), and Equation 4.29, the relationship between the boundary velocity, fluctuation velocity, and stress ratio may be written as

$$\frac{U}{w_1} = \left(\frac{\pi}{2}\right)^{1/2}\left(f + \frac{5}{\sqrt{2J}}\frac{b}{k^2}\right)\frac{S}{N}. \tag{4.31}$$

Next, the expression for pressure (4.22) may be used to determine w_1 in terms of p and ν_1

$$w_1 = \left(\frac{p}{4\rho_1 G_1}\right)^{1/2}.$$

Finally, if desired, the density profile may be obtained by inverting (4.22). Consequently, when p, U, and ν_1 are given, the solution is completely determined.

In Figure 4.3, we show the profiles of normalized fluctuation velocity, average velocity, and volume fraction in half a shear cell with a thickness $L = 14.44\sigma$, a wall bumpiness $\theta = \pi/5.5$, and a wall volume fraction $\nu_1 = 0.25$, for three values of the coefficient of restitution, $e = 0.80, 0.85$, and 0.90. The corresponding values of the stress ratio and pressure, made dimensionless by the product $\rho_s U^2$, are $0.3470, 0.3094, 0.2609$, and $0.0141, 0.0144, 0.0153$, respectively. The values of w/U and u/U at a given height increase with e while those of ν decrease. The assumption of moderate volume fractions is abused at both the wall and the centerline of the flow. At the wall, the volume fraction is somewhat low; at the centerline, the volume fraction for the more dissipative spheres is too high. Near the centerline, a theory for denser shearing flow is required for the more dissipative spheres.

Finally, we note that Equation 4.31 is a quadratic equation in S/p which, when solved for S/p, yields

$$\frac{S}{p} = -\alpha\left(\frac{p}{\rho_1 U^2}\right)^{1/2} + \left[\alpha^2\frac{p}{\rho_1 U^2} + \frac{12J}{5\pi}(1-e)\right]^{1/2} \tag{4.32}$$

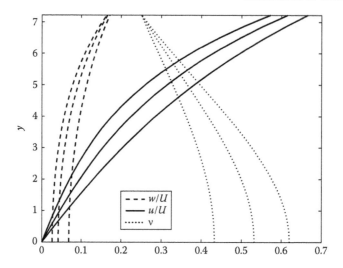

Figure 4.3 Profiles of normalized fluctuation velocity, average velocity, and volume fraction in half a symmetric shear cell for spheres with coefficients of restitution $e = 0.80, 0.85, 0.90$ and $\nu_1 = 0.25$.

where

$$\alpha \equiv \left(\frac{1}{2\pi G_1}\right)^{1/2} \left[\frac{3J}{5}(1-e)f - (1-e_w)(1-\cos\theta)\csc^2\theta\right] \tag{4.33}$$

The coefficient α is proportional to the difference between the rate of slip work in a homogeneous shear flow, in which $S/p = 12J(1-e)/5\pi$, and the collisional dissipation at the boundary. It may be positive or negative, depending on the relative magnitudes of the coefficients of restitution and the bumpiness of the boundary. Typically, α decreases as the bumpiness of the wall increases. When α is positive, fluctuation energy is provided to the flow by the boundaries, and the curves of stress ratio versus normalized boundary velocity are monotone increasing. In this case, the stress ratio at low values of the boundary velocity and/or high values of the normal stress may be substantially lower than its value in simple shear. The steady states corresponding to the curves for which α is negative and the stress ratio is a decreasing function of normalized velocity are likely to be unstable.

Despite the fact that the flow and the geometry are extremely simple and the theory has limited applicability, such an analytical treatment is still useful. It indicates the elements necessary to phrase and solve a boundary-value problem and provides a simple example of how momentum and energy transfer and the production and dissipation of energy in the flow and at the boundaries interact to determine the structure of the flow. This can help to develop intuition for flows of more complex and realistic materials in more complicated geometries. However, because most rapid shearing flows of granular materials involve interactions that are far from being nearly elastic and do not involve small values of sliding friction, it is important to have a kinetic theory for collisions that involve more dissipation in collisions.

4.3.2   Very Dissipative Collisions

For shearing flows of more dissipative spheres at solid volume fractions less than 0.49, we adopt the constitutive relations obtained by Garzo and Dufty (1999) for identical frictionless, inelastic spheres, but do not incorporate the small terms introduced by their function c^* of the coefficient of restitution. The magnitude of c^* is less than 0.4 and terms proportional to it are typically multiplied by a small numerical coefficient. The theory is linear in the first spatial gradients of the fields ρ, u, and T, as is the theory for nearly elastic spheres, and its derivation involves the tacit assumption that the deviatoric part $\widehat{\mathbf{A}}$ of the second moment is a small fraction of T, its trace. However, the determination (4.8) of $\widehat{\mathbf{A}}$ in the simplest theory, used with the solution (4.19) for T in steady, homogeneous shearing, indicates that $\widehat{\mathbf{A}}/T$ can become large as e becomes small. Consequently, the theory has to be used with some caution.

If the x coordinate is taken in the flow direction, the y coordinate taken in the direction of shear, and the z coordinate orthogonal to these, the pressure $p \equiv -\left(t_{xx} + t_{yy} + t_{zz}\right)/3$ is given by

$$p = 4\rho FGT,$$

where, again, $G \equiv \nu g_0$, with g_0 given by (4.5), and

$$F = \frac{(1+e)}{2} + \frac{1}{4G}.$$

The shear stress $S \equiv t_{xy}$ is

$$S = \mu\frac{du}{dy}, \tag{4.34}$$

where

$$\mu = \frac{2J}{5\pi^{1/2}}\frac{p\sigma}{FT^{1/2}}, \tag{4.35}$$

with

$$J = \frac{(1+e)}{2} + \frac{\pi}{32}\frac{[5 + 2(1+e)(3e-1)G][5 + 4(1+e)G]}{\left[24 - 6(1-e)^2 - 5(1-e^2)\right]G^2} \tag{4.36}$$

The energy flux $Q \equiv q_y$ is

$$Q = -\kappa\frac{dT}{dy} - \eta\frac{d\nu}{dy}, \tag{4.37}$$

where

$$k = \frac{M}{\pi^{1/2}} \frac{p\sigma}{FT^{1/2}}$$

(4.38)

and

$$M = \frac{(1+e)}{2} + \frac{\pi}{16} \frac{\left[5 + 3(1+e)^2(2e-1)G\right]\left[5 + 6(1+e)G\right]}{(1+e)\left[16 - 7(1-e)\right]G^2};$$

(4.39)

and, finally, the rate of collisional dissipation γ is given by

$$\eta = \frac{25\pi^{1/2}}{128} \frac{pT^{1/2}}{4FG} \frac{\sigma}{\nu^2} N,$$

(4.40)

with

$$N = \frac{96}{25} \frac{(1-e)}{(1+e)} \frac{\nu}{G} \frac{5 + 6G(1+e)}{16 + 3(1-e)}$$

$$\times \left\{ \frac{20\nu(\ln G)_\nu \left[5 + 3G(1+e)^2(2e-1)\right]}{[48 - 21(1-e)]} - G\left[1 + \nu(\ln G)_\nu\right](1+e)e \right\},$$

(4.41)

in which the subscript indicates a derivative with respect to ν, and

$$\gamma = \frac{12}{\pi^{1/2}} \frac{\rho G}{\sigma} (1 - e^2) T^{3/2}.$$

These relations reduce to those for nearly elastic spheres when the coefficient of restitution e is near unity. They differ in a qualitative way in the presence of the term proportional to the spatial derivative of the volume fraction in the flux of fluctuation energy.

To incorporate the additional dissipation due to friction, Jenkins and Berzi (2010) used the results of a calculation by Herbst et al. (2000) of the rates of dissipation of the rotational and translational energy in a steady, homogeneous shearing flow. Because numerical simulations indicate that the tractions quickly reach limiting values as the coefficient of sliding friction increases above 0.15, they applied their calculation to spheres in the limit of infinite friction. In this case, the ratio R of the energy of the rotational and translation velocity fluctuations is given by

$$R = \frac{2(1+\beta_0)}{14 - 5(1+\beta_0)},$$

where β_0 is the coefficient of tangential restitution in a sticking collision (e.g., Foerster et al. 1994). The rate of dissipation of translation fluctuation energy is

$$\gamma = \frac{48}{\pi^{1/2}} \frac{\rho G}{\sigma} \left[\frac{1-e^2}{4} + \frac{1+\beta_0}{7} - \left(\frac{1+\beta_0}{7}\right)^2 \left(1 + \frac{5}{2}R\right)\right] T^{3/2}.$$

This allows for the definition of an effective coefficient of restitution, e_{eff}:

$$\frac{1 - e_{eff}^2}{4} \equiv \frac{1 - e^2}{4} + \frac{1 + \beta_0}{7} - \left(\frac{1 + \beta_0}{7}\right)^2 \left[1 + \frac{5(1 + \beta_0)}{14 - 5(1 + \beta_0)}\right].$$

From this point onward, we use the symbol e to denote this effective coefficient of restitution.

In the context of a theory for frictionless, inelastic spheres, Woodhouse et al. (2010) have phrased the boundary-value problem for steady, uniform, flows driven by gravity down a bumpy incline and solved it numerically over a range of volume flow rates and angles of inclinations. They find as many as three solutions at a given inclination and volume flux and, in a subsequent analysis (Woodhouse et al. 2010), they characterize the stability of the solutions that they find. The average volume fraction in their solutions ranges from dilute to dense. Those solutions in the intermediate range of volume fraction indicate the features of the solutions, the fact of their multiplicity, and the details of their stability. Flows in their dilute limit are probably better treated by a theory that incorporates anisotropy in the velocity fluctuations; flows in the dense limit are likely to require a theory with a somewhat more complicated structure.

4.4   Very Dense Flows

Recent numerical simulations of dense granular shearing flows (Mitarai and Nakanishi 2007, Reddy and Kumaran 2007) indicate that correlations between the positions and/or the velocities of particles, not considered in simple kinetic theories for collisional flows, play an important role in determining the rheology. In a shear flow of dense, inelastic, compliant particles, the duration of a typical collision may equal or exceed the time between collisions. Then, simultaneous interactions between more than two particles become likely, both in discrete-element simulations that employ compliant spheres and in physical experiments. In this event, small groups of particles overlap (Lois et al. 2005, 2006, Mills et al. 2008) and/or interact through repeated, weak, "chattering" collisions (Mitarai and Nakanishi 2007, Reddy and Kumaran 2007).

Overlaps reduce the frequency of collisions while chattering replaces strong binary collisions with numerous weak ones. In a range of volume fractions between 0.49 and 0.60 and coefficients of restitution greater than 0.60, the influence of the correlations on the fluxes of momentum and energy is compensated for by nonlocal transport associated with the correlated motion. Consequently, overlaps and chattering first influence the collisional rate of dissipation. For denser and/or more inelastic flows, anisotropic and rate-independent contributions to the pressure and shear stress, associated with chains or clusters that eventually span the flow, are anticipated to develop (Goldman and Swinney 2006, Hatano et al. 2007, Schröter et al. 2007, Chialvo et al. 2012).

Models of dense granular shearing flows have attempted to incorporate these correlations in various ways. Ertas and Halsey (2002) introduce a length associated with vorticial structures, Kumaran (2006) employs higher spatial gradients in a kinetic theory for collisional interactions in dense shearing flows, Mitarai and Nakanishi (2007) focus on a modification of the strength of the particle agitation that is associated with correlated velocities, and Jenkins (2006, 2007) introduces a length associated with the size of particle clusters into the expression for the rate of collisional dissipation.

Here, we outline the extension by Jenkins and Berzi (2010, 2012) of the theory to very dissipative, frictional spheres. They incorporate the observed frictional behavior in an effective coefficient of restitution and, because this results in description that involves substantial dissipation, they adopt the kinetic theory of Garzo and Dufty (1999) for frictionless, but very inelastic spheres. When applying the theory to dense flows, they treat overlapping and chattering particles in the same way; assume, as indicated by the numerical simulations (Mitarai and Nakanishi 2007, Reddy and Kumaran 2007) that the singularity in the radial distribution function occurs at volume fractions less than that for random dense packing, and choose the two parameters in Jenkins (2010) theory to provide a reasonable agreement with the results of Pouliquen's (1999) physical experiments on inclined flows. With these material parameters and a simple, algebraic form of the energy balance, they predict relations between volume flux, flow depth, and angle of inclination in dense granular flows between friction sidewalls over the inclined surface of a heap. Their predictions compare favorably with the quantities measured by Jop et al. (2005) and Félix et al. (2007) in their experiments. Consequently, we next outline this extension of the kinetic theory for dense, dissipative shearing flows.

When the volume fraction is greater than 0.49, but less than 0.60, the constitutive relations are taken to be those given earlier in the limit that the terms proportional to $1/G$ are small compared to unity. The volume fraction dependence in G is taken to be a modification of that determined by Torquato (1995) in simulations of dense aggregates of spheres, and the particle diameter σ in the rate of collisional dissipation is replaced by the length L of a typical chain of contacting particles. Then

$$p = 2(1+e)\rho G T, \tag{4.42}$$

where

$$G = \frac{0.63\nu}{(0.60-\nu)};$$

the expressions (4.34) for S and (4.35) for μ are unchanged, but $J = (1+e)/2 + (\pi/4)(3e-1)(1+e)^2/[24-(1-e)(11-e)]$; the expressions (4.37) for Q and (4.38) for κ are unchanged, but $M = (1+e)/2 + (9\pi/8)(2e-1)(1+e)^2/[16-7(1-e)]$; and

$$\gamma = \frac{12}{\pi^{1/2}} \frac{\rho G}{L} (1-e^2) T^{3/2}. \tag{4.43}$$

We expect that the volume fraction at which G becomes singular depends upon the coefficient of restitution; the value of 0.60 seems to be appropriate for a coefficient of restitution of 0.70 (e.g., Mitarai and Nakanishi 2007). At volume fractions greater than 0.60, shear rigidity develops (Hatano et al. 2007, Chialvo et al. 2012) and this contributes a rate-independent term to the shear stress and pressure.

4.4.1 Chain Length

We assume that the spheres are forced into overlapping or chattering contact along the principal compressive axis of the shearing flow and that the random motion of the spheres acts to destroy this order. The principal compressive axis is the eigenvector of the strain rate $\mathbf{D}$ that is associated with its most negative eigenvalue. Then, the magnitude and direction of the vector $\mathbf{L}$ of chain length are determined by a balance between the creation and destruction mechanisms

$$\hat{c}G^{1/3}D_{ik}L_k + \frac{LT^{1/2}}{\sigma^2}L_i = 0, \tag{4.44}$$

where $\hat{c}$ is a constant of order one. The power of G is chosen to be $1/3$ because this power, together with the value $\hat{c} = 0.50$, provides a relatively good fit to the physical experiments of Pouliquen (1999).

The relation (4.44) is a rough, microscopic balance between the effect that is conjectured to create chains or clusters—particles being forced together by the mean shearing into correlated interactions along the principal compressive axis that persist for a time equal to the inverse of the shear rate—and the effect that destroys them—collisions between particles in directions other than that of the principal axis of compression. The balance is phenomenological and crude by the standards of the kinetic theory that it is used with. However, when employed in conjunction with the algebraic form of the energy balance and the constitutive relations of the kinetic theory, the resulting theory has been shown (Jenkins 2006, 2007) to reproduce the qualitative features of inclined flows seen in numerical simulations. With appropriate choices for the constant $\hat{c}$, power of G, and coefficient of restitution e, it reproduces the quantitative relations between volume flow rates, flow depths, and angles of inclination measured in physical experiments.

In a steady, homogeneous shearing flow, (4.44) yields

$$\frac{L}{\sigma} = \frac{1}{2}\hat{c}G^{1/3}\frac{\sigma}{T^{1/2}}\frac{du}{dy}.$$

The shear stress $S \equiv \mu du/dy$ (Equation 4.34) becomes, in the dense limit (where terms proportional to $1/G$ are ignored),

$$\frac{\sigma}{T^{1/2}}\frac{du}{dy} = \frac{5\pi^{1/2}}{2J}\frac{(1+e)}{2}\frac{S}{p}. \tag{4.45}$$

This allows us to express the chain length L/σ in terms of the stress ratio and the volume fraction:

$$\frac{L}{\sigma} = \frac{5\pi^{1/2}}{4J}\frac{(1+e)}{2}\frac{S}{p}\hat{c}G^{1/3}.$$

It is possible to express both L/σ and S/p in terms of the volume fraction ν and coefficient of restitution e by using the algebraic approximation to the energy balance:

$$S\frac{du}{dy} - \gamma = 0,$$

in which γ is assumed to be given by (4.43). Then,

$$\frac{\sigma^2}{T}\left(\frac{du}{dy}\right)^2 = \frac{15}{2J}\frac{\sigma}{L}(1-e^2),$$

and, using (4.45) to eliminate L/σ,

$$\left(\frac{\sigma}{T^{1/2}}\frac{du}{dy}\right)^3 = \frac{15}{J}\frac{(1-e^2)}{\hat{c}G^{1/3}}.$$

With this, L/σ and S/p can be expressed as functions of ν and e as

$$\frac{L}{\sigma} = \frac{1}{2}\left[\frac{15}{J}(1-e^2)\hat{c}^2\right]^{1/3}G^{2/9} \tag{4.46}$$

and

$$\frac{S}{p} = \frac{4J}{5\pi^{1/2}}\frac{1}{(1+e)}\left[\frac{15}{J}\frac{(1-e^2)}{\hat{c}}\right]^{1/3}\frac{1}{G^{1/9}}. \tag{4.47}$$

Inverting (4.47) and using the relation between G and ν results in a determination of the volume fraction in terms of the stress ratio and the coefficient of restitution

$$\nu = \frac{0.60G}{0.63+G}, \tag{4.48}$$

where

$$G = \left[\frac{192}{25\pi^{3/2}}\frac{J^2}{\hat{c}}\frac{(1-e)}{(1+e)^2}\left(\frac{p}{S}\right)^3\right]^3. \tag{4.49}$$

This can also be phrased as a relationship between the stress ratio S/p and the inertial parameter $I \equiv \sigma u'/(p/\rho)^{1/2}$ introduced by GDR MiDi (2004). Upon eliminating the temperature in favor of the pressure in (4.34) and (4.35), we obtain

$$\frac{\sigma}{(p/\rho)^{1/2}}\frac{du}{dy} = \frac{5}{4J}\left(\frac{\pi}{2}\right)^{1/2}\left(\frac{1+e}{G}\right)^{1/2}\frac{S}{p}.$$

With (4.47), this yields the relation

$$\frac{S}{p} = \left\{\frac{5}{4}\left(\frac{\pi}{2}\right)^{1/2}\left[\frac{25\pi^{3/2}}{192}\frac{\hat{c}}{J^{8/3}}\frac{(1+e)^{7/3}}{(1-e)}\right]^{3/2}\right\}^{-2/11}I^{2/11}. \tag{4.50}$$

Equations 4.48 through 4.50 specify the volume fraction and the stress ratio as functions of the inertial parameter, the coefficient of restitution, and the parameter $\hat{c}$. These relations are equivalent to those proposed by GDR Midi (2004) over the range of volume fractions and coefficients of restitution for which exchanges of momentum in collisions dominate the momentum transfer and before force chains span the system. That is, as v increases for a given e, the chain length L given by (4.46) will approach the system size. When it does, there is an additional mechanism for the transfer of momentum in the flow that we do not consider (e.g., Hatano et al. 2007)—ephemeral chains of particles that transfer force across the flow and are responsible for the development of a yield stress. The model of GDR Midi continues to apply above this volume fraction and includes the rate-independent mechanism of momentum transfer, but the model described here does not.

4.4.2 A Boundary-Value Problem

We next use the balance laws and constitutive relations outlined earlier to phrase and solve a boundary-value problem for the steady, fully-developed flow over a rigid, bumpy base inclined at an angle ϕ to the horizontal. In the boundary-value problem, the chain length is determined as in a steady, homogeneous shearing flow.

4.4.2.1 Differential Equations

The balances of momentum parallel and perpendicular to the flow are

$$\frac{dS}{dy} = -\rho g \sin \phi, \tag{4.51}$$

where

$$\frac{du}{dy} = \frac{S}{\mu},$$

with μ given by (4.35) and (4.36), and

$$\frac{dp}{dy} = -\rho g \cos \phi. \tag{4.52}$$

Upon carrying out the differentiation and using (4.42), this first-order equation for p can be employed as a first-order equation for v

$$\left\{ T \frac{d}{dv}[\rho(1+4G)] - \frac{\eta}{\kappa} \right\} \frac{dv}{dy} = \frac{\rho(1+4G)}{\kappa} - \rho g \cos \phi, \tag{4.53}$$

in which η is given by (4.40) and (4.41) and k is given by (4.38) and (4.39).

The energy balance is

$$\frac{dQ}{dy} = S \frac{du}{dy} - \gamma,$$

where γ is given by (4.43), with $L = \sigma$, if $15(1 - e^2)\hat{c}G^{2/3}/(8J) \leq 1$, and $L = \sigma(5\pi^{1/2}/8J)(1 + e)\hat{c}G^{1/3}S/p$, otherwise; and

$$\frac{dT}{dy} = -\frac{(Q + \eta v')}{\kappa},$$

where v' is given in (4.53).

The specification of the volume V of the material over a unit area of the base, or volume holdup, is implemented as a boundary condition to a first-order differential equation for the partial holdup, $v(y) \equiv \int_0^y v(\xi)d\xi$,

$$\frac{dv}{dy} = v,$$

with $v(0) = 0$ and $v(H) = V$.

4.4.2.2 Boundary Conditions

We assume that the base consists of a flat wall to which spheres identical to those in the flow have been fixed and apply the boundary conditions derived for nearly elastic collisions to this more dissipative flow. Then,

$$\frac{u}{T^{1/2}} = \left(\frac{\pi}{2}\right)^{1/2} f \frac{S}{p}, \qquad (4.54)$$

where, with $\bar{\sigma} = \sigma$,

$$f = \frac{3}{4\sqrt{2}J} \frac{2\sqrt{2}J - 5F(1 + B)\sin^2\theta}{2\csc^2\theta(1 - \cos\theta) - \cos\theta} + \frac{5F}{\sqrt{2}J}$$

and

$$B = \frac{\pi}{12\sqrt{2}}\left(1 + \frac{5}{8G}\right),$$

and

$$D = 2\left(\frac{2}{\pi}\right)^{1/2} p(1 - e_w)T^{1/2}(1 - \cos\theta)\csc^2\theta. \qquad (4.55)$$

We take the top of the flow to be the point where the free flight trajectory of a particle ejected normal to the flow with velocity $T^{1/2}$ first equals the mean free distance between collisions. At this point (Pasini and Jenkins 2005),

$$p = 4\rho GFT = 0.037\frac{\rho g \sigma}{v}.$$

The flow momentum and energy flux there, associated with the acceleration of the particle under gravity (Jenkins and Hanes 1993), are

$$S = p\tan\phi$$

and

$$Q = -pT^{1/2}\tan^2\phi.$$

4.4.2.3 Numerical Results

We take $e = e_w = 0.60$, $\theta = \pi/3$, and $\hat{c} = 0.50$ and use σ, $(g\sigma)^{1/2}$, $\rho_s g\sigma$, and $\rho_s(g\sigma)^{1/2}$ to make dimensionless lengths, velocities, stresses, and the energy flux, respectively. We then employ the two-point MATLAB® boundary-value problem solver bvp4c to determine solutions at three values of inclination angle for different values of dimensionless volume holdup. The inclinations are among those employed in Pouliquen's (1999) experiments, and the range of volume holdup provides depths within the range of those measured. The resulting profiles of volume fraction are shown in Figure 4.4 while those of mean velocities are shown in Figure 4.5. The profiles of volume fraction exhibit the uniformity observed, for example, by Silbert et al. (2001) in numerical simulations, with the essentially constant value of volume fraction decreasing with increasing inclination and flows of any dimensionless volume holdup possible at a given angle of inclination. These predictions are in contrast to those of a kinetic theory that does not involve the additional length scale. Such a theory predicts a nonuniform profile of volume fraction, with volume fraction increasing with inclination, and essentially only one steady depth possible at any angle of inclination.

4.4.3 Algebraic Theory

We next show that in relative deep flows of the type just considered, the constant value of ν and the profiles of T and u can be obtained using a much simpler theory. In such flows, the influence of the top and bottom boundaries is minimized and

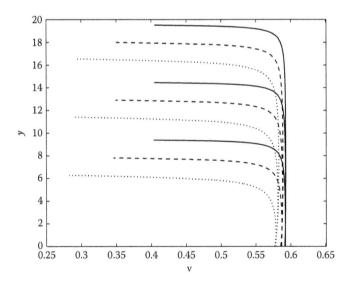

Figure 4.4 Solid volume fraction versus dimensionless depth for three angles of inclination and three values of the dimensionless volume holdup at each angle: $\phi = 26°$, $V = 3.5, 6.5, 9.5$ (dotted lines); $\phi = 25°$, $V = 4.5, 7.5, 10.5$ (dashed lines); $\phi = 24°$, $V = 5.5, 8.5, 11.5$ (solid lines).

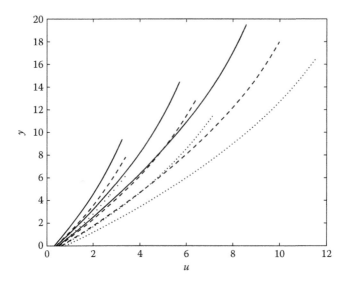

Figure 4.5 Dimensionless mean velocity versus dimensionless depth for the inclinations and dimensionless volume holdups of Figure 4.3.

there is an algebraic balance between the rates of production and dissipation of fluctuation energy in much of the flow Mitarai and Nakanishi 2005).

The balances of momentum (4.51) and (4.52) may be combined and integrated to give

$$S = p \tan \phi,$$

and, because the mass density is nearly constant, a good approximation to the pressure is

$$p \doteq \rho g (H - y) \cos \phi.$$

Then, with (4.42),

$$T = \frac{1}{2(1+e)} \frac{p}{\rho G} = \frac{1}{2(1+e)} \frac{g(H-y)\cos\phi}{G}, \tag{4.56}$$

where, from (4.47),

$$G = \left[\frac{192}{25\pi^{3/2}} \frac{J^2}{\hat{c}} \frac{(1-e)}{(1+e)^2} \frac{1}{\tan^3\phi} \right]^3. \tag{4.57}$$

Finally, the mean velocity satisfies

$$\frac{du}{dy} = \frac{5\pi^{1/2}}{4J}(1+e)\frac{T^{1/2}}{\sigma}\tan\phi,$$

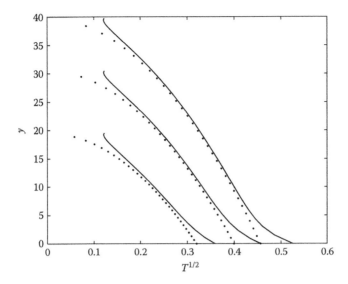

Figure 4.6   Comparison of the profiles of dimensionless fluctuation velocity versus dimensionless depth for $\phi = 24°$ and $V = 11.5, 18.0, 23.5$. The solid lines are the numerical solution of the full equations; the dots are the approximate algebraic solution.

so, upon integration,

$$u = u_0 + \frac{2}{3}\frac{5\pi^{1/2}}{4\sqrt{2}J}\frac{(1+e)^{1/2}}{G^{1/2}}\left[\left(\frac{H}{\sigma}\right)^{3/2} - \left(\frac{H}{\sigma}-\frac{y}{\sigma}\right)^{3/2}\right](g\sigma\cos\phi)^{1/2}\tan\phi, \qquad (4.58)$$

where u_0 is given by (4.54).

The values of ν predicted by (4.48) and (4.57) for the angles 26°, 25°, and 24° are 0.585, 0.590, and 0.593, respectively, compared to values away from the boundaries in the full solutions of 0.571, 0.586, and 0.592, respectively. In Figure 4.6, we compare the predictions of the distribution of fluctuation velocity $T^{1/2}$ by (4.56) with the result of the integration of the full theory for an angle of 24° and different dimensionless holdups that correspond to depths of approximately 20, 30, and 40 particle diameters, and, in Figure 4.7, we do the same for the mean velocity u. The agreement is good, despite the number of approximations that have been made in the analysis. The indication is that the associated algebraic theory used with the same parameters as the full kinetic theory reproduces the profiles of density, average velocity, and fluctuation velocity in the core of relatively deep dense flows, but that there are modest deviations near the upper and lower boundaries.

4.5   Conclusion

We have outlined the derivation of the simplest possible kinetic theory for collisional flows of identical, nearly elastic spheres and sketched the analogous derivation of the boundary conditions at a flat, bumpy wall. The constitutive

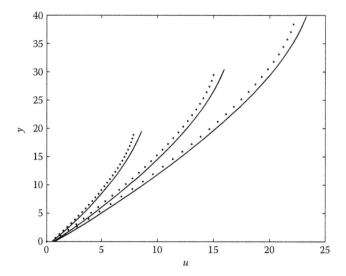

Figure 4.7   Comparison of the profiles of dimensionless mean velocity versus dimension-less depth for $\phi = 24°$ and $V = 11.5, 18.0, 23.5$. The solid lines are the numerical solution of the full equations; the dots are the approximate algebraic solution.

relations and the energy balance of the kinetic theory were used to indicate how the quadratic dependence of the shear stress and pressure on the shear rate emerges in a steady, homogeneous shearing flow. The balance laws and constitutive relations were then employed with the boundary conditions on the flow momentum and flux of fluctuation energy to phrase and solve a boundary-value problem for a steady, uniform, moderately dense, shearing flow between parallel boundaries in relative motion. We next introduced more elaborate constitutive relations of a theory for more dissipative particle interactions. For very dense flows, we incorporated a phenomenological extension of this theory that involved an additional length scale in the rate of collisional dissipation that is associated with repeated and/or multiple collisions. This extended theory was used to formulate a boundary-value problem for steady, uniform flow down an incline. Solutions of this problem were obtained numerically, and profiles of volume fraction and mean velocity were exhibited. Finally, we showed that for deep, dense, inclined flows a simple version of the theory that involves an algebraic balance between the rate of production of fluctuation energy and the modified rate of dissipation had the capacity to predict the profiles of volume fraction, fluctuation velocity, and mean velocity in good agreement with the numerical solution of the full theory.

References

Adler, B. J. and T. E. Wainwright. 1957. Phase transition for a hard sphere system. *J. Chem. Phys.* 27, 1208–1209.

Bagnold, R. A. 1954. Experiments on a gravity-free dispersion of solid spheres in a Newtonian fluid under shear. *Proc. Roy. Soc. Lond.* A225, 49–63.

Brilliantov, N. and T. Pöschel. 2004. *Kinetic Theory of Granular Gases*. Oxford, U.K.: Oxford University Press.

Carnahan, N. F. and K. E. Starling. 1969. Equation of state of non–attracting rigid spheres. *J. Chem. Phys.* 51, 635–636.

Chapman, S. and T. G. Cowling. 1970. *The Mathematical Theory of Nonuniform Gases*. 3rd edn., Cambridge, U.K.: Cambridge University Press.

Chialvo, S., J. Sun, and S. Sundaresan. 2012. Bridging the rheology of granular flows in three regimes. *Phys. Rev. E* 85, 021305.

Chou, C.-S. and M. W. Richman. 1998. Constitutive theory for homogeneous granular shear fows of highly inelastic spheres. *Phys. A* 259, 430–448.

Da Cruz, F., S. Emem, M. Prochnow, J.-N. Roux, and F. Chevoir. 2005. Rheophysics of dense granular materials: Discrete simulation of plane shear flows. *Phys. Rev. E* 72, 021309.

Ertas, D. and T. C. Halsey. 2002. Granular gravitational collapse and chute flow. *Europhys. Lett.* 60, 931–937.

Félix, G., V. Falk, and U. D'Ortona. 2007 . Granular flows in a rotating drum: the scaling law between velocity and thickness of the flow. *Eur. Phys. J. E* 22, 25–31.

Foerster, S. F., M. Y. Louge, H. Chang, and K. Allia. 1994. Measurements of collision properties of small spheres. *Phys. Fluids* 6, 1108–1115.

Garzo, V. and J. W. Dufty. 1999. Dense fluid transport for inelastic hard spheres. *Phys. Rev. E* 59, 5895–5911.

Goldhirsch, I. 2003. Rapid granular flows. *Ann. Rev. Fluid Mech.* 35, 267–293 (2003).

Goldman, D. I. and H. L. Swinney. 2006. Signatures of glass formation in a fluidized bed of hard spheres. *Phys. Rev. Letts.* 96, 145702.

Haff, P. K. 1983. Grain flow as a fluid mechanical phenomenon. *J. Fluid Mech.*, 134, 401–433.

Hanes, D. M. and D. L. Inman. 1985. Observations of rapidly flowing granular-fluid materials. *J. Fluid Mech.* 150, 357–380.

Hatano, T., M. Otsuki, and S. Sasa. 2007. Criticality and scaling relations in a sheared granular material. *J. Phys. Soc. J.* 76, 023001.

Henderson, J. R. and F. van Swol. 1984. On the interface between a fluid and a planar wall: Theory and simulation of a hard sphere fluid at a hard wall. *Mol. Phys.* 51, 991–1010.

Herbst, O., M. Huthmann, and A. Zippelius. 2000. Dynamics of inelastically colliding spheres with Coulomb friction: Dynamics of the relaxation of translational and rotational energy. *Granular Matter* 2, 211–219.

Hungr, O. and N. R. Morgenstern. 1984. Experiments in high velocity open channel flow of granular materials. *Géotechnique* 34, 405–413.

Jenkins, J. T. 2001. Boundary conditions for collisional grain flows at bumpy, frictional walls. In *Granular Gases*, edn. T. Poschel and S. Luding, 125–138, Berlin, Germany: Springer.

Jenkins, J. T. 2006. Dense shearing flows of inelastic disks. *Phys. Fluids* 18, 103307.

Jenkins, J. T. 2007. Dense inclined flows of inelastic spheres. *Granular Matter* 10, 47–52.

Jenkins, J. T. and D. Berzi. 2010. Dense inclined flows of inelastic spheres: Tests of an extension of kinetic theory. *Granular Matter* 12, 151–158.

Jenkins, J. T. and D. Berzi. 2012. Kinetic theory applied to inclined flows. *Granular Matter* 14, 79–84.

Jenkins, J. T. and D. M. Hanes. 1993. The balance of momentum and energy at an interface between colliding and freely flying grains in a rapid granular flow. *Phys. Fluids A* 5, 781–783.

Jenkins, J. T. and M. W. Richman. 1985. Grad's 13-moment system for a dense gas of inelastic spheres. *Arch. Rat'l. Mech. Anal.* 87, 355–377.

Jenkins, J. T. and S. B. Savage. 1983. A theory for the rapid flow of identical, smooth, nearly elastic, spherical particles. *J. Fluid Mech.* 130, 187–202.

Jenkins, J. T. and C. Zhang. 2002. Kinetic theory for identical, frictional, nearly elastic spheres. *Phys. Fluids* 14, 1228–1235.

Johnson, P. C., P. Nott, and R. Jackson. 1990. Frictional–collisional equations of motion for particulate flow and their applications to chutes. *J. Fluid Mech.* 210, 501–535.

Jop, P., Y. Forterre, and O. Pouliquen. 2005. Crucial role of sidewalls in granular surface flows: Consequences for the rheology. *J. Fluid Mech.* 451, 167–192.

Kumaran, V. 2006. The constitutive relation for the granular flow of rough particles and its application to the flow down an incline plane. *J. Fluid Mech.* 561, 1–42.

Lois, G., J. Carlson, and A. Lemaitre. 2005. Numerical tests of constitutive laws for dense granular fows. *Phys. Rev. E* 72, 051303.

Lois, G., A. Lemaitre, and J. Carlson, 2006. Emergence of multi-contact interactions in contact dynamics simulations. *Europhys. Lett.* 76, 318–324.

Lun, C. K. K. 1991. Kinetic theory for granular flow of dense, slightly inelastic, slightly rough spheres. *J. Fluid Mech.* 233, 539–559.

Lun, C. K. K., S. B. Savage, D. J. Jeffrey, and N. Chepurniy. 1984. Kinetic theories for granular flow: Inelastic particles in couette flow and lightly inelastic particles in a general flow field. *J. Fluid Mech.* 140, 223–256.

MiDi, G. D. R. 2004. On dense granular flows. *Eur. Phys. J. E* 14, 341–365.

Mills, P., P. G. Rognon, and F. Chevoir. 2008. Rheology and structure of granular materials near the jamming transition. *Europhys. Lett.* 81, 64005.

Mitarai, N. and H. Nakanishi. 2005. Bagnold scaling, density plateau, and kinetic theory analysis of dense granular flow. *Phys. Rev. Lett.* 94, 128001.

Mitarai, N. and H. Nakanishi. 2007. Velocity correlations in dense granular shear flows: Effects on energy dissipation and normal stress. *Phys. Rev. E* 75, 031305.

Pasini, J. M. and J. T. Jenkins. 2005. Aeolian transport with collisional suspension. *Phil. Trans. Roy. Soc.* 363, 1625–1646.

Pouliquen, O. 1999. Scaling laws in granular flows down a bumpy inclined plane. *Phys. Fluids* 11, 542–548.

Reddy, K. A. and V. Kumaran. 2007. Applicability of constitutive relations from kinetic theory for dense granular flow. *Phys. Rev. E* 76, 061305.

Reif, F. 1965. *Fundamentals of Statistical Mechanics and Thermal Physics.* New York: McGraw-Hill.

Richman, M. W. 1988. Boundary conditions based upon a modified Maxwellian velocity distribution function for flows of identical, smooth, nearly elastic spheres. *Acta Mech.* 75, 227–240.

Savage, S. B. 1972. Experiments on shear flows of cohesionless granular materials. In *Continuum Mechanical and Statistical Approaches in the Mechanics of Granular Materials*, edn. S. C. Cowin and M. Satake, pp. 241–254. Tokyo, Japan: Gakujutsu Bunken Fukyu-kai.

Savage, S. B. and S. McKeown. 1983. Shear stresses developed during rapid shear of dense concentrations of large spherical particles between concentric rotating cylinders. *J. Fluid Mech.* 127, 453–472.

Savage, S. B. and M. Sayed. 1984. Stresses developed by dry cohesionless granular materials sheared in an annular shear cell. *J. Fluid Mech.* 142, 391–430.

Schröter, M., S. Nägle, C. Radin, and H. L. Swinney 2007. Phase transition in a static granular system. *Europhys. Lett.* 78, 44004.

Sela, N., I. Goldhirsch, and S. H. Noskowicz. 1996. Kinetic theoretical study of a simple sheared two-dimensional granular gas to Burnett order. *Phys. Fluids* 8, 2337–2353.

Silbert, L. E., D. Ertas, G. S. Grest, T. C. Halsey, D. Levine, and S. J. Plimpton. 2001. Granular flow down an inclined plane: Bagnold scaling and rheology. *Phys. Rev. E* 64, 51302.

Snook, I. K. and D. Henderson. 1978. Monte carlo study of a hard sphere fluid near a hard wall. *J. Chem. Phys.* 68, 2134–2139.

Torquato, S. 1995. Nearest-neighbor statistics for packings of hard spheres and disks. *Phys. Rev. E* 51, 3170–3182.

Walton, O. R. and R. L. Braun. 1986. Stress calculations for assemblies of inelastic spheres in uniform shear. *Acta Mech.* 63, 73–86.

Woodhouse, M. J. and A. J. Hogg. 2010. Rapid granular flows down inclined planar chutes. Part 2. Linear stability analysis of steady flow solutions. *J. Fluid Mech.* 652, 461–488.

Woodhouse, M. J., A. J. Hogg, and A. A. Sellar. 2010. Rapid granular flows down inclined planar chutes. Part 1. Steady flows, multiple solutions and existence domains. *J. Fluid Mech.* 652 , 427–460.

Yoon, D. K. and J. T. Jenkins. 2005. Kinetic theory for identical, frictional, nearly elastic disks. *Phys. Fluids* 17, 083301.

CHAPTER 5

Statistical Mechanics of Dry Granular Materials: Statics and Slow Dynamics

Bulbul Chakraborty

CONTENTS

5.1   Introduction

This chapter introduces tools from statistical mechanics that are useful for analyzing the behavior of static and slowly driven granular media. (For fast dynamics, refer to the previous chapter on Kinetic Theory by Jenkins.) These tools encompass techniques used to predict emergent properties from microscopic laws, which are the analogs of calculations in equilibrium statistical mechanics based on the concept of statistical ensembles and stochastic dynamics. Included, for example, are the Edwards approach to static granular media, and coarse-grained models of granular rheology. The discussion will focus on a set of generic tools that can be honed to study particular problems.

This chapter is organized as follows. Section 5.2 is devoted to a discussion of the statistical ensemble framework applied to static granular media, and Section 5.3 discusses statistical approaches to modeling the rheology of slow, dense flows.

5.2 Statistical Tools for Static Packings

For static granular packings, which are end states of some protocol, statistical techniques are needed for analyzing properties such as stress transmission and porosity. I first introduce the concept and utility of statistical ensembles in the context of equilibrium statistical mechanics and then generalize the approach to static granular assemblies.

5.2.1 Statistical Ensembles in Equilibrium Statistical Mechanics

A statistical ensemble is a set of microscopic states that are consistent with a given set of fixed macroscopic variables. For example, the microcanonical ensemble of equilibrium statistical mechanics refers to all microscopic states of a system at a given energy. A remarkable feature of systems in thermal equilibrium is that specification of an ensemble determines the probability distribution of microscopic states. Once the probability distribution is known, in principle, one can calculate all statistical properties through the evaluation of the partition or generating function [20]. In practice, exact analytic calculation of the partition function is possible only for a handful of models. There are, however, numerous numerical techniques for calculating statistical averages defined in terms of the partition function. The Metropolis Monte Carlo method is one of the most widely used numerical approaches [20] for calculating equilibrium properties. The three common ensembles correspond to macroscopic observables that are most amenable to being controlled in experiments and simulations. I will provide a brief discussion of each of these later with the intent of highlighting the fundamental assumptions to set the stage for discussing differences between thermal and granular systems.

There are two ways of constructing the probability distributions, P_v of microscopic states in different ensembles. One approach is based on conservative, Hamiltonian dynamics and the fundamental postulate (in addition to energy conservation) that, in an isolated system, all states that have the same energy are equally likely. This *a priori* uniform measure on the canonical variables is justified by the Liouville theorem in classical systems [41]. Starting from these postulates, one can construct P_v for the different ensembles by positing one additional assumption: that for two systems in contact with each other, the effects of boundaries can be neglected [20]. This assumption implies that the density of states of the combined system is $\Omega_{12}(E = E_1 + E_2) = \Omega_1(E_1)\Omega_2(E_2)$. The entropy is defined as $S(E) = \ln \Omega(E)$.

In an alternative approach, P_v is determined from a variational principle that is based on the Gibbs formulation of the entropy: $S[P_v]$ [20]. The entropy is a functional of the probability distribution, which for particles is a function of the positions and momenta. The postulate of maximum entropy then translates to a maximization of S in the space of functions, P_ν, subject to macroscopic constraints. In the Jaynesian formalism [37], these constraints are whatever macroscopic quantities are held fixed in experiments. In the standard formulation of equilibrium statistical mechanics, the constraints relate to energy, volume, and number of particles. Whether these are constrained to fixed values or allowed to fluctuate leads to different ensembles and different probability distributions.

5.2.1.1 Microcanonical Ensemble

The microcanonical ensemble describes an "isolated" system. The dynamical approach says all states for a given energy are equally likely and the energy is fixed in an isolated system. From conservation of probability, $P_\nu = 1/\Omega(E)$ if $E_\nu = E$ and is zero otherwise. In the variational approach, one maximizes $S[P_\nu]$ subject to the constraints that $\sum_\nu P_\nu = 1$ and that $E_\nu = E$. The result is the same from both approaches. There is a microcanonical temperature, which is a response function and is an intensive quantity. The inverse temperature $\beta = \frac{\partial S}{\partial E}|_{V,N}$ measures how the entropy changes as the energy is varied.

5.2.1.2 Canonical Ensemble

The canonical ensemble is one in which the system is in contact with a heat bath. In the dynamical picture, one envisions that the system of interest is part of a much larger system that is isolated such that the total energy of the composite system is fixed. The energy of the subsystem (the system of interest) can fluctuate. The probability of a given microstate in this subsystem can be calculated by using energy conservation (of the total system) and $\Omega_{12}(E = E_1 + E_2) = \Omega_1(E_1)\Omega_2(E_2)$, where 1 and 2 refer to the subsystem and the bath. The result is the famous Boltzmann distribution: $P_\nu = e^{-\beta E_\nu}/Z$, where β is the temperature of the bath obtained by using the expression from the previous paragraph and evaluated at the energy of the total system (throughout it is assumed that E_ν is much smaller than the bath energy). The partition function, $Z = \sum_\nu e^{-\beta E_\nu}$, is the generating function of all correlations of fluctuations, and all response functions through the fluctuation dissipation relations [20].

In the variational approach, the entropy, $S[P_\nu]$, is maximized as in the microcanonical, but instead of the strict requirement that $E_\nu = E$, there is the weaker condition that the *average* energy: $\sum_\nu P_\nu E_\nu = \langle E \rangle$ is fixed. We again obtain the Boltzmann distribution, and the relationship between β and $\langle E \rangle$ is obtained from the constraint equation on E_ν.

5.2.1.3 Grand Canonical Ensemble

Another commonly used ensemble in statistical mechanics is one in which the energy and the number of particles are allowed to fluctuate. The distribution is now $P_\nu = \frac{e^{-\beta(E_\nu - \mu N_\nu)}}{Z_G}$, where μ is the chemical potential. The intensive quantity $\beta\mu$ describes how the density of states changes with the number of particles, and Z_G

is the grand canonical partition function which generates all correlation functions in this ensemble.

The statistical ensemble framework of equilibrium statistical mechanics gives us the tools to analyze experimental data and to make theoretical predictions. The concept of entropy, its maximization, and the ensuing definition of intensive quantities such as pressure and temperature reduces the complexity of a statistical system of 10^{23} particles to manageable proportions. In principle, the problem of predicting the collective behavior of equilibrium systems starting from microscopic interactions is solved. In practice, exact calculations are rarities. Computational tools such as the Monte Carlo Metropolis method can, however, fill in this void: a priori knowledge of the probability distribution of microstates is at the heart of the Metropolis algorithm.

Even if we cannot "solve" for equilibrium properties exactly, there are universal relations that apply to all equilibrium systems that impose powerful constraints on the behavior of equilibrium systems. The relationship between fluctuations and response follows immediately from the form of the Boltzmann distribution [51]. This is a remarkable feature of equilibrium statistical mechanics and its existence is often used as the proof that a system is in thermal equilibrium. What the fluctuation-dissipation relations say is that the response of a system to small external perturbations (in the linear regime) can be computed from a knowledge of the fluctuation spectrum in the absence of the external perturbation. For example, the specific heat, which describes how the average energy of a system changes when the temperature is varied, is simply related to the fluctuations of the energy at the original temperature. The existence of intensive quantities such as temperature and chemical potential that become equal when two systems that are in contact reach equilibrium and the existence of the relations between fluctuations and response are the distinctive features of equilibrium systems. In the rest of this section, we will explore which of these concepts can be generalized to granular materials and under what conditions. It is instructive to remember that it is precisely because they are liberated from the constraints of equilibrium, that out-of-equilibrium systems can exhibit a much richer array of phenomena.

5.2.2 Statistical Ensembles in Static Granular Assemblies

The ubiquitous presence of fluctuations and well-defined statistical distributions [19] observed in static granular systems suggests that statistical ensembles, derived from microscopic descriptions, could prove to be an important tool for predicting the emergent macroscopic properties. The spectrum of fluctuations exhibited by granular materials differs from those in thermal systems, an indicator of the fact that they are intrinsically out of equilibrium. Unlike systems that are driven out of equilibrium or glassy systems that fall out of equilibrium [36], systems of grains are too large to be sensitive to thermal fluctuations and are inherently out of equilibrium. Fluctuations can be induced by driving and are nonnegligible. For static assemblies of grains, the "fluctuations" we refer to are variations under repeated experiments and protocols such as tapping [15]. The focus of this chapter is on these static systems and their behavior under slow driving.

5.2.2.1 Nature of Ensemble: Non-Ergodic, Non Thermal

The objective of a statistical ensemble framework is to predict the probability of occurrence of a microscopic configuration, given a set of macroscopic constraints. As discussed in the previous section, this can be achieved for thermal systems and the framework leads to universal principles. The statistical ensemble approaches to granular media are seeking these types of universal guiding principles that can aid in our prediction of the collective properties of grains such as the response to external stresses. One crucial difference between granular and thermal ensembles is the lack of ergodicity in the former. In thermal systems, ergodicity relates time averages to ensemble averages [20]. There is a natural dynamics generated by a Hamiltonian and averages over these dynamical trajectories are equal to ensemble averages calculated with P_v.

The microscopic state of a static granular assembly is completely specified by the positions of the grains and the forces (including frictional forces) at inter-grain contacts. As in thermal systems, and other jammed (or glassy) systems, a granular packing can exist in many different microscopic states for a fixed set of macroscopic variables such as volume or applied stress. In addition, there is a microscopic indeterminacy arising from the Coulomb condition for static friction, which imposes an inequality constraint: $|f_T| \leq \mu |f_N|$ that the magnitude of the tangential force, $\mathbf{f_T}$, cannot exceed the static friction coefficient μ times the magnitude of the normal force $\mathbf{f_N}$. Thus, there is a range of tangential forces that are allowed for any *microscopic* grain configuration.

Dry granular materials living at $T = 0$ have some distinctive features that set them apart from other amorphous solids such as metallic glasses or colloidal structures. Since there is no attractive interaction between grains, and there is no thermal motion to generate a kinetic stress, the zero stress state of an assembly of dry grains is ill-defined [17]. Cohesion in these systems arises purely from applied stress. Since the states are defined at $T = 0$, each grain has to satisfy the constraints of force and torque balance, and each contact has to satisfy the Coulomb condition. Such states have been named "blocked" states by Edwards [50].

Blocked states, by definition, are not ergodic. These configurations of grains are end points of some dynamical protocol such as shaking, pouring, or slow shearing of a collection of grains. The granular ensemble is thus closer to the ensemble of metastable states of a glassy system [15]. Ensemble averages are explicitly not equal to time averages in these systems. The ensemble is generated by repeated "experiments" with the macroscopic variables held fixed and the relevant probability is that of the occurrence of a microscopic grain configuration under this type of protocol. Quoting from [15], "The natural question is then: with which statistical weight the different equilibrium (blocked) states appear in a given experiment? To what extent are these weights dependent on the dynamics that leads to the blocked states? Is the ensemble of native (as-built) packings identical to the ensemble of packings obtained under tapping?"

5.2.2.2 Edward's Ensemble

The first theoretical approach to address this question was that of the Edwards ensemble, which asserts that the total volume (V) of a mechanically stable grain packing plays the role of energy and that there is a temperature-like quantity called the compactivity [25–27,50]. The mechanically stable grain packings of infinitely rigid objects have been termed blocked states [27] and are related to strictly jammed states [24]. The basic hypothesis underlying the Edwards ensemble is the analog of the microcanonical [20] hypothesis and states that all blocked states with the same total volume V are equally likely. This hypothesis has been tested in numerical simulations, in experiments, and in some exactly solvable models [2,3,21,22,42,45]. The consensus seems to be that the hypothesis is not universally valid, but does hold under some conditions. The Edwards hypothesis involves the definition of the Edwards entropy: $S(V) = \ln \Omega(V)$, where $\Omega(V)$ is the density of blocked states, that is, the number of blocked states between V and $V + \delta V$. Evaluating this function has been a challenging problem, although progress has been made recently [12]. The compactivity is defined as $1/X = \partial S/\partial V$. The parallel with the microcanonical ensemble is complete, with energy replaced by volume. There is, however, no analog of the Liouville theorem that provides a scaffold for this equiprobability hypothesis in blocked states.

In a more recent development, it has been shown that the "equally likely" hypothesis of Edwards is not essential for the definition of a temperature-like quantity and the much weaker condition of factorizability of distributions (the density of states factorization discussed in the previous section, e.g.) is sufficient [5–7]. The necessary conditions for being able to define a temperature-like variable and a statistical ensemble based on this variable are (I) the existence of a physical quantity that is conserved by the natural dynamics of the system (in thermal systems energy is conserved but in dissipative granular media, it is not) and (II) the frequency of finding different states with the same value of the conserved quantity is factorizable: $\omega_{\nu_1 + \nu_2} = \omega_{\nu_1} \omega_{\nu_2}$. The latter condition implies that if one creates a configuration by bringing together two configurations ν_1 and ν_2, then the frequency of occurrence of this joint configuration is a product of the frequencies ω_ν of the individual configurations. This is a much weaker condition than the equiprobable hypothesis that ω_ν is independent of the microstate, ν. As long as these conditions are satisfied, one can define ensembles based on the dynamically conserved quantity and its analogous intensive variables. These intensive variables are the derivatives of a generalization of $\ln \Omega(E)$. Using the notation U for the conserved variable, we define $G(U) = \sum_\nu \omega_\nu \delta(U_\nu - U)$, which reduces to the density of states under the equiprobable hypothesis. The intensive quantities are then defined as the derivatives of $G(U)$.

To apply these general ideas to the Edwards ensemble based on volume, we must first restrict ourselves to dynamics that conserves volume. A relevant question for an ensemble created by tapping is whether the system can sponta-neously change its volume, a distinctly different question from externally fixing the volume. Recall that when a system is thermally isolated, the energy is not con-strained by definition. Rather, the underlying conservative dynamics never takes the system out of its initial energy shell. If the dynamics to which we subject a granular assembly conserves volume, then we should be able to define com-pactivity by using the general definition given in the previous paragraph. (Note, even if the equiprobable hypothesis may still not be satisfied.) At the end of this section, we will discuss tests of the various granular ensembles using experimen-tal and simulation results. Another important note is that, unlike $\Omega(E)$, $G(V)$ is not determined by any Hamiltonian. A pragmatic approach is to assert that dif-ferent protocols lead to different $G(V)$. The universal principles of an ensemble approach would then still be valid, but each protocol would be like choosing a new Hamiltonian. Even this seemingly wide-open scenario results in concepts that are valid independent of protocol. The state variables and equations of state will vary from protocol to protocol. Obviously, the ensemble approach would be most use-ful if there are large classes of protocols that lead to the same $G(V)$.

5.2.2.3   Generalized Ensemble Based on Stress

In the context of mechanically stable packings of grains (soft, deformable, or rigid), a conserved quantity that has been identified is the force moment tensor [1,11,35], which is related to the Cauchy stress tensor

$$\hat{\sigma} = \left(\frac{1}{V}\right) \sum_{ij} \vec{r}_{ij} \vec{F}_{ij}. \tag{5.1}$$

The summation in Equation 5.1 is over all contacts $\{ij\}$ in an assembly of grains, occupying a volume V, with contact vectors $\vec{r}_{ij}$ and contact forces $\vec{F}_{ij}$ (for grains with friction $\vec{F}_{ij}$ does not lie along the direction of $\vec{r}_{ij}$).

Unlike volume, which is only globally conserved, the conservation of the force moment tensor follows from *local* constrains of force and torque balance. The microscopic force moment tensor for a grain is given by

$$\hat{\kappa}_i = \sum_{j} \vec{r}_{ij} \vec{F}_{ij}, \tag{5.2}$$

where the sum is over all grains j that contact grain i. For grains in mechanical equilibrium, it can be shown, using a generalized Stoke's theorem [25,33,34], that the *total* force moment tensor

$$\hat{\Sigma} = \sum_{i} \kappa_i$$

becomes a boundary integral for systems with open boundaries and is a topologi-cally conserved quantity for systems with periodic boundary conditions [8,35]. In two dimensions, this property can be explicitly demonstrated by introducing a set of auxiliary variables [1] which are the analog of the vector potential in electromagnetism. The connection to electromagnetism is natural if one remembers that for mechanically stable packings, the divergence of the stress tensor is zero, just as the divergence of the magnetic field is zero in electromagnetism.

To change the components of the force moment tensor, while maintaining force and torque balance on every grain, one must change the forces along a string of grains that spans the system [18,47]. Local perturbations, such as tapping, can and will change forces and torques on individual grains but cannot change the sum of the grain-level force moment tensors. The only way to do this is to *simultaneously* change the forces on grains along a line that spans the system. This situation is closer to the dynamical conservation of global energy in conservative systems than the purely global conservation of volume.

The phase space of all mechanically stable packings of grains can, therefore, be grouped into sectors characterized by the value of $\hat{\Sigma}$. These sectors are not connected by any local, tapping type, dynamics. The dynamics can change forces and positions (including breaking of contacts). The only restriction is that the rearrangement be local; under this type of dynamics, $\hat{\Sigma}$ is a conserved quantity that meets criterion (I) of the previous section on the Edwards ensemble. If we now assume criterion (II), then a statistical ensemble can be constructed to describe the end states of processes such as tapping grains. The role of energy is played by $\hat{\Sigma}$ which is an extensive quantity (scales with system size) and the analog of temperature is a tensorial quantity, $\hat{\alpha}$, which has been named angoricity by Edwards [25]. This tensor is defined by

$$\hat{\alpha}(\hat{\Sigma}) = \partial \ln(G(\hat{\Sigma}))/\partial \hat{\Sigma}|_{\hat{\Sigma}} \tag{5.3}$$

where

$$G(\hat{\Sigma}) = \sum_{\nu} \omega_{\nu} \delta(\hat{\Sigma}_{\nu} - \hat{\Sigma})$$

and the sum is over all mechanically stable packings ν. The constraints on the microstates are that (1) forces and torques are balanced on every grain, (2) the Coulomb condition for static friction is satisfied, and (3) all forces are positive.

The stress or generalized Edwards ensemble has much in common with statistical frameworks developed for glassy systems [15]. Collective properties in both are determined by the complexity of the landscape of metastable states. Complexity in spin glasses is measured by the logarithm of the number of local free energy minima that have a given free energy density [15,48]. An analogous definition applies to the potential energy landscape of supercooled liquids [56,64]. For granular systems, complexity is measured by the logarithm of the number of mechanically stable states with a given force moment tensor, which is related to the entropy $S(\hat{\Sigma})$. As illustrated by the example of spatial correlations of stress, discussed earlier, stress fluctuations are controlled by the configurational entropy, $S(\hat{\Sigma})$. Unlike thermal ensembles, the glassy and granular systems have broken ergodicity with barriers separating metastable states. In meanfield spin glass models, these barriers diverge with system size. Within the ensemble of mechanically stable states, the barriers between sectors with different $\hat{\Sigma}$ diverge with systems size because of the constraints of local force and torque balance.

Defining the ensemble for static grains is the conceptually difficult part. Calculating the partition function is technically difficult because of the constraints that one still needs to take into account. Some progress has been made in this regard [8,13,23,34] but this field is still in its infancy. The remainder of this section is devoted to short descriptions of tests of the existence of angoricity and compactivity in granular assemblies and an example of application of the equiprobability ansatz to construct a model for force transmission.

5.2.2.4 Tests of the Statistical Ensemble Framework

The ability to define an ensemble does not necessarily mean that real granular packings conform to this ensemble. The assumption of factorizabilty could break down or the frequency of occurrence of a packing, ω_ν could be history dependent, making the temperature or the ensemble concept not very useful. It is, therefore, essential to check the predictions of this ensemble against experiments and simulations.

5.2.2.5 Volume Ensemble

The existence of compactivity and the validity of the Edwards volume ensemble [50] without the restriction to the equiprobable hypothesis can be tested by measuring the analog of the canonical thermal distribution. According to Edwards, the probability to access the state ν with volume v_ν in a subsystem that is part of a larger system with volume V_{tot} is

$$P_\nu = \exp\left(\frac{-v_\nu}{X}\right), \tag{5.4}$$

where X is the *compactivity*, $\frac{1}{X} = \frac{\partial(\ln G(V_{tot}))}{\partial V_{tot}}$. If Equation 5.4 applies, then the probability of finding a packing with volume V at a compactivity X is given by the canonical distribution,

$$P(V) = \frac{1}{Z(X)}G(V)\exp\left(\frac{-V}{X}\right), \tag{5.5}$$

the analog of the Boltzmann distribution. In this equation, $G(V)$ is the generalized density of states and $Z(X)$ is the partition function or generating function that generates all correlations functions at a given X. It should be emphasized that the partition function involves a sum over mechanically stable states *only*.

In a recent work [46], experiments and simulations have been used to directly test the form of the volume distribution, Equation 5.5. The results show that the distribution can indeed be written in this very special form, which is a nontrivial result that supports the entropic principles behind the Edwards hypothesis. As the authors point out, however, testing Equation 5.5 does not test the equiprobable hypothesis. The analog of density of states, $G(V)$ involves the microscopic ω_ν,

which do not have to be equal [7,43], and therefore, the microcanonical probability distribution, $P_v(V) = \omega_v/G(V)$, is no longer uniform. The extended definition introduces the possibility of a protocol dependence of $G(V)$ since the weights
associated with the microscopic states, ω_v could depend on how the packing is
prepared. The utility of the Edwards framework depends on how sensitively $G(V)$
depends on history. If the sensitivity is large, then the utility is limited since the
full history will be needed for any calculation and the framework will not have
much predictive power.

5.2.2.6 Stress Ensemble: Simulations

When a body reaches thermal equilibrium, the temperature must be everywhere
the same. In the same way, the stress ensemble framework implies that the
angoricity of a packing must be the same everywhere inside a packing. In addition,
one should demand that the angoricity is independent of the size of the regions
into which the packing is subdivided, as long as the size is large compared to a
grain size, and that there should be a unique relationship between the angoricity
$\hat{\alpha}$ and the force moment tensor $\hat{\Sigma}$ of the packing, providing an equation of state
analogous to temperature energy equations of state.

The equality of $\hat{\alpha}$ inside a packing has been tested in isotropically compressed
packings of frictionless grains, generated by simulations of systems with Hertzian
and harmonic contact forces [34]. The test was designed to (1) check that the distri-
bution of stresses followed the Boltzmann-like distribution: $P_\alpha(\Gamma) = \frac{G(\Gamma)}{Z(\alpha)} \exp(-\alpha\Gamma)$,
where the scalar α is the component of the inverse angoricity that is conjugate to
$\Gamma = Tr\hat{\Sigma}$, (2) measure the α of packings, and (3) measure the effective density of
states, $G(\Gamma)$.

Results were obtained by measuring the distribution of the *local* values of Γ_m
in an m grain cluster obtained from simulations of frictionless grains [34], and it
was demonstrated that α is approximately independent of the number m of grains
included in the subsystem and that there is a linear relation: $\alpha \simeq \frac{2N}{\Gamma}$, where N is
the total number of grains in the packing.

5.2.2.7 Stress Ensemble: Experiments

Until very recently, there had been no experiments to test the underlying framework of the stress ensemble. Nontrivial experimental validation has, however, now
emerged from experiments performed on 2D arrays of photoelastic disks that have
been jammed under compression [54] or shear [10].

In the first set of experiments, repeated compression is used to generate a large
number of independent configurations, each of which are in mechanical equilibrium. Each configuration has a smaller subsystem with a different inter-particle
friction coefficient than the "bath." The experiment tests the validity of both the
Boltzmann-like distribution characterized by the compactivity and the one characterized by angoricity. The testing adopts the same approach as described in the
previous subsection on testing the volume ensemble, that is, testing the overlap of
distributions. The experiments demonstrate that the Boltzmann–like distributions

do provide a good description of those observed in the experiments. In addition, the experiments test whether the compactivity and angoricity of the subsystem and the bath are equal. This is a nontrivial test of the idea of statistical ensembles for granular media. As shown in Figure 5.1, the experiments show an interesting result: the compactivity of the bath and systems are not equal but the angoricity does equalize. Presumably the local conservation principle underlying the stress ensemble makes it more robust than the volume ensemble, where volume

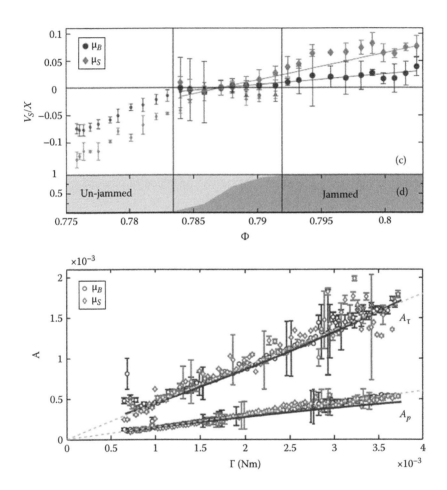

Figure 5.1   Compactivity and angoricity equations of states. (Adapted from Puckett, J.G. et al. 2012.) The top figures hows the comp activity X as a function of the packing fraction ϕ for the particles in the bath with friction coefficient μ_B, and particles in the system, which have a friction coefficient μ_S. The packing fraction range includes jammed (solids) and unjammed states. The comp activity should only be defined for the jammed states. The bottom panel shows two components of the angoricity, A_p and A_τ corresponding to the pressure and shear components of the stress tensor. The angoricities are shown as a function of the overall compression, Γ, and indicates a linear equation of state, as was observed in the simulations. (From Henkes, S. et al., *Phys. Rev. Lett.*, 99, 038002, 2007.)

conservation is only a global property. The reason for the differences is, however, not well understood.

In the second set of experiments, jammed states are created through a quasi-static shearing process [8]. (For more on experimental shear-jammed states, see Chapter 7 by Behringer.) Unlike the compression protocol, this is a strictly nonequilibrium process for creating a solid and the full tensorial form of the stress ensemble framework is invoked. The applicability of the stress ensemble in shear-jammed (SJ) states was tested by computing the force moment tensor, $\hat{\Sigma}_m$ of a m grain subregion of a jammed configuration. The stress ensemble predicts the following form for the distribution:

$$P(\hat{\Sigma}_m) = \frac{1}{Z(\hat{\alpha})} \exp\left[G_m(\hat{\Sigma}_m)\right] \exp\left(-\hat{\alpha} : \hat{\Sigma}_m\right). \tag{5.6}$$

In order to apply Equation 5.6 to the experimentally generated SJ states, we need a measure of $\hat{\alpha}$. Unfortunately, there is no "thermometer" for directly measuring angoricity and we must base its measurement on postulated or measured equations of state relating $\hat{\alpha}$ to $\hat{\Sigma}$. In isotropically compressed states, the equation of state deduced from both experiments [54] and simulations [35] shows a linear relationship between the angoricity and the trace of the force moment tensor. A straightforward generalization of this equation of state was postulated for the equation of state for SJ states

$$\hat{\alpha} \propto M\left(\hat{\Sigma}_M\right)^{-1} = \frac{M}{\det\left(\hat{\Sigma}_M\right)} \begin{pmatrix} \Sigma_{22,M} & -\Sigma_{12,M} \\ -\Sigma_{12,M} & \Sigma_{11,M} \end{pmatrix}. \tag{5.7}$$

Using this equation of state, each m grain SJ state can be labelled by an angoricity tensor, as the force moment tensor is measured in the experiments.

Equation 5.6 predicts the distribution of stress in a subregion containing m grains at the angoricity given by $\hat{\alpha}$. As has been pointed out earlier [46], this Boltzmann-like distribution has a very special form. It has a term that depends purely on $\hat{\Sigma}$, a term that depends purely on $\hat{\alpha}$, and the combination of these two variables appears only in the exponential. The stress ensemble would, therefore, predict that multiplying the distribution by $e^{\hat{\alpha}:\hat{\Sigma}}$ and taking its logarithm would yield a function that depends on $\hat{\Sigma}$ and distributions corresponding to different $\hat{\alpha}$s would differ only by an additive constant, or

$$\log\left[P(\hat{\Sigma}_m)e^{\hat{\alpha}:\hat{\Sigma}_m}\right] = S_m\left(\hat{\Sigma}_m\right) - \log Z(\hat{\alpha}) \tag{5.8}$$

We directly test these predictions as follows. All SJ states, regardless of the value of the shear strain or the packing fraction ϕ, are categorized by their global force moment tensor $\hat{\Sigma}_M$. Through the equation of state (Equation 5.7), the value of the angoricity tensor is determined. Then for each SJ state at a particular $\hat{\alpha}$, random contiguous clusters containing m grains (Figure 5.2a) are chosen to form an ensemble of subregions with different values of $\hat{\Sigma}_m$, giving a probability distribution $P_{\hat{\alpha}}(\hat{\Sigma}_m)$. To simplify the analysis, the reduced distributions for

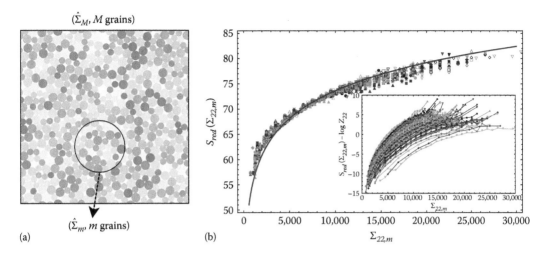

$(\hat{\Sigma}_M, M \text{ grains})$

$(\hat{\Sigma}_m, m \text{ grains})$

(a)   (b)

Figure 5.2   (a) Schematic demonstration of a subregion of size m and $\hat{\Sigma}_m$ inside a packing with $(M, \hat{\Sigma}_M)$. Here, we define subregions as any contiguous cluster of m grains roughly circular in shape. (b) Equation 5.6 predicts that when the distribution of local force moment tensor is multiplied by $e^{\hat{\alpha}:\hat{\Sigma}}$ (shown in inset for $m=15$ and the 22–component of the local force moment tensor in SJ states), its logarithm should be related to the entropy of microstates G_m, and for different $\hat{\alpha}$'s this function should only differ by an additive constant. The data can be collapsed onto the universal curve in the main figure. The functional form of the collapsed data also shows good agreement with the postulated form of entropy, given by the solid line, which is based on the ideal gas model. (From Bi, D. et al., *Europhys. Lett., 203.*)

each component of the local force moment tensor were considered instead of the multidimensional distribution function.

Figure 5.2b shows that the data collapse implied by Equation 5.8 works remarkably well for a subregion containing 15 grains. Similar data collapse was observed for the other components of the force moment tensor and for m as small as 4. For smaller m, there is more of a spread in the data; however, there is remarkably fast convergence as a function of m to a universal functional form, which can be interpreted as the thermodynamic entropy $G(\Sigma_{kl})$ for $kl = 11, 22, 12$. The solid line in Figure 5.2b is the functional form deduced from the assumed equation of state and the definition of of $\hat{\alpha}$ (Equation 5.3)

$$G_M(\hat{\Sigma}_M) = A\, M \log\left[\det \hat{\Sigma}_M\right]. \tag{5.9}$$

Here, $A = 0.5$ is a constant of proportionality that was determined from the best fit to the collapsed data. A few features of Figure 5.2b are of note: there is a small but systematic difference between the entropy deduced from the equation of state, and the form of the collapsed data, especially at small values of Σ_{kl}. In spite of these small differences, the fact that a data collapse exists demonstrates that the Boltzmann-like distribution defined by an angoricity tensor (Equation 5.6) works remarkably well for SJ states.

The stress ensemble analysis of the SJ states also tested whether the ideal gas–like equation of state captures detailed features of the stress distributions. The results showed that, although the ideal gas model captures the mean and overall shape of the experimental distributions, theory and experiment differed significantly in the width of the distributions [10]. Direct measurement of the variance of the stress for m grain subregions showed that the variance scaled as $m^{1.5}$ rather than m, which would be expected if the stress correlations did not extend beyond the subregion. This observation indicates the presence of long-range stress correlations in shear-jammed states. The variance is thus much more sensitive to correlations than the equation of state. This is a well-known feature of thermal equilibrium distributions [40].

Fluctuation-response relations relate the variance to two-point correlations. It can be shown [40], however, that two-point correlations enter calculations of entropy (or free energy) through logarithmic corrections. It is, therefore, reasonable to expect that the entropy and the equation of state are much less sensitive to the existence of correlations than the variance of the force moment tensor. The emergence of the same feature in the granular systems is reassuring since fluctuation-response relations are a direct consequence of a Boltzmann-like distribution. The relationship between variance and two-point correlations has now been directly verified in SJ states [55]. This test of the fluctuation relations is a bit complicated in granular media since the correlations have to be measured only between force-bearing grains, and in just-jammed systems, there can be complicated paths, weaving through non-force-bearing grains, that connect force-bearing grains.

To conclude, recent research indicates that it is possible to construct a robust statistical ensemble framework for static granular media. The ensemble includes all grain packings that are in mechanical equilibrium. The ensemble is generated by either repeatedly generating such configurations or by shearing quasi statically, allowing the systems enough time to relax to mechanical equilibrium. The ensemble is, therefore, very different from a thermal ensemble and there is no ergodicity and there is no time average analog of ensemble average. Within this ensemble, however, intensive quantities exist that are equalized when two grain assemblies that are brought in contact with each other reach mechanical equilibrium. Since a Boltzmann-like distribution exists, there is an analog of fluctuation dissipation relations and response is directly related to fluctuations. The experiments on photoelastic disks [54] test this relationship in measuring compactivity and angoricity.

A straightforward, important application of the statistical ensemble approach is to construct phenomenological equations of state: the analogs of $PV = NRT$ or the van der Waals equations for ideal gases and gas-liquid transitions, respectively. As in thermal systems, equations of state can be used to predict the macroscopic behavior of a system that can have different microsopic interactions. Equations of states capture the relationships between macroscopic variable that emerge from the collective behavior of the microscopic objects. The angoricity-stress equations of state obtained from experiments [54], for example, predict the behavior of shear stress with increasing pressure. Without statistical ideas, the only way to obtain such relations would be to solve for the full dynamics of approach to mechanically stable configurations. The advantage of knowing the a priori distribution of microscopic states, which is what a statistical ensemble give us, is not to be underestimated.

5.2.3 Force Network Ensemble: A Model Based on the Edwards Equiprobability Hypothesis

The Edwards picture was originally based on the assumption of completely rigid particles. However, all physical particles have finite stiffness, and in the usual case, interact via frictional contact forces. The conditions of finite vs. infinite stiffness and frictionless vs. frictional interactions has some important consequences for granular packings. Relaxing the infinite stiffness constraint introduces couplings between the positions of grains and the contact forces. If the grains are rigid enough that small changes in compression lead to large changes in force, then under weak tapping, it is reasonable to assume that the configurations sampled differ only in their contact force, not in their real space geometry. This assumption forms the basis of a widely used model for force transmission in granular solids, the force network ensemble. This is a statistical ensemble that adopts the Edwards equiprobability hypothesis for blocked states. All force, and torque-balanced configurations corresponding to a given geometry are, therefore, sampled with equal weight.

A mechanically stable packing of frictionless, deformable, spherical particles must satisfy the equations of force balance. In addition, there is the force law relating the positions of the particles to the forces [57,66]. For grains interacting through purely repulsive, short-range forces, there are dN equations of force balance for N grains in a space of dimensionality d

$$\text{Force - balance} \quad dN \text{ eqs}: \quad \sum_j F_{ij} \frac{\mathbf{r}_{ij}}{|\mathbf{r}_{ij}|} = 0 \tag{5.10}$$

$$\text{Force - law} \quad \langle z \rangle N/2 \text{ eqs}: \quad F_{ij} = f(\mathbf{r}_{ij}) \tag{5.11}$$

Here,

$\langle z \rangle$ is the average number of contacts per grain

F_{ij} is the magnitude of the contact force between grains i and j, the angles of the contacts being fixed by the geometry

$f(\mathbf{r}_{ij})$ is a function specifying the intergrain force law

For a given geometry (*i.e.*, fixed $\frac{\mathbf{r}_{ij}}{|\mathbf{r}_{ij}|}$), the equations of force balance involve $\langle z \rangle N/2$ unknowns. The number of force balance equations cannot be greater than the number of unknowns, otherwise the forces are overdetermined, and, therefore, $\langle z \rangle N/2 \geq dN$.

For rigid, nondeformable grains, the force law becomes a constraint on the positions of the grains [66]. There is one constraint for each contact, which leads to $\langle z \rangle N/2$ equations for the dN positions, the unknowns. Since the number of constraint equations must be larger than the number of unknowns, $\langle z \rangle N/2 \leq dN$. The force law and the force balance constraints can, therefore, be satisfied only if $\langle z \rangle = z_{iso} = 2d$. This enumeration argument applies only to disordered packings, since for ordered, crystalline packings, angles of the lines connecting the grains are not independent, and therefore, not all the constraint equations are linearly independent. Considering such disordered packings, the enumeration argument leads to the conclusion that packings of rigid, isotropic, frictionless particles are *isostatic*, with $\langle z \rangle = z_{iso}$.

For deformable particles, mechanically stable packings can exist for $\langle z \rangle \geq z_{iso}$. If the particles are very stiff, the magnitude of the forces change greatly with small changes in the separation between grains. There is, therefore, an effective separation of scales [57,62] in Equations 5.10 and 5.11, and one can consider the ensemble of forces that satisfy force balance for a fixed geometry of the packing, that is, a packing with fixed grain positions. The properties of such force-ensembles have been studied [57–62,67], and it has been shown that $P(F)$, the PDF of contact forces, evolves to an exponential form as the particles are made increasingly rigid [57,62]. Work on sheared and isotropically compressed packings has shown that there are more extended spatial correlations of the forces in the force ensembles for sheared packings [65]. The force ensemble approach captures many features of packing near Point J [44] in the phase diagram of jammed systems (see Chapter 7).

5.3 Statistical Tools for Slowly Driven Granular Media

Granular dynamics can be broadly divided into two categories: fast flows where the kinetic energy of the grains is a significant source of fluctuations and slow, dense flows where the fluctuations are largely a result of plastic events disrupting the elastic response of granular solids. The basic question of whether there is any regime of elastic response has not yet been answered definitively. From the basic definition of a solid, however, we know that they can at least sustain infinitesimal load without suffering complete rearrangement of the grain configuration.

The discussion of fast flow belongs in the realm of kinetic theory and is the subject of Chapter 4 by Jim Jenkins. In this chapter, I discuss some techniques that are applicable to the rheology of granular flows close to jamming. The stochasticity in these flows arises because of the plastic events and various theories have been developed to incorporate this statistical description into the continuum framework of rheology. These approaches share many common features with the theory of glassy rheology [15]. The basic theoretical framework is a time evolution equation for the probability distribution of stresses. The aim of all these models is to describe the rheology of yield stress fluids and they differ in the way yielding events and interactions between yielding events are modeled.

5.3.1   Trap Models and Rheology

A model of granular rheology must account for the disordered nature of the packings and metastability [15]. These are common feature of all soft glassy materials,
and a framework that has been applied is that of soft glassy rheology (SGR) [63].
SGR builds on the trap model of glassy dynamics in thermal systems [49], which
describes activated dynamics in a landscape with a broad distribution of barrier
heights. As discussed earlier, the existence of a large number of different microscopic metastable states that are macroscopically equivalent puts granular materials into a wider class of previously well-studied systems, including gels, glasses,
colloids, emulsions, polymers, and foams. A minimal model that encapsulates
these features and leads to a glass transition is the "trap model" [49]. In this model,
one considers a system made of independent subsystems of a certain size ξ, where
each subsystem acts self-coherently and independent of the others. Their dynamics involve hopping between different metastable states aided by some kind of
fluctuation. This idea of a random walk in a rugged landscape has its roots in the
context of glasses [16]. In what follows, we discuss the trap model and its generalization, which is the framework of soft glassy rheology [63].

Monthus and Bouchaud [49] constructed a one-element model for glasses. In
this model, there exists an energy landscape of traps with various depths E. An
element hops between traps when activated; the fluctuations are assumed to be
thermal. At a temperature $k_B T \equiv 1/\beta$, the probability of being in a trap with depth
E at time t evolves according to

$$\frac{\partial}{\partial t} P(E, t) = -\omega_0 e^{-\beta E} P(E, t) + \omega(t)\rho(E) \tag{5.12}$$

where

$$\omega(t) = \omega_0 \left\langle e^{-\beta E} \right\rangle_{P(E,t)} = \omega_0 \int dE P(E, t) e^{-\beta E} \tag{5.13}$$

The first term on the right-hand side (rhs) of Equation 5.12 is the rate of hopping
out of a trap, where $e^{-\beta E}$ is the activation factor and ω_0 a frequency constant. The
model assumes that choosing a new trap is independent of history, so that a new
trap is chosen from a distribution of trap depths $\rho(E)$ that reflects the underlying disorder in the glassy models. Equation 5.13 gives the average hopping rate.
This rate, multiplied by the distribution of trap depth, Equation 5.12 gives the
probability rate of choosing a new trap.

The existence of a glass transition in this model can be demonstrated as follows: assume the distribution function has a simple exponential tail such as

$$\rho(E) = e^{-\beta_0 E}, \tag{5.14}$$

where β_0 is a fixed parameter describing the disorder in the inherent energy
landscape. The physical justification for having an exponentially decaying $\rho(E)$
is borrowed from systems with quenched random disorder, such as spin glasses,

which use extreme value statistics [49]. Then we can easily solve for the steady-state ($\frac{\partial}{\partial t}P(E,t) = 0$) solution

$$P_{eq}(E) \propto e^{\beta E}\rho(E) = e^{(\beta_0 - \beta)E}. \tag{5.15}$$

Immediately, we see that Equation 5.15 is non-normalizable for $\beta > \beta_0$ or $T < T_0$. This shows that below a temperature T_0, the system is out of equilibrium; it is non-ergodic and ages by evolving into deeper and deeper traps. Therefore, we call $T = T_0$ a point of glass transition [49].

It should be noted that dynamical equations of the trap model do not lead to a Boltzmann distribution at T since the equations do not obey detailed balance and, specifically, the traps are sampled from a quenched distribution. The escape rate from a trap is, however, determined by the Kramers process [29].

The SGR model [63] incorporated an important extra degree of freedom: the strain. As in the trap model, SGR assumes that a macroscopic soft glassy material can be subdivided into a large number of mesoscopic regions, each having a linear size ξ. With each mesoscopic region one can then associate an energy E and a strain variable l, which both evolve with time. The number of mesoscopic regions must be large enough so that ensemble averaging can be performed to yield macroscopic properties. Similar to Bouchaud's trap model, the SGR model assumes the meso-scopic regions "live" in an energy landscape. This is a mean-field energy landscape in that it is not characterized by a metric and is defined only by the distribution $\rho(E)$. The new strain variable describes local elastic deformation of the mesoscopic regions, so that l contributes quadratically to the energy E. Since a strain variable was added to the SGR, the model can now describe material under imposed shear strain, also in a mean-field spirit, with all mesoscopic regions responding uniformly to externally imposed shear strain. Different from Bouchaud's trap model, thermal fluctuations are unimportant in soft materials ($k_B T$ is too small to cause structural rearrangements). In the SGR model, it is the fluctuations in the elastic energy that facilitate structural rearrangements. This fluctuation is determined by a temperature-like quantity x called the "noise level." As a result, the escape rate from a trap becomes $e^{-E/x}$. The fluctuation temperature, x, is a mean-field way of representing the interactions between mesoscopic regions that arise from yielding or plastic events [14].

The trap model has also been applied to granular compaction by adapting the model to describe the time evolution of the probability distribution of free volumes [31]. Trap models capture disorder and intermittency within the simplest, nontrivial framework.

Given the discussion of angoricity in the previous section and increasing evidence that this intensive quantity is the relevant one for describing the equilibration of granular solids [54], it is tempting to adapt the SGR framework to slow granular flows by (1) considering probability distributions of stress rather than energy or free volume and (2) identifying the fluctuation temperature, x, with the angoricity. Such a framework has been constructed [9] and was used to analyze data from packings being sheared in a Couette geometry [30]. The angoricity-driven-activated mechanism provided an explanation for the observed logarithmic strengthening with shear rate [4]. Using the theoretical framework to analyze the temporal fluctuations of the stress results in a relationship between the angoricity, packing fraction, and shear rate [9]. This approach has not seen widespread use but promises to be an attractive starting point for analyzing experiments and connecting to continuum models.

The idea of stress fluctuations playing the role of temperature has been used previously to understand shear zones in granular chute flows [53]. The stress ensemble suggests a connection between the stress temperature in slow flows and angoricity, which needs further exploration.

5.3.2 Stress-Activated Failures and Non Local Rheology

Pouliquen and Gutfraind [62] proposed a model for the quasi-static regime in which the stress fluctuations play the role of a temperature and lead to activated yielding processes. In this picture, the shear zone is a result of noise-driven activation across the Coulomb yield threshold. Theoretical models have been developed [28,62] in which yielding and rearrangement act as a source of noise, activating dynamics elsewhere in the system. These rearrangement events are responsible for the average flow profile when coarse grained in time and space. This picture has been developed further, and a kinetic equation for the probability distribution of stresses in jammed materials has been proposed [14,52]. These approaches have led to the construction of continuum equations with nonlocal constitutive relations [39]. In this section, I will provide an introduction to the Fokker-Planck equation using the example of plastic failure in granular media [14].

There are intriguing connections between the stress-induced fluctuation temperature and angoricity, which is rigorously defined only for static granular media where grains are strictly in force balance. The definition of the latter is based on entropy, whereas the stress temperature in flows has been related to steady-state distributions of the stress [62]. Since entropy is after all related to distributions, it seems plausible that the stress temperature in slow flows and angoricity are related. Exploring this connection can lead to a better understanding of the dynamical origin of angoricity and the challenging issue of protocol dependence.

5.3.2.1 Fokker–Planck Equation Applied to Plastic Failure

As discussed in this chapter, the workhorse of equilibrium statistical mechanics is the statistical ensemble framework. I have discussed at length how this framework has been generalized to granular materials, which are intrinsically out of equilibrium. Nonequilibrium statistical mechanics is concerned with the time evolution of probability distributions. The workhorse in this realm is the framework of Markov processes: generalized random walks. The hallmark of a random walk is that there is no memory. In the simplest example of a walker in one dimension forced to take steps with a fixed length, the walker moves to the right or left with a given predetermined probability. After taking the step, it takes the next step with the same set of probabilities with no memory of where it came from. This is a stochastic process, and the only meaningful question to ask is how does the probability of finding a particle at position x and time t, $P(x,t)$, evolve with time. This concept can be generalized to arbitrary dimensions, and one can construct a probability distribution, $P(v,t)$, where v is a microscopic state of the type used in statistical ensembles. The equation describing the time evolution of $P(v,t)$ is the Master equation [38]

$$\frac{\partial P(v,t)}{\partial t} = -\sum_{v,v'} [W_{v,v'}P_v - W_{v',v}P_{v'}]. \qquad (5.16)$$

In this equation, $W_{v,v'}$ is the transition rate from the current state, v, to a new state v'. The rates do not depend on history. If the probability distribution asymptotically approaches a known equilibrium distribution, then the rates are constrained to fulfill that requirement. In a general case, the rates are defined by the microsopic process that one is trying to model. This framework is broadly applicable and is, therefore, the most widely used scheme for studying nonequilibrium systems. The Fokker–Planck equation is an approximation to a Master equation, which retains only the first two moments of the probability distribution, $P(v,t)$ [38]. In recent work, a Fokker–Planck equation has been constructed to describe plastic flow in disordered solids [14]. In Ref. [14], the authors construct a model based on the SGR philosophy but include elastic interactions between the regions undergoing plastic failure, as illustrated in Figure 5.3. The processes represented in the model (Figure 5.3) can be represented by a Master equation for the local probability distribution of stress (taken to be a scalar quantity here) in a box i: $P_i(\sigma,t)$. The nonlocal process of plastic events in one box, i, modifying stress in a nearby box, j, makes the Master equation nonlinear in the probability distribution, P. These equations are, therefore, much more difficult to solve than the SGR equations. To simplify these equations, the authors draw an analogy with equations that form the basis of kinetic theory: the theoretical framework that describes the evolution of probability distributions in equilibrium systems at low density. In kinetic theory, the *Boltzmann Stosszahlansatz*, or molecular chaos assumption, posits that colliding particles are not statistically correlated before the collision. In the context of the plastic failure model, the ansatz translates to saying that the

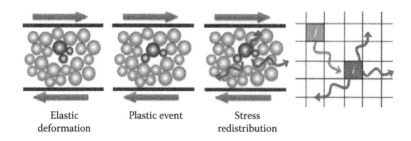

Elastic   Plastic event   Stress
deformation   redistribution

Figure 5.3   Schematic illustration of an SGR-type model of plastic deformation in amorphous media. Deformation occurs via elastic deformation, localized plastic events (in the regions represented by boxes), and nonlocal redistribution of the elastic stress, potentially triggering other plastic events. (From Bocquet, L. et al., *Phys. Rev. Lett.*, 103, 036001, 2009.)

joint probability of finding a stress σ in box i and stress σ' in box j can be written as a product: $P_i(\sigma, t)P_j(\sigma', t)$. The use of the molecular chaos assumption in this context provides a very nice example of how ideas borrowed from classical equilibrium statistical mechanics can aid in solving problems in granular physics. I refer the readers to the original paper [14] for a discussion of the additional approximations that have to be made to obtain a closed-form solution to the probability distribution of stresses in an amorphous material, in this particular model.

The discussion in this section and the previous one emphasizes the utility of stochastic models in elucidating the rheological properties of granular materials. The statistical arguments of Ref. [14] have been used recently to construct a continuum theory of dense flows [32], which captures all essential aspects of experimental observations in highly nontrivial geometries.

References

1. R. C. Ball and R. Blumenfeld. Stress field in granular systems: Loop forces and potential formulation. *Physical Review Letters*, 88:115505, 2002.

2. A. Barrat, J. Kurchan, V. Loreto, and M. Sellitto. Edwards' measures for powders and glasses. *Physical Review Letters*, 85(24):5034, 2000.

3. A. Barrat, J. Kurchan, V. Loreto, and M. Sellitto. Edwards' measures: A thermodynamic construction for dense granular media and glasses. *Physical Review E*, 6305(5), 2001.

4. R. P. Behringer, Dapeng Bi, B. Chakraborty, S. Henkes, and R. R. Hartley. Why do granular materials stiffen with shear rate? a test of novel stress-based statistics. *Physical Review Letters*, 101:268301, 2008.

5. E. Bertin, O. Dauchot, and M. Droz. Temperature in nonequilibrium systems with conserved energy. *Physical Review Letters*, 93, 230601, 2004.

6. E. Bertin, O. Dauchot, and M. Droz. Nonequilibrium temperatures in steady-state systems with conserved energy. *Physical Review E*, 71, 046140, 2005.

7. E. Bertin, O. Dauchot, and M. Droz. Definition and relevance of nonequilibrium intensive thermodynamic parameters. *Physical Review Letters*, 96, 120601, 2006.

8. D. Bi. *Origin of rigidity of marginal solids*. PhD thesis, Brandeis University, Waltham, MA, 2012.

9. D. Bi and B. Chakraborty. Rheology of granular materials: dynamics in a stress landscape. *Philosophical Transactions of Royal Society*, 367:5073–5090, 2009.

10. D. Bi, J. Zhang, R. P. Behringer, and B. Chakraborty. Fluctuations in shear-jammed states: A statistical ensemble approach. *Europhysics Letters*, 102, 34002, 2013.

11. R. Blumenfeld. On entropic characterization of granular materials. to appear in Lecture Notes edited by T. Aste, World Scientific.

12. R. Blumenfeld and S. F. Edwards. Granular entropy: Explicit calculations for planar assemblies. *Physical Review Letters*, 90(11), March 21, 2003.

13. R. Blumenfeld, J. F. Jordan, and S. F. Edwards. Inter-dependence of the volume and stress ensembles and equipartition in statistical mechanics of granular systems. arXiv:1204.2977, 2012.

14. L. Bocquet, A. Colin, and A. Ajdari. Kinetic theory of plastic flow in soft glassy materials. *Physical Review Letters*, 103(3):036001, July 2009.

15. J.-P. Bouchaud. Granular media: Some ideas from statistical physics. In J.-L. Barrat, J. Dalibard, M Feigelman, and J. Kurchan, eds., *Slow Relaxations and Nonequilibrium Dynamics in Condensed Matter*, p. 131. Springer, Berlin, Germany, 2003.

16. J. P. Bouchaud and A. Georges. Anomalous diffusion in disordered media: Statistical mechanisms, models and physical applications. *Physics Reports*, 195(4–5):127–293, 1990.

17. M. E. Cates, J. P. Wittmer, J. P. Bouchaud, and P. Claudin. Jamming, force chains, and fragile matter. *Physical Review Letters*, 81:1841–1844, 1998.

18. B. Chakraborty. Statistical ensemble approach to stress transmission in granular packings. *Soft Matter*, 6:2884, 2010.

19. B. Chakraborty and B. Behringer. Jamming of granular matter. In R. A. Meyers, ed., *Encyclopedia of Complexity and Systems Science*. Springer, 2009.

20. D. Chandler. *Introduction to Modern Statistical Mechanics*. Oxford University Press, New York, 1987.

21. A. Coniglio, A. Fierro, and M. Nicodemi. Applications of the statistical mechanics of inherent states to granular media. *Physica A-Statistical Mechanics and Its Applications*, 302(1–4):193, 2001.

22. A. Coniglio, A. Fierro, and M. Nicodemi. Probability distribution of inherent states in models of granular media and glasses. *European Physical Journal E*, 9(3):219, 2002.

23. E. DeGiuli and J. McElwaine. Laws of granular solids. Geometry and topology. *Physical Review E*, 84, 041310, 2011.

24. A. Donev, S. Torquato, F. H. Stillinger, and R. Connelly. Jamming in hard sphere and disk packings. *Journal of Applied Physics*, 95:989–999, 2004.

25. S. F. Edwards and R. Blumenfeld, in Physics of Granular Materials, Ed. A. Mehta, Cambridge University Press, Cambridge, 2007.

26. S. F. Edwards and D. V. Grinev. Statistical mechanics of stress transmission in disordered granular arrays. *Physical Review Letters*, 82:5397, 1999.

27. S. F. Edwards and D. V. Grinev. Granular materials: Towards the statistical mechanics of jammed configurations. *Advances in Physics*, 51:1669, 2002.

28. Y. Forterre and O. Pouliquen. Flows of dense granular media. *Annual Review of Fluid Mechanics*, 40:1–24, 2008.

29. P. Hanggi, P. Talkner, and M. Borkovec. Reaction-rate theory—50 years after kramers. *Reviews of Modern Physics*, 62(2):251–341, April 1990.

30. R. R. Hartley and R. P. Behringer. Logarithmic rate dependence of force networks in sheared granular materials. *Nature*, 421:928, 2003.

31. D. Head. Phenomenological glass model for vibratory granular compaction. *Physics Review E*, 62:2439, 2000.

32. D. L. Henann and K. Kamrin. A predictive, size-dependent continuum model for dense granular flows. *PNAS*, 110:6730, 2013.

33. S. Henkes and B. Chakraborty. Jamming as a critical phenomenon: A field theory of zero-temperature grain packings. *Physical Review Letters*, 95, 2005.

34. S. Henkes and B. Chakraborty. A statistical mechanics framework for static granular matter. *Physical Review E*, 79:061301, 2009. cond-mat/0810.5715.

35. S. Henkes, C. S. O'Hern, and B. Chakraborty. Entropy and temperature of a static granular assembly: An *Ab Initio* approach. *Physical Review Letters*, 99:038002, July 2007.

36. H. M. Jaeger and A. J. Liu. Far-from equilibrium physics. arXiv:1009.4874, 2010.

37. E. T. Jaynes. Information theory and statistical mechanics. *Phyical Review*, 106:320, 1957.

38. N. G. Van Kampen. *Stochastic Processes in Physics and Chemistry*, 3rd edn. North Holland, 2007.

39. K. Kamrin and G. Koval. Nonlocal constitutive relation for steady granular flow. *Physical Review Letters*, 108:178301, 2012.

40. M. Kardar. *Statistical Physics of Fields*. Cambridge University Press, Cambridge, U.K., 2007.

41. M. Kardar. *Statistical Physics of Particles*. Cambridge University Press, Cambridge, U.K., 2007.

42. J. Kurchan. Recent theories of glasses as out of equilibrium systems. *Comptes Rendus De L Academie Des Sciences Serie Iv Physique Astrophysique*, 2:239–247, 2001.

43. F. Lechenault, F. da Cruz, O. Dauchot, and E. Bertin. Free volume distributions and compactivity measurement in a bidimensional granular packing. *Journal of Statistical Mechanics-Theory and Experiment*, 2006.

44. A. Liu and S. Nagel. Nonlinear dynamics: Jamming is not just cool any more. *Nature*, 1998.

45. H. A. Makse and J. Kurchan. Testing the thermodynamic approach to granular matter with a numerical model of a decisive experiment. *Nature*, 415:614–617, 2002.

46. S. McNamara, P. Richard, S. Kiesgen de Richter, G. Le Caer, and R. Delannay. Measurement of granular entropy. *Physical Review E*, 80:031301, 2009.

47. P. T. Metzger and C. M. Donahue. Elegance of disordered granular packings: A validation of edward's hypothesis. *Physical Review Letters*, 94:148001, 2005.

48. M. Mezard and G. Parisi. Thermodynamics of glasses: A first principles computation. *Physical Review Letters*, 82:747, 1999.

49. C. Monthus and J.-P. Bouchaud. Models of traps and glass phenomenology. *Journal of Physics A Mathematical General*, 29:3847–3869, July 1996.

50. R. B. S. Oakeshott and S. F. Edwards. Theory of powders. *Physica A*, 3:1080–1090, 1989.

51. G. Parisi. *Statistical Field Theory*. Advanced Book Classics Series. Addison-Wesley, 1998.

52. G. Picard, A. Ajdari, F. Lequeux, and L. Bocquet. Slow flows of yield stress fluids: Complex spatiotemporal behavior within a simple elastoplastic model. *Physical Review E*, 71(1), January 2005.

53. O. Pouliquen and R. Gutfraind. Stress fluctuations and shear zones in quasistatic granular chute flows. *Physical Review E*, 53:552–561, January 1996.

54. J. G. Puckett and K. E. Daniels. Equilibrating temperature-like variables in jammed granular subsystems. arXiv:1207.7349, 2012.

55. S. Sarkar, D. Bi, J. Zhang, R. P. Behringer, and B. Chakraborty. Origin of rigidity of sry granular solids. *Physics Review Letters*, May 2013.

56. S. Sastry. Inherent structure approach to the study of glass forming liquids. *Nature*, 409:164, 2001.

57. J. H. Snoeijer, T. J. H. Vlugt, M. van Hecke, and W. van Saarloos. Force network ensemble: A new approach to static granular matter. *Physics Review Letters*, 91:072303, 2003.

58. J. H. Snoeijer, W. G. Ellenbroek, T. J. H. Vlugt, and M. van Hecke. Sheared force networks: Anisotropies, yielding, and geometry. *Physical Review Letters*, 96:098001, 2006.

59. J. H. Snoeijer, M. van Hecke, E. Somfai, and W. van Saarloos. Force and weight distributions in granular media: Effects of contact geometry. *Physical Review E*, 67:030302, 2003.

60. J. H. Snoeijer, M. van Hecke, E. Somfai, and W. van Saarloos. Packing geometry and statistics of force networks in granular media. *Physical Review E*, 70:011301, 2004.

61. J. H. Snoeijer, T. J. H. Vlugt, W. G. Ellenbroek, M. van Hecke, and J. M. J. van Leeuwen. Ensemble theory for force networks in hyperstatic granular matter. *Physical Review E*, 70:061306, 2004.

62. J. H. Snoeijer, T. J. H. Vlugt, M. van Hecke, and W. van Saarloos. Force network ensemble: A new approach to static granular matter. *Physical Review Letters*, 92:054302, 2004.

63. P. Sollich. Rheological constitutive equation for a model of soft glassy materials. *Physics Review E*, 58:738, 1998.

64. F. H. Stillinger. A topographic view of supercooled liquids and glass formation. *Science*, 267:2509, 1995.

65. B. P. Tighe, J. E. S. Socolar, D. G. Schaeffer, W. G. Mitchener, and M. L. Huber. Force distributions in a triangular lattice of rigid bars. *Physical Review E*, 72, 031306, 2005.

66. A. V. Tkachenko and T. A. Witten. Stress propagation through frictionless granular material. *Physics Review E*, 60:687, 1999.

67. A. R. T. van Eerd, W. G. Ellenbroek, M. van Hecke, J. H. Snoeijer, and T. J. H. Vlugt. Tail of the contact force distribution in static granular materials. *Physical Review E*, 75:060302, 2007.

Section II

Current Research

CHAPTER 6

Packings: Static

Mark D. Shattuck and Scott V. Franklin

CONTENTS

Granular materials are ubiquitous in nature, including sands and soils, debris flows, avalanches, and earthquakes. Efficient and reliable processing of granular media is also of paramount importance in the oil, pharmaceutical, food, and personal care industries. Non-cohesive granular materials are composed of discrete, rigid grains that interact via frictional contact interactions. Due to dissipative interactions between particles, granular materials remain static with zero net force and torque on each grain in the absence of external driving such as applied shear or vibration. In addition to net force and torque balance on each grain, mechanically stable (MS) static granular packings are linearly stable to small perturbations. A compelling motivation for studying static granular packings is to develop the ability to predict the response of granular systems to applied deformations by quantifying the static structure. Further, static hard-sphere packings have been shown to capture important structural and mechanical properties of dense liquids [29], colloids [30], and other particulate materials.

Early experimental studies of static packings of nearly monodisperse steel ball bearings [1,2,34–36] identified several characteristic structural properties of amorphous sphere packings. The most likely packing fraction for amorphous, monodisperse sphere packings occurs near random close packing $\phi_{rcp} \approx 0.64$ [3], much below $\phi_{xtal} \approx 0.74$ for face-centered-cubic (FCC) crystals. The radial distribution function $g(r)$, where r is the center-to-center separation between spheres, possesses a split second-neighbor peak and a strong nearest neighbor peak that, when integrated, yields fewer than 12 nearest neighbors, the number obtained for sphere packings with crystalline order.

More recent theoretical and computational studies have quantified the long-wavelength ($q \to 0$) behavior of the static structure factor $S(q)$ [8], fluctuations in the free volume, void statistics [33], and emphasized the sensitive dependence of structural and mechanical properties of static sphere packings on the packing protocol that generates them [46]. For example, hard-sphere systems that are slowly compressed possess more positional and bond-orientational order than those that are rapidly compressed [19]. Very dilute static packings of *frictional* spheres, with tangential as well as normal forces, can be obtained with $\phi = \phi_{\mathrm{rlp}} \approx 0.58$ [11,27,38] much lower than ϕ_{rcp}. In contrast, static packings of particles with anisotropic shapes, such as ellipsoids [6,7,20,32,50], Platonic solids [18,41], and superballs [17], can pack much more densely than frictionless spheres, which are believed to be the worst packing shape in three dimensions.

This chapter will expose the reader to recent computational results for the structural and mechanical properties of MS static packings of frictionless and frictional spherical particles as well as frictionless particles with anisotropic shapes (e.g., ellipsoids). Instead of dwelling on previous studies, we present new, but preliminary simulation results that point to interesting future research topics, focusing on the following important questions: (1) How does the number of distinct MS packings grow with the number of particles in the system? (2) Do MS packings occur as discrete points or do they occupy finite hypervolumes in configuration space? (3) Are there significant differences in the structural and mechanical properties of static packings of spheres versus particles with anisotropic shapes? (4) What is the shape of the distribution of jammed packing fractions for frictional spherical particles?, and (5) Does the likelihood to jam upon the application of shear strain depend on the static friction coefficient or system size? This chapter also serves to highlight computational work to enumerate all distinct MS packings of bidisperse disks in two dimensions and spheres in three dimensions. The motivation of this work is to explicitly test, in small systems, the hypothesis of Edwards and coworkers [10] that MS packings occur with equal probabilities, which forms the basis of thermodynamic and statistical mechanics descriptions of static [5,48] and slowly driven granular media [23,26,43]. See Chapter 7 for a more complete review of the Edwards ensemble.

6.1   Frictionless Disks and Spheres

This section describes the structural and mechanical properties of static packings of frictionless disks and spheres including the distributions of contact number and packing fraction, the number of distinct static packings as a function of the number of particles N and pressure, and the density of vibrational modes in the harmonic approximation. To prevent ordering in two dimensions, bidisperse mixtures (50–50 by number with diameter ratio $\sigma_l/\sigma_s = 1.4$) [13] are used. In three dimensions, monodisperse systems in the fast quench-rate limit are studied. Details of the computational methods employed to generate static packings of spherical particles at jamming onset are given in Chapter 3.

6.1.1 Properties at Jamming Onset

Amorphous static packings of frictionless disks in two dimensions and spheres in three dimensions generated from molecular dynamics methods are typically isostatic, with the minimal number of contacts required to constrain all nontrivial degrees of freedom. In periodic boundary conditions, a necessary condition for mechanical stability of isotropically compressed static packings is that the total number of contacts satisfies

$$N_c \geq N_c^{iso} = d(N' - 1) + 1. \tag{6.1}$$

N' is defined as the total number of particles minus the number of "rattlers," $N' = N - N_r$, where the N_r rattlers are defined as those with fewer than $d+1$ contacts (d is the spatial dimension). For static packings generated using the soft interaction potentials, we can also test whether the static packings are MS by calculating the dN' eigenvalues of the dynamical matrix

$$M_{ij} = \frac{\partial^2 V}{\partial \xi_i \partial \xi_j}\bigg|_{\xi = \xi^0}, \tag{6.2}$$

where $\xi = \{\vec{r}_1, \vec{r}_2, \ldots, \vec{r}_{N'}\}$ and ξ^0 represents the configuration of the static packing. MS packings possess $dN' - d$ nontrivial eigenvalues $m_i > 0$ and eigenvectors $\hat{m}_i$ with $\hat{m}_i^2 = 1$. In systems with periodic boundary conditions, d of the eigenvalues are zero due to translational invariance. The eigenfrequencies are given by $\omega_i = \sqrt{m_i \sigma_s^2 / \epsilon}$, where ϵ is the characteristic energy scale of the repulsive interaction between particles. Distinct static packings are defined by their spectrum of eigenfrequencies. Two static packings α and β are considered the same if $\Delta = \max_i |\omega_i^\alpha - \omega_i^\beta| / (\omega_i^\alpha + \omega_i^\beta) < 10^{-6}$, where the eigenfrequencies are sorted by magnitude. The density of vibrational modes $D(\omega)$ (in the harmonic approximation) can be obtained by binning the eigenfrequencies, $D(\omega) = (N(\omega + \delta\omega) - N(\omega))/\delta\omega$, where $N(\omega)$ is the number of eigenfrequencies less than ω.

We performed exhaustive searches for MS packings in small systems in two dimensions and three dimensions using the soft-particle molecular dynamics packing-generation protocol described in Chapter 3. There are 20 distinct MS packings for 2D bidisperse systems with $N = 6$, as shown in Figure 6.1. All of the packings are isostatic; for 19 packings without rattler particles $N_c = N_c^{iso} = 11$ and for one packing with a single rattler $N_c = N_c^{iso} = 9$. Figure 6.2a displays the eigenfrequency spectrum for MS packings 1, 2, and 3 (*cf.* Figure 6.1). MS packings 1 and 2 possess 10 nonzero eigenfrequencies, whereas MS packing 3 has 8 nonzero eigenfrequencies. Figure 6.2b shows the density of vibrational modes $D(\omega)$ averaged over the 20 distinct MS packings. The high-frequency behavior ($\omega > 1$) is similar to that found previously [25,32]; however, $D(\omega)$ does not possess a plateau that persists to $\omega \to 0$ as found in larger systems.

For 3D monodisperse systems, we identified nearly all distinct MS packings for $N = 2$ to 12 particles (*cf.* Table 6.1). The 19 distinct MS packings for $N = 5$ are shown in Figure 6.3. Figure 6.4 shows the number of distinct MS packings

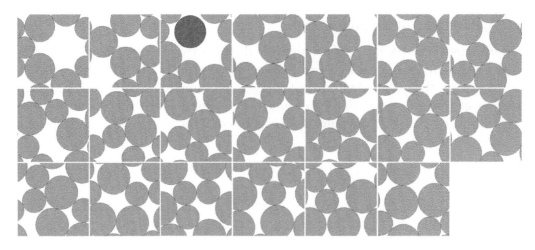

Figure 6.1   Snapshots of the 20 distinct MS packings for $N = 6$ bidisperse disks at jamming onset in a unit square with periodic boundary conditions. The packings are ordered from most dilute to densest with packing fractions $\phi_J = 0.6326, 0.6649, 0.6997, 0.7147, 0.7174, 0.7365, 0.7394, 0.7396, 0.7416, 0.7423, 0.7488, 0.7572, 0.7651, 0.7656, 0.7662, 0.7686, 0.7695, 0.7723, 0.7732$, and 0.7775 from left to right and top to bottom. The dark shaded particle is a "rattler" with fewer than three interparticle contacts. All packings are isostatic with $N_c = 11(9)$ for systems without (with) a rattler particle.

N_s versus N, which contains more structures than that for 2D bidisperse MS packings [49]. We find that the probability for a MS packing to have an excess number of contacts above the isostatic value, $\Delta N_c = N_c - N_c^{\text{iso}}$, is appreciable for several N, for example, 7 and 8 (Figure 6.4c). The available symmetries for systems with these N reduce the number of distinct MS packings. As N increases, the probability for $\Delta N_c > 0$ decreases strongly and nearly all MS packings are isostatic. A rough estimate for the complexity of the potential energy landscape for 3D monodisperse MS packings gives $c = N^{-1} \ln N_s / N_s^0 \approx 1.4 - 1.5$ compared to $1.1 - 1.2$ for 2D bidisperse MS packings [49]. As found in previous studies of 2D bidisperse MS packings, the probabilities for obtaining 3D monodisperse MS packings are highly nonuniform, varying by more than five orders of magnitude for $N = 5$–12 (Figure 6.4b). As shown in the inset to Figure 6.4a, the average packing fraction at jamming onset $\langle \phi_J \rangle$ approaches $\phi_{\text{rcp}} \approx 0.64$ as N increases, albeit more slowly for uniform weighting compared to weighting the packing fraction by the MS packing probabilities [49].

6.1.2   Packings at Nonzero Pressure

In this section, we study MS static packings of 2D bidisperse disks interacting via the purely repulsive linear spring potential ($V(r_{ij})$ given in Equation 3.3 with $\alpha = 2$) at nonzero compression $\Delta \phi$ or pressure,

$$\overline{P} = \frac{P \sigma_s^2}{\epsilon} = \frac{\sigma_s^2}{\epsilon L^2} \sum_{ij} \vec{F}_{ij} \cdot \vec{r}_{ij}, \tag{6.3}$$

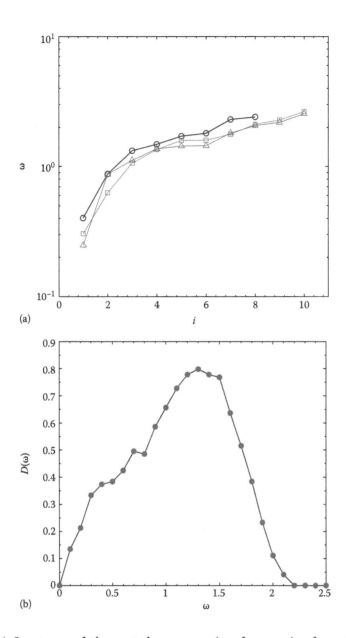

(a)

(b)

Figure 6.2   (a) Spectrum of the sorted nonzero eigenfrequencies from the dynamical matrix for MS packings 1 (squares), 2 (triangles), and 3 (circles) in Figure 6.1 for $N = 6$ bidisperse systems in two dimensions. MS packings 1 and 2 do not have rattlers, and thus possess 10 nonzero eigenvalues, while MS packing 3 contains a single rattler particle and has only 8 nonzero eigenfrequencies. (b) Density of vibrational modes (in the harmonic approximation) $D(\omega)$ obtained by averaging over the 20 distinct MS packings (with equal weights) for 2D bidisperse disks with $N = 6$.

Table 6.1 The Number of Distinct Mechanically Stable
Packings N_s and Total Number of Trials Performed N_t versus
the Number of Particles N for Monodisperse Spheres in
Three Dimensions

N	N_s	N_t
2	1	2×10^5
3	1	1×10^6
4	1	1×10^6
5	19	3×10^6
6	143	1×10^6
7	288	2×10^6
8	167	4×10^5
9	272	7×10^5
10	3,485	5×10^5
11	10,003	4×10^5
12	8,555	3×10^5

where the sum is over all distinct pairs of overlapping particles i and j, and
$\vec{F}_{ij} = -\hat{r}_{ij} \frac{\partial V}{\partial r_{ij}}$, which were obtained using the soft-particle molecular dynamics packing-generation algorithm. For reference, the MS packings of 2D bidis-
perse disks at jamming onset discussed in the previous section were obtained
at $\overline{P} \approx 10^{-8}$.

MS packings in the limit $\Delta\phi \to 0$ (or $\overline{P} \to 0$) possess a given network of interparticle contacts specified by the adjacency matrix A_{ij}, where $A_{ij} = 1$ if particles i and j are in contact, and the entries of A_{ij} are zero otherwise. The network
of interparticle contacts of overcompressed MS packings begins to deviate from
that at jamming onset at a characteristic $(\Delta\phi)_c = \phi_c - \phi_J$. Figure 6.5a shows $(\Delta\phi)_c$
for each of the 20 MS packings for 2D bidisperse systems at $N = 6$. The aver-
age $\langle(\Delta\phi)_c\rangle \sim N^{-\alpha}$ decreases as a power law with increasing system size, where
$\alpha \approx 1.9$ [31] (Figure 6.5b). The exponent α is similar to the system-size scaling
of the minimum separation $r_{ij}/\sigma_{ij} - 1 > 0$ between nearly contacting (not overlapping) particle pairs in MS packings at jamming onset as shown in Figure 6.6.
A power-law exponent near 2 can be obtained by assuming that new contacts are
formed randomly from particles that have separations on the $r_{ij} > \sigma_{ij}$ side of the
first peak in the radial distribution function, which diverges with a power-law
exponent near 0.5 [9,40]. This behavior indicates that nearly contacting particles
play an important role in particle rearrangements induced by isotropic compression near jamming onset.

The complexity of the minima in the potential energy landscape also changes
with increasing pressure. Figure 6.7a shows the number of distinct MS packings versus system size N over five orders of magnitude in pressure, including
$\overline{P} = 10^{-6}, 10^{-4}, 10^{-2}, 8 \times 10^{-2}$, and 10^{-1}. The number of distinct MS packings grows
exponentially for all $\overline{P}$ studied, $N_s \sim e^{c(\overline{P})N}$, where $c(\overline{P})$ is the $\overline{P}$-dependent complexity of the potential energy landscape [28]. However, as shown in Figure 6.7b

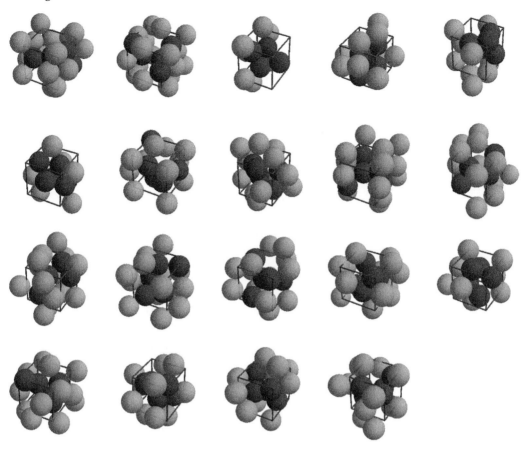

Figure 6.3 Snapshots of the 19 distinct MS packings at jamming onset for $N = 5$ monodisperse spheres in a unit cube with periodic boundary conditions. Spheres with centers located in the central simulation box have darker shading and their images have lighter shading. The packing fractions at jamming onset are $\phi_J = 0.4064, 0.4247, 0.4292, 0.4377, 0.4385, 0.4399, 0.4407, 0.4435, 0.4436, 0.4454, 0.4473, 0.4681, 0.4756, 0.4762, 0.4769, 0.4774, 0.4789, 0.4808$, and 0.4822 from left to right and top to bottom.

the complexity begins to *decrease* strongly for pressures above $\overline{P}^* \sim 10^{-3}$, which coincides with the change in the form of the normal force distribution for MS packings of frictionless spherical particles. Below $\overline{P}^*$ the normal force distribution decays nearly exponentially, while above $\overline{P}^*$ the normal force distribution is Gaussian [22].

6.1.3 Shear-Induced Jamming

Recent experimental studies have shown that the structural and mechanical properties of jammed packings generated via simple and pure shear strain differ from those obtained from isotropic compression [4]. For example, static packings generated via shear can occur at lower packing fractions than those generated via isotropic compression and possess strongly anisotropic networks of particle contacts [21]. In this section, we study static packings of frictionless disks generated

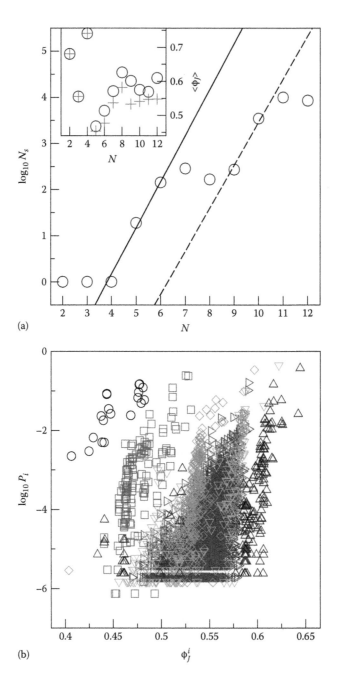

Figure 6.4   (a) Number of distinct mechanically stable packings N_s versus the number of particles N for 3D monodisperse systems. The solid (dashed) line has slope 1.4 (6.4). The inset shows the mean packing fraction at jamming onset $\langle \phi_J \rangle$ assuming a uniform probability distribution (pluses) or weighted by the MS packing probabilities (circles). (b) Scatter plot of the probabilities P_i for each distinct MS packing with ϕ_J^i for 3D monodisperse systems for $N = 5$ (circles), 6 (squares), 7 (diamonds), 8 (upward triangles), 9 (downward triangles), 10 (leftward triangles), and 12 (rightward triangles).   (*Continued*)

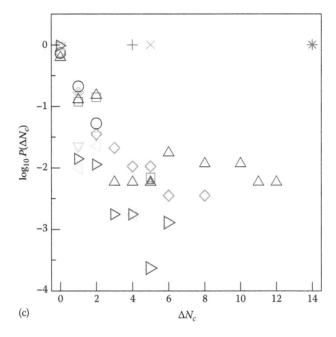

(c)

Figure 6.4 (*Continued*)   (c) Distribution of $\Delta N_c = N_c - N_c^{\text{iso}}$ for $N = 2$ (pluses), 3 (exes), 4 (stars), and 5–12 using the same symbols in (b).

via simple shear and determine whether shear-induced jammed packings of frictionless particles remain stable in the large system limit.

Starting with static packings of frictionless bidisperse disks generated via isotropic compression with a distribution of packing fractions at jamming onset $P(\phi_J)$, the disk packings are decompressed by decreasing the particle sizes uniformly by $\delta\phi = \phi_J - \phi > 0$. Thus, the initial packings are unjammed with zero total potential energy $V = 0$ (i.e., $V/N\epsilon \leq V_{\text{tot}} = 10^{-16}$). An affine simple shear displacement is then applied to all disks of the form

$$x_i' = x_i + \delta\gamma y_i \tag{6.4}$$

in conjunction with shear-periodic (Lees–Edwards) boundary conditions [12], where x_i' is the new strained x-position of the ith particle, $\delta\gamma = 10^{-4}$ is the shear strain increment, and y is the shear gradient direction. (See Chapter 3 for a discussion of shear-periodic boundary conditions.) After applying each incremental affine shear displacement, a conjugate gradient technique is used to minimize the total potential energy V at fixed accumulated shear strain γ and packing fraction. The total potential energy determines whether the system is jammed or not. Packings with $V/N\epsilon < V_{\text{tot}}$ after minimization are assumed to be unjammed. If $V/N\epsilon > V_{\text{tot}}$, the system is assumed to be jammed (with small, but finite interparticle overlaps).

Figure 6.8a shows the distribution $P(\gamma_c)$ of characteristic shear strains γ_c at which isotropic configurations of frictionless bidisperse disks at $\delta\phi$ below ϕ_J first jam under planar shear. The distributions are broad and non-Gaussian.

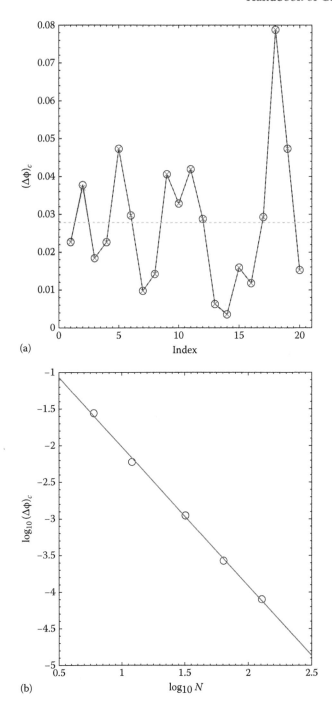

Figure 6.5 The characteristic deviation in packing fraction $(\Delta\phi)_c = \phi_c - \phi_J$ at which the adjacency matrix deviates from that at ϕ_J for each of the 20 MS packings at jamming onset for $N = 6$ bidisperse systems using small successive compressions 10^{-7} (circles) and 10^{-8} (crosses). The dashed line indicates the average $\langle(\Delta\phi)_c\rangle = 0.028$ over the 20 MS packings. (b) The average $\langle(\Delta\phi)_c\rangle$ over 100 2D bidisperse MS packings as a function of N. The solid line has slope -1.9.

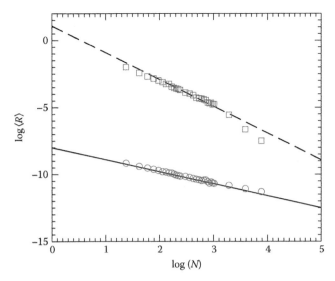

Figure 6.6 The minimum overlap $R_+ = 1 - r_{ij}/\sigma_{ij} > 0$ (circles) between contacting particle pairs and separation $R_- = r_{ij}/\sigma_{ij} - 1 > 0$ (squares) between noncontacting particle pairs averaged over MS packings of bidisperse disks for system sizes $N = 24–780$ at $\Delta\phi = 10^{-8}$. The dashed and solid lines have slopes -2 and -1, respectively.

Figure 6.8b shows the mean strain $\langle \gamma_c \rangle$ averaged over the ensemble of initial conditions versus $\delta\phi$ for system sizes ranging from $N = 32$ to 512. The plateau at small $\delta\phi$ decreases and the divergence near $\delta\phi \approx 10^{-2}$ shifts to larger $\delta\phi$ with increasing system size, which indicates that shear-induced jamming events become more frequent with increasing system size. It is not yet clear whether these trends depend on the number of samples included in the average and an important open question is whether shear-induced jamming in systems composed of frictionless particles is a finite-size effect. One possible scenario is that shear-induced jamming is system-size dependent for $\mu < \mu_c$ (below which the system displays properties of frictionless packings), but only weakly dependent on system size for $\mu > \mu_c$. Another possibility is that $\langle \gamma_c \rangle$ scales inversely with μ. Further studies are necessary to understand the role of friction and compression on shear-induced jamming.

6.2 Frictional Disks

A number of computational studies have shown that the structural and mechanical properties of frictional sphere packings differ significantly from those of frictionless spheres [22,38,39,42,47]. For example, the average packing fraction at jamming onset $\langle \phi_J \rangle$ and contact number $z = 2N_c/N'$ decrease with increasing static friction coefficient μ. Figure 6.9a shows $\langle \phi_J \rangle$ and z as a function of μ for 2D bidisperse static packings obtained using an isotropic compression packing-generation protocol (see Chapter 3.) The packing fraction $\langle \phi_J \rangle$ decreases from ≈ 0.84 near random close packing for 2D bidisperse disks [25] to $\approx 0.76–0.78$

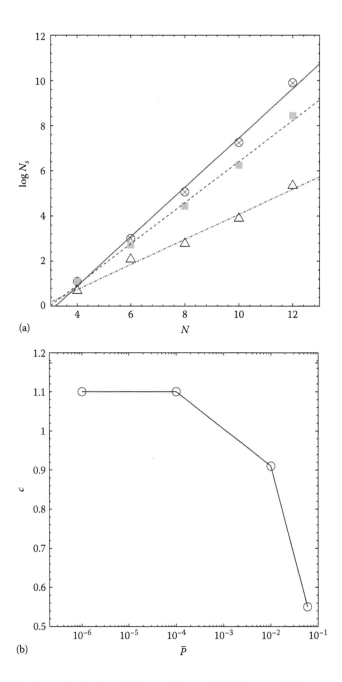

Figure 6.7 (a) Number of distinct MS packings of bidisperse disks as a function of N for pressure $\overline{P} = 10^{-6}$ (circles), 10^{-4} (crosses), 10^{-2} (squares), and 8×10^{-2} (triangles). The slopes of the solid, dashed, and dotted lines are 1.1, 0.9, and 0.5, respectively. (b) Complexity c plotted versus the pressure $\overline{P}$.

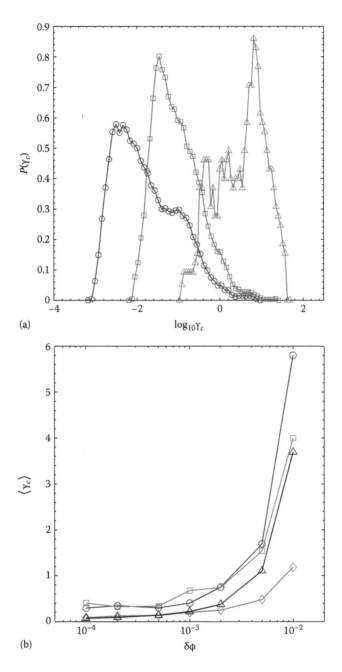

Figure 6.8   (a) Distribution of accumulated planar shear strain $P(\gamma_c)$ required to jam an isotropically compressed configuration of $N = 256$ bidisperse frictionless disks at packing fraction $\delta\phi = \phi_J - \phi = 10^{-4}$ (circles), 10^{-3} (squares), and 10^{-2} (triangles). (b) The mean planar shear strain $\langle\gamma_c\rangle$ required to induce jamming averaged over at least 100 configurations as a function of $\delta\phi$ for $N = 32$ (circles), 64 (squares), 128 (triangles), and 512 (diamonds).

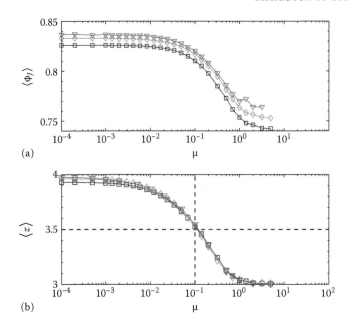

Figure 6.9 (a) Average packing fraction at jamming onset $\langle \phi_J \rangle$ and (b) contact number $z = 2N_c/N'$ as a function of the static friction coefficient μ for $N = 32$ (squares), 128 (diamonds), and 256 (triangles). The vertical dashed line indicates the static friction coefficient $\mu^* \approx 0.1$ at which the contact number crosses over from frictionless to frictional behavior.

for large μ, which is near the random loose packing limit below which amorphous disk packings are not MS [16]. In Figure 6.9b, z decreases from 4 to ≈ 3 contacts per particle for large μ. For 2D bidisperse packings, $\mu^* \approx 10^{-2}$–10^{-1} is the characteristic static friction coefficient above which the structural and mechanical properties begin to deviate from their frictionless values. In contrast, the characteristic $\mu^* < 0.01$ for static packings of monodisperse spheres in three dimensions [38].

Figure 6.10a shows the distribution of packing fractions at jamming onset $P(\phi_J)$ generated via isotropic compression for two static friction coefficients $\mu = 10^{-3}$ (low friction) and 1.0 (high friction). As shown in Figure 6.10b and c, the width σ and the location of the peak $\langle \phi_J(\infty) \rangle - \langle \phi_J(N) \rangle$ display power-law scaling behavior with system size for both low and high friction. The power-law scaling exponents for the widths of the distributions are similar, but the scaling exponent for the peak position is more than 30% lower for high friction than for low friction. The fact that the distribution of packing fractions at jamming onset for high friction narrows to a δ-function in the large system limit is interesting, since it is known that frictional packings can be obtained over a wide range of packing fractions [44].

These results raise several important questions concerning static packings of frictional spheres: (1) What determines the characteristic μ_c that separates frictionless and frictional behavior of the structural and mechanical properties of static sphere packings? (2) Since isotropic compression yields a well-defined

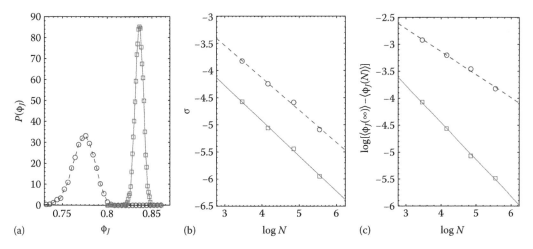

Figure 6.10   (a) Distribution of packing fractions at jamming onset $P(\phi_J)$ for 2D static packings composed of bidisperse frictional disks with $N = 128$ and static friction coefficient $\mu = 10^{-3}$ (squares) and 1.0 (circles). (b) The width σ of the distribution of jamming onsets for $\mu = 10^{-3}$ (squares) and 1.0 (circles) versus N. The slopes of the solid and dotted lines are -0.64 and -0.59, respectively. (c) The scaling of the mean $\langle\phi_J\rangle$ in the distribution of packing fractions at jamming onset for $\mu = 10^{-3}$ (squares) and 1.0 (circles) versus N. $\langle\phi_J(\infty)\rangle \approx 0.843$ for $\mu = 10^{-3}$ and ≈ 0.805 for $\mu = 1.0$. The slopes of the solid and dotted lines are -0.66 and -0.42, respectively.

packing fraction $\langle\phi_J\rangle$ and contact number z for each static friction coefficient μ (in the large system limit), can the protocol be varied to yield different values of $\langle\phi_J(\mu)\rangle$ and $z(\mu)$, and (3) If multiple packing fractions $\langle\phi_J(\mu)\rangle$ and contact numbers $z(\mu)$ can be obtained at a given μ, what physical quantity distinguishes between the different ensembles?

We have also investigated the contact number distribution, eigenvalues of the distance matrix, the mobility distribution, and shear modulus of static packings of bidisperse frictional disks as a function of μ. Figure 6.11a shows the probability for static packings of bidisperse frictional disks to possess $N_c^{iso} - N_c$ contacts for $N = 6$, where $N_c^{iso} = 2(N - N_r) - 1$ is the isostatic number of contacts for frictionless disks and N_r is the number of rattler particles. For $\mu < \mu_c \approx 0.01$, most packings possess $N_c = N_c^{iso}$. As the friction coefficient increases, however, the most probable number of contacts is no longer the isostatic value. For $\mu = 0.2$, $P(N_c^{iso} - N_c)$ possesses a peak at $N_c^{iso} - N_c = 1$ that shifts to 2 for larger μ. Figure 6.11b shows the probability for obtaining static packings of bidisperse frictional disks with $N_c^{iso} - N_c = 0$, 1, and 2 as a function of μ. From this plot, it is clear that most static packings of bidisperse frictional disks possess $N_c \approx N_c^{iso}$ for $\mu \lesssim \mu_c$, whereas higher-order saddle packings become most probable for $\mu \gtrsim \mu_c$. μ_c indicates the crossover in the static friction coefficient below which the static packings behave as frictionless, isostatic systems, and above which frictional forces play a dominant role.

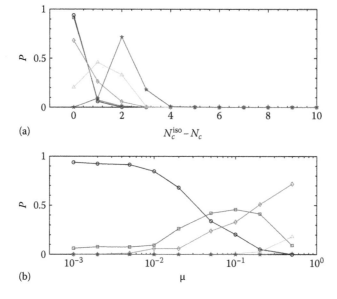

(a)

(b)

Figure 6.11 (a) The probability distribution of the deviation in the number of contacts $N_c^{iso} - N_c$ (where $N_c^{iso} = 2(N - N_r) - 1$ is the isostatic number of contacts for frictionless disks) in static packings of $N = 6$ bidisperse frictional disks for static friction coefficient $\mu = 0.002$ (circles), 0.02 (squares), and 0.2 (diamonds) 1 (triangles), and 2 (stars). (b) The probability for static packings of $N = 6$ bidisperse frictional disks to possess $N_c^{iso} - N_c = 0$ (circles), 1 (squares), 2 (diamonds), 3 (triangles), and 4 (stars). as a function of μ.

Figure 6.12 shows a scatter plot of the second invariant q_2 of the distance matrix, defined as

$$D_{ij} = |\vec{r}_i - \vec{r}_j|, \tag{6.5}$$

where $\vec{r}_i$ is the location of particle i, and its second invariant

$$q_2 = \frac{1}{2}((TrD)^2 - TrD^2), \tag{6.6}$$

versus the packing fraction ϕ. (A similar plot is shown in Figure 3.4a.) Static disk packings with $N_c^{iso} - N_c = 0$ and 1 generated using the "basic" Monte Carlo method and soft-particle molecular dynamics simulations of frictional disks are shown. (See Chapter 3 for descriptions of the Monte Carlo and molecular dynamics packing-generation methods.) Static packings of frictional disks for $\mu \gtrsim \mu_c$ begin populating the geometric families for first-order saddles obtained from Monte Carlo simulations of hard frictionless disks. Preliminary results suggest that frictional packings with small $\mu < \mu_c$ do not preferentially populate geometrical families near frictionless MS packings, that is, the symbols for the $\mu = 0.002$ static packings do not cluster near the frictionless MS packings in Figure 6.12. Frictional packings with larger μ more uniformly sample the geometrical families of the first-order saddle packings, and in particular large-μ packings frequently occur close to MS packings in the space of q_2 and ϕ.

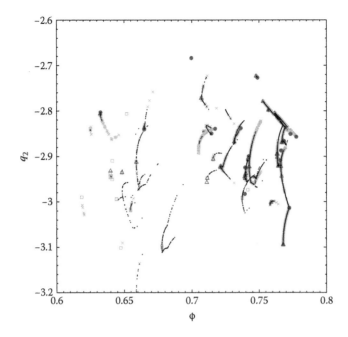

Figure 6.12   The second invariant q_2 of the distance matrix D plotted versus the packing fraction ϕ for $N = 6$ bidisperse disk packings with $N_c^{iso} - N_c = 1$ (first-order saddles) generated via the "basic" Monte Carlo method (black dots) and soft-particle molecular dynamics simulations of frictional particles with static friction coefficient $\mu = 0.002$ (triangles), 0.02 (squares), and 0.2 (crosses). MS packings for frictionless bidisperse disks with $N_c^{iso} - N_c = 0$ are indicated by filled circles.

The fraction of contact forces that are close to the Coulomb sliding threshold plays an important role in determining the response to applied deformations [15]. Figure 6.13a shows the probability distribution of mobilities $P(\xi)$ at each contact, where $\xi = |\vec{F}_{ij}^t|/\mu|\vec{F}_{ij}^r|$, for static packings of bidisperse frictional disks with $N = 64$. For all friction coefficients, $P(\xi)$ possesses a large peak near $\xi = 1$, which indicates that many of the contacts are near sliding. For $\xi < 1$, the distributions are flat for $\mu \lesssim 1$. At large $\mu \gtrsim 1$, a modest peak at small mobilities forms in addition to the large one near $\xi = 1$. The abundance of contacts near the edge of mobility $\xi = 1$ can also be seen clearly as a discontinuity in the accumulated mobility distribution $C(\xi) = \int_0^\xi P(\xi')d\xi'$ in Figure 6.13b, which is robust with respect to varying the system size N as shown in Figure 6.13c. Similar mobility distributions are found when configurations above the onset of jamming (with large overlaps at packing fraction $\phi_i = 0.90$) are successively decompressed (and the energy is minimized). These results for the mobility distributions for static packings of frictional disks and spheres were also found in Refs. [37,38].

The mobility distribution in static packings of bidisperse frictional disks depends on the treatment of the tangential contact forces during the packing-generation procedure. Three variants of the Cundall–Strack model (*cf.* Section 6.2.2) are used. In the first (method 1), the relative tangential displacement $u_{t_{ij}}$

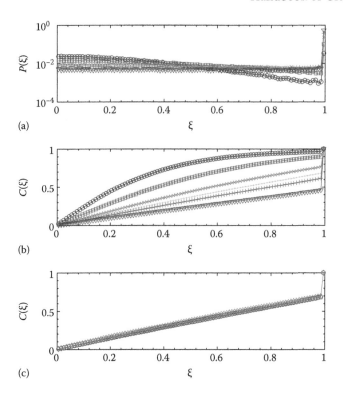

Figure 6.13 (a) Probability distribution $P(\xi)$ of the mobility at each contact, where $\xi = |\vec{F}^t_{ij}|/\mu|\vec{F}^r_{ij}|$, and (b) the cumulative mobility distribution $C(\xi) = \int_0^\xi P(\xi')d\xi'$ for static packings of $N = 64$ bidisperse frictional disks (using method 1 for the treatment of the tangential forces) with static friction coefficient $\mu = 0.01$ (downward triangles), 0.05 (pluses), 0.1 (filled dots), 0.2 (exes), 0.5 (squares), and 1 (circles). (c) $C(\xi)$ for static packings of bidisperse frictional disks with $\mu = 0.1$ and $N = 16$ (circles), 64 (squares), and 128 (diamonds) using method 1 for the treatment of the tangential forces.

between two contacting particles is set to $u_{t_{ij}} = \mu F^r_{ij}/k_s$ whenever the pair tangential force magnitude exceeds $|\vec{F}^t_{ij}| > \mu \vec{F}^r_{ij}$. This enforces the Coulomb sliding condition. When particles are no longer in contact, $u_{t_{ij}}$ is reset to 0. In the second variant (method 2), $u_{t_{ij}}$ is reset to 0 after each compression or decompression step, even if the contact between disks i and j remains intact. Method 3 reverses the compression direction to approach the jamming threshold from above, and begins with systems at high packing fraction $\phi_i = 0.90$ and large overlaps and successively shrinks the particles. If the potential energy per particle falls below a small threshold $\overline{V} < 10^{-16}$ after a given decompression and relaxation step, a smaller decompression step (typically half the previous step) is reapplied to the previous configuration and the energy is minimized. If the relaxed potential energy $\overline{V}$ is now greater than 10^{-16}, we continue the decompression process. If $\overline{V}$ is still below 10^{-16}, the compression step is again reduced and the process continued until the total potential energy reaches the small threshold $\overline{V} = 10^{-16}$ from above.

The cumulative mobility distributions $C(\xi)$ for methods 2 and 3 (Figure 6.14) are quantitatively similar to that for method 1 (Figure 6.13b) for both low and high values of the static friction coefficient, which suggests that the mobility distribution is relatively insensitive to the fraction of contacts that break and reform during the compression and decompression packing-generation method.

Figure 6.15 shows the distribution of the ratio of the tangential to normal pair force magnitudes F_{ij}^t/F_{ij}^r at each contact for $\mu = 1$ and 10, and the limit $\mu \to \infty$. The $\mu \to \infty$ limit is obtained by allowing the tangential component of the force F_{ij}^t to become arbitrarily large while disks i and j are in contact. The distribution of the ratios becomes independent of the static friction coefficient for $\mu \gtrsim 10$. Note that the distributions of $|F_{ij}^t|/F_{ij}^r$ at finite μ can be obtained qualitatively from the $\mu \to \infty$ distribution by placing the ratios that obey $|F_{ij}^t|/F_{ij}^r > \mu$ in the $\mu \to \infty$ distribution at $|F_{ij}^t|/F_{ij}^r = \mu$ to satisfy the Coulomb criterion. A more quantitative comparison of the distribution of force ratios shows that a fraction of the force ratios $|F_{ij}^t|/F_{ij}^r > \mu$ in the $\mu \to \infty$ distribution are spread over the range $0 < |F_{ij}^t|/F_{ij}^r < \mu$ in the finite-μ distribution, while most satisfy the equality $|F_{ij}^t|/F_{ij}^r = \mu$.

We now compare the shear modulus for static packings of frictionless and frictional disks. Figure 6.16 shows the shear modulus G for static packings of bidisperse frictionless disks in response to simple shear as a function of pressure $\overline{P}$ (Equation 6.3) for several system sizes from $N = 32$ to 1024. To measure G, a static

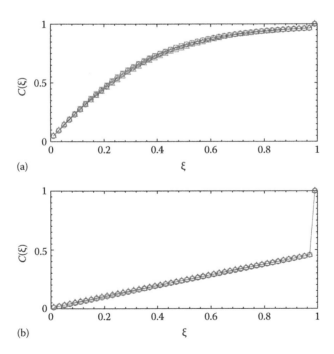

(a)

(b)

Figure 6.14 Cumulative mobility distribution for static packings of $N = 64$ bidisperse frictional disks with (a) $\mu = 1$ and (b) 0.01 using packing-generation methods 1 (circles), 2 (squares), and 3 (triangles).

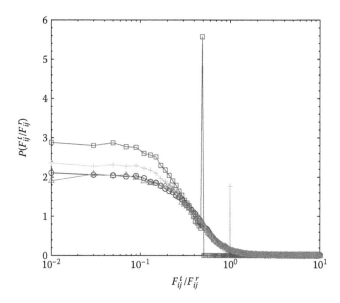

Figure 6.15   Distribution of the ratio of the tangential to normal pair force magnitudes $|F_{ij}^t|/F_{ij}^r$ at each contact for static packings of $N = 64$ bidisperse frictional disks with $\mu = 0.5$ (squares), 1 (pluses), 10 (circles), and the limit $\mu \to \infty$ (triangles) using packing-generation method 1.

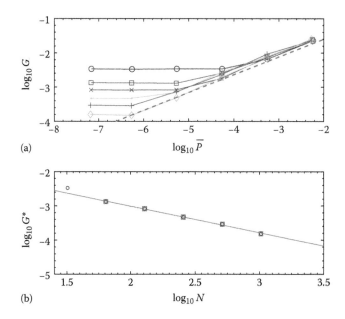

Figure 6.16   (a) Static linear shear modulus G (in response to simple shear) versus pressure $\overline{P}$ for static packings of bidisperse frictionless disks for $N = 32$ (circles), 64 (squares), 128 (exes), 256 (dots), 512 (pluses), and 1024 (diamonds). The value of G at each $\overline{P}$ is averaged over ≈ 100 configurations. The dashed line has slope 0.5. (b) The scaling of $G^* = G(\overline{P} \to 0)$ with system size N. The dashed line has slope -0.8.

packing at jamming onset (with a small initial value of compression $\Delta\phi = 10^{-8}$) is compressed in increments of $\Delta\phi_0 = 10^{-6}$, the total potential energy is minimized, and the pressure and contact number are measured after each compression. Contacts can break and form during the successive compressions, but the dominant effect is that the contact number $z - z_{iso}$ increases with overcompression. Plotting the shear modulus versus pressure (instead of $\Delta\phi$) accounts for particle rearrangements that occur during compression.

A simple shear strain can be applied to each particle in increments of $\Delta\gamma = 10^{-7}$ (coupled with Lees–Edwards boundary conditions) followed by potential energy minimization from total strain $\gamma_{tot} = 0$ to $\gamma_{tot} = 10^{-5}$. The shear stress is

$$\overline{\Sigma}_{xy} = \frac{\sigma_s^2}{\epsilon L^2} \sum_{ij} F_{xij}^r y_{ij}, \tag{6.7}$$

where the sum is over all distinct pairs of overlapping particles i and j. The linear shear modulus G is the proportionality constant $\overline{\Sigma}_{xy} = G\gamma_{tot}$ between the shear stress and total shear strain. For each $\overline{P}$, we average over at least 100 configurations for which the correlation coefficient R_2 of the linear fit of shear stress versus strain is greater than 0.99. Consistent with previous results in Ref. [25], the shear modulus for frictionless MS packings scales as a power law with $\overline{P}$, $G \sim \overline{P}^{0.5}$ (for repulsive linear spring interactions), above a system-size dependent value of the pressure $\overline{P}^*$. The power-law exponent 0.5 for the shear modulus G matches that for the growth of the contact number with $\overline{P}$, $z - z_{iso} \sim \overline{P}^{0.5}$ [25]. For $\overline{P} < \overline{P}^*$, the contact network remains the same as that at jamming onset, and G remains constant [14]. $G^* = G(\overline{P} \to 0) \sim N^{-\lambda}$, where $\lambda \sim 0.8$ for the system sizes considered, as shown in Figure 6.16b.

Figure 6.17 shows the shear modulus in response to simple shear deformations as a function of pressure $\overline{P}$ for static packings of bidisperse, frictional disks at $\mu = 10^{-3}$ and 0.1 using method 1 for the treatment of the tangential contact forces. The shear modulus scales as a power law with pressure, $G \sim \overline{P}^{0.5}$, for $\overline{P} > \overline{P}^*$, as found previously in Refs. [38,42,47]. This scaling is the same as that obtained for frictionless disks. For $\overline{P} < \overline{P}^*$, the contact network does not differ from that at jamming onset, and thus, G is constant.

6.3 Frictionless Non Spherical Particles

Static packings of ellipsoidal particles generated via isotropic compression in two dimensions and three dimensions are MS at finite compression $\Delta\phi > 0$, [20,32] despite the fact that they possess fewer contacts $z < z_{iso}$ than expected from naive contact counting arguments [39]. Thus, the dynamical matrix (i.e., Equation 6.2 with $\vec{\xi} = \{x_1, \ldots, x_{N'}, y_1, \ldots, y_{N'}, l_2\theta_1, \ldots, l_2\theta_{N'}\}$, where $l_2 = \sqrt{a_s^2 + b_s^2}\sqrt{1 + \sigma_l^2/\sigma_s^2}/4\sqrt{2}$, and a_s and b_s are the major and minor axes of the small particles) for each static packing of ellipsoidal particles possesses a full spectrum of nonzero eigenfrequencies. (The number of nonzero eigenvalues is $(2d - 1)N' - d + 1$, where

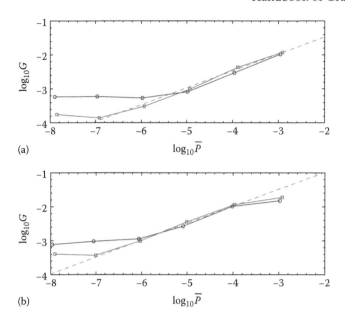

Figure 6.17 Static linear shear modulus G (in response to simple shear) versus pressure $\overline{P}$ for static packings of bidisperse frictional disks for $N = 32$ (circles) and 128 (squares) at (a) $\mu = 10^{-3}$ and (b) 0.1 using method 1 for the treatment of the tangential contact forces. The value of G at each $\overline{P}$ is averaged over ≈ 1000 configurations. The dashed lines in (a) and (b) have slope 0.5.

d is the spatial dimension, for ellipses in two dimensions and spheroids in three dimensions.)

Static packings of ellipsoidal particles also possess $N_q = N(z_{iso} - z)$ "quartic" vibrational modes near jamming onset. For perturbations along these quartic modes,

$$\vec{\xi} = \vec{\xi}_0 + \delta \hat{m}_i^q, \tag{6.8}$$

where

$\hat{m}_i^q$ is one of the N_q quartic mode eigenvectors

δ is the amplitude of the perturbation

$\vec{\xi}_0$ is the point in configuration space corresponding to the original static packing, the change in the potential energy scales as $\Delta V = V(\vec{\xi}) - V(\vec{\xi}_0) \propto \delta^4$ when $\delta > \delta_q$, where $\delta_q \propto \Delta\phi^{1/2}$

In contrast, the change in potential energy scales as $\Delta V \propto \delta^2$ for small perturbation amplitudes $\delta < \delta_q$ along the quartic modes $\hat{m}_i^q$. The change in potential energy scales as $\Delta V \sim \delta^2$ over a wide range of δ for perturbations along the quadratic modes. (See Figure 6.18a.)

The density of vibrational modes (in the harmonic approximation) for static packings of ellipse-shaped particles is plotted in Figure 6.18b and c. Static packings of ellipses with small aspect ratios $\alpha = a/b \lesssim 1.1$ possess N_q quartic and R_2

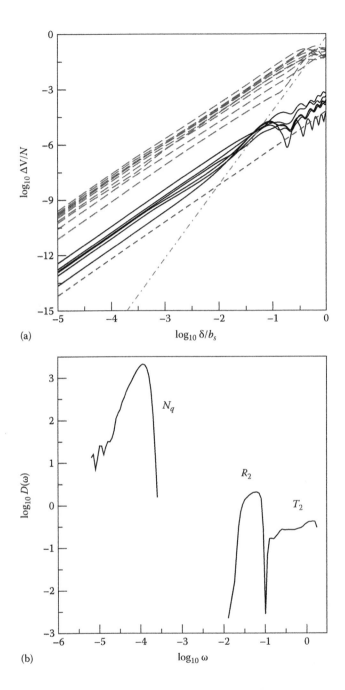

Figure 6.18 (a) The change in potential energy per particle $\Delta V/N$ versus perturbation amplitude δ/b_s for perturbations along the quadratic (dashed line) and quartic (solid line) modes for static ellipse packings with $N = 6$, $\alpha = 1.02$, and $\Delta\phi = 10^{-3}$. The dotted and dot–dashed lines have slopes 2 and 4, respectively. The density of vibrational modes in the harmonic approximation $D(\omega)$ for static ellipse packings with $N = 240$, $\Delta\phi = 10^{-7}$ and (b) $\alpha = 1.05$. (*Continued*)

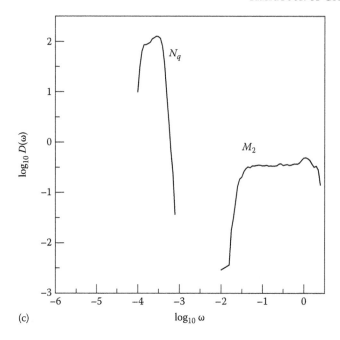

(c)

Figure 6.18 (Continued)   (c) $\alpha = 2$.

quadratic modes with mainly rotational content, and T_2 quadratic translational modes. For larger aspect ratios $\alpha \gtrsim 1.1$, the quadratic modes merge, and thus, only N_q quartic modes with mixed rotational and translational character and M_2 mixed rotational and translational quadratic modes are found. Note that the location of the quartic mode peak in $D(\omega)$ scales as $\omega_q \propto (\Delta\phi)^{1/2}(\alpha-1)^{1/2}$ [32].

The vibrational response of MS packings is strongly nonharmonic near jamming onset $\Delta\phi \to 0$ and for all compressions $\Delta\phi$ in the large-system limit $N \to \infty$ due to the breaking of contacts during vibration [31]. Thus, the vibrational response for packings of ellipsoidal particles is influenced by two sources of nonlinearity: (1) contact breaking and (2) quartic modes. In this section, we determine how these two sources of nonlinearity affect the vibrational response of packings of frictionless ellipse-shaped particles when subjected to thermal fluctuations.

To study the vibrational response, MS packings are perturbed in the direction of eigenvector $\vec{m}_l$ of the dynamical matrix by an amplitude δ, and constant energy MD simulations are run for a time $t > t_{\text{tot}} = 10\tau_1$, where $\tau_1 = 2\pi/\omega_1$ and ω_1 is the smallest nontrivial eigenfrequency from the dynamical matrix. The normalized power spectrum of the particle displacements is $P_{\text{norm}}(\omega^l) = P(\omega^l)/\sum_m P(\omega^m)$ as a function of frequency $\omega^l = \frac{l}{N_t}\frac{2\pi}{\Delta t}$, where

$$P(\omega^l) = \sum_{j=1}^{N}\left(\left|\sum_{k=1}^{N_t}\Delta x_j(k\Delta t)e^{-2\pi ikl/N_t}\right|^2 + \left|\sum_{k=1}^{N_t}\Delta y_j(k\Delta t)e^{-2\pi ikl/N_t}\right|^2\right.$$

$$\left. + l_2^2\left|\sum_{k=1}^{N_t}\Delta\theta_j(k\Delta t)e^{-2\pi ikl/N_t}\right|^2\right),\tag{6.9}$$

$N_t = \frac{t_{tot}}{\Delta t}$, $\Delta t = \frac{1}{3}\frac{2\pi}{\omega_{3N-2}}$, $\Delta x_j(t) = x_j(t) - \langle x_j(t)\rangle_t$, $\Delta y_j(t) = y_j(t) - \langle y_j(t)\rangle_t$, and $\Delta\theta_j(t) = \theta_j(t) - \langle\theta_j(t)\rangle_t$, and the angle brackets indicate time averages. We drop the superscript l on the frequency ω^l for the remainder of the chapter. By calculating the power spectrum, we decompose the particle displacements onto the basis of the dynamical matrix eigenmodes at frequency ω.

We first show results for the vibrational response of disks to illustrate the effects of contact breaking in the absence of quartic modes. Figure 6.19a shows the normalized power spectrum $P_{norm}(\omega)$ for 2D MS packings of bidisperse frictionless disks with $N = 6$ and $\Delta\phi = 10^{-3}$ following perturbations along the sixth eigenmode with eigenfrequency ω_6. The vibrational response of disk packings displays four characteristic regimes as a function of the perturbation amplitude δ/σ_s (Figure 6.19a): (I) Harmonic: For perturbations $\delta/\sigma_s \lesssim \overline{\delta}_c \approx 5 \times 10^{-5}$, no interparticle contacts break, and the system vibrates at frequency ω_6; (II) Nonharmonic 1: For $5\times10^{-5} \lesssim \delta/\sigma_s \lesssim 5\times10^{-4}$, the vibrational response is no longer confined to ω_6, and therefore, the response is not strictly harmonic, but remains roughly confined to the discrete set of eigenfrequencies of the dynamical matrix; (III) Nonharmonic 2: For $5 \times 10^{-4} \lesssim \delta/\sigma_s \lesssim 5 \times 10^{-2}$, the power spectrum has significant weight over a range of frequencies $\omega > 0$; and (IV) Liquid like: For $\delta/\sigma_s \gtrsim 5 \times 10^{-2}$, particle rearrangements occur and the vibrational response displays a continuum of frequencies with a large peak at $\omega = 0$. Note in Figure 6.19a that the time-averaged number of contacts $\langle N_c\rangle_t$ begins to decrease from the isostatic value $N_c^{iso} = 2N'-1 = 11$ for $\delta/\sigma_s > \overline{\delta}_c$.

To calculate the vibrational density of states for perturbation amplitudes in the nonharmonic regime, $\delta/\sigma_s > \overline{\delta}_c$, the displacement matrix

$$C_{kl} = \langle\Delta\xi_k\Delta\xi_l\rangle_t, \tag{6.10}$$

and associated $2N' - 2$ (for frictionless disks in two dimensions) eigenfrequencies c_i, and vibrational frequencies $\omega_i^c = \sqrt{T/c_i}$ are measured, where T is the temperature. Figure 6.19b shows that $D(\omega)$ obtained from the displacement matrix matches that from the dynamical matrix for perturbation amplitudes in regimes I and II. However, for perturbation amplitudes in regimes III and IV, $D(\omega)$ for the displacement matrix has significant weight near $\omega = 0$ in contrast to that for the dynamical matrix.

Vibrations of ellipse packings show both contact breaking and the activation of quartic modes. The power spectrum $P_{norm}(\omega)$ for ellipse packings with aspect ratio $\alpha = 1.02$ in response to perturbations along a mainly translational quadratic mode (T_2 in Figure 6.18b) is shown in Figure 6.20a. Vibrated ellipse packings exhibit behavior similar to that found for disks in regimes I (harmonic) and II (nonharmonic 1), but do not display behavior found in regime III (see Figure 6.19a), where the power spectrum of the displacements is spread over a range of nonzero frequencies that do not correspond to those from the dynamical matrix. In contrast to the behavior in regime III in disk packings, much of the power in vibrated ellipse packings resides in the quartic rotational modes (*cf.* Figure 6.18a), both before ($10^{-6} \lesssim \delta/b_s \lesssim 2 \times 10^{-5}$; regime V) and after contacts break ($10^{-4} \lesssim \delta/b_s \lesssim 8 \times 10^{-3}$; regime VI), but before the system becomes liquid like (regime IV).

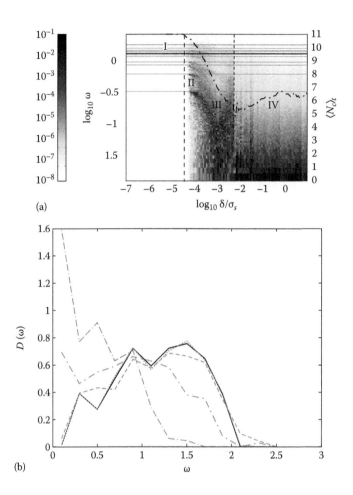

(a)

(b)

Figure 6.19 (a) The gray scale gives the normalized power spectrum $P_{\text{norm}}(\omega)$ obtained by vibrating a $N = 6$ MS bidisperse disk packing with $\Delta\phi = 10^{-3}$ above $\phi_J = 0.7651$ (MS packing 13 in Figure 6.1) as a function of the frequency ω and perturbation amplitude δ/σ_s. The solid gray horizontal lines indicate the eigenfrequencies of the dynamical matrix ω_i of the original static packing, the solid horizontal black line indicates the eigenmode along which the perturbations were applied, and the dashed vertical line indicates the perturbation amplitude $\overline{\delta}_c$ at which the first interparticle contact is broken. The dotted vertical line indicates the perturbation amplitude $\overline{\delta}_l$ above which particle rearrangements occur. On the right axis, we also show the time-averaged number of contacts $\langle N_c \rangle_t$ (dot–dashed line) versus the perturbation amplitude δ/σ_s. The labels I, II, III, and IV locate the "harmonic," "nonharmonic 1," "nonharmonic 2," and "liquid-like" regimes. (b) Density of vibrational frequencies $D(\omega)$ for MS packings of bidisperse disks with $N = 6$ at $\Delta\phi = 10^{-3}$ obtained using the dynamical (solid line) and displacement matrices for vibrated systems with perturbation amplitudes $\delta/\sigma_s \approx 10^{-5}$ (dotted line), 10^{-4} (dashed line), 10^{-3} (dot–dashed line), and 10^{-2} (dot–dash–dashed line).

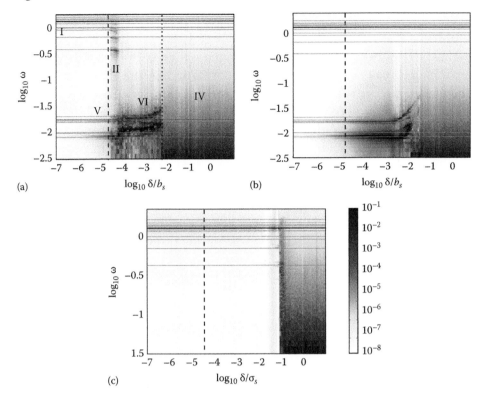

Figure 6.20   (a) The gray scale shows the normalized power spectrum $P_{norm}(\omega)$ obtained by vibrating a $N = 6$ bidisperse ellipse packing with aspect ratio $\alpha = 1.02$ at compression $\Delta\phi = 10^{-3}$ above $\phi_J = 0.771$ as a function of the frequency ω and perturbation amplitude δ/b_s. The solid gray horizontal lines indicate the eigenfrequencies of the dynamical matrix ω_i of the original static packing, the solid black horizontal line indicates the eigenmode along which the perturbations were applied, and the dashed vertical line indicates the perturbation amplitude $\overline{\delta}_c$ at which the first interparticle contact is broken. The labels I and II locate the "harmonic" and "nonharmonic 1" regimes. In regimes V and VI, quartic modes are found in the power spectrum. For perturbations above that given by the vertical dotted line, the system undergoes particle rearrangement events (regime IV). (b) $P_{norm}(\omega)$ for the same system in (a), but with double- not single-sided spring interactions between ellipses that were in contact in the static packing. (c) $P_{norm}(\omega)$ for the MS bidisperse disk packing in Figure 6.19a, but with double-sided spring interactions between disks that were in contact in the static packing.

Figure 6.20b shows the power spectrum for a vibrated packing of bidisperse ellipses (same as that in Figure 6.20a) in which contacting ellipses interact via double-sided spring potentials. The vibrational response is dominated by the quartic modes beyond $\overline{\delta}_c = 2 \times 10^{-5}$, where contacts would break in this system with only single-sided spring potentials. The power spectrum for vibrated ellipse packings with double-sided spring interactions differs strongly with that for vibrated disk packings with double-sided spring interactions, which is confined to the eigenfrequency of the perturbation until the perturbation amplitude reaches the liquid-like regime. (See Figure 6.20c.)

The critical perturbation amplitude above which contacts (that exist in the static packing) break scales as $\bar{\delta}_c \propto \Delta\phi/N$ [31]. Further, the characteristic amplitude beyond which perturbations along quartic modes give rise to quartic nonlinearities in the total potential energy scales as $\bar{\delta}_q \propto \Delta\phi^{1/2}/(\alpha-1)^{1/4}$ in packings of ellipsoidal particles. For ellipse packings with $N = 6$, aspect ratio $\alpha = 1.02$, and $\Delta\phi = 10^{-3}$, $\bar{\delta}_c < \bar{\delta}_q$ as shown in Figure 6.21a. As $\Delta\phi$ increases, $\bar{\delta}_c$ increases above $\bar{\delta}_q$, and quartic nonlinearities are activated before contacts break (Figure 6.21b and c).

Figure 6.22 shows the mean positional ($\Delta_T = \sqrt{\langle \Delta x_i(t)^2 + \Delta y_i(t)^2 \rangle_{i,t}}/b_s$) and angular displacement ($\Delta_R = \sqrt{\langle \langle \Delta\theta_i(t)^2 \rangle_{i,t}}$) together with the spectral response for vibrated bidisperse ellipse packings considered in Figure 6.20 at $\Delta\phi = 10^{-3}$ following perturbations along a quadratic translational mode (panel) and rotational

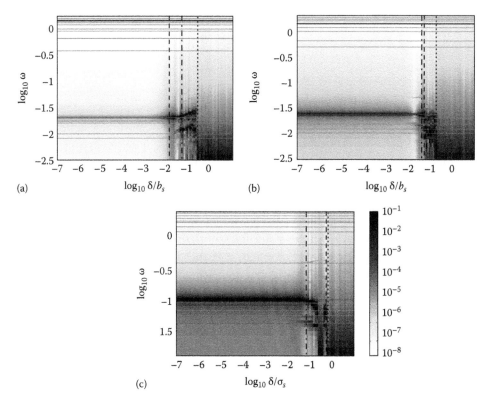

Figure 6.21   Gray scale of the normalized power spectrum $P_{\mathrm{norm}}(\omega)$ obtained for a $N = 6$ bidisperse ellipse packing with aspect ratio 1.02 at compressions (a) $\Delta\phi = 10^{-3}$, (b) $\Delta\phi = 10^{-2}$, and (c) $\Delta\phi = 10^{-1}$ above $\phi_J = 0.771$ as a function of the frequency ω and perturbation amplitude δ/b_s. The solid gray horizontal lines indicate the vibrational frequencies given by the dynamical matrix, the solid black horizontal line shows the frequency of the perturbation, the vertical black dashed line shows the perturbation amplitude $\bar{\delta}_c$ at which the first contact is broken, and the vertical dot–dashed line shows the amplitude $\bar{\delta}_q$ beyond which the quartic modes possess significant weight in the power spectrum. For perturbations above $\bar{\delta}_l$ (dotted line), the system undergoes particle rearrangement events.

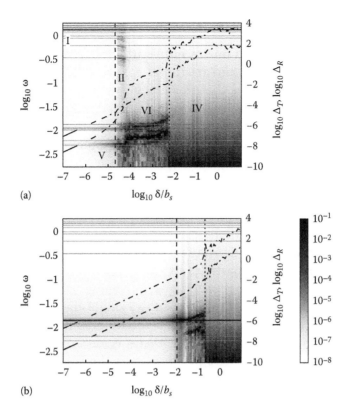

Figure 6.22 The gray scale shows the normalized power spectrum $P_{\text{norm}}(\omega)$ versus perturbation amplitude δ/b_s (a) along a translational quadratic mode and (b) along a rotational quartic mode (indicated by a solid black horizontal line) for the $N = 6$ bidisperse ellipse packing at $\alpha = 1.02$ and $\Delta\phi = 10^{-3}$ above jamming onset ϕ_J considered in Figure 6.20a. The solid gray horizontal lines indicate the vibrational frequencies given by the dynamical matrix of the original static configuration. The regions labeled I, II, IV, V, and VI are the same as those in Figure 6.20a. On the right axes, we plot the mean translational Δ_T (dot–dashed) and rotational Δ_R (dash–dash–dotted) displacement. The dashed and dotted vertical lines indicate $\bar{\delta}_c$ and $\bar{\delta}_I$, respectively.

quartic mode (panel). For the translational quadratic mode perturbation, in region I of the spectral response, $\Delta_T > \Delta_R$, and both Δ_T and Δ_R increase with the perturbation amplitude δ/b_s. In region V, Δ_R begins increasing faster than Δ_T. In region II, contacts break, the vibrational response spreads to all modes, and Δ_R remains below Δ_T. In region VI, the rotational quartic modes dominate the vibrational response and $\Delta_R > \Delta_T$. For sufficiently large perturbations, the system enters the liquid-like regime (region IV), and both Δ_R and Δ_T increase sharply. Note that Δ_R jumps at smaller perturbation amplitudes than Δ_T due to the formation of a "plastic solid" phase, where particles undergo large rotational fluctuations, but do not translate appreciably [24]. In Figure 6.22b, we show the spectral response and displacements after perturbations along a rotational quartic mode. In this case, the rotational displacement is always larger than the translational displacement.

6.4   Conclusions

In this chapter, we highlighted our recent results from computational studies of static packings of frictionless and frictional spherical particles and frictionless ellipsoidal particles. We believe that this chapter will spur new investigations of the structural and mechanical properties of jammed systems, especially systems composed of frictional and nonspherical particles. In this work, we enumerated and counted the MS, isostatic packings in small systems of 3D monodisperse spheres and bidisperse disks at jamming onset as well as over a range of pressure. We also reviewed calculations of the static shear modulus and distribution of mobilities as a function of the static friction coefficient for frictional disks. We showed that MS packings of frictional disks do not occur as single points in configuration space (as found for frictionless spherical particles), but instead as continuous geometrical families in configuration space. Finally, we measured the response of packings of ellipse-shaped particles to vibrations to quantify the effects of the breaking of interparticle contacts and "quartic" modes of deformation. We find that perturbations that are initially confined to a single eigenmode of the dynamical matrix can leak to low-frequency quartic modes even before the occurrence of contact breaking. These results emphasize that it is important to measure directly the vibrational response of particle packings near the jamming transition rather than using the dynamical matrix to infer the linear response. Direct measurements can be obtained, for example, by calculating the power spectrum of the fluctuating particle displacements.

References

1. J. D. Bernal and J. Mason. Packing of spheres: Co-ordination of randomly packed spheres. *Nature*, 188:910, 1960.

2. J. D. Bernal, J. Mason, and K. R. Knight. Radial distribution of the random close packing of equal spheres. *Nature*, 188:910, 1962.

3. J. G. Berryman. Random close packing of hard spheres and disks. *Phys. Rev. A*, 27:1053, 1983.

4. D. Bi, J. Zhang, B. Chakraborty, and R. P. Behringer. Jamming by shear. *Nature*, 480:355, 2011.

5. M. P. Ciamarra, A. Coniglio, and M. Nicodemi. Thermodynamics and statistical mechanics of dense granular media. *Phys. Rev. Lett.*, 97:158001, 2006.

6. A. Donev, I. Cisse, D. Sachs, E. A. Variano, F. H. Stillinger, R. Connelly, S. Torquato, and P. M. Chaikin. Improving the density of jammed disordered packings using ellipsoids. *Science*, 303:990, 2004.

7. A. Donev, R. Connelly, F. H. Stillinger, and S. Torquato. Underconstrained jammed packings of nonspherical hard particles: Ellipses and ellipsoids. *Phys. Rev. E*, 75:051304, 2007.

8. A. Donev, F. H. Stillinger, and S. Torquato. Unexpected density fluctuations in jammed disordered sphere packings. *Phys. Rev. Lett.*, 95:090604, 2005.

9. A. Donev, S. Torquato, and F. H. Stillinger. Pair correlation function characteristics of nearly jammed disordered and ordered hard-sphere packings. *Phys. Rev. E*, 71:011105, 2005.

10. S. F. Edwards and R. B. S. Oakeshott. Theory of powders. *Phys. A*, 157:1080, 1989.

11. G. R. Farrell, K. M. Martini, and N. Menon. Loose packings of frictional spheres. *Soft Matter*, 6:2925, 2010.

12. G.-J. Gao, J. Blawzdziewicz, and C. S. O'Hern. Geometrical families of mechanically stable granular packings. *Phys. Rev. E*, 80:061303, 2009.

13. G.-J. Gao, J. Blawzdziewicz, C. S. O'Hern, and M. D. Shattuck. Experimental demonstration of nonuniform frequency distributions of granular packings. *Phys. Rev. E*, 80:061304, 2009.

14. C. P. Goodrich, A. J. Liu, and S. R. Nagel. Finite-size scaling at the jamming transition. *Phys. Rev. Lett.*, 109:095704, 2012.

15. S. Henkes, M. van Hecke, and W. van Saarloos. Critical jamming of frictional grains in the generalized isostaticity picture. *Europhys. Lett.*, 90:14003, 2010.

16. D. Howell, R. P. Behringer, and C. Veje. Stress fluctuations in a 2d granular couette experiment: A continuous transition. *Phys. Rev. Lett.*, 82:5241, 1999.

17. Y. Jiao, F. H. Stillinger, and S. Torquato. Optimal packings of superballs. *Phys. Rev. E*, 79:041309, 2009.

18. Y. Jiao and S. Torquato. Maximally random jammed packings of platonic solids: Hyperuniform long-range correlations and isostaticity. *Phys. Rev. E*, 84:041309, 2011.

19. A. R. Kansal, T. M. Truskett, and S. Torquato. Nonequilibrium hard-disk packings with controlled orientational order. *J. Chem. Phys.*, 113:4844, 2000.

20. M. Mailman, C. F. Schreck, B. Chakraborty, and C. S. O'Hern. Jamming in systems composed of frictionless ellipse-shaped particles. *Phys. Rev. Lett.*, 102:255501, 2009.

21. T. S. Majmudar and R. P. Behringer. Contact force measurements and stress-induced anisotropy in granular materials. *Nature*, 435:1079, 2005.

22. H. A. Makse, D. L. Johnson, and L. M. Schwartz. Packing of compressible granular materials. *Phys. Rev. Lett.*, 84:4160, 2000.

23. H. A. Makse and J. Kurchan. Testing the thermodynamic approach to granular matter with a numerical model of a decisive experiment. *Nature*, 415:614, 2002.

24. C. De Michele, R. Schilling, and F. Sciortino. Dynamics of uniaxial hard ellipsoids. *Phys. Rev. Lett.*, 98:265702, 2007.

25. C. S. O'Hern, L. E. Silbert, A. J. Liu, and S. R. Nagel. Jamming at zero temperature and zero applied stress: The epitome of disorder. *Phys. Rev. E*, 68:011306, 2003.

26. I. K. Ono, C. S. O'Hern, D. J. Durian, S. A. Langer, A. J. Liu, and S. R. Nagel. Effective temperatures of a driven system near jamming. *Phys. Rev. Lett.*, 89:095703, 2002.

27. G. Y. Onoda and E. G. Liniger. Random loose packings of uniform spheres and the dilatancy onset. *Phys. Rev. Lett.*, 64:2727, 1990.

28. G. Parisi and F. Zamponi. The ideal glass transition of hard spheres. *J. Chem. Phys.*, 123:144501, 2005.

29. J. Russo and H. Tanaka. The microscopic pathway to crystallization in supercooled liquids. *Sci. Rep.*, 2:505, 2012.

30. P. Schall, D. A. Weitz, and F. Spaepen. Structural rearrangements that govern flow in colloidal glasses. *Science*, 318:1895, 2007.

31. C. F. Schreck, T. Bertrand, C. S. O'Hern, and M. D. Shattuck. Repulsive contact interactions make jammed particulate systems inherently nonharmonic. *Phys. Rev. Lett.*, 107:078301, 2011.

32. C. F. Schreck, M. Mailman, B. Chakraborty, and C. S. O'Hern. Constraints and vibrations in static packings of ellipsoidal particles. *Phys. Rev. E*, 85:061305, 2012.

33. G. E. Schröder-Turk, W. Mickel, M. Schröter, G. W. Delaney, M. Saadatfar, T. J. Senden, K. Mecke, and T. Aste. Disordered spherical bead packings are anisotropic. *Europhys. Lett.*, 90:34001, 2010.

34. G. D. Scott. Packing of spheres: Packing of equal spheres. *Nature*, 188:908, 1960.

35. G. D. Scott. Radial distribution of the random close packing of equal spheres. *Nature*, 194:956, 1962.

36. G. D. Scott and D. M. Kilgour. The density of random close packing of spheres. *J. Phys. D: Appl. Phys.*, 2:863, 1969.

37. K. Shundyak, M. van Hecke, and W. van Saarloos. Force mobilization and generalized isostaticity in jammed packings of frictional grains. *Phys. Rev. E*, 75:010301, 2007.

38. L. E. Silbert. Jamming of frictional spheres and random loose packing. *Soft Matter*, 6:2918, 2011.

39. L. E. Silbert, D. Ertas, G. S. Grest, T. C. Halsey, and D. Levine. Geometry of frictional and frictionless sphere packings. *Phys. Rev. E*, 65:031304, 2002.

40. L. E. Silbert, A. J. Liu, and S. R. Nagel. Structural signatures of the unjamming transition at zero temperature. *Phys. Rev. E*, 73:041304, 2006.

41. K. C. Smith, M. Alam, and T. S. Fisher. Athermal jamming of soft frictionless platonic solids. *Phys. Rev. E*, 82:051304, 2010.

42. E. Somfai, M. van Hecke, W. G. Ellenbroek, K. Shundyak, and W. van Saarloos. Critical and non-critical jamming of frictional grains. *Phys. Rev. E*, 75:020301, 2007.

43. C. Song, P. Wang, and H. A. Makse. Experimental measurement of an effective temperature for jammed granular materials. *Prof. Natl. Acad. Sci. U.S.A.*, 102:2299, 2004.

44. C. Song, P. Wang, and H. A. Makse. A phase diagram for jammed matter. *Nature*, 453:629, 2008.

45. A. V. Tkachenko and T. A. Witten. Stress propagation through frictionless granular material. *Phys. Rev. E*, 60:687, 1999.

46. S. Torquato, T. M. Truskett, and P. G. Debenedetti. Is random close packing of spheres well-defined? *Phys. Rev. Lett.*, 84:2064, 2000.

47. M. van Hecke. Jamming of soft particles: Geometry, mechanics, scaling, and isostaticity. *J. Phys. Cond. Matt.*, 22:033101, 2010.

48. K. Wang, C. Song, P. Wang, and H. A. Makse. Edwards thermodynamics of the jamming transition for frictionless packings: Ergodicity test and role of angoricity and compactivity. *Phys. Rev. E*, 86:011305, 2012.

49. N. Xu, J. Blawzdziewicz, and C. S. O'Hern. Reexamination of random close packing: Ways to pack frictionless disks. *Phys. Rev. E*, 71:061306, 2005.

50. Z. Zeravcic, N. Xu, A. J. Liu, S. R. Nagel, and W. van Saarloos. Excitations of ellipsoid packings near jamming. *Europhys. Lett.*, 87:26001, 2009.

CHAPTER 7

Forces in Static Packings

Robert P. Behringer

CONTENTS

7.1 Introduction

7.1.1 Forces and Stresses for Static Packings

This chapter focuses on packings and forces in dense non-cohesive granular materials. To maintain such a material in a steady static state requires the application of stresses at the boundaries. The forces and torques on each grain must add up to zero, and this condition (plus Newton's third law) results in forces being sustained across a material. Equivalently, in the absence of body forces, the divergence of the stress tensor must vanish. This effect has sometimes been referred to as "propagation," but it refers to a static, not dynamic, property. This is a strong mechanical constraint, and the way in which this constraint is satisfied for random packings is still at best a partially resolved issue. This chapter reviews some of the models and experiments that touch on this question; the interested reader can also find useful material in several reviews [1,2].

The vanishing of the divergence of the stress tensor (in the absence of body forces) is a direct result of the divergence theorem, similar to the vanishing of the divergence of the velocity field in an ideal fluid (in the absence of sources or sinks) or that of the electric field (in the absence of charges). The integral of the force acting on a body in a given direction can be written as $\int F_i dV = \int \partial \sigma_{ij}/\partial x_j dV = \int \sigma_{ij} da_j$, where the first two integrals are over a volume, V, of the body, the last integral is over the corresponding surface, A, of the body, and the repeated index j implies a summation [3]. For the static case of no body forces, the divergence of the stress tensor implies that the surface integral of the stresses vanishes. Imagine now that V is an infinitesimally thin slice of material. The surface integral of the stress on either side of the slice must be equal and opposite, implying that the forces on either side of the slice are equal and opposite.

To formulate the problem of force response, imagine a static granular material subject to some set of forces/stresses at its boundaries. Applying a small additional force to this system, either at a boundary or within, changes the stresses within the material and, in particular, the forces acting on each grain. The basic question then is what are the grain scale, and ultimately macroscale, stress changes that result from the additional force. While these changes must propagate throughout the system, presumably by a sound-like process, here we are concerned with the final state after any acoustic transients have vanished. This change constitutes the response of the system.

7.1.2 Jamming

A clearly identified feature of *static* packings is encapsulated in the idea of jamming, present (with different terminology) in soil mechanics models [14] but brought to the fore by Liu and Nagel [4]. A wide range of systems, foams, colloids, glasses, granular materials, etc., can be characterized in the context of a single universal "jamming diagram," as sketched in Figure 7.1a. In this figure, states to the right of the sloping line in a space of packing fraction ϕ (or effectively density) and shear stress are mechanically stable, hence, jammed, and those to the left are "unjammed," that is, not mechanically stable. The packing fraction is the volume in three dimensions or the area in two dimensions that is occupied by particles, normalized by the total system volume (or area). If the grains are formed from a single type of material that has a bulk density, ρ_b, then the density of the granular material is $\rho = \rho_b \phi$. The idea is that jammed states can resist small applied forces without deforming irreversibly, whereas unjammed states deform irreversibly under such actions. This idea has been compelling, and it has generated a substantial effort, as reviewed recently [5,6]. Of particular note are numerical simulations by O'Hern et al. [7]. These indicated that frictionless spherical particles satisfy such a picture, and that point-J, the point of lowest density, ϕ_J, with shear stress $\tau = 0$, for which jamming can occur is a special point that is well defined and, in principle, independent of protocol. More recently, Schreck et al.

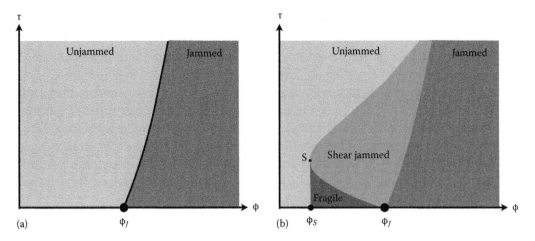

Figure 7.1 Jamming diagrams in a space of shear stress, τ, and packing fraction, ϕ. These diagrams pertain to $T = 0$ systems, which include granular materials. (a) is a sketch of the Liu-Nagel diagram [4], in which states at high enough ϕ and low enough τ exist in a jammed state, the darker region. (b) is the diagram that characterizes frictional particles, as seen by Bi et al. [10]. In addition to the jammed region of the Liu-Nagel diagram, there is a region below ϕ_J, extending to a lowest density, ϕ_S, where it is possible to start from a stress-free state, apply shear, and arrive at a jammed state, that is, green region marked shear jammed. Note that there is an intermediate region between the ϕ-axis and the shear-jammed region where fragile states can exist.

[8] have argued that jamming is protocol dependent and that there may not in fact be a unique jamming density.

Recently, Zhang et al. [9] and Bi et al. [10] have shown that for frictional particles, the situation is more complex, as sketched in Figure 7.1b. For the frictional case, it is possible to create stress-free states below ϕ_J and, by shearing these states, produce jammed states that are characterized by stress anisotropy. That is, they are characterized by $\tau \neq 0$. In addition, the network of contacts, as characterized by the fabric (discussed later), is anisotropic. Intermediate between the zero-stress and shear-jammed states are highly anisotropic fragile states, as originally described by Cates et al. [11], in a paper that was roughly contemporaneous with the Liu-Nagel jamming proposal. The details of the fragile and shear-jammed states are discussed later. The remainder of this chapter deals with granular materials that are jammed (or perhaps shear jammed). Although the original work of Bi et al. [71] was carried out with frictional particles, Kumar and Luding [12] have since found that shear jamming can also occur in frictionless systems. See also Chapter 6 by O'Hern for a description of computational studies of jammed systems.

7.1.3   Continuum Models of Granular Materials

The manner in which forces are described in jammed granular materials has its roots in soil mechanics, dating to the time of Coulomb [13]. More recent soil mechanics is contained in sophisticated models, including critical state soil mechanics [14], and its descendants. These models have been constructed to address the obvious needs of engineers to build granular handling devices, to ensure the stability of foundations, and to carry out many other practical operations involving granular materials. The physical approach is to assume that under shear loads, granular materials are elastic like (with the possibility of having a rigid material, hence infinite stiffness) but that, for large enough shear strains, the material deforms irreversibly (plastically) along planes where the stresses reach the yield curve. In the Coulomb picture, this curve is given by the condition $|\tau| = \mu P$. In general, the yield curve is not a straight line, and modern soil mechanics models contain considerable complexity [15–17].

In considering force in jammed granular materials, an intermediate step that deserves mention is the Janssen model [18,19]. This model was developed to understand how stresses are carried in tall granular storage devices, such as silos, and it emphasizes the facts that (1) stresses in granular materials are frequently not isotropic and (2) friction can play an important role that differentiates granular solids from more conventional continua. Figure 7.2 illustrates the key ideas behind this model. A tall container, such as a cylindrical silo, filled with a granular material. In an incompressible fluid, the pressure P increases linearly with depth z (measured from the container's top), $P \propto z$. Janssen's

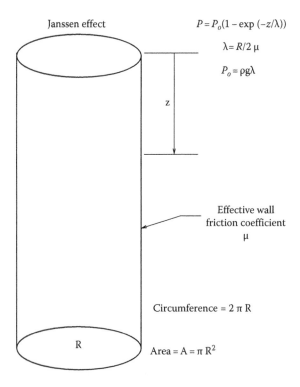

Figure 7.2 Sketch of container of granular material, as described by the Janssen effect.

analysis yields a very different result, with the pressure saturating exponentially with depth,

$$P = P_o \left(1 - \exp\left(\frac{-z}{\lambda}\right) \right) \tag{7.1}$$

where $\lambda = R/2\mu$, μ is an effective friction coefficient, and $P_o = \rho g \lambda$, with ρ the density of the granular material.

This result is obtained as follows. We average over thin horizontal slices of thickness dz and assume a proportional relation between the vertical stresses P acting on a horizontal plane and the corresponding normal stresses on the outer wall. These normal stresses result in a proportional tangential frictional stress σ_t from the outer wall (acting upward) $\sigma_t = \mu P$. Force balance on a horizontal disk of material of thickness dz is then given by

$$\rho g A dz + P(z - dz)A = P(z)A + C dz \mu P. \tag{7.2}$$

Here, $A = \pi R^2$ and the circumference, $C = 2\pi R$. This can be rearranged to give

$$\frac{P(z) - P(z - dz)}{dz} = \rho g A - C\mu P. \tag{7.3}$$

The solution to this differential equation (in the limit $dz \to 0$) is a vertical pressure profile that exponentially approaches an asymptote P_o. Note that these are averages and the stresses are in general not isotropic.

7.1.4 Experimental and Computational Evidence for Heterogeneity

Some time ago, an upsurge of interest in force structures was triggered by a number of studies, including discrete element method (DEM)/molecular dynamics (MD) simulations [20], acoustic experiments [21] and photoelastic and time-resolved force measurements [22]. The important aspect of these works was the idea of force heterogeneity for dense granular materials (i.e., jammed materials). Forces in these materials are carried primarily on structures known as force chains. In fact, this phenomenon was discovered at least half a century ago by Dantu [23] and Wakabayashi [24], using photoelastic particles, with more recent experimental studies by Dresher and Joselin de Jong [25,26], and computational innovations by Cundall et al. [27,28]. Force chains can be isotropic or anisotropic, depending on the stress state and preparation of a sample, as shown in Figure 7.3 (from Majmudar and Behringer [29] and Zhang et al. [9]). Figure 7.3 shows networks that resulted from isotropic compression, part (a), and from pure shear, starting from an unjammed state, part (b). These two different types of networks are not equivalent, and jamming/force response for anisotropic vs. isotropic states need to be carefully differentiated. Additional details on how photoelastic materials can be used to obtain quantitative force data are given at the end of this chapter.

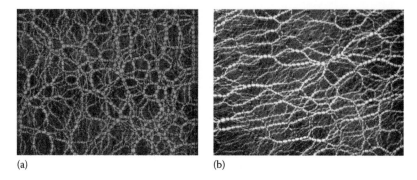

(a)   (b)

Figure 7.3 Contrasting photoelastic images of an isotropically compressed quasi-2D system of photoelastic disks (a) and a shear-jammed system of the same photoelastic disks (b). Note the long quasi-linear force chains in the sheared system, vs. the rather short entangled force network of the isotropically compressed system. More details on the use of photoelastic materials to obtain quantitative force data are given later.

7.2 Models for Force Transmission in Dense Granular Materials

7.2.1 Overview

It is clear, however, that regardless of whether states are isotropic or not, forces show fluctuations at the microscale. The statistical properties at all scales, and how the microscale physics is manifested at the macroscale, are important questions. The idea that forces might have a stochastic description tied to force chains leads to a flowering of novel models, including the grain-scale Q-model [31], the three-legged model and a continuum limit convection-diffusion model [32], the oriented stress linearity (OSL) model [33,34,85], the Double-Y model [35,38], and an elastic-grain picture [36,37]. The mathematical classes of proposed models are particularly surprising, whether based directly on these models if they are partial differential equations (PDEs), or on their continuum limit: essentially, all the standard PDE types, hyperbolic, elliptic, parabolic, as well as nonstandard PDE types, are represented [14,31,32,35,37–43]. A number of experiments [45–47,59,62] were designed to test these models to determine which class of solutions was appropriate.

A brief review of the different classes of second-order partial differential equations is appropriate. Parabolic PDEs are typified by the heat equation $\partial u/\partial x = D\partial^2 u/\partial y^2$, hyperbolic equations by the wave equation $\partial^2 u/\partial x = c^{-2}\partial^2 u/\partial y^2$, and elliptical equations by Laplace's equation $\partial^2 u/\partial x + \partial^2 u/\partial y^2 = 0$. x and y are spatial coordinates, and u is representative of forces, stresses, or similar quantities. The diffusion coefficient D in the heat equation must have units of length, and the "sound speed" in the wave equation is dimensionless.

A number of models addressed the impact of disorder on the force networks. Disorder in frictional granular materials has two different origins, one source of randomness, which can be seen in Figures 7.3 and 7.5, is rooted in the disordered spatial structure of grain packings. This effect can be determined quantitatively through measures such as the radial distribution function, $g(r)$. (cf. Luding et al. [48] on the effect of polydispersity in the context of granular gases.) A different source of disorder comes from the fact that, due to the fact that the Coulomb frictional constraint $|F_t| \leq \mu_s F_N$ is an inequality, the set of contact forces that satisfy force and torque balance may not be (and, in fact, are most likely not) unique. (F_t and F_N are the tangential and normal forces at the contact and μ_s the coefficient of static friction.) This means that the contact forces in granular materials with frictional interactions may be assumed to, in fact, be random values, even if the grains exist in a perfectly ordered lattice. The goal of the lattice models was to introduce essential randomness by making the forces at contacts random variables, subject to mechanical stability, but without introducing the complexity of spatial disorder. Other source of complexity for granular packings are also worth mentioning. These include the fact that above jamming, particles typically have more contacts than they need for mechanical stability. Forces between grains are often nonlinear in the relative displacement of contacting grains. And, grains need not have simple "spherical" shapes, that is, spheres in three dimensions or disks in two dimensions. As a consequence, if experiments or simulations to measure force response (or many other local properties) are repeated under nominally identical conditions, the actual measured response typically varies substantially. The next several subsections explore some of the models that were proposed to address the issues that are raised earlier.

7.2.2   The Stress-Dip Puzzle and the OSL Model

The OSL model was constructed in order to explain curious observations reporting that the maximum pressure P in a sandpile was not necessarily directly below the pile's peak but, rather, could occur on a ring of nonzero radius [49–52] (see also Savage [53]). In some cases, the pressure at the base was actually reported to have a local minimum under the peak, the so-called stress dip phenomenon. The 2D OSL model has a Janssen-like constitutive relation of the form $\sigma_{xx} = \eta\sigma_{zz} + \mu\sigma_{xz}$, where z is the vertical and x is the horizontal direction. When coupled with the constraint of stress balance, this leads to the proposal that (static) stresses within a granular material satisfy a hyperbolic PDE in the spatial variables, x and z. Bouchaud et al. then showed that this model could predict a stress dip. Savage [53] argued that soil mechanics models [14] can also account for a stress dip. Elasto-plastic soil mechanics models [14] are elastic below yield and are described in this case by elliptic equations (above yield, they are characterized by hyperbolic equations). Hence, the OSL and soil mechanics approaches are inherently different types of models.

Initially, it was not clear whether the stress dip was real or an experimental artifact, as early experiments on heaps were inconclusive. Eventually, Vanel et al. [54] and Geng et al. [55] showed that the stress dip was intrinsically tied to how the system was prepared. If the preparation history induced a non-isotropic fabric (defined later), it was possible to create a stress dip. The idea is that a fabric, that is, the network of contacts, that leads to a strong force network which is inclined along the slope of the heap can effectively create "buttresses" that shield the more central part of the heap. In fact, this picture can be captured by an anisotropic elasto-plastic model. To demonstrate that preparation and fabric was key to the presence/absence of the stress dip, Vanel et al. and Geng et al. prepared heaps of ordinary sand and quasi-2D photoelastic particles by different protocols that were designed to produce fabrics that spread the weight of the sandpile with varying degrees of isotropicity. When the granular material was deposited on the supporting surface by pouring the material through a sieve, the pressure had its maximum at the center (Figure 7.4). When the material was poured onto the surface with a funnel and allowed to avalanche down the slope of the heap as it formed, the pressure was clearly at a local minimum at the center of the heap. Photoelastic studies

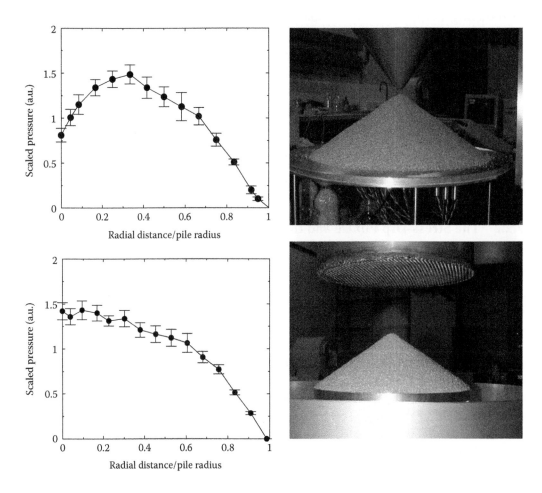

Figure 7.4 Photographs of different heap preparation protocols (right) and data for the corresponding base pressure profiles as a function of radial distance from the center of the heap, after Vanel et al. [54]. The top photograph is for a heap formed by the flowing material out of a funnel onto the top of the heap, and the lower photograph is for a heap that was formed by sieving sand onto the base and then lifting the seive. The top (bottom) data pertain to the funnel (seive) protocol. Note the obvious stress did for the funnel protocol that is absent in the seive protocol.

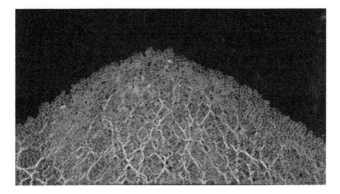

Figure 7.5   Photoelastic image of a heap formed from elliptical photoelastic particles via a protocol that is roughly similar to the seive method mentioned earlier [56]. Note that the force chains arriving at the base are roughly uniformly distributed in this, unlike what might occur when a force dip occurs.

clearly showed that the internal force structures also depended sensitively on the preparation protocol. Figure 7.5 shows the structure and fabric that occurs for a heap of photoelastic ellipses [56]. Interestingly, for ellipses, the shape of the heap (e.g., the angle of repose) is not necessarily affected very much by the preparation protocol.

The key difference between the two preparation protocols concerns the stress tensor and contact anisotropy, as represented by the fabric tensor F. These quantities are computed [11,57] formally from the branch vectors, $\vec{r}_{ij}$, that is, the vectors from the center of mass of a particle i to a contact from particle j, and from the force at that contact, f_{ij}:

$$\hat{\sigma} = \frac{1}{V} \sum_{i \neq j} \vec{r}_{ij} \otimes \vec{f}_{ij},$$

$$\hat{F} = \frac{1}{N} \sum_{i \neq j} \frac{\vec{r}_{ij}}{\|\vec{r}_{ij}\|} \otimes \frac{\vec{r}_{ij}}{\|\vec{r}_{ij}\|}. \qquad (7.4)$$

The sums are computed by first summing over all contacts for a given particle and then by summing over all particles ($\otimes$ refers to the exterior product of the vectors). Note that a continuum stress field can be computed by appropriate coarse-graining strategies [30]. The pressure is then $P = (\sigma_2 + \sigma_1)/2$ and the shear stress is $\tau = (\sigma_2 - \sigma_1)/2$ for the 2D case, where the σ_i are the eigenvalues of $\hat{\sigma}$.

7.2.3   The Q-Model and the Three-Leg Model

The Q-model of Coppersmith et al. [31] was the first lattice model to try to understand force propagation in granular materials from microscopic point of view. Its important features are sketched in Figure 7.6. One imagines that there is an ordered packing of the grains, with random transmission of the force from one grain in the packing to the next two nearest-neighbor particles. The force, F, that

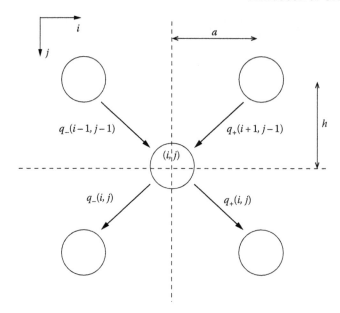

Figure 7.6   Sketch showing how forces are carried, via a probabilistic scheme, for Q-model. (After Coppersmith, S.N. et al., *Phys. Rev. E*, 53, 4673, 1996.)

"arrives" at a given particle from above is transferred to the left lower neighbor by the fraction qF, with the fraction $(1-q)F$ being transferred to the right lower neighbor. Here, the eponymous q is a random variable in the unit interval, $0 \leq q \leq 1$. If the distribution for q is flat, then the force "random walks" from the top to the bottom of the packing, effectively a diffusive process that leads to a parabolic description in the long-wavelength, or continuum limit. Other behavior can occur for different distributions of q.

The Q-model was the first attempt at using a lattice description and is an inherently scalar model that can only describe forces in a single direction. A different lattice model, the "three-leg" model, that is tensorial in nature was proposed by Claudin et al. and is sketched in Figure 7.7. This microscopic tensoral model was constructed to justify their proposition that force transmission within a granular

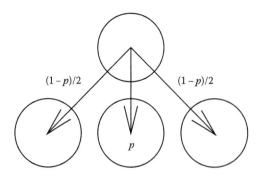

Figure 7.7   Sketch of forces for three-legged model. (After Claudin, P. et al., *Phys. Rev. E*, 57, 4441, 1998.)

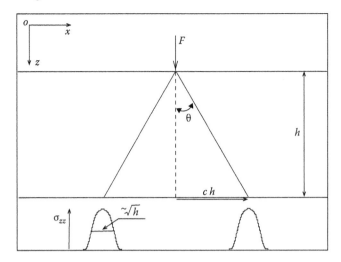

Figure 7.8 Sketch showing features of convection–diffusion model.

material might be described by a wave equation. They then considered the possibility of weak disorder and argued that force transmission would be wavelike, with additional diffusive terms. A point force applied to such a model granular system would lead to forces propagating along rays, while broadening at the same time, as suggested by Figure 7.8. The idea of a wave equation with diffusion was also pursued by Kenkre [58].

7.2.4 Double-Y Force Chain Model

Bouchaud et al. [38] and Socolar et al. [35] have taken a rather different approach to addressing the inherent disorder in granular materials in a Boltzmann-like force chain splitting model that is illustrated in Figure 7.9. Their double-Y model considers stochastic processes whereby forces, carried on chain-like structures, redistribute stresses as force chains bifurcate or merge. The idea behind this model is to incorporate as much of the relevant physics of these splitting and merging process which give their name to the double-Y model.

7.2.5 Elastic Grain Models

Goldenberg and Goldhirsch [36,37] have considered the possibility, which is at least conceptually consistent with elasto-plastic models, that grain packings respond elastically for small applied forces. A sketch in Figure 7.10 suggests the nature of their model, in which "grains" in a 2D packing of disks interact via linear springs that may be bi- or unidirectional. These models yield, on microscales, behaviors associated with granular materials, including force chains. Goldenberg and Goldhirsch pursued this idea [36] by arguing that the addition of friction at contacts would tend to sustain an elastic-like response. They concluded that, depending on the scale of the observation, both force chain like and elastic behavior could be observed.

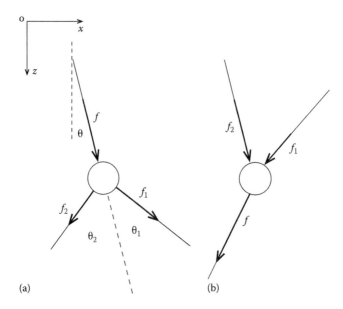

Figure 7.9 Sketch showing force chain splitting and merging, from the double-Y model of Bouchaud et al. [38] and Socolar et al. [35].

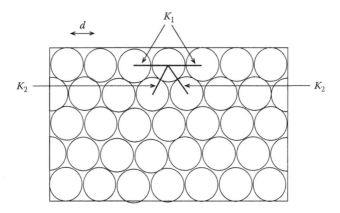

Figure 7.10 Sketch showing the granular spring model of Goldenberg and Goldhirsch [36,37]. Note that individual grains are connected by springs that may have variable spring constants, K_i.

7.3 Experiments

There has been considerable interplay between experimental tests, stimulated for instance by the Q-model and the OSL model, and the development of new models in response to experiments. Both experiments and numerical simulations [54,55, 59,60,81] have shown that the existence of non-isotropic textures due to different deposition procedures of sandpiles or other packing procedures can determine the way forces are transmitted and produce different stress distributions.

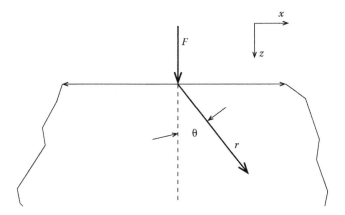

Figure 7.11   Sketch of coordinates for describing an elastic response to a point force.

Many of these experiments involved the idea of a granular Greens function in which experimenters applied a small local force to a granular sample and observed the response. The idea of determining a Greens function is, in fact, fairly natural. To understand how to implement such an approach, consider the response that one would expect from a material that is subject to a small local force at its upper surface, as in the schematic shown in Figure 7.11. The solution [3] for a point force applied to the upper boundary of a (semi-infinite) plane of a linear elastic material (thickness t) is given by a stress tensor that, in cylindrical coordinates, has only one nonzero element, namely, $\sigma_{rr} = -2F\cos(\theta)/(\pi tr)$. It is not hard to show that if we define a width of the response to a point force by $W^2 = \int x^2 \sigma_{zz}(x,z)dx / \int \sigma_{zz}(x,z)dx$, then $W \propto z$ for an elastic material. Here, z is the distance below a point contact and x is the horizontal direction of the material. By contrast, for a model that is diffusive in character, such as the Q-model, this width grows as $z^{1/2}$. For a convection–diffusion model, one expects a response that consists of two rays, along the characteristics, that broaden diffusively, that is, as $z^{1/2}$, with depth.

Several examples of the earliest work using granular Greens function approaches include studies by Reydellet and Clément [46], by Rajchebach et al. [47], by Mueggenburg et al. [55], and by Geng et al. [45,62]. Reydellet and Clément [46] used weak localized driving on a 3D container of grains. They found that the response as a function of depth of the granular layer consisted of a single "peak" as a function of radial distance from a line directly beneath the driver. The width of this peak grew linearly with depth, indicating an elastic-like model. Other experiments by Da Silva and Rajchenbach [47] used rectangular photoelastic blocks that were densely packed and stacked in ordered rows. These authors found a response that indicated diffusive behavior. Mueggenburg et al. [59] carried out 3D measurements using glass spheres by creating ordered and disordered packings. For disordered packings, these authors found a broad central response to a point force, although they could not resolve the difference between an elastic and a diffusive response. For ordered packings, Mueggenburg et al. found that forces were carried predominantly along preferred directions that reflected the packing structure.

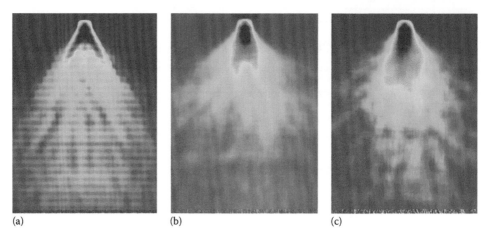

(a)   (b)   (c)

Figure 7.12   Response for increasing amounts of spatial disorder: (a) regular hexagonal packing of monodisperse photoelastic disks; (b) polydisperse disk packing; (c) packing of pentagons [62]. In this and the next several figures, the response is shown in color for an ensemble of multiple experiments where a point force was applied to the upper boundary of the indicated stacking of photoelastic particles. More details about photoelastic techniques are given at the end of this chapter.

For example, the response for face centered cubic (FCC) packings followed along three well-defined lines, intrinsic to the FCC structure, with moderate broadening with depth. By contrast, hexagonal close packed (HCP) packings showed a response along cones that broadened more rapidly with depth than the FCC packings. Interestingly, HCP packings did not show the simple connected paths (i.e., a fabric issue) seen in FCC packings. Geng et al. used a Greens function approach on 2D packings of photoelastic disks and pentagons [62], thus probing the effect of order–disorder on force transmission. In addition, they changed the friction of their particles by coating them with Teflon tape, thus reducing the possibility of disorder due to random tangential forces [45]. Figures 7.12 and 7.13 show the effect of increasing disorder on the local mean photoelastic response. There is indeed preferred propagation reminiscent of wavelike behavior, but the preferred direction is determined by the packing crystalline order, not independently, as suggested by Claudin et al. [32]. For the disordered packings of pentagons, the force response consists of a broad central peak that grows linearly with depth, indicating an elastic response. Figures 7.14 and 7.15 show the mean photoelastic response for square crystalline packings of photoelastic monodisperse disks that have nominal friction coefficients of $\mu \sim 0.9$ (1) and $\mu \sim 0.5$ (2). Note that these data show a deeper penetration of the response along the crystalline axes than for the regular hexagonal packing. In addition, packings with reduced inter-grain friction show deeper penetration and less widening of the peaks, demonstrating that frictional disorder plays an important role in force transmission.

The data of Figures 7.14 and 7.15 are reasonably well modeled by a convection–diffusion equation. As noted, the variances of the central peaks do not

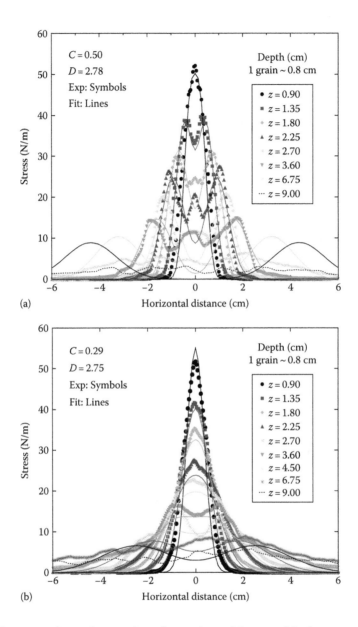

Figure 7.13 Same results as the previous figure, but with a graphical representation of the data. Note that for the ordered monodisperse disk packing (a), forces are carried primarily down the principal directions of the hexagonal lattice. For the bidisperse packing (b), there is still a remnant of the regular packing behavior, but there is now a dominant central peak that was only barely visible for the ordered packing. For the pentagonal packing (c), there is only a central response, with a width, W, that grows linearly with depth, z. (*Continued*)

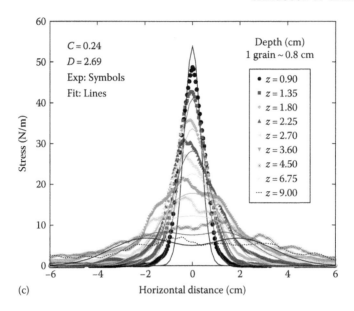

(c)

Figure 7.13 (*Continued*) Same results as the previous figure, but with a graphical representation of the data. Note that for the ordered monodisperse disk packing (a), forces are carried primarily down the principal directions of the hexagonal lattice. For the bidisperse packing (b), there is still a remnant of the regular packing behavior, but there is now a dominant central peak that was only barely visible for the ordered packing. For the pentagonal packing (c), there is only a central response, with a width, W, that grows linearly with depth, z.

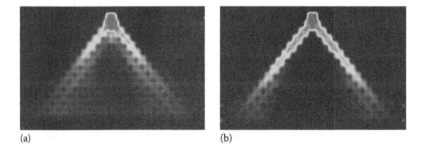

Figure 7.14 Photoelastic response for a point force response for square packings of photoelastic disks, where part (a) has a higher interparticle friction coefficient than part (b), as discussed in [45]. These results are ensemble averages over multiple experiments.

vary with depth z as $z^{1/2}$, but rather are more suggestive of a linear scaling. The proposal of Claudin et al. that ordered packings might have wavelike responses and that disorder causes such responses to broaden with depth is consistent in these studies although, on balance, disordered packings behave more nearly like elastic media, in terms of their response to weak point forces.

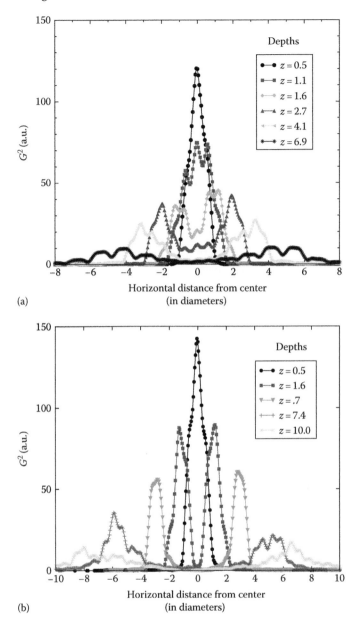

Figure 7.15 Same results as the previous figure, but with graphical representation of the data. Note that for the lower-friction particles, the strong responses penetrate deeper and broaden more slowly.

It is also possible to probe the response of a granular system by considering applied forces that include tangential components. This provides a good test of the ideal elastic model for disordered systems, and the effect of the lattice structure of spatially ordered packings. Figures 7.16 through 7.18 show data for nonnormal point forces applied to systems of pentagonal (hence disordered) and hexagonal (ordered) particles. For pentagonal particles, there is a qualitative agreement with

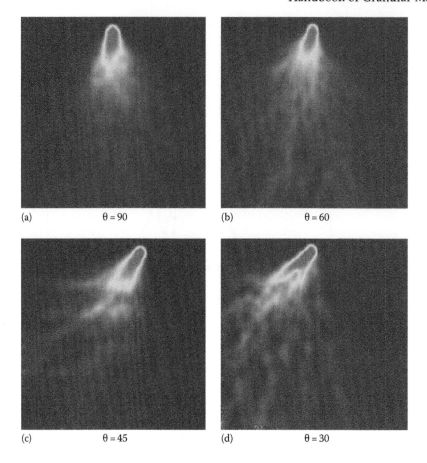

(a)   θ = 90   (b)   θ = 60

(c)   θ = 45   (d)   θ = 30

Figure 7.16   Point force response for a force applied at the indicated angle for a layer formed from pentagonally shaped photoelastic particles. (From Geng, J. et al., *Phys. D*, 182, 274, 2003.)

what one would expect from an elastic material, but the solid lines show that as the angle of the applied force from vertical increases, the agreement with an ideal elastic picture worsens. For the hexagonal packing, Figure 7.18, the response follows the two obvious principal lattice directions for small inclinations of the applied force from vertical. Interestingly, for large enough tilts from vertical, the forces are also carried along a different lattice direction that is obvious in Figure 7.18.

7.4   Granular Materials Near Jamming

Much of the work discussed earlier applies to systems well above the threshold for jamming. However, it is important to consider at least briefly dense granular materials near the jamming point. There has been an extensive range of studies of systems near jamming, with much of this work focused on idealized models of particulate systems, such as frictionless spherical particles. Several reviews [5,6] provide a resume of these studies, and this chapter will focus instead on systems of particles where friction does play a role. In particular, we describe the *shear-jamming transition* that is more noticeable in frictional systems (although it does occur, more weakly, with frictionless particles [12]).

Figure 7.3b shows an example of a system of photoelastic disks, which is in a shear-jammed state. In order to understand where such states come from, we consider Figure 7.1b. In the range of packing fractions, $\phi_S \leq \phi \leq \phi_J$, it is possible to generate completely stress-free states, that is, states for which both P and τ are zero. If a shear strain, γ, is applied in that density regime, starting from a $\tau = P = 0$ state, the system first passes through a regime where the system remains

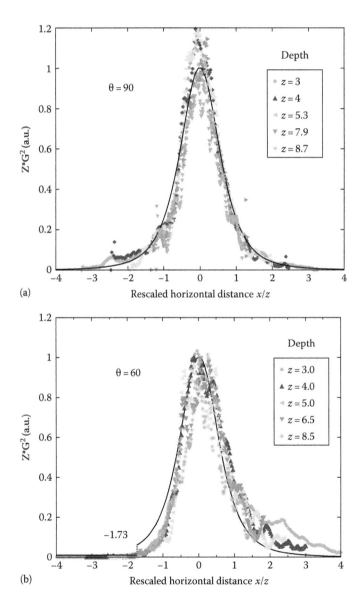

Figure 7.17   Same results as the previous figure, but with graphical representation of the data.   (*Continued*)

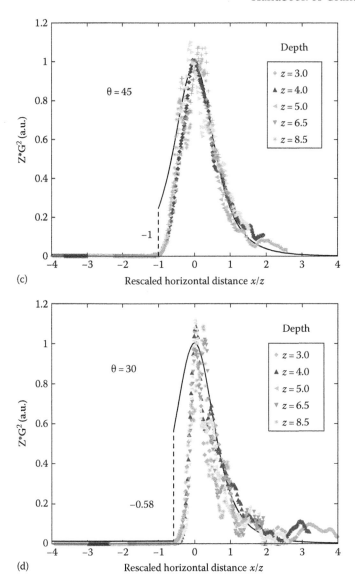

Figure 7.17 (*Continued*)   Same results as the previous figure, but with graphical representation of the data.

essentially stress free, but where the contact network, or fabric, becomes increasingly anisotropic. The reason for the growth of anisotropy is easy to understand in the context of Figure 7.19. Note that the effect of shear on a 2D rectangular region is to compress it in one direction and to expand or dilate it in the other direction. In the case of simple shear, there is an additional rotation, but in all cases, shear strain is volume preserving. Hence, particle separations are decreased in the compressive direction and increased in the dilational direction, assuming that the shear is applied uniformly across the material, something that is, in fact, nontrivial to achieve experimentally. At some point during shearing of frictional

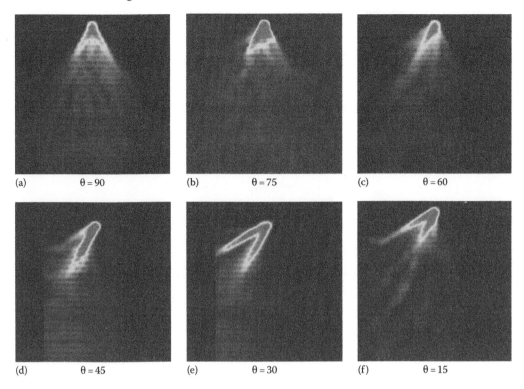

(a) $\theta = 90$ (b) $\theta = 75$ (c) $\theta = 60$

(d) $\theta = 45$ (e) $\theta = 30$ (f) $\theta = 15$

Figure 7.18   Response to a nonnormal point force applied to a hexagonally packed layer of monodisperse disks [45]. Note that for smaller angles (from vertical), the force are carried primarily along the two most vertical crystal axes of the packing. For larger angles from vertical, the forces are carried on one of the previous crystal axes, plus a new crystal axis that is more nearly horizontal.

particles, a network of strong force contacts propagates from one boundary to the other along the compressional direction, but there are at best only weak contacts along the dilational direction. This is a fragile state, which is solid like along the compressive direction, but "squishy" along the dilational direction. Such states were proposed in the context of colloids by Cates et al. [11]. If the shear strain is continued past the onset of the fragile states, something very interesting happens: the strong network eventually percolates in both directions, and the system arrives at a "shear-jammed" state that is solid like in all directions. In Figure 7.20, we show examples from the photoelastic shear studies of Ren et al. [63] of various states in the shear-jammed region through, and past, the shear-jamming transition. Note that as the system is sheared further and further past the onset of shear jamming, the system tends to become increasingly isotropic.

At least in two dimensions, one can see how friction facilitates this transition to shear jamming. Consider particles that are part of a force chain that has formed due to shear strain. Particles that are members of a force chain typically experience two strong, roughly diametric contacts. Idealizing this situation, we consider a particle that is in force and torque balance and that experiences exactly two

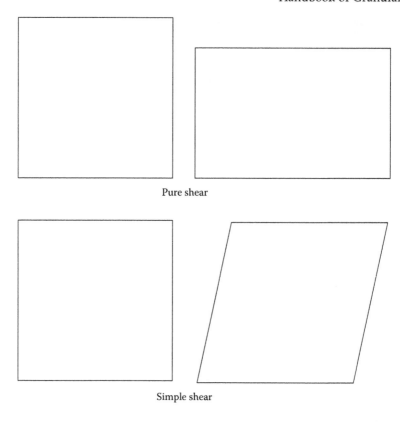

Pure shear

Simple shear

Figure 7.19   Sketch contrasting pure and simple shear strains. Note that in both cases, there is one direction that expands (dilates) under shear strain and another that is compressed. In both cases, the area occupied by the material is constant.

contacts, as sketched in Figure 7.21. The force at each contact for the middle particle has both a normal and a tangential (hence frictional) component. We choose a coordinate system such that x-axis bisects the angle between the two contacts and such that the contacts are inclined at an angle θ relative to the y-axis, as sketched. Torque balance requires that the two tangential forces are equal in magnitude and have an opposite sense of rotation. Force balance then requires that the two normal components must be equal and that the ratio of the tangential to normal components of any pair of these two contact forces satisfies $F_t/F_n = \tan(\theta)$. Note that the ratio F_t/F_n is constrained by Coulomb friction to be $F_t/F_n = \pm\tan(\theta) \leq \mu$, where μ is the coefficient of friction between the two particles. Thus, the smaller the friction coefficient, the more diametric the pairwise contacts must be. Force chains for low-friction particles must be nearly straight for pairwise contacts, whereas for higher-friction particles, force chains can bend. Of course, it is still possible to have force chains for low or zero friction, but the chains must be buttressed by additional particles. In any event, this argument indicates that as a system of frictional grains is sheared past the onset of fragility, force chains can spread by bending in the dilational direction without breaking, thus transmitting stress in all directions. Continued shear can distort these force chains, cause them to

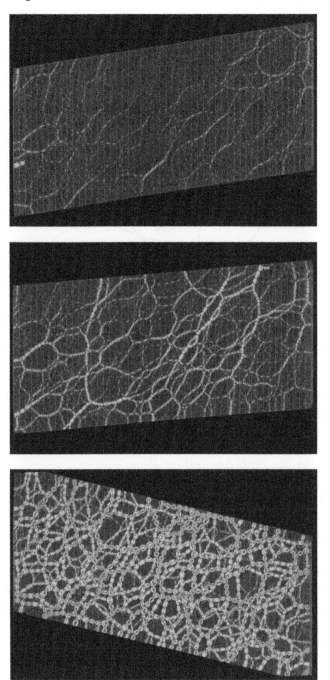

Figure 7.20 Images showing the photoelastic response of disks that are subject to simple shear, after Ren et al. [63]. Top to bottom shows states that are, respectively, fragile, near shear jamming, and well above shear jamming. In these experiments, a layer of particles is sheared, using a special apparatus that allows for uniform shear strain across the whole layer. The initial state is one that is stress free and that has a parallelogram shape that is "tilted down" at the lower left corner. The final state is one where the tilt is reversed. Note that throughout the shearing process, the density of the system remains constant.

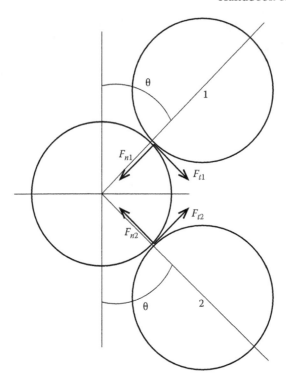

Figure 7.21   Sketch of possible force- and torque-balanced state for a disk that has exactly two contacts with two other disks, labeled "1" and "2." Note that the total force, that is, the vector sum of the normal and tangential forces, is along the y-axis.

buckle [64], or cause individual contacts to fail. These processes lead to added lateral contacts that make the system more homogeneous.

Note too that strong pairwise contacts imply anisotropy in the stress tensor, and if there are more contacts associated with the compressive direction of shear, there also is anisotropy in the fabric. Again referring to Figure 7.21, the fabric tensor associated with the central particle is diagonal for this coordinate system, with components $F_{11} = 2\sin^2(\theta)$ and $F_{22} = 2\cos^2(\theta)$. Hence, the difference, $F_{11} - F_{22} = 2(\cos^2(\theta) - \sin^2(\theta)) = 2\cos(2\theta)$, is nonzero for any pair of frictional contacts with finite μ. The situation for the stress tensor contribution from a particle with pairwise contacts is "maximally anisotropic" in the sense that one diagonal component of the stress or force moment tensor is nonzero, and all other components are zero (in the coordinate system shown here). In this case, the one nonzero component is σ_{yy}. Then the contribution to the particle-scale "shear stress," $\tau = (\sigma_2 - \sigma_1)/2$ and the particle-scale "pressure," $P = (\sigma_2 + \sigma_1)/2$ are $\tau = \sigma_{yy} = P$.

There are several other aspects of the shear-jamming transition reported in Bi et al. [10] that are of interest. First, the amount of shear that is needed to reach a shear-jammed state from an initial stress-free state depends on ϕ, as shown in Figure 7.22. If ϕ is just below ϕ_J, then the amount of strain to reach shear jamming is small. If ϕ is just above ϕ_S, then the amount of strain needed to reach a

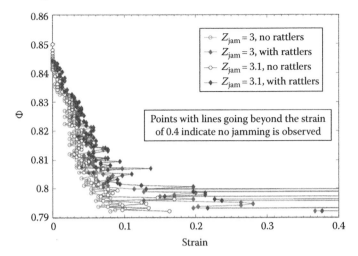

Figure 7.22   Amount of shear strain needed for a given φ to reach a shear-jammed state. (After Bi, D. et al., *Nature*, 480, 355, 2011.)

shear-jammed state is rather large and appears to diverge, that is, there is a lowest $\phi = \phi_S$ such that shear jamming cannot be achieved with finite strain.

It is interesting to ask whether there are preferred "state" variables for the shear-jammed states. Bi et al. [10] found that, in the shear-jamming regime, the pressure P and the shear stress τ collapse quite well when expressed as functions of the fraction of particles that are non-rattlers f_{NR} (Figure 7.23). In fact, this probably is not true at higher densities, that is, above ϕ_J, since f_{NR} can be increased by changing ϕ and by shearing, unlike what happens in the range $\phi_S \le \phi < \phi_J$.

Given the interesting jamming behavior of frictional systems that are subject to shear jamming, it is reasonable to ask about the stress/force response in anisotropic materials. A number of studies have addressed elastic behavior for anisotropic systems. A key point is that, in the limit of strong anisotropy, it may be very difficult to distinguish between hyperbolic wavelike and anisotropic elastic behavior [77–79]. Figures 7.24 and 7.25 show the response that is obtained by applying a localized vertical force to the top of a layer of pentagons that has been previously subject to simple shear strain. In this case, the strong force network is roughly oriented at an angle of 45°, sloping from top left to lower right. This direction, in the compressive direction of shear (see, e.g., Figure 7.19), is favored even though the force is applied vertically.

7.5   Force Ensemble Approaches

An aspect of forces that deserves mention, but that is somewhat aside of the main focus of this review, concerns new statistical ensemble approaches, which were first proposed by Edwards and collaborators [66,67], and explained in detail in Chapter 5 by Chakraborty. These ensemble approaches address the fact that granular materials are many-body systems which, on the one hand, are highly

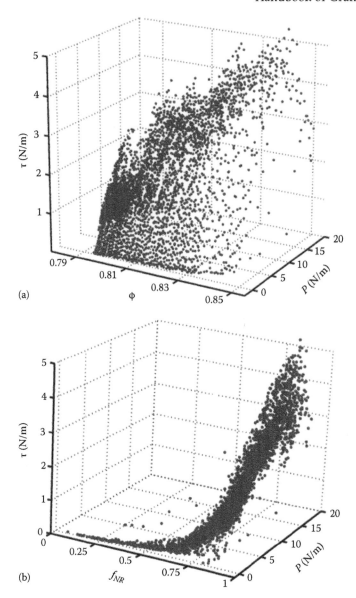

Figure 7.23 Scaling collapse for the pressure and shear stress. Part (a) shows P and τ as a function of ϕ for a number of experiments in which a stress-free sample of photoelastic disks was subject to pure shear strain. Part (b) shows the same data plotted as a function of the non-rattler fraction, f_{NR}. (From the work of Bi, D. et al., *Nature*, 480, 355, 2011.)

fluctuating—hence best described by a statistical approach, yet, they are not described in a sensible way by the ordinary Boltzmann approach, since temperature in the usual sense is not relevant. To appreciate this last point, note that the exchange of thermal energy between a granular system and a thermal bath does not appreciably change the granular system. And, energy is generally not conserved in granular interactions. However, there are quantities that are conserved for granular systems, including volume (density) for hard grains, and

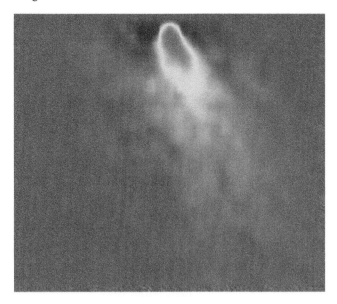

Figure 7.24 Photoelastic response of a layer of pentagons that has first been sheared and then has been subjected to a local normal force at the top of the layer [45]. Note that the preferred propagation direction for the force response deviates substantially from normal. The compressive direction of the applied shear is from top left to lower right.

stress for static collections of deformable grains. The idea of these ensembles is a generalization of equal probability of microstates, where a microstate must be consistent with whatever property (volume or stress, for instance) that is conserved for a given system. This has led to a number of different studies, of which several are cited here [68–76].

Perhaps the simplest approach to these generalized ensembles is the one taken originally by Edwards for the case of rigid grains. The idea is to base the ensemble on conserved properties, which in the rigid particle case are the number of particles, N, and the volume, V, that they occupy. Edwards then proposed that over an ensemble of blocked or jammed states, there would be an entropy, corresponding to the number of states with N particles occupying a volume V such that $S = S(V,N)$ and such that the probability of observing a state of volume V would be $P(V) = [\Omega(V)/Z(X)]\exp(-V/X)$, where $\Omega(V)$ is a density of states. Z is an appropriate partition function, and X is an analogue to the normal thermodynamic temperature. For particles of finite stiffness, the stress tensor is an additional global system property (for static systems), so that there is a parallel entropy and Boltzmann-like factor proportional to $\exp(-\hat{\alpha} : \hat{\Sigma})$. Here, $\hat{\Sigma}$ is the force moment tensor, an extensive version of the stress tensor (the stress times the system area in two dimensions or volume in three dimensions) and $\hat{\alpha}$ is a tensoral field, whose inverse is known as the angoricity: $\alpha_{ij} = \partial S(V,\hat{\Sigma})/\partial \Sigma_{ij}$, where the derivative is carried out at constant V. Here, a carat is used to distinguish a tensor.

One of the important aspects of this formulation is a method to account for microscale fluctuations between different realizations of systems with the

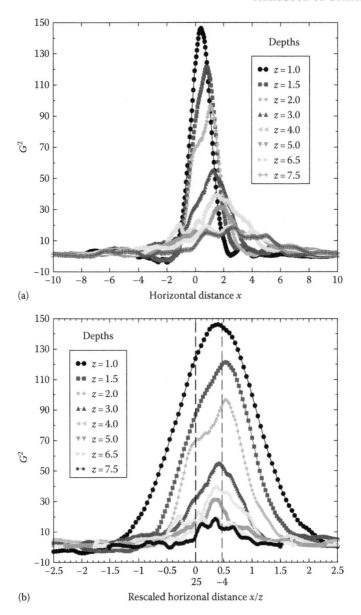

(a)

(b)

Figure 7.25 Quantitative representation of the data from the previous figure, showing the photoelastic response for various depths below the point of contact [45]. Part (a) shows the data vs. the horizontal coordinate. Part (b) shows the same data in terms of a normalized horizontal coordinate, x/z, where z is the depth. Note that this implies a clearly preferred direction that does not vary appreciably with depth.

same macroscopic parameters such as volumes and stresses. One theoretical approach to investigating force fluctuations has been carried out using ensembles of particles that live on a lattice, but that have different interparticle forces, subject to global force constraints [72,76]. This approach allows an exploration of contact force fluctuations without the complication of geometric complexity.

Alternatively, other approaches involve methods for finding the generalized partition functions [68,69,73,74]. Two recent experimental studies have shown that fluctuations in systems of photoelastic particles prepared under isotropic and anisotropic conditions are described well in the context of the force ensemble [70,71].

7.6 Outlook

This chapter has provided a brief summary of recent models for how forces should behave in dense granular materials. Much of the work from roughly a decade ago focused on granular systems that had densities well above jamming. For such systems, it seems as though an elastic-like picture for force response is usually reasonable, although the data are, to some extent, limited by the fact that "point-force" experiments show significant variability and avoiding strains that are large enough to rearrange the contacts or even the packing structure may be difficult. Given the range of interesting behavior that occurs for both frictional and frictionless systems, the true nature of how forces are carried in granular material remains open to debate.

Appendix 7A

7A.1 Using Photoelasticity to Obtain Forces

Since quite a few researchers have used photoelasticity to obtain information on the local response of granular materials [9,10,23–26,29,45,47,55,56,62–66,70,78, 84,86,87], it seems useful to discuss this approach briefly. There are essentially two ways to get such local information, one empirical and the other "exact" in the sense that in an appropriate limit, and within experimental error, it is possible to obtain precise vector forces at interparticle contacts of photoelastic disks. Both methods use the fact that if a disk or other quasi-2D object is illuminated by an incident beam of intensity I_0 and placed between crossed circular polarizers, then the light intensity that emerges [52] from this system is given by

$$I = I_0 \sin[(\sigma_2 - \sigma_1)CT/\lambda]. \tag{7A.1}$$

Here
   the σ_i are the principal stresses in the object
   C is the material-dependent stress-optic coefficient
   T is the object thickness along the direction of transmitted light
   λ is the wavelength of the light

The empirical approach was first proposed by Howell et al. [84]. The idea is that contact forces acting on a grain create stress fields inside the grain. In a photoelastic image of such a grain, increasing applied contact forces increases the stress within the grain and leads to an increasing density of light/dark fringes.

Since the pressure is a reflection of the mean normal forces on a particle, hence the internal stress, it is possible to calibrate a measure of the fringe density against the pressure, P. One measure of the fringe density is the square magnitude of the photoelastic image intensity, g^2 or G^2. Thus, G^2, integrated over a particle, provides an empirical connection to the local P.

It is also possible to solve an inverse problem, as first done in Majmudar [29] and then in several other works [9,10,63,64,70]. This method uses an exact solution of the stress field within a disk that is subject to vector point contact forces. The algorithm involves solving the inverse problem, particle-by-particle, to obtain the contact forces. The underlying idea is that if the contact forces acting on a disk are known, then the stresses inside the disk are known; and if the stresses inside the disk are known, then the photoelastic pattern for the disk is given by the equation given earlier. The algorithm runs this logic in reverse, starting from the photoelastic pattern of a disk, to yield the contact forces acting on it.

References

1. For a broad perspective see Focus Issue on Granular Materials, *Chaos* **9**, 509–696 (1999); *Physics of Dry Granular Media*, H. J. Herrmann, J.-P. Hovi, and S. Luding, eds., NATO ASI Series, Kluwer, 1997; *Powders and Grains 97*, R. P. Behringer and J. T. Jenkins, eds., Balkema, 1997; H. M. Jaeger, S. R. Nagel, and R. P. Behringer, *Rev. Mod. Phys.* **68**, 1259 (1996).

2. P.-G. de Gennes, *Rev. Mod. Phys.* **71**, 374 (1999).

3. L. D. landau and E. M. Lifshitz, *Theory of Elasticity*, 3rd edn., Butterworth–Heinemann, 1995.

4. A. J. Liu and S. R. Nagel, *Nature* **396**, 21–22 (1998).

5. M. van Hecke, *J. Phys. Cond. Matt.* **22**, 033101 (2010).

6. B. Chakraborty and R. Behringer, Jamming of granular matter, *Encyclopedia of Complexity and Systems Science*, **39**, 4997–5021 (2009).

7. C. S. O'Hern, L. E. Silbert, A. J. Liu, and S. A. Langer, *Phys. Rev. E* **68**, 011306 (2003).

8. Carl F. Schreck, Corey S. OHern, and Leonardo E. Silbert, *Phys. Rev. E* **84**, 011305 (2011).

9. J. Zhang, T. Majmudar, A. Tordesillas, and R. P. Behringer, *Granular Matter* **12**, 159–172 (2010).

10. D. Bi, J. Zhang, B. Chakraborty, and R. P. Behringer, *Nature* **480**, 355–358 (2011).

11. J. D. Goddard, *Physics of Dry Granular Media*, H. J. Herrmann, J.-P. Hovi, and S. Luding, eds., NATO ASI Series, Kluwer, pp.1–24, (1998).

12. N. Kumar, Ph.D. Thesis, Micro-Macro and Jamming Transition in Granular Materials, Multi Scale Mechanics, University of Twente, Enshede, The Netherlands, (2014).

13. C. A. Coulomb, *Mémoires de Mathématique & de Physique présentés à l'Acadèmi Royale des Sciences par divers Savans, & lŝ dans ses Assemblées*, Paris 7, 343–82, (1776).

14. D. M. Wood, *Soil Behaviour and Critical State Soil Mechanics*, (Cambridge University, Cambridge, U.K., 1990); R. M. Nedderman, *Statics and Kinematics of Granular Materials*, Cambridge University Press, Cambridge, U.K., 1992.

15. R. Jackson, in *Theory of Dispersed Multiphase Flow*, R. E. Meyer, ed., Academic Press, New York, pp. 291–337, (1983).

16. D. G. Schaeffer, *J. Differential Equations*, **66**, 19–50 (1987).

17. G. Gudehus, *Soils Foundations* **36**, 1–12 (1996).

18. H. A. Janssen, Versuche über Getreidedruck in Silozellen, *Zeitschr. d. Vereines deutscher Ingenieure*, **39**, 1045–1049 (1895).

19. M. Sperl, *Granular Matter* **8**, 59–65 (2006).

20. F. Radjai, M. Jean, J. J. Moreau, and S. Roux, *Phys. Rev. Lett.* **77**, 274 (1996).

21. C.-H. Liu and S. R. Nagel, *Phys. Rev. Lett.* **68**, 2301–2304 (1992).

22. G. W. Baxter, R. Leone, G. A. Johnson, and R. P. Behringer, *Eur. J. Mech. B: Fluids* **10**, 181 (1991).

23. P. Dantu, *Géotechnique* **18**, 50 (1968).

24. T. Wakabayashi, *J. Phys. Soc. Jpn* **5**, 383–385 (1950).

25. A. Drescher and G. De Josselin De Jong, *J. Mech. Phys. Solids* **20**, 337 (1972).

26. A. Drescher, *Géotechnique* **26**, 591–601 (1976).

27. P. A. Cundall and O. D. L. Strack, *Géotechnique* **29**, 47–65 (1979).

28. P. A. Cundall and O. D. L. Strack, Modeling of Microscopic Mechanisms in Granular Materials, in *Mechanics of Granular Materials: New Models and Constitutive Relations*, J. T. Jenkins and M. Satake, eds., Elsevier, Amsterdam, the Netherlands, pp. 137–149 (1983).

29. T. S. Majmudar and R. P. Behringer, *Nature* **435**, 1079 (2005).

30. C. Goldenberg and I. Goldhirsch, in *Handbook of Theoretical and Computational Nanotechnology*, M. Rieth and W. Schommers, eds., American Scientific Publishers, Vol. 1, pp. 1–58. (2005).

31. S. N. Coppersmith, C.-h. Liu, S. R. Nagel, D. A. Schecter, S. Majumdar, O. Narayan, and T. A. Witten, *Science* **269**, 513 (1995), *Phys. Rev. E* **53**, 4673 (1996).

32. P. Claudin, J. P. Bouchaud, M. E. Cates, and J. P. Wittmer, *Phys. Rev. E* **57**, 4441 (1998).

33. J.-P. Bouchaud, M. E. Cates, and P. Claudin, *J. Phys. I* **5**, 639–656 (1995).

34. J. P. Wittmer, P. Claudin, M. E. Cates, and J.-P. Bouchaud, *Nature* **382**, 336–338 (1996).

35. J. E. S. Socolar, D.G. Schaeffer, and P. Claudin, *Eur. Phys. J. E* **7**, 353 (2002).

36. C. Goldenberg and I. Goldhirsch, *Nature* **435**, 188–191 (2005).

37. C. Goldenberg and I. Goldhirsch, *Phys. Rev. Lett.* **89**, 084302 (2002).

38. J.-P. Bouchaud, P. Claudin, D. Levine, and M. Otto, *Eur. Phys. J. E* **4**, 451 (2001).

39. D. A. Head, A. V. Tkachenko, and T. A. Witten, *Eur. Phys. J. E* **6**, 99 (2001).

40. A. V. Tkachenko and T. Q. Witten, *Phys. Rev. E* **60**, 687 (1999).

41. A. V. Tkachenko and T. Q. Witten, *Phys. Rev. E* **62**, 2510 (2000).

42. J.-P. Bouchaud, M. E. Cates, and P. Claudin, J. Phys. I, France **5**, 639 (1995); J. P. Wittmer, M. E. Cates, and P. Claudin, *J. Phys. R.*, France **7**, 39 (1997).

43. M. E. Cates, J. P. Wittmer, J.-P. Bouchaud, and P. Claudin, *Phys. Rev. Lett.* **81**, 1841 (1998).

44. Junfei Geng, D. Howell, E. Longhi, R. P. Behringer, G. Reydellet, L. Vanel, E. Clément, and S. Luding, *Phys. Rev. Lett.* **87**, 035506 (2001).

45. J. Geng, R. P. Behringer, and G. Reydellet, E. Clément *Phys. D* **182**, 274–303 (2003).

46. G. Reydellet and E. Clément, *Phys. Rev. Lett.* **86**, 3308 (2001).

47. M. Da Silva and J. Rajchenbach, *Nature (Lond.)* **406**, 708 (2000).

48. S. Luding, O. Strauß, in *Granular Gases*, T. Pöschel, and S. Luding eds., Springer-Verlag, Berlin, Germany (2000).

49. F. H. Hummel and E. J. Finnan, *Proc. Instn. Civil Engn.* **212**, 369–392 (1921).

50. Tomosada Jotaki and Ryuichi Moriyama, *J. Soc. of Powder Technol.*, Japan **16**, 184–191 (1979).

51. R. B. Heywood, *Designing by Photoelasticity*, Chapman and Hall Ltd., London, (1952).

52. R. Brockbank and J. M. Huntley and R. C. Ball, *J. Phys. II France* **7**, 1521–1532 (1997).

53. S. B. Savage, in *Powders & Grains 97*, R. P. Behringer and J. T. Jenkins, eds., Balkema, Rotterdam, pp. 185–194 (1997).

54. L. Vanel, D. W. Howell, D. Clark, and R. P. Behringer, *Phys. Rev. E Rapid Comm.* **60**, R5040 (1999).

55. J. Geng, E. Longhi, R. P. Behringer and D. Howell, *Phys. Rev. E* **64**, 060301(R), 2001.

56. J. Krim and R. P. Behringer, *Phys. Today* 66–67 (2009).

57. J. D. Goddard, in *Physics of Dry Granular Media*, H. J. Herrmann, J.-P. Hovi, and S. Luding, eds., NATO ASI Series, Kluwer, pp. 1–24 (1998).

58. V. M. Kenkre, J. E. Scott, E. A. Pease, and A. J. Hurd, *Phys. Rev. E* **57**, 5841 (1998).

59. Nathen W. Mueggenburg, heinrich M. Jaeger, and Sidney R. Nagel, *Phys. Rev. E* **66**, 030304 (2002).

60. K. Liffman, M. Nguyen, G. Metcalfe, and P. Cleary, *Granular Matter* **3**, 165 (2001).

61. L. Landau and E. Lifshitz, *Elasticity Theory*. Pergamon, New York, (1986).

62. J. Geng, D. Howell, E. Longhi, R. P. Behringer, G. Reydellet, L. Vanel, E. Clément, and S. Luding, *Phys. Rev. Lett.* **87**, 0335506 (2001).

63. J. Ren. J. Dijksman, and R. P. Behringer, *Phys. Rev. Lett.* **110**, 018302 (2013).

64. A. Tordesillas, J. Zhang, and R. P. Behringer, *Geomech. Geoeng.* **4**, 3–16 (2009).

65. T. S. Majmudar, M. Sperl, S. Luding, and R. P. Behringer, *Phys. Rev. Lett.* **98**, 058001 (2007).

66. S. F. Edwards and R. B. S. Oakeshott, *Phys. A* **157**, 1080 (1989).

67. S. F. Edwards and C. C. Mounfield, *Phys. A* **226**, 1 (1996).

68. S. Henkes, C. S. O'Hern, and B. Chakraborty, *Phys. Rev. Lett.* **99**, 038002 (2007).

69. S. Henkes and B. Chakraborty, *Phys. Rev. E* **79**, 061301 (2009).

70. James G. Puckett and Karen E. Daniels, *Phys. Rev. Lett.* **110**, 058001 (2013).

71. D. Bi, J. Zhang, R. P. Behringer, and B. Chakratorty, *Europhys. Lett.* **102**, 34002 (2013).

72. J. H. Snoeijer, T. J. H. Vlugt, M. van Hecke, and W. van Saarloos, *Phys. Rev. Lett.* **92**, 054302 (2004).

73. C. Song, P. Wang, and H. A. Makse, *Nature*, **453**, 629–632 (2008).

74. C. M. Song, P. Wang, and H. A. Makse, *Proc. Natl. Acad. Sci. USA* **102**, 2299 (2005).

75. G. D'Anna, P. Mayor, A. Barrat, V. Loreto, and F. Nori, *Nature* **424**, 909 (2003).

76. B. P. Tighe, J. E. S. Socolar, D. G. Schaeffer, W. G. Mitchener, and M. L. Huber, *Phys. Rev. E* **72**, 031306 (2005).

77. M. Otto, J.-P. Bouchaud, P. Claudin, and J. E. S. Socolar, *Phys. Rev. E* **67**, 031302 (2003).

78. A. P. F. Atman, P. Brunet, J. Geng, G. Reydellet, P. Claudin, R. P. Behringer, and E. Clément, *Eur. Phys. J. E* **17**, 93–100 (2005).

79. D. Serro, G. Reydellet, P. Claudin, E. Clément, and D. Levine, *Eur. Phys. J. E* **6**, 169–179 (2001).

80. S. Henkes and B. Chakraborty, *Phys. Rev. E* **79**, 061301 (2009).

81. H. G. Matuttis, Gran. Mat. **1**, 83 (1998); H. G. Matuttis and A. Schinner, Gran. Mat. **1**, 195 (1999); H. G. Matuttis, S. Luding, and H. J. Herrmann, *Powder Technol.* **109**, 278 (2000).

82. R. B. Heywood, *Designing by Photoelasticity*. Chapman and Hall Ltd., London, U.K. (1952).

83. M. Latzel, S. Luding, and H. J. Herrmann, *Granular Matter* **2**, 123 (2000).

84. D. Howell, C. Veje, and R. P. Behringer, *Phys. Rev. Lett.* **82**, 5241 (1999).

85. J. P. Wittmer, M. E. Cates, and P. Claudin, *J. Phys. I* **7**, 39–80 (1997).

86. D. S. Bassett, E. T. Owens, K. E. Daniels, and Mason A. Porter, *Phys. Rev. E* (2012).

87. B. Utter and R. P. Behringer, *Phys. Rev. Lett.* **100**, 208302 (2008).

CHAPTER 8

Sheared Dense Granular Flows

Andreea Panaitescu, Ashish V. Orpe, and Arshad Kudrolli

CONTENTS

This chapter reviews 3D sheared granular flows in simple laboratory systems. The review is confined to flow of particles that are monodisperse in size, shape, and density to avoid complexities induced by segregation dynamics (which are covered in detail in Chapter 10). We further focus on the slow flow regime where the particles (essentially) remain in contact, as opposed to the kinetic regime where particles are agitated significantly and undergo binary collisions. Kinetic theory approaches have been well developed in the latter regime [1]; a review of kinetic theory is found in Chapter 4.

Granular media often experience shear when a boundary is moved at a prescribed speed, as in a Couette cell, or when flowing past a stationary boundary wall, as in chutes or in converging flow from a silo. The formal study of sheared granular materials goes back to at least Osborne Reynolds [2], a pioneer of fluid dynamics, who showed that a compacted granular packing dilates when sheared. Conversely, a loose packing may be compacted through shear [3,4]. The resulting change in the packing fraction of the particles can have profound effect on the response of the granular material, leading to shear thinning or thickening behavior depending on the initial packing fraction. Thus, a systematic understanding of the structure of initial granular packing, its evolution under shear, and its effect on the related flow dynamics are all needed in order to understand granular flows.

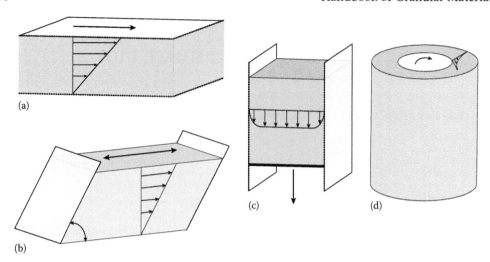

Figure 8.1   Examples of various kinds of laboratory systems in which effect of shear on granular materials has been investigated. (a) A planar cell in which the top plate is moved parallel to the bottom plate. (b) A cyclic shear cell used to obtain bulk shear. (c) Gravity-driven flow in a tall silo with a uniform rectangular cross section. (d) A Couette cell in which the inner cylinder is rotated by the outer cylinder is held at rest.

Typical systems used to study quasi-static shear in granular materials are illustrated in Figure 8.1. As indicated by the representative velocity profiles, depending on the geometry the shear may be localized in narrow shear bands (often near the boundaries) or be broadly distributed across the system. For example, in a planar geometry as in Figure 8.1a, the shear may be linear, provided gravitational stresses are absent, but localized along the inner cylinder in a Couette cell shown in Figure 8.1d.

Randomly prepared assemblies of frictionless colloidal particles (subject to thermal motion) show random packing structure for fractions below ≈ 0.64 [5], but, when sheared, rearrange to form disordered hexagonal structures that have higher configurational entropy at the same volume fraction [6]. It appears that, by ordering, the grains can more easily move past one other (at least in frictionless systems). In granular systems, however, friction and their athermal nature can prevent particles from uniformly exploring the local phase space, leading to different responses to shear. The initial packing fraction of the grains are typically between 0.55 and 0.63, depending on both the amount of friction and the sample preparation technique (e.g., slowly sedimented through a viscous fluid [7] or compacted by repeated tapping [8]).

Significant progress has been made in 2D systems that allow all the particles to be visualized, for example, using photoelastic disks (cf. Chapter 7). To overcome the inherent difficulty of visualizing particle motion inside a 3D packing, various advanced techniques such as x-ray imaging and optical index matching have been developed [9,10]. At the same time, increased computation power has led to more realistic simulations using event-driven (ED) method or discrete element method (DEM) which can be tested directly against experimental data (cf. Chapters 2 and 3) [11,12]. In the following sections, we discuss some of the important advances in elucidating the impact of shear on the structure and dynamics of granular materials. The rheology of sheared granular flows has been extensively reviewed recently [13–15], and we, therefore, do not duplicate that discussion here.

8.1 Simple Shear

The classic linear shear system is arguably the simplest in which to study shear. Spherical particles with diameter d are sheared between two large horizontally moving parallel plates separated by distance h, as shown in Figure 8.2. The horizontal extent L is considered to be much larger than h, rendering the flow translationally invariant along the x- and z-axes. In a classic fluid, $d \ll h$ and the shear scale can be considered to be much larger than d. This is not necessarily true in case of granular medium, where considerable slip can occur between the boundary and grains and shear bands can be several grain diameters wide. We consider a situation where the applied normal stress is constant and a constant shear stress is applied to the top plate parallel with constant velocity V to the right.

Simple shear is, in fact, difficult to implement in 3D experiments due to the presence of gravity. (Using density matching liquids, for example, significantly changes the physics of the system.) One method is to study gravitationally driven flow past a rough wall aligned with the gravitational field (cf. the experiments of [16] and DEM simulations of [12]). These rough boundaries lead to particle ordering in a region $4 - 5d$ adjacent to the boundary [12,16]. This ordering has a significant effect on the flow, with shear being localized adjacent to the walls, as shown in Figure 8.3. If Lees–Edwards boundary conditions are used, a linear velocity profile and homogeneous density develops [17]. The feasibility of implementing such a boundary condition in an experimental laboratory system, however, remains unclear.

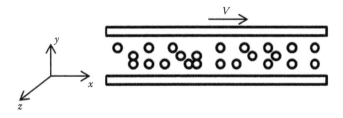

Figure 8.2   Schematic of simple shear being applied to a granular medium. The top plate is moved with constant velocity V relative to the granular medium to the right. The thickness of the granular layer is h. Shear banding occurs with flow confined near the top boundary if gravity is directed along the y-axis.

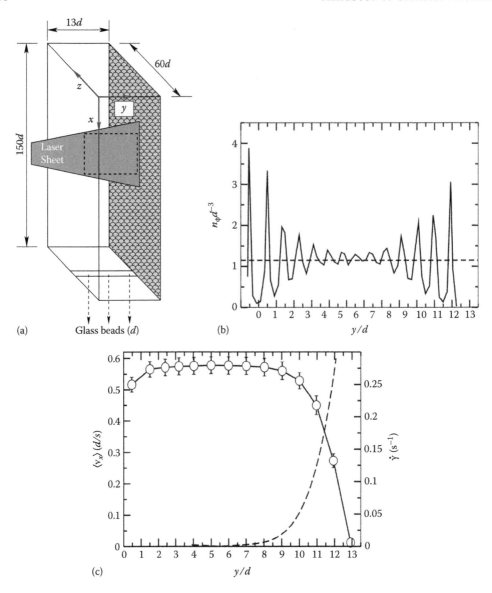

Figure 8.3   (a) Schematic of simple shear being applied by one of the sidewalls, which is made rough by gluing a layer of glass beads. (b) The number density of particles across the shear plane. (c) Mean velocity and the corresponding strain rate across the shear plane. (Adapted from Orpe, A.V. et al., *Europhys. Lett.*, 84, 64003, 2008.)

8.2   Diffusion and Mixing Due to Shear

Shear also has significant impact on the diffusion and mixing of grains, as measured by the evolution of mean squared displacement (MSD) (see Figure 8.4a, and b). A transition from super-diffusive motion (in which the MSD scales faster than $t^{1/2}$) to normal diffusion (in which the MSD $\propto t^{1/2}$) is observed [12,16,18], although the Peclet number remains large indicating that advection dominates. The diffusion coefficients are anisotropic, with greater magnitude in the direction

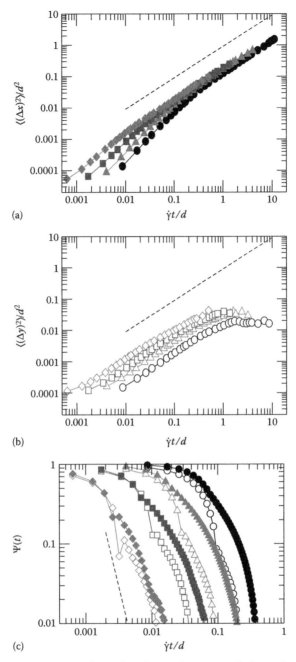

Figure 8.4 Fluctuations measured in the sheared region of the velocity profile. Scaled mean squared displacements in (a) flow direction and (b) gradient direction plotted against the scaled shear rate. (c) Velocity autocorrelation function in the flow and gradient direction. Different colours represent measurements made at different positions (shear rate) along the mean velocity profile shown in Figure 8.3c. Open symbols represent data in gradient direction while filled symbols represent data in the flow directions. Dashed lines in (a) and (b) are of slopes of 1 (i.e., diffusive motion) and dashed line in (c) is of slope 7/2 (i.e., a faster decay). (Adapted from Rycroft, C.H. et al., *Phys. Rev. E*, 80, 031305, 2009; Orpe, A.V. et al., *Europhys. Lett.*, 84, 64003, 2008.)

of flow [12,16]. It appears that the ordering of particles in the shear region inhibits diffusion across the layers in gradient direction [16]. The related velocity autocorrelation function decays with time as $t^{-7/2}$, faster is found in unsheared media and elastic fluids, which decay as $t^{-3/2}$ [19]. This observed faster decay, shown in Figure 8.4c, supports the prospect of applying kinetic theory calculations to dense sheared granular media [20], which hitherto have been thought to be applicable only for dilute systems.

Next, let us consider flow at an angle with respect to gravity. Siavoshi et al. [21] pushed a plate at constant velocity over a static granular bed to measure the effective friction and used index-matching techniques to measure complementary information on the packing structure. The experimental results show that the friction encountered by a sliding plate is sensitive to the roughness characteristics of the shearing plate and the bed thickness (see Figure 8.5a.) For a sufficiently large bed thickness and for sliding surface roughened with a layer of particles of the same kind as the granular bed, a thin shearing layer develops with a velocity which decays as a function of depth (see Figure 8.5b) as

$$v(y) = V \exp\left[-a\left(\frac{h-y}{d}\right) - b\left(\frac{h-y}{d}\right)^2 \right] \tag{8.1}$$

where h is the granular layer thickness and the decay constants a and b experimentally measured for glass beads to be $a = 0.6$ and $b = 0.03$. The effects of gravity are quite noticeable in the localization of shear near the sliding surface, where the gravitational normal stress is minimal. The normalized density profile across the bed thickness shows oscillations, suggesting that the particles slide in layers past one another (Figure 8.5c). In subsequent work, Divoux and Géminard [22] used a similar shear apparatus to measure the friction coefficient for thicker layers ($>10d$), for a range of particle diameters, and also for a granular bed immersed in liquids of varying viscosities. The effective friction coefficient for a dense granular medium depended neither on the particle size nor on the interstitial fluid, provided the strain rate is small.

Experiments using parallel plate shear, simple as they may be to carry out, cannot be implemented for extended period of time or strain due to the finite length of the shearing surface. One way to overcome this limitation is through simulations with periodic boundary conditions. In experiments, a spatially periodic Couette cell or cyclic shear cell can be used to reduce the effect of confining boundaries and apply a prolonged shear.

8.3   Periodic Couette Flow

The Couette cell is a classical geometry used to study the rheological properties of the complex fluids (see Figure 8.1d). One typically imposes a shear rate and measures the resulting shear stress and velocity profile [23]. In a 3D Couette cell, the granular material is confined between two coaxial cylinders. The wall friction can be controlled by coating the surface of each cylinder with a layer of randomly

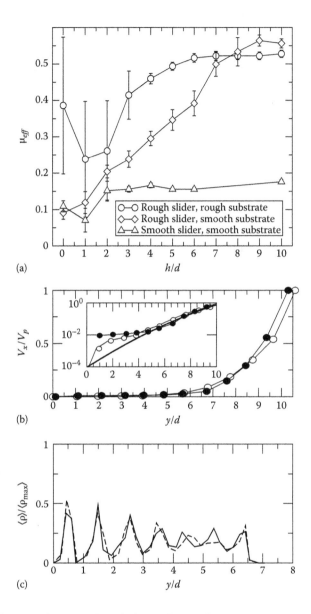

Figure 8.5 (a) Effective friction (μ_{eff}) for different boundary conditions of the sliding surface and the bed bottom. (b) Normalized mean velocity of the grains as a function of bed depth for two different boundary conditions and slider velocity of $v_p = 0.3d/s$. Rough slider on a rough base (filled symbols) and rough slider on a smooth base (open symbols). Inset: Corresponding plot in log-linear scale. The thick solid line represents fit (see text). (c) Normalized density of the grains as a function of bed depth for a rough slider with rough base. Solid lines represent $v_p = 0.3d/s$ and dashed lines represent $v_p = 1.3d/s$. (Adapted from Siavoshi, S. et al., *Phys. Rev. E*, 73, 010301(R), 2006.)

packed particles. A steady-state flow profile is set up by keeping the outer cylinder stationary while rotating the inner cylinder at constant velocity. With the upper surface free to expand, the packing fraction can adjust freely in response to the shear-induced dilation and gravity. In contrast, in a 2D Couette cell the granular material is arranged in a single layer, resulting in a constant volume and packing fraction.

In the Couette geometry, a shear band about $5d - 10d$ wide is located next to the inner cylinder (see Figure 8.6), regardless of which cylinder is rotated, as a result of the curvature of the system [24]. The initial flow properties depend strongly on the history [25–27]. If the motion of the rotating boundary is reinitiated in the same direction, the system very quickly reaches the steady state, indicating the quasi-static nature of the flow. If, however, the shear direction is reversed, the shear band becomes wider and the shear stress decreases. In a 3D system, the material compacts by a fraction of a particle diameter over a characteristic strain of $0.5d$ upon shear reversal [26]. The behavior of the material during the transient period can be explained by the reorganization of the force network presented in a dense granular material [27].

Using photoelastic disks, Howell et al. [28] showed that the forces in a dense granular material are carried preferentially along a network of force chains in which a minority of the grains carries the majority of the load. As the system is sheared, the force network aligns at 45° to resist the shear [28,29]. When the system is stopped these forces relax to some extent, but the contact network remains

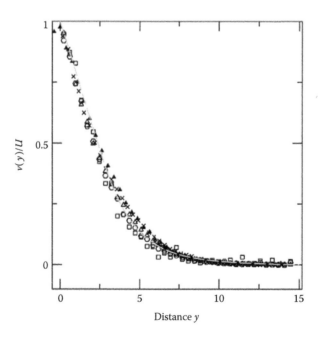

Figure 8.6   The measured velocity as a function of distance from the inner cylinder of a Couette cell, which is moving at a tangential speed U. (From Losert, W. et al., *Phys. Rev. Lett.*, 85, 1428, 2000.)

unchanged and the system immediately returns to the steady-state behavior if the shear is re-applied in the same direction. When the shear is restarted in the opposite direction, the particles outside the initial shear band acquire mobility since they are pushed along an initially fragile direction and the shear band widens. After reversal, the previously aligned force network breaks, making the stress within the material to be negligible initially, and then reforms in a perpendicular orientation to resist the new direction of the shear.

8.4 Velocity Profiles in Sheared Granular Flows

Particle dynamics in dense granular Couette flows have been extensively studied in two [30] and three dimensions [9]. Studies of the surface flow of a 3D Couette system showed that the time average azimuthal velocity $v_\theta(r)$ decays as a function of the radial distance from the inner cylinder r, showing a behavior somewhere between exponential and Gaussian [24,31]. For polydisperse materials, $v_\theta(r)$ decays more slowly and is more Gaussian than for monodisperse materials. Bulk measurements in three dimensions were taken by Mueth et al. [9] using magnetic resonance imaging (MRI) and x-ray techniques to track the average mass flow. The velocity profile does not vary with the height of the system, including the regions near the top and the bottom surfaces, and $v_\theta(r)$ can be expressed as the product of an exponential decay and a Gaussian

$$v_\theta(r) = v_0 \exp\left[-\left(\frac{r}{\alpha}\right) - \left(\frac{r}{\sigma}\right)^2\right] \tag{8.2}$$

where the fitting parameters α and σ describe the decay lengths of the exponential and Gaussian terms, respectively. The exponential term is associated with the slip between the layers formed in packings of smooth monodisperse spheres. For systems of nonspherical, rough, or polydisperse grains the exponential term vanishes and the azimuthal velocity decays as a Gaussian. In both the cases, in slow shearing regime, the velocity profile normalized by the applied shear rate was independent of the magnitude of the shear rate [9,24,32]. It is quite noteworthy that the form of the decay is similar as in planar shear case (Equation 8.1) in spite of the curvature of the Couette geometry.

The packing fraction is not uniform throughout the system, increasing slightly with distance from the inner cylinder [9,33]. Layering of the material close to the inner wall can be seen in systems composed of smooth, monodisperse particles, similar to that seen in simple shear systems (cf. Figure 8.3). For small r, the packing volume fraction varies almost linearly with the shear rate. This trend is not preserved for larger r, however, as both the packing fraction and the shear rate oscillate strongly as r increases.

In order for the particles to flow under applied shear, the packing must dilate, which implies particle motion perpendicular to the flow direction. There can be no average net motion, however, so the relevant information is embedded in the root mean square (rms) velocity fluctuations. The rms velocity fluctuations decay more slowly as a function of r than the mean velocity profile [24,31,33] and increase with shear rate as $\sim \dot{\gamma}^{0.4-0.5}$ [31,33]. Calculating the cross-correlation between the velocities of two particles at a given distance r from the moving boundary, Mueth [33] determined a characteristic length scale $\psi(r)$ over which the particles are correlated. The correlation length is almost constant away from the inner cylinder, which corresponds to a uniformly disordered packing, and ψ increases in the vicinity of the inner wall, showing that in this region particles move in clusters. Correlations between the velocity fluctuations decay over a timescale that varies with r. This is due to discrete motion events becoming increasingly rare, and occuring over shorter times, at large r.

8.5   Effect of Shear on Compaction

As mentioned earlier, the global volume fraction is fixed in 2D experiments, leading to a strong dependence of the flow properties on the average packing fraction. Studies of quasi-static 2D Couette flows [29,32] show that the packing remains at rest below a characteristic packing fraction ($\phi_c \sim 0.77$). Just above ϕ_c, sporadic displacements of the particles can be observed at isolated and temporary fluctuating locations. Well above the close packing ϕ_c, particle displacements occur simultaneously at several angular positions. This transition to shear at ϕ_c is characterized by some important features: the stress increases from zero as ϕ increases to ϕ_c playing the role of an order parameter and the system slows dramatically. The transition near ϕ_c is also characterized by a change from completely slipping to increasingly nonslip dynamics. This transition is not visible in 3D systems because gravity compacts the packing. For monodisperse systems, above ϕ_c, the velocity profile is exponential and approximately independent of the shear rate [28,29], while for a polydisperse packing the velocity profile decays strongly with r [30]. Utter et al. [30] investigated the velocity fluctuations in slowly sheared granular flow and the connection between packing microstructure and flow properties. In dense shear flow the particle motion is diffusive, with the self-diffusivity proportional to the local shear rate ($D \approx d^2\dot{\gamma}$) and an anisotropic diffusion tensor. As a result of the anisotropy of the velocity field, the tangential diffusion is about twice the radial diffusion. The local diffusivity outside the immediate vicinity of the shearing surface is dominated by anisotropic force chains, leading to a minimum diffusivity approximately along the mean force chain direction.

8.6   Broad Flow versus Shear Banding

An important characteristic of dense granular flow in Couette geometries is the localization of the shear in a narrow band next to the inner cylinder. Fenistein et al. [34] modified the Couette geometry to obtain an arbitrary wide shear band in the bulk of the packing, away from the sidewalls of the container. In the modified Couette geometry, the bottom of the cell is split at radius $r = R_s$ and shear is created by rotating both the outer ring and the cylindrical boundary while keeping the central disk fixed. (For more discussion of the "split-bottom" cell, see

Chapter 10.) For shallow layers, a narrow shear zone centered at $r = R_s$ was observed. As the filling height H is increased, the shear zone broadens continuously and shifts away from R_s toward the center of the shear cell. For a large enough H, the shear zone reaches the inner cylinder where it approaches the asymptotic regime of wall-localized shear band reported earlier. The evolution of the center of the shear zone with the height appeared to be independent of the grain properties, while the shear zone width W depends on the particle size and type. Analyzing the flow at the top surface, Fenistein et al. showed that, after rescaling, the velocity profile collapses onto a universal curve given by an error function

$$\omega(r) = \frac{1}{2} + \frac{1}{2}\mathrm{erf}\frac{r - R_c}{W}. \tag{8.3}$$

Using MRI and large-scale simulations, Cheng et al. [35] investigated the properties of the shear band in the bulk as a function of filling height. They found that the modified Couette geometry produces two distinct forms of shear: a radial component whose width increases with height and an axial component with a small constant width that appears only when the filling height exceeds a threshold. The radial velocity profile is described by an error function, while the axial profile is best fit by a Gaussian. The different character of the shear bands in the radial and axial directions shows the fundamental influence that boundary conditions have on shear localization.

8.7 Ordering Due to Shearing

To investigate the effect of prolonged shear on granular materials, Tsai et al. [10,36] conducted experiments in which a granular material was continuously sheared under a fixed normal load for an extended period of time. Under seemingly identical unidirectional shear conditions, some experiments evolved toward a crystalline structure (hexagonal lattice planes oriented perpendicular to the imposed velocity gradient), while others ended up in a random stable configuration. Interestingly, applying a cyclic shear made it possible to select the final crystallized state. The structure and the stability of these two states depended both on the volume fraction of the initial packing and on the boundary condition of the stationary substrate.

If the initial packing was compacted, say by first applying a unidirectional shear with a fixed normal load, then the disordered state becomes highly stable and the cyclic shear becomes ineffective at generating partial ordering. If the substrate was flat, or a monolayer of particles arranged in a quasi-hexagonal arrangement (a favorable boundary condition for crystallization) glued to the surface, the system (under unidirectional shear) always evolved toward a crystalline state. The ordered and disordered states also had different flow rheologies: in the ordered state, hexagonal planes slide with nearly uniform motion parallel to the shearing velocity, while the disordered flow is Poiseuille like with some slip near the boundary, and the packing volume fraction and the boundary shear force were much smaller in the crystalline state. Polydispersivity prevented the evolution toward an ordered structure, similar to the behavior observed with colloidal suspensions [6].

8.8   Cyclic Shear

A complementary method to apply extended shear is with a cyclic shear apparatus, illustrated in Figure 8.7, in which lateral confining walls are tilted periodically in opposite directions. A large confining pressure can be also applied, reducing, if not completely eliminating, the effects of gravitational gradients. Under periodic shear in these systems, granular materials evolve slowly towards a more compact configuration [37,38], accompanied by spatial ordering [38–40]. Contrary to the shear flows inside a Couette cell or a silo, the packing is sheared in the bulk and localization is not seen [41]. The compaction process is more efficient for larger shear amplitudes, and the evolution of the packing is sensitive to initial random structure of system, exhibiting short-term memory [38]. A sudden change in the shear amplitude is followed by a rapid change in packing volume fraction [38], independent of the packing history. The slow and rapid dynamics are found to be uncorrelated, suggesting that the particle rearrangements associated with the two dynamics are of a fundamentally different nature.

Mueggenburg [39] studied the transient behaviors of the granular shear flows in a cell of independently movable slats, essentially creating a well with sidewalls that could deform in response to the packing. The flow profile was obtained by monitoring the position of the walls. The initial velocity profile was found to vary significantly from packing to packing, ranging from a linear profile across the entire width of the cell to an exponential profile with a width of about six particles diameter. After repeated shear cycles, however, the profile became exponential with a width of approximately $3d$. Further shearing caused the velocity profile to deviate from exponential to one more closely resembling a Gaussian or error function.

The velocity profiles are not affected if the flow is suddenly stopped and then restarted in the same direction. A wide velocity profile, however, is observed when

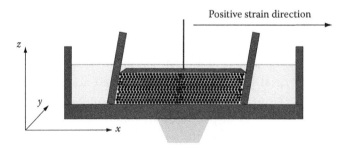

Figure 8.7   Schematic of a cyclic shear apparatus used to obtain homogeneous shear. This geometry also allows quasi-static deformation making it possible to obtain position and velocity information on the particles inside the bulk using particle index-matching techniques.

the direction of the shear is reversed. These results are consistent with experiments in 2D and 3D Couette geometries [25,27] and confirm that the shear history is stored in the interparticle contacts, with reversal of shear breaking and rebuilding the force network. Interestingly, the changes in the force network do not disrupt the large-scale crystalline structure induced by the shear.

8.9   Compaction and Particle Diffusion

The link between the shear flow properties and the microscopic structure of the packing has been studied at constant density in 2D systems [42,43] and during compaction in 3D systems [41,44] as shown in Figure 8.8. Investigation of the tra-

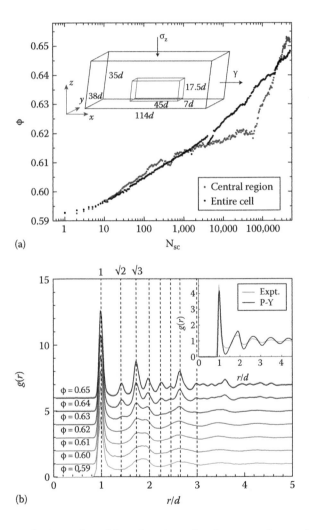

(a)

(b)

Figure 8.8   (a) The evolution of packing fraction of spheres under cyclic shear. Inset: A schematic of the apparatus used to measure the packing. (b) The radial correlation function $g(r)$ develops sharp peaks as the system compacts. The inset shows that $g(r)$ of the initial packing is described approximately by the Percus–Yevick equation.   (*Continued*)

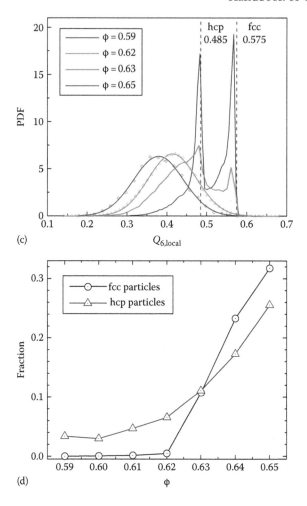

(c)

(d)

Figure 8.8 (*Continued*)   (c) The distribution of the local orientation order parameter Q_6 shows the kinds of packing symmetries observed at various volume fractions. (d) Fraction of particles with hcp and fcc symmetry as a function of volume fraction. (Adapted from Panaitescu, A. et al., *Phys. Rev. Lett.*, 108, 108001, 2012.)

jectories of individual particles has confirmed that particles spend most of their time trapped in cages defined by their nearest neighbors, and occasionally escape through rare, brief, events. In 2D bidisperse systems, the motion of the particles presents two regimes: at short times the dynamics are subdiffusive (the MSD scales as $\langle r^2 \rangle \sim t^{1/4}$), and are result of cage trapping, at long times the dynamics become diffusive ($\langle r^2 \rangle \sim t^{1/2}$)), when particles have succeeded in escaping from the cages [42]. The cage rearrangements is a highly cooperative phenomena, and the self-intermediate scattering function displays slower than exponential relaxation, suggesting dynamic heterogeneities [42,43].

In a 3D system undergoing continuous compaction, the particle motion is not diffusive [41,44]. The cumulative MSD over the first hundreds of shear cycles shows a linear growth, but the overall fluctuations are small compared with the particle size [41]. The particles seem to settle in place, and fast or large cage-breaking events are not observed. The growth of the MSD becomes fairly homogeneous with increasing shear cycles and the packing volume fraction approaches random close packing. The random motion within confined volumes is directly proportional to the shear amplitude [44], and the probability for a particle to escape from its cage decreases as the system becomes more compact. The cage changes are associated with the slow dynamics of compaction during which irreversible structural rearrangements take place [44], while the sudden change of packing volume fraction that accompanies a change in shear amplitude is correlated with cage rearrangements. These rearrangements, however, are reversible and therefore not related to permanent structural rearrangements. These results have been found in both monodisperse and bidisperse materials [44].

The onset of flow, when material goes from a jammed to a flowing state, was addressed recently in a 3D granular suspension [45]. The material is first pre-sheared to a reproducible initial state and the applied strain increased from 0% to 18.2%. Analyzing the trajectories of the individual particles revealed that the flow is inhomogeneous at the particle level: some particles remained trapped in a cage during the entire shear interval while others moved freely. These trajectories also showed a flow field with linear shear gradient, pressure-dependent dilation, and super-diffusive motions around the mean flow. At small shear intervals, particle displacements are correlated, with these correlations increasing with increasing confining pressure (at small strains).

8.10   Growth of Order

As noted in the previous sections, shear can induce spatial ordering in a monodisperse packing of spheres. The timescale for the formation of ordered regions under cyclic shear seems to be faster than that under unidirectional shear [39]. Under cyclic shear, a globally disordered granular packing exhibits small clusters with hexagonal close-packed (hcp) symmetry [40], the clusters appearing and disappearing frequently, but becoming more stable with increasing packing volume fraction. The crystallites are nonspherical in shape and oriented preferentially along the shear axis. Once the nucleating clusters reach a critical size of 10–60 particles, they start to grow rapidly and crystallites with face-centered cubic (fcc) order are observed with increasing probability. After hundreds of shear cycles, a polycrystalline structure with well-separated regions of hcp and fcc emerges (see Figure 8.9.)

As the ordered structure is formed, the velocity profile undergoes repeated changes between periods of time during which there is large shear within the ordered region and periods of time where there is virtually no shear within that region [39]. After prolonged application of periodic shear deformations, however, the fcc phase becomes more abundant [39,40], and no shear is observed within this region. The fcc and hcp stackings have identical close-packed volume, but are structurally distinct. While some calculations of the free energy phase indicate that the fcc structure is more stable, the difference is extremely small [46], and so the fcc stacking observed in shear-induced crystallization experiments may arise from another mechanism. One possibility is related to the unique way in which fcc packings support stress. Experiments on static granular packings have shown that forces are supported along straight lines in fcc packings, whereas in hexagonally packed layers the forces branch or split between each close-packed plane [47]. As a consequence, under shear the regions with symmetry other than fcc will experience slip between the horizontal layers and rearrange, while the regions with fcc configuration are more stable and move together as a solid block.

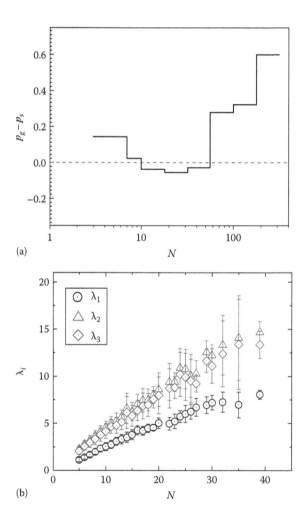

(a)

(b)

Figure 8.9   (a) The difference between the probability of a crystalline nucleus to grow and shrink as a function of the number of particles in the nucleus. The nucleus reaches a critical size when these two probabilities become equal. (b) The square root of the eigenvalues of the moments of inertia tensor as a function of the number of particles in the nucleus shows anisotropy of the nucleus.   (*Continued*)

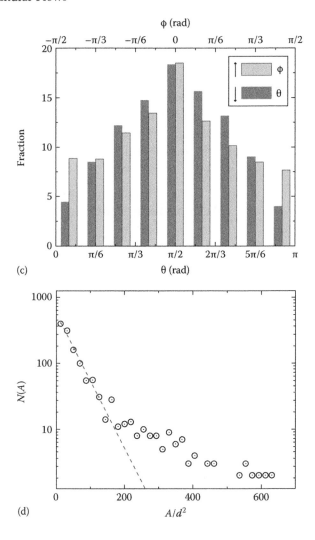

Figure 8.9 (*Continued*)   (c) The histogram of the polar angle and the azimuthal angle made by the longest axis of each cluster shows alignment with shear. (d) The decay of number of nuclei $N(A)$ versus nucleus surface area A is related to surface tension. (Adapted from Panaitescu, A. et al., *Phys. Rev. Lett.*, 108, 108001, 2012.)

8.11   Discussion

We have presented a brief overview of the experimentally observed shear profiles that result when monodisperse granular materials are sheared by linear or periodic displacements of the boundary. The typical velocity profiles that result in the case where the shear flow is driven by interactions with the boundary have a stretched exponential form. Such profiles have been discussed in terms of an order parameter description [48] and can be potentially described by nonlocal constitutive laws [49]. While the shear is localized in these cases, it may be noted that this is not always the case. For example, flow is found to be sheared over a wide area in the convergent zone of a silo or in the split bottom Couette cell.

We have also considered the impact of shear on the variations observed in the packing structure. It appears that granular packings undergo ordering upon prolonged shear similar to transitions that appear in colloidal systems consisting of thermally excited frictionless particles. Further, shearing has a significant effect on velocity fluctuations in the granular flow, leading to an anisotropic temperature distribution. The velocity autocorrelation decays faster than $t^{-3/2}$, which leads to theoretical descriptions of granular flows using modified kinetic theory approaches being applicable in the dense regime. [20,23] (cf. Chapter 4).

Acknowledgments

This review is based on the research of many previous members of the group including S. Siavoshi and A. Reddy and was supported by funds from the National Science Foundation under grant number CBET-0853943.

References

1. C. S. Campbell. *Ann. Rev. Fluid Mech.*, 22:57, 1990.

2. O. Reynolds. On the dilatancy of media composed of rigid particles in contact, with experimental illustrations. *Phil. Mag. Ser.*, 5, 20:469, 1885.

3. R. M. Nedderman. *Statics and Kinematics of Granular Materials*. Nova Science, 1991.

4. A. Lemaitre. Rearrangements and dilatancy for sheared dense materials. *Phys. Rev. Lett.*, 89:195503, 2002.

5. S. Torquato, T. M. Truskett, and P. G. Debenedetti. Is random close packing of spheres well defined? *Phys. Rev. Lett.*, 84:2064, 2000.

6. U. Gasser. Crystallization in three- and two-dimensional colloidal suspensions. *J. Phys.: Cond. Matter*, 21:203101, 2009.

7. G. R. Farrell, M. Martini, and N. Menon. Random loose packings of frictional spheres. *Soft Matter*, 6:2925, 2010.

8. E. R. Nowak, J. B. Knight, E. Ben-Naim, H. M. Jaeger, and S. R. Nagel. Density fluctuations in vibrated granular materials. *Phys. Rev. E*, 57:1971, 1998.

9. D. E. Mueth, G. F. Debregeas, G. S. Karczmar, P. J. Eng, S. R. Nagel, and H. M. Jaeger. Signatures of granular microstructure in dense shear flows. *Nature*, 406:385–388, 2000.

10. J. P. Gollub, G. Voth, and J. C. Tsai. Internal granular dynamics, shear induced crystallization, and compaction steps. *Phys. Rev. Lett.*, 91:064301, 2003.

11. L. E. Silbert, D. Ertaş, G. S. Grest, T. C. Halsey, D. Levine, and S. J. Plimpton. Granular flow down an inclined plane: Bagnold scaling and rheology. *Phys. Rev. E*, 64(5):051302, October 2001.

12. C. H. Rycroft, A. V. Orpe, and A. Kudrolli. Physical test of a particle simulation model in a sheared granular system. *Phys. Rev. E*, 80:031305, 2009.

13. G. D. R. Midi. On dense granular flows. *Euro. Phys. J. E.*, 14:341–365, 2004.

14. P. Jop, Y. Forterre, and O. Pouliquen. A constitutive law for dense granular flows. *Nature*, 441:727–730, 2006.

15. Y. Forterre and O. Pouliquen. Flows of dense granular media. *Ann. Rev. Fluid Mech.*, 40:1, 2008.

16. A. V. Orpe, V. Kumaran, K. A. Reddy, and A. Kudrolli. Fast decay of the velocity autocorrelation function in dense shear flow of inelastic hard spheres. *Europhys. Lett.*, 84:64003, 2008.

17. V. Kumaran. Dynamics of dense sheared granular flows. part 1. structure and diffusion. *J. Fluid Mech.*, 632:109–144, 2009.

18. J. Choi, A. Kudrolli, R. R. Rosales, and M. Z. Bazant. Diffusion and mixing in gravity driven dense granular flows. *Phys. Rev. Lett.*, 92:174301, 2004.

19. A. V. Orpe and A. Kudrolli. Velocity correlations in dense granular flows observed with internal imaging. *Phys. Rev. Lett.*, 98:238001, 2007.

20. V. Kumaran. Velocity autocorrelations and viscosity normalization in sheared granular flows. *Phys. Rev. Lett.*, 96:258002, 2006.

21. S. Siavoshi, A. V. Orpe, and A. Kudrolli. Friction of a slider on a granular layer: Nonmonotonic thickness dependence and effect of boundary conditions. *Phys. Rev. E*, 73:010301(R), 2006.

22. T. Divous and J. C. Géminard. Friction and dilatancy in immersed granular matter. *Phys. Rev. Lett.*, 99:258301, 2007.

23. R. A. Bagnold. Experiments on a gravity free dispersion of large solid spheres in a newtonian fluid under shear. *Proc. R. Soc. Lond. Ser. A*, 225, 1954.

24. W. Losert, L. Bocquet, T. C. Lubensky, and J. P. Gollub. Particle dynamics in sheared granular matter. *Phys. Rev. Lett.*, 85:1428–1431, 2000.

25. W. Losert and G. Kwon. Transient and steady state dynamics of granular shear flows. *Adv. Compl. Syst.*, 4:369–377, 2001.

26. M. Toiya, J. Stambaugh, and W. Losert. Transient and oscillatory granular shear flow. *Phys. Rev. Lett.*, 93:088001, 2004.

27. B. Utter and R. P. Behringer. Transients in sheared granular matter. *Eur. Phys. J. E*, 14(4):373–380, 2004.

28. D. W. Howell, R. P. Behringer, and C. T. Veje. Fluctuations in granular media. *Chaos*, 9:559, 1999.

29. C. T. Veje, D. W. Howell, and R. P. Behringer. Kinematics of a two-dimensional granular couette experiment at the transition to shearing. *Phys. Rev. E*, 59:739–745, 1999.

30. B. Utter and R. P. Behringer. Self-diffusion in dense granular shear flows. *Phys. Rev. E*, 69:031308, 2004.

31. L. Bocquet, W. Losert, D. Schalk, T. C. Lubensky, and J. P Gollub. *Phys. Rev. E*, 65:011307, 2001.

32. D. Howell, R. P. Behringer, and C. Veje. Stress fluctuations in a 2d granular couette experiment: A continuous transition. *Phys. Rev. Lett.*, 82:5241–5244, 1999.

33. D. M. Mueth. Measurements of particle dynamics in slow, dense granular couette flow. *Phys. Rev. E*, 67:011304, 2003.

34. D. Fenistein and M. van Hecke. Wide shear zones in granular bulk flow. *Nature*, 425:256, 2003.

35. X. Cheng, J. B. Lechman, A. Fernandez-Barbero, G. S. Grest, H. M. Jaeger, G. S. Karczmar, M. E. Möbius, and S. R. Nagel. Three-dimensional shear in granular flow. *Phys. Rev. Lett.*, 96:038001, 2006.

36. J.-C. Tsai and J. P. Gollub. Slowly sheared dense granular flows: Crystallization and nonunique final states. *Phys. Rev. E*, 70:031303, 2004.

37. A. M. Scott, G. D. Charlesworth, and M. K. Mak. On the random packing of spheres. *J. Chem. Phys.*, 40:611, 1964.

38. M. Nicolas, P. Duru, and O. Pouliquen. Compaction of a granular material under cyclic shear. *Eur. Phys. J. E*, 3:309–314, 2000.

39. N. W. Mueggenburg. Behavior of granular materials under cyclic shear. *Phys. Rev. E*, 71:031301, 2005.

40. A. Panaitescu, A. Reddy, and A. Kudrolli. Nucleation and crystal growth in sheared granular sphere packings. *Phys. Rev. Lett.*, 108:108001, 2012.

41. A. Panaitescu and A. Kudrolli. Experimental investigation of cyclically sheared granular particles with direct particle tracking. *Prog. Theor. Phys.*, 184:100, 2010.

42. G. Marty and O. Dauchot. Subdiffusion and cage effect in a sheared granular material. *Phys. Rev. Lett.*, 94:015701, 2005.

43. O. Dauchot, G. Marty, and G. Biroli. Dynamical heterogeneity close to the jamming transition in a sheared granular material. *Phys. Rev. Lett.*, 95:265701, 2005.

44. O. Pouliquen, M. Belzons, and M. Nicolas. Fluctuating particle motion during shear induced granular compaction. *Phys. Rev. Lett.*, 91:014301, 2003.

45. K. A. Lorincz and P. Schall. Visualization of displacement fields in a sheared granular system. *Soft Matter*, 6:3044–3049, 2010.

46. P. G. Bolhuis, D. Frenkel, S.-C. Mau, and D. A. Huse. Entropy difference between crystal phases. *Nature*, 388:235–236, 1997.

47. N. W. Mueggenburg, H. M. Jaeger, and S. R. Nagel. Stress transmission through three-dimensional ordered granular arrays. *Phys. Rev. E*, 66:031304, 2002.

48. D. Volfson, L. S. Tsimring, and I. S. Aranson. Order parameter description of stationary partially fluidized shear granular flows. *Phys. Rev. Lett.*, 90:254301, 2003.

49. K. Kamrin and G. Koval. Nonlocal constitutive relation for steady granular flow. *Phys. Rev. Lett.*, 108:178301, 2012.

50. M. Otsuki and H. Hayakawa. Unified description of long-time tails and long-range correlation functions for sheared granular liquids. *Eur. Phys. J. Special Top.*, 179:179–195, 2009.

CHAPTER 9

Avalanches in Slowly Sheared Disordered Materials

Karin A. Dahmen and Robert P. Behringer

CONTENTS

9.1 Introduction

The term "avalanche," as applied to granular and disordered material, has a rich and varied span. The conventional term brings to mind destructive flows of snow, soil, or mud down the surfaces of mountains and hillsides. Indeed, researchers around the world [1,2] are concerned with what happens in such complex flows. Several aspects of these phenomena are key, including the idea of a buildup of stress (here, shear stress), failure, and a dynamical response that typically involves flow until a new stable state is attained. These features are not only unique to avalanches on hillsides, but occur also in polycrystalline magnetic systems, in earthquake fault zones, and in a variety of more idealized granular and frictional systems. These kinds of phenomena are sometimes referred to as stick-slip and can occur at the microscopic scale of atoms, mesoscopic scale of foams or colloids, and macroscopic scale of granular materials.

Understanding the statistical properties of avalanches is of particular interest. Distributions of avalanche sizes and durations, power spectra of slip-velocity time traces, and acoustic emissions can provide information about the system parameters. For example, the statistics of magnetization avalanches in magnetic materials, measured as Barkhausen noise, can be used for nondestructive materials testing. The avalanche size distributions and power spectra of Barkhausen noise provide information about the disorder in the material, the average grain size in polycrystalline materials, the range of the interactions, the dimensionality of the system, and many other properties [55]. For granular materials, the avalanche statistics are expected to reflect parameters such as the packing fraction, applied stress, friction, amount of inter-grain cohesion—for example, resulting from humidity in the material—and other important system parameters [41,73,81–83]. Measuring the avalanche statistics can thus be used as a tool to gain information about microscopic properties of the system from macroscopic measurements.

In applying avalanche theory to granular materials, one seeks relationships between statistical quantities and underlying, fundamentally granular, parameters and to establish which statistics are universal, that is, independent of the microscopic details [73]. For example, while the size of the largest avalanche depends on the packing fraction in the material and the system size, the exponent of the power law that describes the size distribution for smaller-than-maximum avalanches is expected to be universal. Furthermore, the average temporal profile of the avalanche propagation velocity may be described by a predicted universal scaling function [73], as was previously established for magnetic avalanches [55]. Another basic question concerns the size of the underlying universality class, that is, which granular materials and other systems display the same scaling behavior of the avalanche statistics on long length scales. In short, studying avalanche statistics provides a new approach to understanding the complex nonequilibrium dynamics of granular materials and its relation to other dynamical systems.

An important aspect of granular systems that has parallels in stick-slip systems is jamming [3], as reviewed by several authors [4–8]. For instance, in the Liu–Nagel proposal for a jamming diagram, Figure 9.1a, there exist jammed and "unjammed" regions in a space of shear stress and packing fraction, ϕ. (A temperature axis has been omitted here, since by and large, we will focus on the zero-temperature regime that applies to granular systems.) Here, packing fraction is the volume (area in two dimensions) of a system that is occupied by grains, divided by the total system volume (area). In the unjammed regime, to the left of the sloping line in Figure 9.1a, granular-like systems are proposed to have a fluid-like response to shear, and to the right of the line, systems are expected to have a solid-like resistance to shear stress. If a system with $\phi > \phi_J$ experiences a strong enough shear stress, the system begins to flow.

A simple system that connects to the concept of jamming/unjamming is a granular avalanche occurring in a channel filled with grains. If the channel is tilted by a small amount, grains experience a shear stress due to the component of

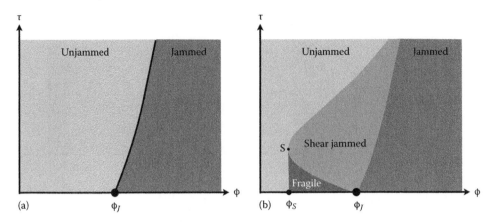

Figure 9.1 Jamming diagrams. (a) The shear stress–packing fraction plane from the Liu–Nagel scenario. (b) Shear jamming. (Diagram from Bi, D. et al. *Nature*, 480, 355, 2011.)

gravity along the downward incline of the channel. The system remains jammed (i.e., it does not flow) until the channel is tilted to a sufficiently high angle, at which point the material becomes unjammed and flows, and this is the initiation of an avalanche. Even for relatively ideal systems, there is a rich phenomenology associated with what happens near the onset of avalanching, the nature of avalanching flows, and the way in which avalanches stop and start [10–26].

Once initiated, avalanche flows, triggered by the buildup of shear stress, typically have relatively high-speed dynamics. Slow shear flows, such as Couette flow, also exhibit buildup of stress, with a release that may be much slower than in avalanches.

In several papers, starting with O'Hern [27], it was shown, for frictionless spherical particles, that there is a jamming transition [4–8], as encapsulated by the Liu-Nagel jamming diagram of Figure 9.1a. For frictional granular materials, the jamming diagram is qualitatively different from the Liu–Nagel picture. Zhang et al. [28] and Bi et al. [9] have shown experimentally for the case of frictional "spherical" particles in two dimensions (i.e., disks) that there is a regime of densities spanning $\phi_S \leq \phi \leq \phi_J$ where there exist stress-free states which, when subject to shear, pass through a regime of fragile states (in the sense of Cates et al. [29]), followed by a regime of fully jammed states. The fragile states, which are characterized by long "force chains," as in Figure 9.2, form along the compressional direction of shear. Shear strain (in say two dimensions) consists of compression in one direction, dilation in the other direction, and possibly rotation, all without change of area. In the fragile regime, the force chain networks that form are robust in the compressive direction, but they cannot resist strain in the direction of dilation. But, for enough shear strain, providing $\phi_S \leq \phi \leq \phi_J$, it is possible to arrive at a state that is robust against strains in both directions. There is less information on the situation in three dimensions but, importantly, Kumar and Luding [30] have found that shear jamming can occur in systems of frictionless 3D spheres. This implies that avalanching/flowing

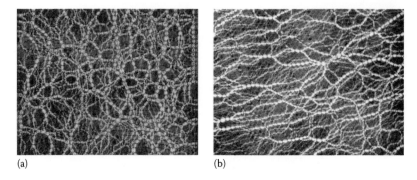

(a)   (b)

Figure 9.2   Force networks for systems of photoelastic disks that have been compressed isotropically (a) and sheared (b). In these images, brighter particles experience higher mean force. The sheared system on the right is in a regime of densities for which shear jamming occurs, and the resulting force networks, or force chains, are highly anisotropic.

phenomena, which inherently involve shear, may be strongly affected by the granular behavior that is associated with shear jamming.

This chapter is laid out as follows. We first discuss several common geometries in which avalanches have been studied, including chute flows, granular piles and rotating drums, and slowly sheared systems. We then develop a simple analytic model that provides insight into the control parameters found to be relevant in the study of avalanches and compare the model predictions with a number of experimental findings. Finally, we discuss experiments on slowly sheared systems and (briefly) discuss other related experiments and models.

9.2   Chute Flow

The flowing regime in granular avalanches has been extensively investigated over a number of years, and here we note a sampling of studies, both experimental and numerical [10–23,31]. The details of geometry, bounding surface properties, material polydispersity, and so forth can have a significant effect on the flow details. We focus on the basics of flow down a chute with a rough base. Pouliquen [19] considered flows of glass spheres down a wide chute with a rough base. For a given material, the key parameters are the height of the flowing layer h and the angle of the incline θ. There exist two boundaries in the parameter space of h/d (where d is the particle diameter) and θ. One of these, $h_{stop}(\theta)$, separates states where no flow occurs, that is, for $h < h_{stop}(\theta)$. The other separates states where steady flow can be obtained, from regimes (at steeper angles for a given height) where flows continually accelerate. In the steady flow regime, much of the data for u, the mean speed of the flow, can be empirically expressed in a scaled form by

$$u/(gh)^{1/2} = \beta h/h_{stop}(\theta), \qquad (9.1)$$

where $\beta = 0.136$ is a dimensionless constant. This result implies that $u \propto h^{3/2}$, which was observed in numerical results [18,20,32]. Specifically, Silbert et al. [20]

showed similar scaling behavior, and they characterized in detail the nature of flow, including density and velocity profiles as a function of θ and h. Their results for the steady flow regime are consistent with a depth-averaged flow speed that varies as $h^{3/2}$. They also showed that the flow consists of a core region where ϕ is nearly constant, and below the value that one would anticipate for ϕ_J, which is bounded from above and below by boundary layers.

Slow granular shear flows in channels, such as those that occur for Couette shear, have been extensively probed. There is necessarily an extensive literature on such flows, and we note only a handful of studies that are of interest to the present review. Miller et al. [33] and later Daniels et al. [34] focused on the nature of stress fluctuations and on order/disorder in channels filled with spherical particles. In order to better understand the nature of these fluctuations, Howell et al. [35] carried out studies using quasi-2D systems of photoelastic particles in a Couette geometry. These studies actually show the effect of shear jamming, and are, in some sense, a prelude to work by Hartley et al. on avalanches and fluctuations in Couette flow, discussed later [36,37]. Mueth et al. [38] focused particularly on flow profiles using very powerful tools of magnetic resonance imaging. In channel flows of fluid saturated particles, Gollub et al. [39] showed highly novel in-plane ordering of shear granular materials. Corwin et al. [40] considered the forces that appeared at the base of sheared cylindrical systems of grains. In these experiments, there was a radial boundary between evolving and jammed particles, which allowed these authors to explore the nature of force distributions in different (e.g., jammed and unjammed) states. Finally, we note that there are related systems, including foams, colloids, and glasses where avalanche phenemena also occur. A complete review would be beyond the scope of the current chapter; more information can be found in Chapter 8 of this book.

9.3 Granular Piles and Rotating Drums

Granular materials can behave like a liquid—with the ability to flow, or like a solid—with a mechanically stable structure [41]. This concept was pursued in early sandpile models, which predict broad distributions of avalanche sizes. In these models, site topplings are triggered when grains are slowly rained onto a sandpile in the absence of cohesion between the grains [42–45]. The distribution, $D(S)$, of avalanche sizes, follows a power law $D(S) \sim S^{-\tau'}$ with $\tau' = 1.05$ in two dimensions (2D), and $\tau' = 1.37$ in three dimensions (3D). Here, S is the number of sites that topple in an avalanche. Bak, Tang, and Wiesenfeld [44] used their sandpile model to introduce the concept of self-organized criticality (SOC). The key proposal of SOC is that a system self-organizes to a critical point without the need for parameter tuning. Soon, more realistic models were developed [45–48], and SOC was studied in models of other systems [36,49–51]. The distribution, $D(s)$, of sizes, s, for avalanches that result from grains being slowly added, and where the avalanche material falls off the pile edge, also follows a power-law relation, $D(s) \sim s^{-\tau}$. Here, $\tau = 1.5$ [52,53] in 3D models and experiments [52,53].

The critical power-law exponents (τ and τ') are expected to be "universal," independent of the system details [42,48,54], and dependent only on fundamental properties such as symmetries, dimensions, interaction ranges, etc. [42,55]. (For more on self-organized criticality, see Chapter 1 by Franklin.)

Experiments on avalanche dynamics in real systems [29,54,56–70] have given a range of different results. Experiments on quasi-1D piles (a 2D pile with 1D flow) of spheres or rice sometimes show power-law behavior depending on the structure or randomness of the base [56] and on the aspect ratio of the particles [57,58]. Rice piles with shorter grains deviate from SOC behavior, while piles with elongated grains produce SOC exponents over a large range of sizes [57,58]. Three-dimensional conical sandpiles [62–64] yield power-law behavior over a range of avalanche sizes; in addition, they also display quasiperiodic large avalanches when the base diameter is larger than 75 grain diameters. These experiments [62–64] use small grains less than 0.8 mm in diameter; grains of this size may be subject to cohesion due to capillary effects from room humidity [71]. In contrast, experiments on conical piles of glass beads with diameter 3.0 mm, in the absence of cohesion, yield a power law over three decades of avalanche size, even for base diameters exceeding 90 bead diameters [72]. In this case, the cutoff (i.e., the maximum avalanche size) has been shown to depend on the drop height [70,72].

Experiments using granular materials in a partially filled, rotating cylindrical drum [54,65–67] report results that are not consistent with SOC predictions. Note, however, that the drum rotation in these experiments makes them fundamentally different from pile experiments. A rotating drum advances a whole line of material past the angle of maximum stability and eventually takes more tightly packed material from the bottom of the pile and places it at the top, routinely rupturing force chains that support the grains [54,65], significantly altering the system dynamics. Overall, real sandpiles show a wide variety of behavior; there is no real consensus yet in the literature on a unifying "dynamical phase diagram" containing all these observations.

9.4 Models and Experiments on Slip Avalanche Statistics in Slowly Sheared Granular Materials

The shear response of granular materials depends on parameters such as the packing fraction of the material [4,5], the shear rate [36,73], and the boundary conditions [36,74–76] (e.g., fixed volume versus fixed pressure). At lower packing fractions, grains tend to flow in a manner that is roughly analogous to fluid flows in similar geometries. At higher packing fractions, grains jam, and the material develops shear rigidity, similar to amorphous solids and glasses [29,77]. At low shear rates, the shear stress is relaxed through local failure avalanches that are similar to cracks or small earthquakes. Recent theoretical studies have focused on the statistical distributions of these avalanches in the quasistatic limit (where avalanches are well separated in time) at high packing fractions [73,78–80].

Simulations [81–83], analytic work [73], and experiments [36,74–76,78,79] show that in certain parameter regimes, these avalanches can span a wide range of scales, often with a power-law distribution, similar to the Gutenberg–Richter distribution of earthquake magnitudes. In other parameter regimes, stick-slip behavior and mode switching between stick-slip and power-law-distributed event sizes occur [73–76,81–85]. Experiments yield similar behavior at lower packing fractions, near the jamming transition [3,86], where some of the key assumptions of the models for high packing fractions are expected to break down [36,73,87].

9.5 Avalanches in Experiments and Comparison with the Predictions from a Simple Analytic Theory

Recent simulations of sheared granular materials show solid-like behavior at high packing fractions, with power-law-distributed avalanches or stick-slip behavior, depending on the system parameters. Some simulations show mode switching between the two regimes, that is, time periods with power-law avalanche statistics followed by sudden switches to stick-slip behavior and vice versa [81–83]. Similar behavior has also been discovered in experiments on sheared granular matter [74–76]. In order to gain an intuition for the relevant control parameters, simple analytic models are often helpful. One such model is presented in the following.

9.5.1 Model

A simple analytic mean field theory predicts avalanche statistics at different packing fractions, at different shear rates, and for different frictional properties [73]. It models a coarse-grained version of the system, where the granular material is represented as a lattice of weak spots where grains may slip. Each weak spot has its own randomly chosen failure stress threshold $\tau_{s,i}$. The threshold randomness reflects the disorder in the granular material. Each lattice point remains stuck until the local shear stress is larger than the local threshold to sliding. When the shear stress exceeds this threshold, the system slides until the stress is reduced to some arrest stress $\tau_{a,i}$. At the highest packing fraction (e.g., modeling randomly closely packed granular materials in experiments), the released stress is redistributed equally to all other points in the lattice. For densely packed materials, the scalar mean field equation for the shear stress $\tau_i(t)$ at each site i in a lattice with N sites is given by [73]

$$\tau_i = J \sum_m \frac{1}{N}[u_m(t) - u_i(t)] + K_L(wt - u_i(t)). \qquad (9.2)$$

where
   J is the mean field coupling
   $u_i(t)$ is the total amount of slip (the displacement discontinuity) at site i and time t

The coarse-grained bulk acts like a soft spring (with spring constant $K_L \sim 1/L$), coupling the lattice to the boundary. L is the linear size of the system. The system is slowly sheared by a boundary that moves at a slow parallel velocity, w, which is proportional to the strain rate. If the local stress τ_i exceeds the local failure stress $\tau_{s,i}$, the site slips during one time step to relax the local stress to a local arrest stress $\tau_{a,i}$, and then resticks. The slip weakens the local failure threshold [88] to a dynamic value $\tau_{d,i} \equiv \tau_{s,i} - \epsilon(\tau_{s,i} - \tau_{a,i})$. Here, $0 \leq \epsilon \leq 1$ is a dynamic weakening parameter that quantifies the difference between effective static and dynamic "friction" [73,89–92]. Weakening could be caused, for example, by local dilation as has been shown in simulations [81–83]. In principle, the amount of weakening could thus depend on the packing fraction.

Since the sites are elastically coupled, a slipping site can trigger other sites to slip in the next time step of the avalanche, etc. The avalanche stops when the stresses at all sites are relaxed below their current failure thresholds. All failure thresholds then fully re-heal to their static values $\tau_{s,i}$. Afterward, the loading stress is slowly increased until a new avalanche is started. (Shear rates faster than the re-healing rate lead to shear band formation.) The system can also be diluted to model more loosely packed grains and to study the effect of packing fraction on avalanche statistics [73]. Essentially, for lower packing fraction, not all of the released stress is redistributed to the other sites in the system, but some of the stress is effectively absorbed by the loading mechanism, in the presence of voids in the material, as shown in [73]. As a result, the model predicts a smaller maximum avalanche size for lower packing fractions.

9.5.2 Predictions

Avalanche statistics can be computed analytically in mean field theory [73,89]. Since in the jammed region there are system spanning force chains, mean field theory with infinite range interactions is expected to give accurate results in the physical dimension. This argument should work especially well at high packing fractions, where the system spanning force chains are quite dense. In such a case, grains are highly confined, and plastic slip events are rare. Effectively, higher density causes the system to become more elastic. In mean field theory, every lattice point is coupled equally to every other point in the lattice. Mean field theory has been shown to give correct predictions for dislocation slips in crystals, due to the long-range character of the elastic interactions [90,92,93]. Our initial comparisons to experiments on granular materials indicate that mean field theory indeed describes densely packed granular materials well.

9.5.2.1 Predicted Phase Diagram

Figure 9.3 shows the dynamical phase diagram for the slip avalanche statistics in granular materials, obtained analytically from this model [73]. The y-axis effectively corresponds to rescaled packing fraction $v \equiv \Phi/\Phi_c$, such that the upper limit on the y-axis $y = 1$ corresponds to the densest possible packing, $\Phi = \Phi_c$.

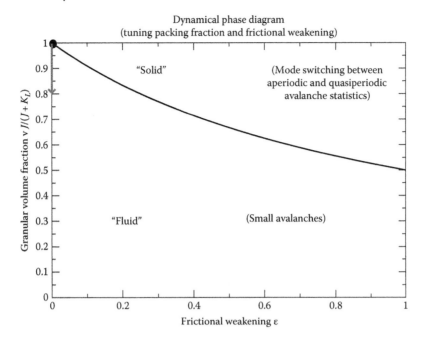

Figure 9.3 Dynamical phase diagram for sheared granular materials according to [73]. The red arrow indicates lowering the packing fraction starting from random close-packed materials. The granular volume fraction v is defined as $v = \Phi/Phi_c$, where Φ is the packing fraction and Φ_c is the maximum packing fraction (like random closed packed). The black line indicates a critical volume fraction $v*$ that separates fluid-like dynamics from solid-like dynamics with larger slip avalanches (see also Table 9.1).

The x-axis corresponds to the effective frictional weakening, which could result from dilation effects in response to local slips, or from dynamic friction between the grains, which is lower than the static friction [73]. At low packing fractions, below the curved line $v \equiv v^*$ in Figure 9.3, the model predicts only small avalanches [73,89]. Above this line for $\epsilon > 0$, the model predicts episodic switching between two fundamentally different types of behavior. The first involves quasiperiodically recurring, large, system-spanning slip events separated by much smaller precursors. The second consists of aperiocially recurring, large slip events with power-law-distributed events, with the size distribution $D(S) \sim S^{-\kappa}\exp(-AS/S_{\max})$. The persistence time for each dynamical "phase" depends on both the packing fraction and the weakening [73,89]. The dependence of these persistence times on the packing fraction qualitatively agrees with experiments reported in [74].

The black dot at the highest packing fraction $v = 1$ and $\epsilon = 0$ denotes a "critical point" of the model, that is, a point where—for adiabatically slow shear rate, such that the avalanches are separated in time—the model predicts power-law distributions of avalanche sizes, durations, power spectra of stress drop rate time series, $V(t)$, etc.

Table 9.1 Initial Results Comparing Predictions from Our Recent Theory to Experiments

Power-law Exponent or Other Universal Quantity	Mean Field Theory see [73]	Granular Experiment [36,74,77–79,84,85]	Granular Simulations [81–83]
Avalanche size distribution $D(S) \sim S^{-\kappa}$	$\kappa = 1.5$	$\kappa = 1.5$	
Avalanche duration distribution $D(T) \sim T^{-\alpha}$	$\alpha = 2$	$\alpha = 2$ or exponential?	
Stress drop rate distribution $D(V) \sim V^{-\psi}$	$\psi = 2$	$\psi = 2$ [36]	
Power spectrum $P(\omega)\ \omega^{-\phi}$ of stress drop rate $V(t)$	$\phi = 2$ if $v \approx 1$ $\phi = 0$ if $v \ll 1$	$\phi = 1.8 - 2.5, 2$ for high time resolution	$\phi = 2$ if solid $\phi = 0$ if fluid
Temporal slip-rate profile	Symmetric (parabola)	Symmetric (parabola)	Symmetric (sine function?)
can see periodic stick-slip	Yes, if $\epsilon > 0$ and $v > v^*$	Yes, sometimes [74]	Yes (mode switching)
Mode switching (between aperiodic and periodic stuck-slip)	Yes, if $\epsilon > 0$ and $v > v^*$	Yes, sometimes [74]	Yes, in solid regime

Note: Many additional properties that are yet to be tested against experiments are listed in [73]. The volume fractions v and v^* are defined as in the caption of Figure 3 with $v = \Phi/\Phi_c$, where Φ_c can be chosen in experiments, for example, as the packing fraction of random close-packed materials. Here, S is the total slip or the total stress drop during an avalanche, T is the avalanche duration, V is the stress drop rate, and is the frequency in the power spectra, which is defined as the absolute square of the Fourier transform of the time series of the stress drop rates $V(t)$.

9.5.2.2 Predictions for the Avalanche Size Distribution

The respective power laws for these properties are given in Table 9.1. As the packing is decreased, as indicated by the red arrow in Figure 9.3, the model predicts that the power-law avalanche size distributions are terminated at a maximum cutoff size, S_{max}, that decreases with decreasing packing fraction. The model predicts [73] that S_{max} also decreases as the shear rate is increased. The reason is that for higher shear rates the fluctuations around the average stress become effectively smaller. At higher strain rates, the avalanche sizes quantify the size of the fluctuations of the stress around its mean, and therefore, the maximum avalanche size decreases with shear rate. The power-law exponent for the dependence of the maximum avalanche size on the strain rate has been computed in mean field theory [73].

This is illustrated in Figure 9.4, which shows a typical prediction for how avalanche size distributions (or related properties) change with packing fraction or shear rate [73]. The black arrow shows how the avalanche size distributions change if either the packing fraction is reduced or the shear rate is increased. Figure 9.4 strongly resembles experimental results on avalanche statistics at different shear rates from [36], which are closely related to theory [73]. The model predicts that, for slow shear rates, the distribution $D(S)$ of avalanche sizes S (which is the total slip in an avalanche, also called the "potency" for earthquakes) scales as

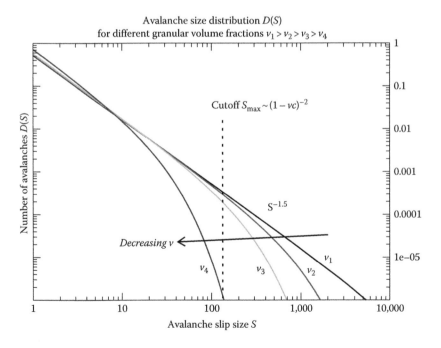

Figure 9.4 Normalized slip avalanche size distributions as predicted for samples with decreasing packing fractions. The black arrow indicates the trend for either increasing shear rate Ω or decreasing packing fraction, similar to the red arrow in the phase diagram in (Dahmen, K.A. et al., *J. Geophys. Res.*, 7, 554, 2011.)

$D(S) \sim S^{-\kappa} \exp(-AS/S_{max})$, where A is a nonuniversal constant. At highest packing fraction, $\Phi \to \Phi_c$, if the shear rate Ω is increased starting from the adiabatic case $\Omega \to 0$, the model predicts that the maximum slip avalanche size scales as $S_{max} \sim \Omega^{-1}$. Here, the slip size at nonzero shear rates can be defined as the total slip that occurs during an event that starts when the slip rate exceeds the applied average slip rate (imposed by the moving boundary), and stops when the total slip rate drops below the imposed average rate. Similarly, for adiabatically slow shear rates ($\Omega \to 0$), the largest avalanche size scales with the packing fraction Φ as $S_{max} \sim (1-\Phi/\Phi_c)^{-2}$, as indicated by the black arrow in Figure 9.4. At high packing fraction, the slip size of the large avalanches is proportional to the total stress drop during the avalanche, so that $D(S)$ is also the prediction for the expected stress drop size distribution. For experiments in two dimensions that can visualize the displacement of each grain, as a function of time, the total amount of slip, S, in an avalanche can also be computed from the curl of the displacement variable, integrated over an entire avalanche, and the dilation can be extracted from the divergence of the displacement field, as discussed in [95].

Table 9.1 lists some of the scaling predictions from mean field theory and their comparison to experiments and simulations. The agreement seen in these initial results is encouraging. Additional predictions, to be studied and compared with experiments, include the scaling behavior of the cutoffs (to the power-law scaling regime) of avalanche size and duration distributions, of power spectra, and of other statistical distributions. The mean field theory predicts how these cutoffs scale with packing fraction, strain rate, and other control parameters. The corresponding scaling predictions are given in the supplementary material of [73].

The predictions of [73] include the predicted average temporal slip-rate profile of avalanches of the same duration, shown in Figure 9.5 [73,88,96–99]. This scaling function is obtained by averaging the slip rate (or the stress drop rate) as a function of time, of all avalanches of the same total duration T. Similar average slip-rate profiles for avalanches of different durations are predicted to be related by simple rescalings of the x-axis by $1/T$ and of the y-axis by $T^{1-1/\sigma v z}$, where $1/\sigma v z = 2$ in MFT. Similar scaling collapes have been obtained for earthquakes [100], magnetic avalanches [101,102], and neural avalanches [103]. This parabolic scaling function is just as universal as the scaling exponents of the avalanche size distributions. Initial comparison with experiments and simulations look encouraging (see Table 9.1 [73]).

9.5.2.3 Extensions

The model can also be extended [41] to include cohesion as a tuning parameter so as to model slip statistics in wet granular materials, which can be different from those in dry granular materials. In the wet case, cohesion can locally be reduced by crack formation. The effect of these local cohesion drops is similar to the effect of the local weakening ϵ in Figure 9.3. If the shear rate is sufficiently fast relative to the healing rate of the cohesion, the system can show shear localization. See Ref. [41] for a full discussion of the effects of cohesion on the slip avalanche

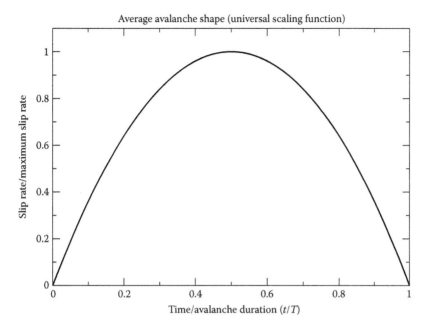

Average avalanche shape (universal scaling function)

Figure 9.5 Predicted universal average temporal slip-rate profile for avalanches of the same duration T (see text). (Dahmen, K.A. et al., *J. Geophys. Res.*, 7, 554, 2011.)

statistics in the model, with predictions for experiments, and a dynamical phase diagram with a third axis added to Figure 9.3 for the ratio of shear rate to healing rate.

9.6 Experiments in Slowly Sheared Granular Materials

Hartley et al. [36,37,104] have carried out extensive studies of fluctuations, rate dependence, and avalanches for slow shear in a Couette geometry. In these experiments, layers of quasi-2D photoelastic particles were sheared between a fixed outer ring and a rotating inner wheel. The authors considered two different shapes, studying particles with (respectively) circular and pentagonal cross sections. Several key parameters, in addition to the particle shape, characterized the resulting flows: the shear rate, the packing fraction, and the system size, that is, the radial span between the shearing wheel and the outer confining ring. This last parameter is particularly interesting because it identifies a characteristic spatial scale that may be small or large compared to the typical length of a force chain. In order to vary this system length, the experimenters varied the diameter of the inner rotating wheel. For smallest, middle-size and largest inner wheel, the radial span of the annulus containing the sample was roughly 17, 15, and 11 particles in width.

Here, we will focus on effects due to changes in the shear rate and the packing fraction. In these experiments, the packing fraction ϕ ranged from that at which it was just possible to sustain stress in the system ($\phi = \phi_S$) to packing fractions where the system was likely shear jammed. In these latter states, the system tended toward more stress-isotropic states. Since ϕ_S likely depended on details of particle distributions and preparation, the results presented in terms of the percentage deviation from ϕ_S. The shear rate was not uniform across the sample, since the system shows a strong shear band, with much of the motion confined to the first half-dozen grains. However, the motion at any give radial position was controlled by Ω, the rate of rotation of the inner wheel.

9.6.1   Strain-Rate Dependence of the Stress

The space-time averaged stress in the system depends logarithmically on the strain rate, as shown in Figure 9.6. Note that the pressure or normal stress, here represented in terms of σ/σ_{min}, where σ_{min} is a background pressure, has several features that change with Ω. It is clear that the scale of pressure fluctuations, both in terms of the size of local stress peaks and in terms of mean pressure, grows with rate. The frequency and duration of positive-going pressure fluctuations also grow as the rate increases. For the largest Ω shown in Figure 9.6, there are essentially no times when the system is quiescent. It is worth emphasizing that the data are presented in dimensionless terms of shearing wheel rotations. If there were no rate dependence, then the data in the different frames of this figure would look statistically similar, which is clearly not the case.

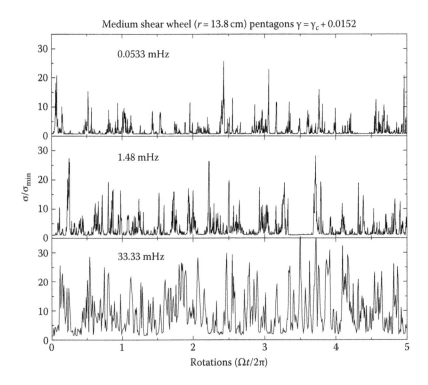

Figure 9.6   Time series for the pressure observed in a region of Couette experiments by Hartley et al. [36,37] for different rotation rates of the inner shearing wheel, expressed in mHz.

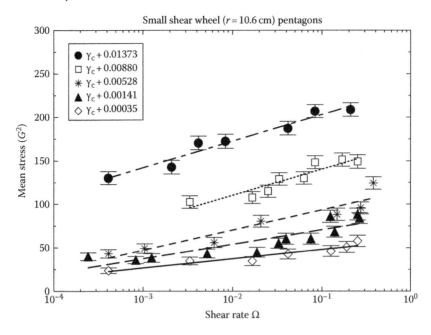

Figure 9.7 Mean stress vs. rate for various densities, here for a small shearing wheel (hence large radial gap) for pentagonal particles.

This rate dependence is made clear by computing the space-time average of the pressure, which shows logarithmic variation with rate (Figure 9.7) over many decades (Figure 9.8). A similar logarithmic variation of the variance of the pressure signal also occurs.

In fact, the increase in stress associated with an increase in rate can cause the distribution of stresses, as measured photoelastically, to change substantively. We show this effect in Figure 9.9. For low shear rates and low density, the distribution of stresses is nearly exponential for a broad range of stresses. However, this distribution narrows substantially as the rate is increased. An initial roughly exponential distribution is cut off at a decreasing value of the stress as the rotation rate increases.

The nature of the stress buildups and drops (avalanches) also depends sensitively on both rate and density. Figure 9.10 shows experimental data for the distribution of slopes associated with stress buildups and drops for a very low and a very high rotation rate. Here, the data are presented on semi-log and log–log axes to highlight the change from distributions that are power laws at low rates and are exponential at high rates.

It is possible to consider the rate-dependent behavior of sheared granular materials as evidence of activated processes, something that has been recently explored by Reddy et al. [105] and Behringer et al. [104]. This type of approach dates to Eyring [106], although the recent granular work has focused on force fluctuations, rather than the energy activation of early work. A key motivation for reformulating the activation model in terms of forces/stresses comes from a

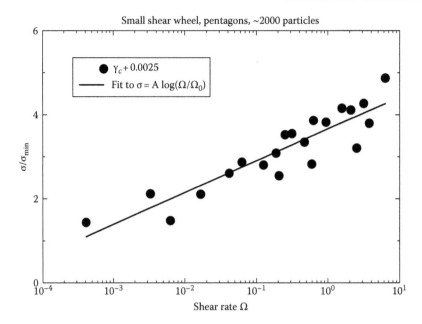

Figure 9.8   Mean stress vs. rate for a small density of pentagonal particles. Note the large range of rotation rates for which there is logarthmic variation of the stress with rate.

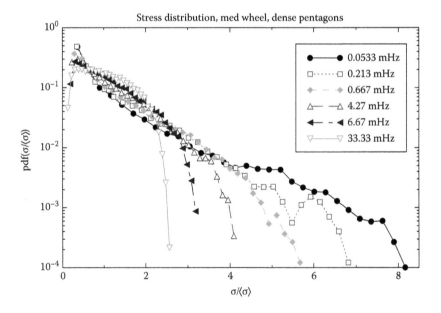

Figure 9.9   Distribution of stresses for a range of shearing rates.

proposal by Edwards et al. and later developed by others [107–109]. The idea is that the statistical behavior of dense granular materials might best be described in terms of a stress ensemble where the force moment tensor replaces the energy of the canonical statistical ensemble, and the "angoricity" replaces the temperature. Specifically, for granular systems, the ordinary temperature is not relevant, since

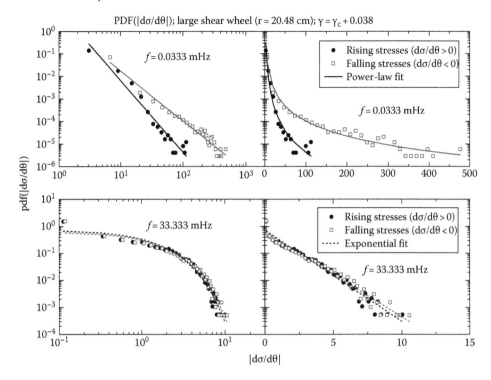

Figure 9.10 Distributions for the slope of stress buildups and decays for low and high rotation rates. Note that the data are plotted on log–log and log–lin axes to emphasize changes between power-law (at low rotation rate) and exponential (at high rotation rate) distributions.

energy is not generally conserved for granular interactions, and the grains "do not care" about a normal thermal bath. But, for (nearly) force-/torque-balanced states, such as those that are generated by quasistatic shear, there is a "conserved quantity," that is, the stress. See Chapter 5 by Chakraborty for a more detailed discussion of statistical methods in general and angoricity in particular.

9.6.2 Slip Avalanche Size Distributions

Hartley, Behringer, et al. report power-law distributions of slip sizes in [36,84,85] for frictional photoelastic disks that are slowly sheared in a Couette geometry, or with a slider. As shown in Figure 9.5 for the Couette geometry, the results are consistent with the simple theory of [73]. Hartley and Behringer find larger mean avalanche sizes for higher packing fractions and a widening range of power-law scaling, in agreement with the model of [73]. They also find shear-rate dependence, with smaller fluctuations for faster shear rates, which is consistent with the theory results of [73].

Experiments motivated by the study of earthquake statistics were performed by Daniels, Hayman, Ducloue, and Foco [74–76] on slowly sheared (at constant volume) photoelastic granular aggregates. The experiments visualized both particle-scale kinematics and interparticle forces (in the form of force chains, see Chapter 7 by Behringer for more discussion). They observed stick-slip deformations and a strengthening of the shear zone. Interesting mode-switching between quasiperiodically recurring large slips and aperiodically occurring slips is seen, similar to the mode-switching region in Figure 9.3 of the simple mean field model discussed in [73]. As in the model, higher packing fractions favor quasiperiodic behavior, with a longer time interval between aperiodic events than between quasiperiodic events. The dependence of the persistence time of each mode on the packing fraction agrees with the predictions of [73]. The authors also discuss similarities with a driven harmonic oscillator model with damping, and insights from shear transformation zone (STZ), jamming, and crackling noise theories for the effective stiffness and patterns of shear localization during deformation. The authors report power-law distributions of slip avalanche sizes for lower packing fractions (which have less hardening) and exponential distributions for higher packing fractions (with more hardening). This also is consistent with the model [73] when hardening is included, as discussed in [110].

9.6.3   Other Studies

Other studies [85,111–119] have focused on the intermittent motion of plates, rods, beads, or other objects that are slowly pulled or pushed through jammed granular materials. Granular friction experiments show reorganizations prior to macroscopic sliding [120,121]. A recurrent theme for such experiments is the fluctuating nature of the forces involved in driving the object through the material, even though there is a well-defined average behavior. For instance, Albert et al. [115] probed the force exerted on rods that were pushed through a bucket of glass spheres. Geng and Behringer [114] carried out experiments in two dimensions that involved pushing a disk through an annular trough that was filled with photoelastic disks. Fluctuations in the force needed to drive the intruder at constant speed are related to buildup and failure of force chains and can be characterized by a spring-failure model that is reminiscent of that used to study avalanches. Indeed, a similar spring-failure model has been used to reproduce experiments in a slider geometry [85,117,119]. Takehara et al. [116] measured force fluctuations for an intruder driven through a 2D bead of grains and found that deformations occur in a fluctuating region around the intruder (which they associate with force chains that formed and failed.) More work is needed to fully understand and predict the spatial structure of the intermittent dynamics on a wide range of scales.

A recent study by Chikkadi, Wegdam, Bonn, Nienhuis, and Schall reports interesting scaling properties of long-range strain correlations in sheared colloidal glasses [122]. The authors report direct experimental evidence of long-range correlations during the deformation of a dense colloidal glass. An imposed external stress forces structural rearrangements. Long-range correlations in the fluctuations of microscopic strain are observed, and their scaling and spatial symmetry are measured. The authors observe a transition from homogeneous to inhomegeneous flow at a critical shear rate and investigate the role of strain correlations in this transition region.

Related avalanche statistics are seen in a broad range of other systems [123], including dry friction [124], crystal plasticity [94], Barkhausen noise in magnetic materials [55,99], and others.

9.7   Related Models

Other numerical models for slowly sheared granular materials have been proposed in various contexts. In an early study, Baldassari et al. [125,126] modeled the stochastic dynamics of a sheared granular medium with a model that has the same scaling behavior as the Alessandro–Beatrice–Bertotti–Montorsi (ABBM) model that was originally developed for the study of magnetic domain wall motion and Barkhausen noise in soft magnets. This model belongs to the interface depinning universality class. They found good agreement between their model predictions for the slip avalanche size distributions ($\kappa \approx 1.5$) and the power spectra of frictional torque time series (which corresponds to a stress time series in our model) for experiments in a Couette cell geometry.

Aharony and Sparks [81–83] performed 2D discrete element simulations of shear in a gravity-free layer of circular grains to study systems such as layers of fault gouge. Their simulations treat individual grains as inelastic disks that undergo linear and rotational accelerations due to grain interactions and external forces. Energy is lost in the system through viscous damping, which models inealastic collisions between grains and the surface friction between the grains. Their simulated granular layers can exhibit stable (steady state) or unstable (stick-slip) motion, depending on the shear rate and the applied confining stress. This is consistent with predictions of the mean field model [73] that predicts steady flow at higher shear rates, and slip avalanches that decrease in size as the shear rate is increased. This is also consistent with the Hartley and Behringer's experiments of [36], which indicate that smaller and smaller fluctuations are seen turning into "steady flow" at higher and higher shear rates. Aharony and Sparks also compare their simulation results to a spring block model.

Bi and Chakroborty studied the rheology of dense driven granular materials as a stress-based ensemble [127], building on studies by Behringer [128] that model the sheared granular material as an activated process driven by the angoricity. The granular material is modeled by a stress-based trap landscape. The angoricity plays the role of fluctuation temperature in soft glassy rheology models. The authors calculate the constitutive equation, the yield stress, and the distribution of stress drops dissipated during granular shearing experiments, and find agreement with experiments [37].

Durian proposed a model [129–131] for slip avalanches in wet, disordered foams at nonzero shear rate. His model treats the foam as consisting of spherical bubbles that can overlap, where the overlap determines the pairwise repulsive interactions between neighboring bubbles. The model also contains a viscous drag term. Simulations using this model yield a broad power-law distribution of slip avalanche energy drops, with a cutoff that depends on the area fraction of the bubbles. For progressively wetter foams, the event size distribution broadens into a power law that is cut off only by the system size.

Related recent studies of slip statistics in molecular dynamics simulations of amorphous materials found that the critical exponents for slip avalanche statistics depend on whether the system is overdamped or underdamped [132]. The highly successful theory of STZs for the deformation of amorphous materials [133–136] has also been discussed in the context of simulations of slowly sheared granular materials, with a particular focus on effective temperature [137], shear localization, and the formation of shear bands [138]. "Soft glassy rheology" [23,139–141] models use an internal "fluidity" variable to model the internal energy relaxation processes that may be spatiotemporally coupled. If the fluidity is coupled nonlocally to the material rheology [142,143], the models can mimic fully developed flows with localized shear bands that may coexist with quasijammed regions. Goyon et al. [144,145] suggested a relation between fluidity and localized structures in an analysis of experiments on confined emulsions. A similar interpretation was applied to sheared bubble rafts [146]. In foams, localized plastic events may be the foam T1 events [147]. In colloidal glasses, they may be observed, for example, by confocal microscopy [148]. Recently, Kamrin and Koval [23] have reformulated fluidity models to account for the yield stress behavior of granular materials. With this model, they have been able to describe a number of shear flow simulations with considerable precision.

9.8 Conclusions

Avalanches are seen in slowly sheared granular materials under a wide range of conditions. Experiments in rotating drums, slowly sheared Couette cells, earthquake fault geometries, chute flows, and particles pulled through granular materials show intermittent shear behavior, spanning a wide range of slip avalanche sizes. Recent theories suggest that the power-law scaling behavior seen in the avalanche statistics at high packing fractions is reflecting an underlying nonequilibrium phase transition above the jamming transition [73]. Experimental studies show similar scaling behavior of the avalanche statistics near the jamming transition. Current models range from coarse-grained analytic models to soft glassy rheology, to STZ models with effective temperatures, and to detailed simulations that compute the motion of every grain. Much experimental and theoretical progress has been made toward understanding and predicting the avalanche statistics as a function of tuning parameters such as packing fraction, particle properties, particle interaction properties, shear rate, boundary conditions, and frictional properties. Intriguing connections to slowly sheared glassy materials and colloidal suspensions help identify relevant and irrelevant tuning parameters. Connections between force chain statistics and intermittent dynamics link the results of different experimental probes. Yet, a precise phase diagram for the avalanche statistics for granular materials at packing fractions across the jamming transition is still missing. For more detailed comparison with the model predictions, experiments at higher time resolutions and spatial imaging experiments are needed. Such measurements, particularly those that allow us to extract spatial correlations between slips, should be sufficient to distinguish between current competing models. Imaging experiments and spatiotemporal measurements of stress evolutions in two and three dimensions are also relevant for the study of earthquake statistics, since direct observation of these properties is not possible for real geological faults. A study of the effect of boundary conditions is necessary to compare results from different experimental geometries. Such experiments will lead to the development of a complete phase diagram of the slip statistics as a function of the common tuning parameters and boundary conditions.

References

1. R. M. Iverson, M. Logan, R. G. LaHusen, and M. Berti. *J. Geophys. Res.* **115**, F03005 (2010).

2. M. J. Woodhouse and A. J. Hogg, *J. Fluid Mech.* **652**, 461 (2010).

3. A. J. Liu and S. R. Nagel, *Nature* **396**, 21–22 (1998).

4. Liu and S. R. Nagel, The jamming transition and the marginally jammed solid *Annl. Rev. Condens. Matter Phys.* **1**, 347–369 (2010) and references therein.

5. Liu and S. R. Nagel, Nonlinear dynamics: Jamming is not just cool any more, *Nature* **396**, 21–22 (1998); doi:10.1038/23819 and references therein.

6. A. J. Liu and S. R. N. *Jamming and Rheology*, Taylor & Francis, New York (2001).

7. B. Chakraborty and R. P. Behringer. Jamming of granular matter. *Ency. Compl. Syst. Sci.* **39**, 4997–5021 (2009).

8. M. van Hecke *J. Phys. Condens. Matter* **22**, 033101 (2012).

9. D. Bi, J. Zhang, B. Chakraborty, and R. P. Behringer. *Nature* **480**, 355–358 (2011).

10. D A. Augenstein and R. Hogg. *Powder Technical* **19**, 205 (1978).

11. S. B. Savage. *J. Fluid Mech.* **92**, 53 (1979).

12. C. S. Campbell, C. E. Brennen, and R. H. Sabersky. *Powder Technol.* **41**, 77 (1985).

13. S. B. Savage and K. Hutter. *J. Fluid Mech.* **199**, 177 (1989).

14. T. G. Drake. *J. Geophys. Res.* **95**, 8681 (1990).

15. P. C. Johnson, P. Nott, and R. Jackson. *J. Fluid Mech.* **210**, 501 (1990).

16. K. G. Anderson and R. Jackson. *J. Fluid Mech.* **241**, 145 (1992).

17. C. S. Campbell, P. W. Cleary, and M. Hopkins. *J. Geophys. Res.* [Solid Earth] **100**, 8267 (1995).

18. E. Azanza F. Chevoir and P. Moucheron. *J. Fluid Mech.* **400**, 199 (1999).

19. O. Pouliquen, *Phys. Flu.* **11**, 542 (1999).

20. L. E. Silbert, D. Ertas, G. S. Grest, T. C. Halsey, D. Levine, and S. J. Plimpton. *Phys. Rev. E* **64**, 050802 (2001).

21. M. Y. Louge. *Phys. Rev. E* **67**, 061303 (2003).

22. G. Midi. *Eur. Phys. J. E* **14**, 341 (2004).

23. K. Kamrin and G. Koval. *Phys. Rev. Lett.* **108**, 178301 (2012).

24. S. Deboeuf, O. Dauchot, L. Staron, A. Mangeney, and J.-P. Vilotte. *Phys. Rev. E* **72**, 051305 (2005).

25. A. Daerr and S. Douady. *Nature* **399**, 241 (1999).

26. J. Rajchenbach. *Phys. Rev. Lett.* **89**, 074301 (2002).

27. C. S. O'Hern, L. E. Silbert, A. J. Liu, and S. A. Langer. *Phys. Rev. E* **68**, 011306 (2003).

28. J. Zhang, T. Majmudar, A. Tordesillas, and R. P. Behringer. *Granular Matter* **12** 159–172 (2010).

29. M. E. Cates, J. P. Wittmer, J.-P. Bouchaud, and P. Claudin, *Phys. Rev. Lett.* **81**, 1841 (1998).

30. N. Kumar and S. Luding, preprint (2013).

31. C. Ancey. *Phys. Rev. E* **65**, 011304 (2002).

32. G. Berton, R. Delannay, P. Richard, N. Taberlet, and A. Valance. *Phys. Rev. E* **68**, 051303 (2003).

33. B. J. Miller, C. O'Hern, and R. P. Behringer. *Phys. Rev. Lett.* **77**, 3110 (1996).

34. K. E. Daniels and R. P. Behringer. *Phys. Rev. Lett.* **94**, 168001 (2005).

35. D. Howell, R. P. Behringer, and C. Veje, *Phys. Rev. Lett.* **82**, 5241–5244 (1999).

36. R. Hartley, PhD thesis, Physics Duke University, Durham, NC, (2005) and references therein.

37. R. Hartley and R. P. Behringer. Logarithmic rate dependence of force networks in sheared granular materials *Nature* **421**, 928–931 (2003).

38. D. Mueth, G. Debregeas, G. Karczmar, P. Eng, Sidney, R. Nagel, and H. M. Jaeger. *Nature* **406** (6794), 385 (2000).

39. P. Gollub, G. Voth, and J. C. Tsai. *Phys. Rev. Lett.* **91**, 064301 (2003).

40. E. Corwin, H. Jaeger, S. Nagel. *Nature* **435**, 1075–1078 (2005).

41. Y. Ben-Zion, K. Dahmen, and J. Uhl. A unifying phase diagram for the dynamics of sheared solids and granular materials, *Pure Appl. Geophys.* **168**, 2221–2237 (2011), doi:10.1007/s00024-011-0273-7 (2011).

42. M. Paczuski and S. Boettcher. Universality in sandpiles, interface depinning, and earthquake models, *Phys. Rev. Lett.* **77**, 111–114 (1996).

43. P. Bak. *How nature works: The Science of Self-organized Criticality*. Copernicus New York (1996).

44. P. Bak, C. Tang, and K. Wiesenfeld. Self-organized criticality: An explanation of the 1/f noise, *Phys. Rev. Lett.* **59**, 381–384 (1987).

45. P. Bak, C. Tang, and K. Wiesenfeld, Self-organized criticality, *Phys. Rev. A* **38**, 364–374 (1988).

46. V. Frette. Sandpile models with dynamically varying critical slopes, *Phys. Rev. Lett.* **70**, 2762–2765 (1993).

47. H. J. Jensen. Self-Organized Criticality: Emergent Complex Behavior in Physical and Biological Systems. Cambridge University Press, Cambridge, U.K. (1998).

48. L. P. Kadanoff, S. R. Nagel, L. Wu, and S. M. Zhou, Scaling and universality in avalanches, *Phys. Rev. A* **39**, 6524–6537 (1989).

49. J. M. Carlson and J. S. Langer. Properties of earthquakes generated by fault dynamics, *Phys. Rev. Lett.* **62**, 2632–2635 (1989).

50. B. Drossel and F. Schwabl. Self-organized critical forest-fire model, *Phys. Rev. Lett.* **69**, 1629–1632 (1992).

51. Z. Olami, H. J. S. Feder, and K. Christensen. Self-organized criticality in a continuous, nonconservative cellular automaton modeling earthquakes, *Phys. Rev. Lett.* **68**, 1244–1247 (1992).

52. K. B. Lauritsen, S. Zapperi, and H. E. Stanley. Self-organized branching processes: Avalanche models with dissipation, *Phys. Rev. E* **54**, 2483–2488 (1996).

53. S. Zapperi, K. B. Lauritsen, and H. E. Stanley. Self-organized branching-processes - mean-field theory for avalanches, *Phys. Rev. Lett.* **75**, 4071–4074 (1995).

54. P. Evesque, D. Fargeix, P. Habib, M. P. Luong, and P. Porion. Pile density is a control parameter of sand avalanches, *Phys. Rev. E* **47**, 2326–2332 (1993).

55. J. P. Sethna, K. A. Dahmen, and C. R. Myers. Crackling noise, *Nature* **410**, 242–250 (2001).

56. E. Altshuler, O. Ramos, C. Martinez, L. E. Flores, and C. Noda. Avalanches in one-dimensional piles with different types of bases, *Phys. Rev. Lett.* **86**, 5490–5493 (2001).

57. K. Christensen, A. Corral, V. Frette, J. Feder, and T. Jossang. Tracer dispersion in a self-organized critical system, *Phys. Rev. Lett.* **77**, 107–110 (1996).

58. V. Frette, K. Christensen, A. Malthe-Sorenssen, J. Feder, T. Jossang, and P. Meakin. Avalanche dynamics in a pile of rice, *Nature* **379**, 49–52 (1996).

59. M. Bretz, J. B. Cunningham, P. L. Kurczynski, and F. Nori. Imaging of avalanches in granular materials, *Phys. Rev. Lett.* **69**, 2431–2434 (1992).

60. J. Feder. The evidence for self-organized criticality in sandpile dynamics, *Fractals* **3**, 431–443 (1995).

61. S. K. Grumbacher, K. M. McEwen, D. A. Halverson, D. T. Jacobs, and J. Lindner. Self-Organized Criticality—An experiment with sandpiles, *Am. J. Phys.* **61**, 329–335 (1993).

62. G. A. Held, D. H. Solina, D. T. Keane, W. J. Haag, P. M. Horn, and G. Grinstein. Experimental study of critical-mass fluctuations in an evolving Sandpile, *Phys. Rev. Lett.* **65**, 1120–1123 (1990).

63. J. Rosendahl, M. Vekic, and J. Kelley. Persistent self-organization of sandpiles, *Phys. Rev. E* **47**, 1401–1404 (1993).

64. J. Rosendahl, M. Vekic, and J. E. Rutledge. Predictability of large avalanches on a Sandpile, *Phys. Rev. Lett.* **73**, 537–540 (1994).

65. P. Evesque. Analysis of the statistics of sandpile avalanches using soil-mechanics results and concepts, *Phys. Rev. A* **43**, 2720–2740 (1991).

66. H. M. Jaeger, C. H. Liu, and S. R. Nagel. Relaxation at the angle of repose, *Phys. Rev. Lett.* **62**, 40–43 (1989).

67. E. Morales-Gamboa, J. Lomnitz-Adler, V. Romero-Rochin, R. Chicharro-Serra, and R. Peralta-Fabi. Two-dimensional avalanches as stochastic Markov processes, *Phys. Rev. E* **47**, R2229–R2231 (1993).

68. C. M. Aegerter, R. Gunther, and R. J. Wijngaarden. Avalanche dynamics, surface roughening, and self-organized criticality: Experiments on a three-dimensional pile of rice, *Phys. Rev. E* **67**, 051306 (2003).

69. C. M. Aegerter K. A. Lorincz, M. S. Welling, and R. J. Wijngaarden. Extremal dynamics and the approach to the critical state: Experiments on a three dimensional pile of rice, *Phys. Rev. Lett.* **92**, 058702 (2004).

70. S. Y. Lehman, E. Baker, H. A. Henry, A. J. Kindschuh, L. C. Markley, M. B. Browning, M. E. Mills, R. M. Winters, and D. T. Jacobs.

71. R. Albert, I. Albert, D. Hornbaker, P. Schiffer, and A.-L. Barabsi. Maximum angle of stability in wet and dry spherical granular media, *Phys. Rev. E* **56**, R6271–R6274 (1997).

72. R. M. Costello, K. L. Cruz, C. Egnatuk, D. T. Jacobs, M. C. Krivos, T. S. Louis, R. J. Urban, and H. Wagner. Self-organized criticality in a bead pile, *Phys. Rev. E* **67**, 041304 (2003).

73. K. A. Dahmen, Y. Ben-Zion, J. T. Uhl. A simple analytic theory for the statistics of avalanches in sheared granular materials, *Nat. Phys.* **7**, 554–557 (2011).

74. K. E. Daniels and N. W. Hayman. Force chains in seismogenic faults visualized with photoelastic granular shear experiments. *J. Geophysi. Res.* **113**, B11411/1–13 (2008).

75. K. E. Daniels and N. W. Hayman. Boundary conditions and event scaling of granular stick-slip events. In: Nakagawa, M., Luding, S. *AIP Conference Proceedings. AIP*; 567–570 (2009).

76. N. W. Hayman, L. Duclou, K. L. Foco, and K. E. Daniels. Granular controls on periodicity of stick-slip events: kinematics and force-chains in an experimental fault preprint (2010).

77. H. S. R. Jaeger and R. P. Nagel Behringer. Granular solids liquids and gases, *Rev. Mod. Phys.* **66**, 1259–1273 (1996) and references therein.

78. A. Baldassari et al. Brownian forces in sheared granular matter. *Phys. Rev. Lett.*, **96**, 118002/1–4 (2006).

79. A. Petri, F. Baldassarri, G. Dalton, L. Pontuale, S. Pietronero, and S. Zapperi. Stochastic dynamics of a sheared granular medium. *Euro. Phys. J. B* **64**, 531–535 (2008).

80. C. Maloney and A. Lemaitre. Amorphous systems in athermal quasistatic shear, *Phys. Rev. E* **74**, 016118 (2006) and references therein.

81. E. Aharonov and D. Sparks. Shear profiles and localization in simulations of granular materials, *Phys. Rev. E* **65**, 051302/1–12 (2002) and references therein.

82. E. Aharonov and D. Sparks. Rigidity phase transition in granular packings, *Phys. Rev E* **60**, 6890–6896 (1999).

83. E. Aharonov and D. Sparks. Shear profiles and localization in simulations of granular materials, *Phys. Rev. E* **65**, 051302/1–12 (2002) and references therein and *J. Geophys. Res* **109**, B09306–18 (2004).

84. J. T. Zhang, S. Majumdar, A. Tordesillas, and R. P. Behringer et al. Statistical Properties of a 2D Granular Material Subjected to Cyclic Shear preprint (2009).

85. Y. Peidong, T. Shannon, B. Utter, and R. P. Behringer. Stick-slip in a 2D granular medium preprint (2009).

86. M. Van Hecke. *J. Phys. Condens. Matter* **21**, 033101 (2010).

87. R. Behringer private communication 2010.

88. J. P. Sethna, K. A. Dahmen, and C. R. Myers. Crackling noise. *Nature* **410**, 242–250 (2001) and references therein.

89. K. Dahmen, D. Ertas, and Y. Ben-Zion. Gutenberg-Richter and characteristic earthquake behavior in simple mean-field models of heterogeneous faults. *Phys. Rev. E* **58**, 1494–1501 (1998).

90. K. A., Dahmen, Y. Ben-Zio, and J. T. Uhl. Micromechanical model for deformation in solids with universal predictions for stress-strain curves and slip avalanches. *Phys. Rev. Lett.* **102**, 175501/1–4 (2009) and references therein.

91. Y. Ben-Zion. Collective behavior of earthquakes and faults: Continuum-discrete transitions, progressive evolutionary changes, and different dynamic regimes. *Rev. Geophys.* **46**, RG4006, 21 (2008).

92. D. S. Fisher, K. Dahmen, S. Ramanathan, and Y. Ben-Zion. Statistics of earthquakes in simple models of heterogeneous faults, *Phys. Rev. Lett.* **78**, 4885–4888 (1997).

93. M. Zaiser. Scale invariance in plastic flow of crystalline solids. *Adva. Phys.* **55**, 185–245 (2006).

94. N. Friedman, A. T. Jennings, G. Tsekenis, J.-Y. Kim, J. T. Uhl, J. R. Greer, and K. A. Dahmen, *Phys. Rev. Lett.* **109**, 095507 (2012).

95. C. E. Maloney and M.O. Robbins. *Phys. Rev. Lett.* **102**, 225502 (2009).

96. M. Kuntz and J. P. Sethna. Noise in disordered systems: the power spectrum and the dynamic exponents in avalanche models. *Phys. Rev. B* **62**, 11699–11708 (2000).

97. S. Papanikolaou, F. Bohn, R. L. Sommer, G. Durin, S. Zapperi, J. P. Sethna. Beyond power laws: Universality in the average avalanche shape preprint (2009) arXiv:09112291.

98. A. P. Mehta, K. A. Dahmen, and Y. Ben-Zion. Universal mean moment rate profiles of earthquake ruptures. *Phys. Rev. E* **73**, 056104 (2006).

99. K. Dahmen and Y. Ben-Zion. The physics of jerky motion in slowly driven magnetic and earthquake fault systems encyclopedia of complexity and system science, C. Marchetti, and R. Meyers (Eds.) Springer vol. **5**, pp. 5021–5037 (2009).

100. A. P. Mehta, K.A. Dahmen, and Y. Ben-Zion. *Phys. Rev. E* **73**, 056104 (2006).

101. J. P. Sethna, K. A. Dahmen, and C. R. Myers. *Nature* **410**, 242–250 (2001).

102. A. P. Mehta, A. C. Mills, K. A. Dahmen, and J. P. Sethna. *Phys. Rev. E* **65**, 046139 (2002).

103. N. Friedman, S. I. Braden, A. W. Brinkman, M. Shimono, R. E. Lee DeVille, K. A. Dahmen, J. M. Beggs, and T. C. Butler. *Phys. Rev. Lett.* **108**, 208102 (2012).

104. R. P. Behringer, D. Bi, B. Chakraborty, S. Henkes, and R. R. Hartley. *Phys. Rev. Lett.* **101**, 268301(2008).

105. K. A. Reddy, Y. Forterre, and O. Pouliquen. *Phys. Rev. Lett.* **106**, 108301 (2011).

106. H. Eyring, *J. Chem. Phys.* **4**, 283 (1936).

107. S. F. Edwards and R. B. S. Oakeshott, *Phys. A* **157**, 1080 (1989).

108. S. Henkes and B. Chakraborty. *Phys. Rev. E* **79**, 061301 (2009).

109. J. H. Snoeijer, T. J. H. Vlugt, M. van Hecke, and W. van Saarloos. *Phys. Rev. Lett.* **92** 054302 (2004).

110. K. A. Dahmen, Y. Ben-Zion, and J. T. Uhl, Micromechanical model for deformation in solids with universal predictions for stress-strain curves and slip avalanches, *Phys. Rev. Lett.* **102**, 175501 (2009).

111. J.-F. Mayer, D. J. Suntrup III, C. Radin, H. L. Swinney, and M. Schroeter, Shearing of frictional sphere packings, *EPL* **93**, 64003/1–5 (2011).

112. A. Amon, V. B. Ngyen, A. Bruant, J. Crassous, and E. Clement. Hot-spots in an athermal system. *Phys. Rev. Lett.* **108**, 135502 (2012).

113. R. Harich, T. Darnige, E. Kolb, and E. Clement. Intruder mobility in a vibrated granular packing. *Europhys. Lett.* **96**, 54003 (2011).

114. J. Geng and R. P. Behringer, *Phys. Rev. E* **71**, 011302 (2005).

115. R. Albert, M. A. Pfeifer, A. L. Barabási, and P. Schiffer. *Phys. Rev. Lett.* **82**, 205 (1999).

116. Y. Takehara, S. Fujimoto, and K. Okumura. *Europhys. Lett.* **92**, 44003 (2010).

117. Y. Piedong. Ph thesis, Physics, Duke University, Durham, NC (2008) and references therein.

118. Evelyn Kolb private communication (2012).

119. J. Krim, Y. Peidong, and R. P. Behringer. *Pure Appl. Geophys.* **168**, 2259–2275 (2011).

120. S. Nasuno, A. Kudrolli, and J. P. Gollub. *Phys. Rev. Lett.* **79**, 949 (1997).

121. S. Nasuno, A. Kudrolli, A. Bak, and J. P. Gollub. *Phys. Rev. E* **58**, 2161 (1998).

122. V. Chikkadi, G. Wegdam, D. Bonn, B. Nienhuis, and P. Schall. *Phys. Rev. Lett.* **107**, 198303 (2011).

123. J. T. Uhl, G. Tsekenis1, N. Friedman1, B. A. W. Brinkman, T. M. Earnest, A. Jennings, P. K. Liaw, Y. Ben-Zion, J. R. Greer, and K. A. Dahmen, preprint 2013, submitted for publication.

124. T. H. W. Goebel, T. W. Becker, D. Schorlemmer, S. Stanchits, C. Sammis, E. Rybacki, and G. Dresen, Identifying fault heterogeneity through mapping spatial anomalies in acoustic emission statistics, *J. Geoph. Res.* **117**, B03310, doi:10.1029/2011JB008763 (2012).

125. A. Petri, A. Baldassarri, F. Dalton, G. Pontuale, L. Pietronero, and S. Zapperi, Stochastic dynamics of a sheared granular medium. *Eur. Phys. J. B* **64**, 531–535 (2008).

126. A. Baldassari, F. Dalton, A. Petri, S. Zapperi, G. Pontuale, and L. Pietronero. Brownian forces in sheared granular matter. *Phys. Rev. Lett.* **96**.

127. B. D. Bi and B. Chakroborty, Rheology of granular materials: dynamics in a stress landscape. *Phil. Trans. R. Soc. A* **367** 5073–5090 (2009).

128. R. P. Behringer, D. Bi, B. Chakroborty, S. Henkes, and R. R. Hartley. Why do granular material stiffen with shear rate? Test of novel stress-based statistics. *Phys Rev. Lett.* **101**, 268301 (2008).

129. D. J. Durian. Foam mechanics at the bubble scale. *Phys. Rev. Lett.* **75**, 4780–4783 (1995).

130. D. J. Durian. Bubble-scale model of foam mechanics: Melting nonlinear behavior and avalanches. *Phys. Rev. E* **55**, 1739–1751 (1997).

131. S. Tewari, D. Schiermann, D. J. Durian, C. M. Knobler, S. A. Langer, and A. Liu. Statistics of shear-induced rearrangements in a two-dimensional model foam. *Phys. Rev. E* **60**, 4185–4396.

132. K. M. Salerno, C. E. Maloney, and M. O. Robbins. Avalanches in strained amorphous solids: Does inertia destroy critical behavior? *Phys. Rev. Lett.* **109**, 105703 (2012).

133. F. Spaepen. *Acta Metall.* **25**, 407 (1977).

134. A. Argon. *Acta Metall.* **27**, 47 (1979).

135. V. V. Bulatov and A. S. Argon, *Model. Simul. Mater. Sci. Eng.* **2**, 1994 (1994).

136. M. Falk and J. S. Langer. *Phys. Rev. E* **57**, 7192 (1998).

137. J. S. Langer and M. L. Manning. Steady-state effective-temperature dynamics in a glassy material, *Phys. Rev. E* **76**, 056107 (2007).

138. M. L. Manning, J. S. Langer, and J. M. Carlson. Strain localization in a shear transformation zone model for amorphous solids, *Phys. Rev. E* **76**, 056106 (2007).

139. P. Sollich, F. Lequeux, P. He braud, and M. E. Cates. *Phys. Rev. Lett.* **78**, 2020 (1997).

140. P. Sollich. *Phys. Rev. E* **58**, 738 (1998).

141. C. Derec, A. Ajdari, and F. Lequeux. *Eur. Phys. J. E* **4**, 355 (2001).

142. G. Picard, A. Ajdari, L. Bocquet, and F. Lequeux. *Phys. Rev. E* **66**, 051501 (2002).

143. L. Bocquet, A. Colin, and A. Ajdari. *Phys. Rev. Lett.* **103**, 036001 (2009).

144. J. Goyon, A. Colin, G. Ovarlez, A. Ajdari, and L. Bocquet. *Nature (Lond.)* **454**, 84 (2008).

145. J. Goyon, A. Colin, and L. Bocquet. *Soft Matter* **6**, 2668 (2010).

146. G. Katgert, B. P. Tighe, M. E. Mobius, and M. van Hecke. *Europhys. Lett.* **90**, 54 002 (2010).

147. A. Kabla and G. Debregeas. *Phys. Rev. Lett.* **90**, 258303 (2003).

148. P. Schall, D. A. Weitz, and F. Spaepen. *Science* **318**, 1895 (2007).

CHAPTER 10

Segregation in Dense Sheared Systems

Kimberly M. Hill

CONTENTS

Most of the chapters to this point have focused on the behavior of granular materials as a single phase. While polydispersity is often introduced in simulations to prevent crystallization, the polydispersity sufficiently narrows so that the dynamics and statistics are typically still treated as a single system, without concern of the response of the individual constituents. Most granular materials in industry and nature, however, are composed of multiple constituents—particles differing in size, shape, and material properties such as density and elastic response. When subjected to shear, jostling, or other excitation, the constituents tend to unmix by particle property. This manifests in a range of striking patterns in nature, from fingering in pyroclastic flows to gravel patches in riverbeds. It also confounds issues in particle processing industries, such as in the pharmaceutical industry when powders need to be well mixed on the scale of a pill diameter. Furthermore, the behavior of granular materials depends on a representative particle size and density, so segregation can affect the local and global behavior of a granular mixture.

In this chapter, we review some basic segregation phenomenology common to a wide range of systems of granular materials. Then, we focus on segregation in dense sheared systems, where a rotating drum has proven an effective test bed for studying the driving factors that give rise to a wide range of segregation patterns. While there are some common segregation patterns that appear in many experiments, simulations, and field observations, perhaps one of the strongest underlying themes is the extreme sensitivity of segregation patterns to boundary conditions. We explore the strengths and limitations of some frameworks used to model segregation in dry sheared granular flows, and we conclude the chapter with a discussion of the effects of segregation on granular flow and also segregation in systems with interstitial fluid.

10.1   Some Complexities of Segregating Systems

Segregation behaviors and associated dynamics in granular mixtures give rise to a wide range of complex segregation patterns. The pictures in Figure 10.1 illustrate just a few. The origin of patterns such as the "petal" patterns in Figure 10.1a results from relatively simple segregation behaviors in combination with chaotic advection and diffusional mixing (Hill et al., 1999a). Other patterns, such as the striped segregation pattern shown in Figure 10.1b (Hill et al., 1999a) and the swirl pattern shown in Figure 10.1c (Rietz and Stannarius, 2008), may also have their origins in relatively simple segregation behaviors, though there is currently not a universally agreed upon physics-based predictive model for these systems. Surprisingly, some of the more complex segregation patterns in nature, such as fingering in pyroclastic flows shown in Figure 10.1d (Pouliquen and Vallance, 1999) and the coarsening of the snout and levees of a debris flow in Figure 10.1e (Bardou, 2003), appear to have a relatively simple underlying physical framework

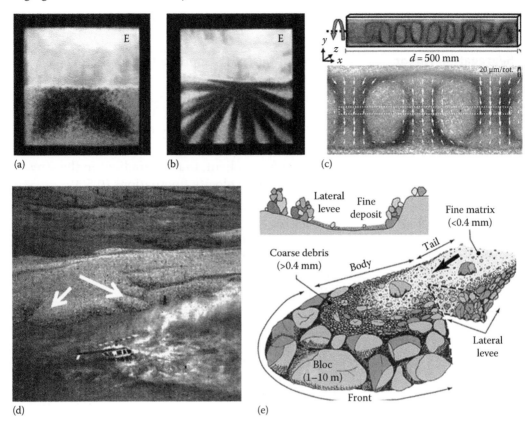

Figure 10.1   Some examples of somewhat complex patterns in (a and b) rotating drums. (From Hill, K.M. et al., *Proceedings of the National Academy of Sciences*, 96, 11701, 1999.), (c) boxes. (From Rietz and Stannarius, *Phys. Rev. Lett.* 100, 078002, 2008) and (d and e) natural settings associated with spontaneous segregation or unmixing of mixtures of different types of particles. (d) "Fingering" in pyroclastic flows. (From Pouliquen, O. and Valance, J.W., *Chaos*, 9, 621, 1999.) The two arrows indicate two "fingers," and the helicopter provides a sense of scale. (e) Coarsening of the snout of a debris flow. (From Bardou, Thèse EPFL, n° 2479, 2002.)

as reflected by some recent modeling efforts (Pouliquen and Vallance, 1999; Gray and Ancey, 2009).

We distinguish four factors that are associated with segregation phenomenology in granular materials: (1) segregation *mechanisms*, or *drivers*, referring to the physical drivers directly responsible for the unmixing of particles; (2) the *properties* that distinguish the species of particles in the mixture and by which they have been reported to have segregated (but, generally, in and of themselves do not give rise to the act of segregating or unmixing the different species); (3) segregation *patterns*, that is, the spatial arrangement of higher concentrations of one or more of the constituents according to particle property; and (4) *other mitigating dynamics* such as advection that may modify the evolving segregation pattern but are not in and of themselves segregating dynamics. Table 10.1 lists some examples

Table 10.1 Summary of some important segregation factors with references to examples in this chapter.

Segregation Factor	Examples (with Some Relevant Section(s))
Apparent segregation mechanism	Gravity (nearly all)
	Pressure/stress gradient (10.5.2, 0.5.3, 0.7.4)
	Granular temperature gradient, that is, gradient in the kinetic energy of velocity fluctuations (0.4.3, 0.5.2, 0.5.3)
	Kinetic stress gradient, that is, gradient in the stress associated with velocity fluctuations (0.5.3)
	Porosity (solid fraction) gradient (not discussed here)
Particle properties (observed associated with segregation)	Particle size (0.2.1, 0.2.3, 0.4.1, 0.4.3, 0.7.2, 0.7.4)
	Particle density (0.2.2, 0.2.3, 0.4.2, 0.4.3, 0.7.1, 0.7.3)
	Particle inertia (0.2.4, 0.4.3)
	Surface roughness/friction (not discussed here)
	Shape (not discussed here)
Mitigating factors	Advection (0.4, 0.5.1, 0.6, 0.7.3)
	Diffusion (0.5.2, 0.7.4)
	Total porosity or solid fraction (0.5.2, 0.5.3)
	Boundary conditions, for example, geometric constraints (0.7.3)
	Absolute pressure (0.4.1)
	Air or other interstitial fluids (0.8)

of mechanisms, particle properties, and associated mitigating dynamics that have been reported to give rise to the wide variety of patterns observed in granular mixtures. We use this as a context as we discuss some commonly and not-so-commonly observed segregation patterns in experiments, in industry, and in nature. In particular, the segregation *patterns* can vary dramatically from one system to the next; in some cases, this is because the segregation *mechanism* varies, but in other cases the differences are due to some mitigating factor such as the boundary conditions that have changed the details of the advection.

10.2 Ups and Downs of Segregation in Simple Shaken Systems

One of the simplest segregation patterns that arise in shaken mixtures is a simple sorting (upward or downward) of particles by size and/or density.

10.2.1 Vertical Segregation by Size in Shaken Systems: The Brazil Nut Problem

While segregation is often governed by a combination of mechanisms, a relatively common observation is that when mixtures of particles differing only in size are sheared, shaken, or otherwise disturbed, the larger particles tend to segregate to the top. For example, in a can of mixed nuts, the largest—often, the Brazil nuts—are found highly concentrated near the top. Correspondingly, the phenomenology associated with this segregation associated with particle size is often referred to as the "Brazil nut problem (BNP)."

The situation is not always this simple. Larger particles may segregate upward, downward, neither, or both, depending on boundary conditions, particle properties, and other details. However, we start by looking at the relatively simple case where the segregation *pattern* involves different-sized particles sorted according to their size, from smallest at the bottom to largest at the top.

While this size-dependent segregation might trivially occur if the smaller particles are sufficiently smaller to travel through pore spaces around the large particles, this also happens if the particles are only modestly different in size. Additionally, if a single large intruder particle is placed in a system of smaller particles, the large particle will often rise, indicating that there is some collective phenomenology driving even a single large particle to the top. This segregation pattern extends to a wide variety of systems, from simply shaken or sheared mixtures to complex systems such as sediment transport along mountain streams, where an armored layer of larger particles is typically found at the surface of the bed.

The mechanism(s) that govern(s) the processes leading to a disproportionately high fraction of large particles on the top of a system of mixed particle sizes vary depending on the system and the type and degree of excitation.

Williams (1963) reported one of the first quantitative studies of the Brazil nut phenomenon in shaken systems—he performed systematic experiments involving a single large sphere in a bed of sand that oscillated vertically. He attributed the large sphere's upward trajectory to two mechanisms: (1) a pressure gradient that set up an asymmetry in the mobility of all particles and (2) geometric constraints, whereby smaller particles could fit into more openings than large particles. Williams suggested that an asymmetry arises in this system because the material is relatively sparse above a large particle, while below the material is "locked" due to the overburden pressure of the larger particles. The net effect is that the large particles may easily move up but not down. Additionally, as small particles progressively fill in gaps beneath a large particle, the large particle is ratcheted upward. The net effect is that large particles are forced upward and not allowed downward, the combination of which segregates different-sized particles vertically (Figure 10.2).

Computational simulations by Rosato et al. (1986, 1987) confirmed the importance of a local geometric segregation mechanism for driving the large particles to the top. In these simulations, Rosato and colleagues observed that small particles were more likely to fill in voids beneath large particles than *vice versa*, and these local geometric rearrangements progressively transported a larger particle upward. Further, Rosato et al. (1986, 1987) demonstrated that this geometric mechanism could be responsible for segregating either an isolated single particle or a conglomeration of large particles.

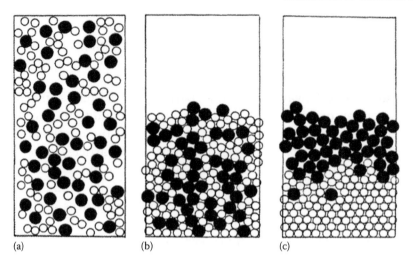

Figure 10.2 Results from Monte Carlo simulations of different-sized particles shaken vertically. (a and b) An initially randomly arranged 50/50 mixture is dropped into a container. The container is shaken with several discrete taps, and the large (dark) particles segregate to the top (c). (From Rosato, A.D. et al., *Phy. Rev. Lett.*, 58, 1038, 1987.)

Knight et al. (1993) demonstrated that a vertical size-dependent segregation pattern could emerge from an entirely different mechanism, one associated with convection. They showed that when a system was shaken by a series of discretely controlled vertical taps, a system-wide circulation occurred. When shaken in a cylindrical container with vertical walls, they found that the majority of the mixture moved gradually upward while a thin cylindrical cell of particles adjacent to the wall traveled more quickly downward; very little local random jostling or rearrangement occurred in either region. The large particles traveled upward *with* the smaller particles in the middle and outward to the walls at the top. However, the large particles could not fit into the thin stream of particles down the outside and were trapped at the top. Knight and colleagues pointed out that in these systems, a system-wide convection roll raises all the particles from the central region away from vertical walls to the top but does not allow for reentry of the larger particles in the downstream direction. Knight and colleagues demonstrated the effectiveness of this explanation through experiments with different boundary conditions that reversed the direction of the convection rolls. When the direction of the convection rolls reversed, so did the segregation; that is, the larger particles moved downward and became trapped at the bottom, reversing the Brazil nut effect!

These studies demonstrate that, for the Brazil nut pattern in shaken systems, there are at least two possible segregation drivers, and which is most important depends on the boundary conditions and detailed manner of the excitation. Recently, additional segregating mechanisms relative to the BNP have been proposed in the context of segregating mixtures of particles differing in both size and density, some of which will be discussed shortly.

10.2.2 Vertical Segregation by Density in Shaken Systems: Effective Buoyancy

When mixtures of same-sized particles differing in density are sheared, shaken, or otherwise disturbed, the lighter particles often segregate to the top although, again, exceptions exist.

In some cases, this segregation by density can be modeled by considering the effective relative density of the particles. Similar to an analogous segregation of immiscible fluids of different densities, upon excitation, buoyancy brings the less dense constituent to the surface (e.g., Khakhar et al., 1997b). However, the segregation pattern often indicates that the situation is not as simple as a buoyancy argument implies.

Shi et al. (2007) performed experiments that demonstrated that what might be called an "intensity of segregation"—the degree to which a shaken system segregates particles of different densities—depends on the relative density of the particles as well as the amplitude and frequency of vibration. As might be expected, the greater the density difference, typically, the more complete the final segregation pattern appears (Figure 10.3). However, they found that the intensity of segregation differed in different regions of their shaken systems. For example, for intermediate shaking amplitudes and frequencies, Shi and colleagues found that each mixture typically evolves to a partially segregated state: the region at the bottom contained a higher concentration of dense particles but remained partly mixed with some less dense particles, while the region at the top contained only less dense particles. The asymmetry of this pattern suggests that the segregation

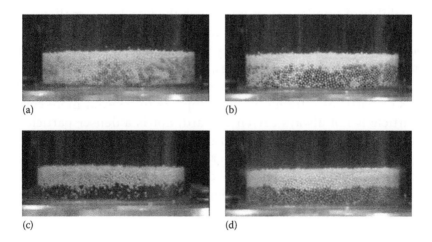

(a)						(b)

(c)						(d)

Figure 10.3 Photographs of the segregated state of binary granular mixtures of particles differing only in density after being shaken repeatedly until a steady segregation state is reached. The lighter particles are made of aluminum oxide (density $\rho \sim 1.31$ g/cm^3) and the heavier ones are made of (a) zirconium oxide (density $\rho \sim 2.87$ g/cm^3), (b) titanium alloy (density $\rho \sim 4.45$ g/cm^3), (c) cobalt–chromium–molybdenum alloy (density $\rho \sim 8.37$ g/cm^3), and (d) tungsten alloy (density $\rho = 18.0$ g/cm^3), respectively. (Shi, Q. et al., *Phys. Rev. E*, 061302/1–4, 2007.)

dynamics may also be dependent on the asymmetry of the dynamics associated with shaking a granular material under gravity.

Other mechanisms associated with particle density have been proposed. One such mechanism, the *pseudo-thermal buoyancy force* (e.g., Alam et al., 2006), appears important for particles with different *granular temperatures* (kinetic energy of velocity fluctuations or variances). This and other dynamics associated with segregation have been investigated in more detail for particles differing simultaneously in density and size, as discussed in the next section.

10.2.3 Vertical Segregation by Size and Density in Shaken Systems: The Reverse Brazil Nut Problem and Related Issues

A number of research efforts have focused on the segregation and mixing dynamics that arise in mixtures of particles differing in both size and density. Based on isolated "kinetic sieving" and "effective buoyancy" conceptual arguments alone, one might expect that smaller denser particles should always sink in a mixture of less dense larger particles, and the behavior should reverse for mixtures where the smaller particles are sufficiently less dense. This could manifest in one of three ways: (1) if larger particles are sufficiently dense, then the density difference dominates and the larger particles sink; (2) if denser particles are sufficiently large, then the size difference dominates and the denser particles rise; (3) if the larger particles are somewhat, but not too much, denser, the two effects could cancel and agitation could produce a well-mixed system. For example, in this last case, the tendency for the denser (but slightly larger) particles to sink could be moderated by the availability of smaller spaces into which the less dense smaller particles are able to fall. Alternatively, the tendency for the smaller (but less dense) particles to sink may be moderated by a buoyancy force propelling them upward among a denser medium of larger denser particles.

In fact, the dependence of segregation behavior on relative size and density in mixtures of particles differing in both appears to be even more complicated: a larger particle is not always driven upward, nor is a denser particle necessarily driven downward. Shinbrot and Muzzio (1998) investigated the behavior of a single large tracer particle among smaller bulk particles undergoing large amplitude shaking in what they referred to as "deep beds," characteristic of commercial transport of granular mixtures. They found that a single "intruder" particle significantly larger than the bulk particles would migrate to the bottom if it was significantly less dense than the smaller bulk particles. Alternatively, if the large intruder particle was significantly denser than the smaller bulk particles, the intruder would migrate to the top, exactly opposite of what one would expect from considering combined effects of density and size.

Shinbrot and Muzzio (1998) suggested that intruder inertia (effectively, its total mass rather than size or density alone) may play an important role in what they called a "reverse buoyancy" effect in shaken systems. They suggested that upon each shake, a heavy intruder thrown upward tends to continue on a parabolic arc, sustained by its larger inertia that provides more space below for the smaller bulk particles to fill in below the particle, ratcheting the intruder upward. In contrast, a light intruder thrown upward is rapidly forced back down by collisions with particles above, unable to maintain the void space below required for the Brazil nut dynamics to be effectual.

This appears to be another situation where the asymmetry of the shaking dynamics associated with "simple" shaken systems may play a significant role in the segregation pattern. Other dynamics appear important as well, such as the relative *granular temperature* or kinetic energy of velocity fluctuations for different species within a mixture, such as that discussed in more detail by Alam et al. (2006) and others. In any case, a simpler way of "jostling" the system helps with the identification and isolation of significant segregation effects, as we describe in the next section.

10.3   Ups and Downs of Segregation in Condensing Systems

Hong et al. (2001) performed event-driven molecular dynamics (MD) simulations to investigate combined effects of size and density differences on segregation in a condensing system. Rather than the intermittent excitation of a shaken system, they initially mixed their system under a relatively uniform granular temperature (i.e., a typical particle kinetic energy was independent of location) and then quenched the system to some lower temperature in the presence of gravity. They reported that this more uniform manner in which the particles are excited and then "cooled" produced some very systematic results.

As one might expect based on discussions earlier, under some conditions the large particles segregate to the top, and increasing the density of the larger particles can reverse this behavior (Figure 10.4a). Figure 10.4b presents a phase diagram from their results illustrating the transition between the BNP and what has been referred to as the reverse Brazil nut problem (RBNP). Specifically, a segregation crossover occurs when $d_A/d_B = (m_A/m_B)^{1/2}$, that is, when $d_A/d_B = \rho_B/\rho_A$. This prediction is supported by a segregation model by Jenkins and Yoon (2002) based on kinetic theory, discussed earlier in this book. It is interesting to note that these results indicate that in the mixtures of equal-density different-sized particles, larger particles should actually sink, that is, mixtures of equal-density different-sized particles exhibit RBNP. This is in contrast to studies described in Section 10.2.1 where large particles have been observed to rise among equal-density smaller particles. Hong et al. (2001) and Jenkins and Yoon (2002) found that for their systems, the transition point between the BNP and RBNP behaviors in different-sized particles occurs when the large particles are sufficiently less dense than the small particles, that is, large particles sink when $\rho_{large}/\rho_{small} \leq d_{small}/d_{large}$.

Hong et al. (2001) argue that their results are associated with competition between two effects: (1) a difference in critical condensation temperature between mixture constituents and (2) the dynamic sieving effect associated with the space filling of small particles beneath larger ones. The analogous results derived by Jenkins and Yoon (2002) were obtained for systems that satisfy certain

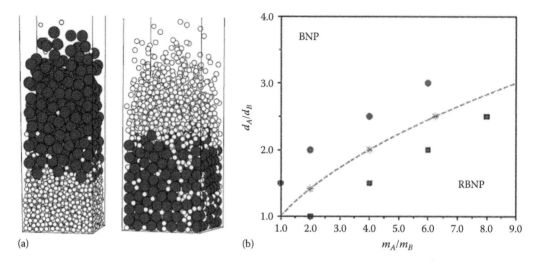

(a)   (b)

Figure 10.4 Simulation results from Hong et al. (2001) for a condensing mixture of particles differing in both size and density. (a) Left: Results from a mixture of particles of size and density ratio $m_A/m_B = d_A/d_B = 2$. Right: Results from a mixture of particles of the same size ratio $(d_A/d_B = 2)$ but a greater mass ratio, $m_A/m_B = 4$ illustrating the transition from BNP to RBPNP for increasing mass (density) ratios in, otherwise, the same particle mixture. (b) A plot with more results, where ● indicates mixtures where the large particles rise to the top, typical of the traditional BNP, and ■ indicates mixtures that exhibit the "reverse Brazil nut problem" (RBNP) where the larger particles rise to the top. ∗ indicates mixtures where the large particles remain dispersed throughout the mixtures, and the dashed line represents $d_A/d_B = (m_A/m_B)^{1/2}$.

assumptions inherited from kinetic theory. For example, particle collisions must be binary and chaotic, that is, not dependent on previous collisions, conditions typically satisfied in sparser systems. When comparing these results to the classic BNP described in Section 10.2.1, it seems reasonable to assume that the nature of even some of the most basic segregation behavior—whether large particles segregate upward or downward—is dependent on the state of the system. The apparent contradiction among results from different modeling frameworks, experiments, and simulations underlines the importance of boundary conditions, including manner and degree of excitation, in determining trends in segregation and mixing. This theme continues when considering vertical segregation in sheared systems in the next sections.

10.4 Ups and Downs of Segregation in Simple Sheared Systems

Sheared granular systems are as prevalent as shaken systems in industry, and more common in nature (e.g., in debris flowing down steep slopes and in bedsheets in riverbeds). The boundary conditions can be significantly more complex than a simple shearing system, as particles are typically advected as they are sheared, which can significantly alter the segregation patterns. Initially, we consider the "simple segregation effects" in two systems: rotated drums and inclined chute flows.

10.4.1 Vertical Segregation by Size in Sheared Systems: The Brazil Nut Problem Revisited

Experiments and simulations with mixtures of different-sized particles flowing down chutes and rotating in drums also exhibit the Brazil nut segregation. A rotated circular drum is particularly appealing for systematic studies of segregation in sheared layers as it allows one to isolate the effect of segregation in one region and study the results in another.

Figure 10.5a illustrates basic flow patterns in a circular drum partway filled with a mixture of particles and rotated at intermediate speeds. The majority of particles move in solid-like rotation with the drum walls, and a thin free-surface layer flows in laminar-like shear flow over the top surface. Segregation occurs in this thin layer and, as in shaken systems, large particles typically rise to the top,

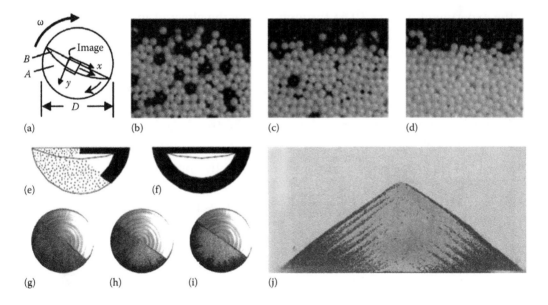

Figure 10.5   Images illustrating the Brazil nut effect in (a–j) a rotated drum and (k) heap flow. (a) A sketch of the circulation of particles rotated in a drum halfway filled with beads. At any time, most beads rotate with the drum A while a thin layer B slides quickly over the rest. (b–d) Images taken as indicated in (a) of 3 mm (dark) and 2 mm (light) plastic beads. (From Hill, K. M. and Zhang, J. Phys. Rev. E 77, 061303, 2008.). (e–f) Sketches and (g–i) photographs illustrating how the segregation in the thin top flowing layer is recorded into a pattern in the solid-like granular materials in the majority of the drum. (From Hill, K.M. et al., *Phys. Rev. Lett.*, 93, 224301, 2004.) (j) An analogous segregation pattern seen in heap flow experiments with small (white) and large (black) particles. (From Williams, J.C., *Powder Technol.*, 2, 13, 1968.)

that is, toward the free surface. Figure 10.5b through d are photographs taken from the center of the flowing layer illustrating the progression of segregation, analogous to those from shaken systems in Figure 10.2b and c.

The overall segregation pattern in the drum is determined in part from the circulation of the particles. In the latter part of the flowing layer, the particles are progressively frozen into the solid-like region of the mixture. The lower part of the flowing layer, dominated by small particles, is frozen first and, therefore, moves into solid-like rotation at smaller radii. The layers close to the free surface, dominated by large particles, are frozen last and therefore at larger radii near the outside of the drum. Subsequently, the vertical segregation in the flowing layer is "frozen" into a place as a radially symmetric segregation pattern in the drum, recording of some of the segregating dynamics of the flowing layer. This process is illustrated in Figure 10.5e through i.

Cantelaube and Bideau (1995) and Clément et al. (1995) performed systematic experimental studies to investigate the segregation dynamics of a single large or small "intruder" particle in a drum of otherwise uniform particles. Figure 10.6 shows results from experiments by Clément et al. (1995); each picture shows all locations of tracer particles from thousands of snapshots taken over the duration of the experiment. Figure 10.6a through c show the effect of increasing the tracer particle diameter relative to that of particles in the bulk. Above each picture is the first iteration map of "reinjection" of the tracer particle where each point represents the radial location of the tracer particle entering (r_i) and leaving (r_{i+1}) the flowing layer. The results help illustrate that the segregation of large particles outward in a drum and the segregation of small particles inward is not necessarily a collective behavior for mixtures of particles; the trend applies to tracer particles in a mixture as well.

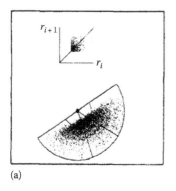 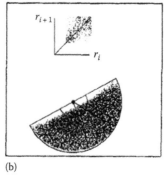 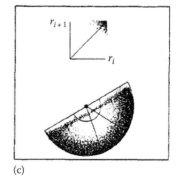

(a)   (b)   (c)

Figure 10.6 Results from experiments by Clément et al. (1995) using one tracer particle among 1.5 mm particles in a nearly half-filled drum. Each picture shows all locations of the tracer particles from 12,000 snapshots taken every 5 s, indicating the position of the tracer particle centre of mass. The diameter of the tracer particle in each case is as follows: (a) 1 mm, (b) 1.5 mm, (c) 2.0 mm. Above each of these pictures is the first iteration map of "reinjection" of the tracer particle where each point represents the radial location of the tracer particle entering (r_i) and leaving (r_{i+1}) the flowing layer.

The dynamics are similar when a mixture of different-sized particles is poured into a heap (e.g., Williams, 1963, 1968), as in Figure 10.5j. At any given moment, only particles in a thin layer near the free surface exhibit any substantial movement, segregation, or mixing. In the sheared surficial layer of flowing particles, particles are free to move, mix, and segregate relative to their neighbors until they subsequently deposit onto the bed where segregation is recorded. In the sheared region, the larger particles migrate upward toward the free surface, and the small particles migrate downward away from the free surface until they are captured into the layers below. The lower part of the layer is frozen "upstream," higher in the heap, and the larger particles are frozen last, transported toward the bottom of the heap.

Drahun and Bridgwater (1983) investigated these dynamics in a depositional heap through using experiments with tracer particles of different properties from the bulk. They injected tracer particles with different-sized bulk particles at the top of the heap and measured where the tracers were deposited along the downslope. They found that tracer particles of different sizes were captured at different distances from the top of the flow. The larger the tracer particles, the closer to the bottom they were captured; the smaller the tracer particles, the closer to the top they were deposited. Furthermore, they found that the relationship between tracer/bulk particle size ratio (for moderate particle size ratios ranging from 0.4 to 1.4) and normalized distance deposited from the top is linear.

While these experiments illustrate a relatively predictable behavior, more recent experiments with sheared granular materials illustrate yet another manner in which the "Brazil nut" segregation phenomenology is not as simple as it seems initially. Thomas (2000) and Félix and Thomas (2004) performed experiments with tracers in a 2D drum (Figure 10.7) and other sheared systems and found that the segregation of large particles toward the free surface increases with particle size ratio only for modest particle size ratios, up to approximately a ratio of 2. For larger particle size ratios, they observed a less intense segregation of large particles to the top and for larger tracer particles still, the large tracer particles can actually sink. In other words, they found that there is not a simple monotonic dependence on the segregation behavior with particle size ratio. Golick and Daniels (2009) and May et al. (2010a) found similar results in an annular shear cell (Figure 10.8). They used the shear cell to investigate primarily how the dynamics of mixing and segregation in a shear flow depend on the applied pressure. Their annular cell confined the particles between a top plate, free to move vertically, and a rotating bottom plate (Figure 10.8a). The confining pressure was adjusted via the top plate. As the bottom plate rotated, friction between the particles and the plates sheared the granular materials in layers (from bottom to top). Independent of initial conditions, the steady state of the experiments showed segregation, with larger particles at the top and smaller particles at the bottom, though they found the segregation rate is a nonmonotonic function of the particle size ratio with a maximum for intermediate particle size ratio of approximately two. They also found that the segregation rate increased with increasing pressure.

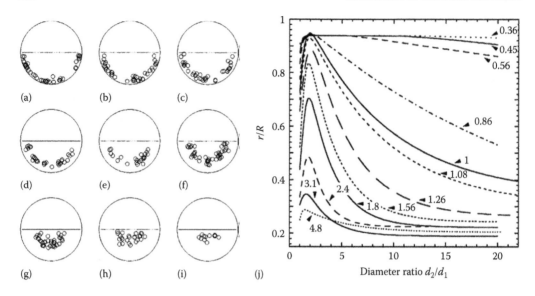

Figure 10.7   (a–i) Sketch of experimental results from rotating larger spherical glass tracer particles (3% by volume) among smaller glass beads in a half-filled drum from Thomas (2000). Tracer positions are superposed from several time steps. The small beads are (a) 1.5 mm, (b) 0.85 mm, (c) 0.60 mm, (d) 0.50 mm, (d) 0.425 mm, (f) 0.30 mm, (g) 0.212 mm, (h) 0.150 mm, and (i) 0.090 mm# (#performed with 1% 3 mm beads,) The location of the segregated beads evolves continuously from the periphery to the center when increasing the diameter ratio. (j) Summary of experimental results from analogous experiments to those of (a–i), but allowing the density to vary as well (Félix and Thomas, 2004). The vertical axis is the distance of the average tracer particle location r normalized by the radius of the drum R. Each curve is labeled with the density ratio ρ_1/ρ_2.

In many ways, models for the segregation of particles by size in sheared systems use mechanisms similar to those proposed for shaken systems. A foundational model proposed by Savage and Lun (1988) evoked an effect analogous to the void filling mechanisms for shaken systems (e.g., Williams, 1963; Rosato et al., 1987). In the Savage and Lun (1988) model, the shearing layers are modeled as "random fluctuating sieves," regularly supplying voids into which particles may fall from the layer immediately above. The void sizes are randomly distributed, and smaller particles have a higher likelihood of falling into the layer below. The voids are not fixed in size or location, hence the names "random fluctuating sieve" or "kinetic sieve." One key difference between this model and similar particle-size-dependent segregation models for shaken systems is the importance of shear rate in supplying voids for segregation and therefore in controlling the rate of segregation. The framework proposed by Savage and Lun explicitly takes into account particle-scale interactions and, as such, the details of the model provide a useful context for thinking critically about the segregation process.

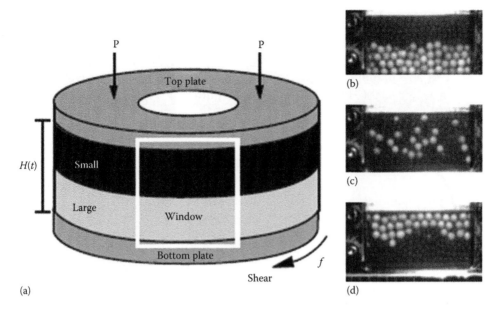

Figure 10.8 (a) Schematic of the experimental annular shear cell used by Golick and Daniels (2009) and May et al. (2010a). (b–d) images taken of particles through window (shown in (a)) illustrating the progression of a mixture of 4 mm (dark) and 6 mm (light) particles in a typical experiment used to study mixing and segregation rates of different-sized particles, (b) initial configuration of particles, (c) mixed state reached partway through the experiment, (d) final segregated state.

More recent models include other considerations such as a condensation mechanism associated with the difference between the energies of different-sized particles as well as separate compressive and tensile geometric forces (e.g., Trujillo et al., 2003). Typically, these are considered in the context of particles that can vary in both size and density, some results of which we describe in Section 10.4.3.

10.4.2   Vertical Segregation by Density in Sheared Systems: Effective Buoyancy Revisited

When a mixture of particles differing only in density is sheared, the behavior is similar to when such systems are shaken or quenched, with denser particles typically sinking relative to the bulk. Figure 10.9 shows such a mixture partially filling a thin drum; the heavier particles segregate downward in the thin surficial layer away from the free surface, and the lighter particles segregate upward, toward the free surface. As described in Section 10.4.1, the lower layers freeze into the solid-like rotation first. As the lower layers become concentrated with heavier particles and the upper layers with lighter particles, the overall segregation pattern in the drum becomes a radially symmetric pattern with the lighter particles on the outside of the drum and the denser particles in the middle (Figure 10.9f).

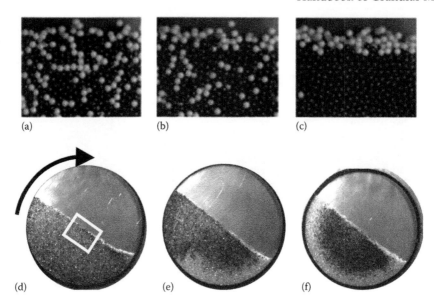

Figure 10.9   Segregation of equal-sized particles of different densities in a drum. The first
row illustrates segregation of equal-sized plastic (gray) and steel (black) particles in the
top sheared layer from (a) mixture to (b) 1/2 rotation of the drum to (c) final steady-state
pattern. The second row illustrates the segregation pattern that emerges in a drum. In this
case, the particles are equal-sized glass (light gray) and steel (dark gray) particles. In (d)
the particles are well-mixed and rotation has just commenced; the segregation happens
quickly as can be seen after just (e) 1/4 of a rotation and (f) 1/2 of a rotation of the drum.
The box in (d) illustrates the region of the drum where pictures in (a) - (c) were taken for
a different mixture.

Drahun and Bridgwater (1983) investigated the effect of density ratio on
the paths of tracer particles in their heap flow experiments. Their results are
somewhat analogous to those for different-sized particles segregating in their
heap flow experiments: denser particles play the role of "sinkers" among less
dense (same-sized) counterparts, and lighter particles are "floaters" among denser
(same-sized) counterparts. They observed that denser tracer particles sank away
from the free surface and were frozen closer to the top of the slope; less dense
tracer particles rose to the free surface and traveled further down the slope toward
the bottom of the pile before being captured. The final locations of the tracer particles in these experiments indicated that for modestly different density particles
(i.e., for particle density ratios between 0.5 and 1.5) the travel distance downslope
depended monotonically on the particle density ratio.

As with shaken systems, this segregation by particle density has been commonly attributed to the Archimedean buoyancy force, essentially, the force on an
object due to the difference between the weight of the object and the weight of the
displaced particles. On the other hand, as detailed in the next section, mass, rather
than density, is sometimes a better indicator of segregation trends.

10.4.3   Vertical Segregation by Size and Density in Sheared Systems: The Reverse Brazil Nut Problem and the Intermediate Brazil Nut Problem

For many industrial applications (e.g., pharmaceutical pills), where particles often differ in both size and density, the degree to which a particle system is well mixed can determine the quality of a final product. In these cases, the search for a crossover point where density and size segregation effects balance one another is compelling. Efforts have typically focused on combinations of relative size and density in a mixture of particles that are smaller and less dense compared with another constituent consisting of particles that are larger and denser.

There have been rather limited and scattered reports of such mixtures that remain uniformly mixed under shear. Three sets of experiments in the 1980s and 1990s in particular stand out (Drahun and Bridgwater, 1983; Alonso et al., 1991; Metcalfe and Shattuck, 1996). In their heap flow experiments described earlier in this chapter, Drahun and Bridgwater (1983) found one particular system mixed well: glass spherical tracers of 78% diameter and 43% density of steel spherical particles in the bulk. Metcalfe and Shattuck (1996) studied the segregation, mixing, and transport behaviors of mixtures of particles differing in both size and density rotated in a drum using a magnetic resonance imaging (MRI) technique. While the material choices were rather limited—it is helpful if one component in a binary mixture is MRI sensitive, typically moist like a seed or oil-filled sphere—they found one mixture that did not segregate. They found that a 50/50 mixture of brown mustard seeds and larger denser glass spheres—diameter and density ratios approximately 0.7 and 0.65, respectively—mixed rather uniformly in a drum.

Alonso et al. (1991) performed perhaps the first systematic experiments in a rotating circular drum that included some theoretical considerations to directly address the competition between size and density effects using combinations of 13 different material types with densities of 1.08–7.78 g/cm^3 and particle diameters of 0.7–5.0 mm (Figures 10.10 and 10.11). Among all the mixtures they studied, two particle mixtures remained well mixed in the drum: (1) smaller glass beads, 30% by volume, and larger steel beads, 70% by volume—diameter and density ratios of 0.32 and 0.43, respectively, and (2) smaller plastic beads, 70% by volume, and larger glass beads, 30% by volume—diameter and density ratios of 0.50 and 0.42, respectively. The dependence of these results on the fraction of large particles in the mixture is shown in Figure 10.10. They found that density effects dominated the segregation behavior for lower fractions of coarse particles in the mixture and size effects dominated for higher concentrations in the mixture.

Alonso et al. (1991) derived a parameter (S in Figure 10.11) to determine when density differences or size differences dominate the segregation processes. The derivation of this parameter involved two effects. The first was segregation by particle size, which they modeled by considering the relative likelihood of the different-sized particles to find a void in the layer below large enough to descend (similar to the classic "kinetic sieving" argument of Savage and Lun, 1988). The second effect they considered was segregation by particle mass, rather

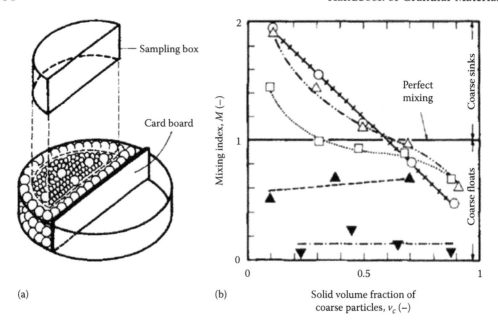

(a)   (b)   Solid volume fraction of
                                                             coarse particles, v_c (–)

Figure 10.10 Some results from segregation experiments performed by Alonso et al. (1991) of segregation and mixing behavior of particles differing in both size and density rotated in a thin drum. (a) Illustration of the sampling used to determine a "mixing index," which is essentially a measure of the difference between the fraction of coarse particle in the core, within the sampling box illustrated, and the fraction of coarse particles in the rest of the drum. (b) Experimental results for five different sets of binary mixtures using different concentrations of coarse particles. In the following, d refers to the mean diameter and ρ refers to the mean density of each component in the five mixtures shown. ▼ d = 2 mm glass beads (ρ = 2.49 g/cm^3) and d = 1.12 mm ZrO$_2$ spheres (ρ = 6.04 g/cm^3) ▲ d = 4.00 mm chocolate balls (ρ = 1.42 g/cm^3) ○ d = 1 mm plastic beads (ρ = 1.05 g/cm^3) and d = 3 mm steel balls (ρ = 7.78 g/cm^3) □ d = 1 mm plastic beads (ρ = 1.05 g/cm^3) and d = 2 mm glass beads (ρ = 2.49 g/cm^3) Δ d = 3 mm steel balls (ρ = 7.78 g/cm^3) and d = 1 mm plastic beads (ρ = 1.05 g/cm^3).

than density. For this, they considered the relative ability of a heavier particle (due to combined effects of size and density) to push open a space below it large enough to descend. They reasoned that this capability should vary not only with relative size and density of the components in a mixture but also in their relative concentration. With these considerations they derived a segregation parameter S (see Alonso et al., 1991).

$$S = \left\{ \frac{\rho_r \times [1 + v_c \times (d_r - 1)]}{d_r \times [1 + v_c \times (\rho_r - 1)]} \right\} \times \left\{ \frac{1 - P(d_r, v_c, \varepsilon)}{1 - P(d_r = 1, \varepsilon)} \right\} \tag{10.1}$$

where

$\rho_r = \rho_{large}/\rho_{small}$ and $d_r = d_{large}/d_{small}$ are the density and size ratios, respectively, of the large and small particles

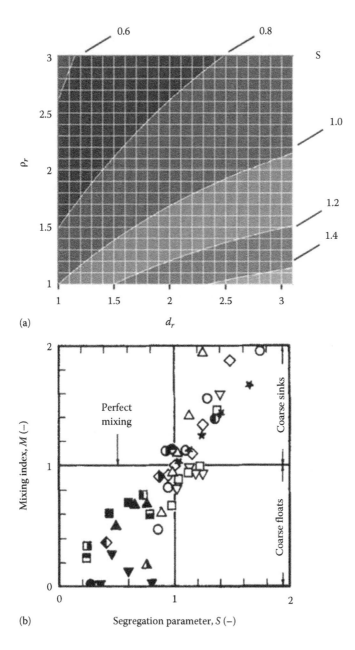

Figure 10.11   (a) A contour plot of segregation parameter S (Equation 10.3) proposed by Alonso et al. (1991) for a range of size and density ratios. For these calculations, porosity ϵ = 0.55 and the fraction of large particles v_c = 0.5. The theory predicts that when S = 1, density and size segregation effects cancel; otherwise either segregation by density ($S > 1$) or size ($S < 1$) should dominate. (b) Comparison between theoretical predictions (S) and experimental observations (in terms of M) of segregation trends. Each symbol type represents a different mixture described in detail in Alonso et al. (1991).

ν_c is the fraction of large particles

$P(d_r, \nu_c, \varepsilon)$ is the probability of finding a void of diameter equal to or greater than d_{small} in an assembly of randomly packed particles of particle ratio d_r, where the fraction of large particles is ν_c, and the void ratio is ε

$$P(d_r, \nu_c, \varepsilon) = \varepsilon \times \exp\left\{ -\frac{1-\varepsilon}{\varepsilon} \times \left[1 + \nu_c \times \left(\frac{1}{1 + \nu_c \times (d_r - 1)} \right)^2 - 1 \right] \right\}. \tag{10.2}$$

The theory predicts that when $S < 1$, small particles should sink independent of relative density and when $S > 1$ dense particles should sink independent of relative size. When $S = 1$, the theory predicts that there is no segregation. Figure 10.11 contains a contour plot for S for $\nu_c = 0.5$ and $\varepsilon = 0.45$ over a range of density and size ratio.

Alonso et al. (1991) compared the predictions from this theory to segregation trends they observed in a drum, characterized by a segregation mixing index M. Here, M is essentially a measure of the difference between the fraction of large particle in the core of the drum ν_c^i, within the sampling box illustrated in Figure 10.10a, and the fraction of large particles in the rest of the drum, ν_c^o:

$$M = 1 + \frac{\nu_c^i - \nu_c^o}{\Delta\nu_c} \tag{10.3}$$

where

$$\Delta\nu_c = \left(\nu_c^i\right)_{max} - \left(\nu_c^i\right)_{min} = \begin{cases} 2\nu_c, & \text{for } \nu_c \le 0.5 \\ 2(1-\nu_c), & \text{for } \nu_c \ge 0.5 \end{cases} \tag{10.4}$$

They found reasonable agreement in the segregation trends predicted by their model and observed in their experiments (Figure 10.11b). Comparing results with MD simulations from Hong et al. (2001), these results emphasize the importance of boundary conditions, particularly in determining the crossover between density- or mass-dominated segregation effects and size-dominated effects.

In all of the segregation discussion to this point, one constituent had the tendency to move *up or down* throughout a system. Thomas (2000) and Félix and Thomas (2004) showed that the segregation direction was not always uniform for a particular system. As mentioned in Section 10.4.2, for equal-density particles, Thomas (2000) observed a transition between the Brazil nut segregation where the large particles rise to a gradual reversal in the segregation direction for a diameter ratio of approximately two. When a mixture of equal-density particles of relatively small diameter ratios (from just over one to five) was rotated in a drum, the large particles rose to the top of the flowing layer (e.g., see Figure 10.7) and formed a segregated ring of particles near the outside of the drum (Figure 10.7a and b). As the size ratio increased beyond that, the larger particles segregated lower in the flow. This was reflected in the segregation pattern, where the ring of larger particles became smaller in diameter until, for the largest particle ratios, the segregation direction had nearly entirely reversed (Figure 10.7c through i).

Félix and Thomas (2004) simultaneously varied the density ratio and the size ratio of the particles and qualitatively reproduced some of the phenomenology of the RBNP reported by Hong et al. (2001) in shaken systems. For a particular particle size ratio, if the density of the larger particles was significantly larger than those of the smaller particles, the large particles sank to the bottom of the flowing layer. The "transition" observed by Félix and Thomas (2004) in the drum was not abrupt; however, it was a significant difference from that observed by Hong et al. (2001). As illustrated in Figure 10.7j, as the larger particles increased in density, the average steady-state location of the larger particle gradually shifted inward, whereas in the condensing systems reported by Hong et al. (2001) the segregation pattern changed abruptly.

Hong et al. (2001) did not report any conditions under which the larger particles segregated to an intermediate height in the system.

These results—what one might call an "intermediate Brazil nut problem"—do not appear unique to tracer particle experiments (Figure 10.12a). Experiments and simulations (Figure 10.12b and c) (from Jain et al., 2005a; Zhang, 2009, respectively) show that certain 50/50 mixtures of different-sized steel and glass beads can exhibit this "intermediate Brazil-nut" segregation behavior as well, though the segregation pattern is not nearly as sharp as when fewer large dense tracer particles are used.

This intermediate Brazil nut pattern has been relatively infrequently reported—almost nonexistent in the literature until the 2000s—yet, there is at

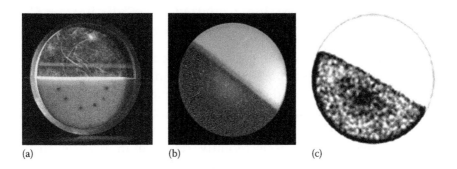

(a)   (b)   (c)

Figure 10.12   Images of the "intermediate Brazil nut problem" observed in experiments (a and b) and simulations (c) of particulate mixtures rotated in drums. (a) Photograph from an experiment (Thomas, 2000) using 0.150 mm small glass beads and ~0.3% 1.5 mm glass tracers showing the clear segregated ring pattern that appears for relatively few large tracer particles. (b) Photograph from an experiment performed by Jain et al. (2005a) showing a similar phenomenon in a 50/50 mixture of 2 mm glass beads (white) and 4 mm steel beads (darker but also shiny) rotated in a drum. (c) Image from a soft-sphere discrete element method (DEM) simulation performed by Zhang (2008) of a 50/50 mixture of larger denser particles (white) and smaller less dense particles (black). ($\rho_1/\rho_2 = 3.1$; $\rho_1/\rho_2 = 1.5$). (b and c) The ring of smaller particles around the periphery in addition to the high concentration near the center of rotation contains a wide band highly concentrated in large dense beads at an intermediate distance from the center of rotation.

least one isolated report of a similar intermediate segregation behavior of a large tracer particle in a shaken system in the 1960s. Williams (1963) reported that he could tune the height of a large intruder particle in a deep bed of smaller particles of the same density by varying the frequency of vibration. Williams described a series of experiments where he initially placed a large particle on the bottom of the bed and filled the rest of the container with smaller particles. Provided the agitation was sufficiently high to induce some movement of the large particle relative to the rest of the bed, he found that only low vibration frequencies brought the particle to the top of the bed. Increasing the frequency further would increase the steady-state depth of the large particle. The process was repeatable and reversible, so each intermediate frequency was associated with a unique intruder particle height in the bed. Williams (1963) attributed the behavior he observed to pressure variations in the bed trapping the large particle at specific locations that depended on the mass of the large particle and the local density variation. Thomas (2000) and Félix and Thomas (2004) proposed a balance between size, where the dynamics tended to ratchet larger particles upward, while more massive particles had the ability to push down into the bed (not unlike that of Alonso et al. (1991). Hill et al. (2010) suggested that dynamics associated with a shear-rate gradient apparent in many gravity-driven free-surface sheared mixtures may also play a role, a conjecture supported in simulations in a chute flow (Fan and Hill, 2011a) and associated theoretical work (Fan and Hill, 2011b).

10.5   Shear-Driven Segregation: Couette Cells, Split-Bottom Cells, and Vertical Chutes

The discussion of segregation to this point has focused on segregation patterns primarily driven by gravity and possibly associated pressure gradients. Other segregation mechanisms have been identified in more energetic systems. For example, a gradient in granular temperature—the kinetic energy of velocity fluctuations—has been shown to drive segregation (e.g., Jenkins and Mancini, 1989) and shear-driven segregation has been recently reported in particle suspensions (Barentin et al., 2004), mitigated somewhat by curvature in the mean particle path (Krishnan et al., 1996). In typical experimental systems (e.g., chutes and rotated drums), gravity, velocity gradients, and porosity gradients coexist in the direction of segregation (Hill et al., 2003). In this section, we consider geometries that have the ability to isolate segregation effects associated with a shear-rate gradient. In other words, in contrast with the previous sections, we consider the effects of shear-rate gradients *isolated from segregating effects of gravity*.

10.5.1   Shear-Driven Segregation in a Split-Bottom Cell

A compelling geometry for studying shear-driven segregation is a "split-bottom-cell" (e.g., Fenistein and vanHecke, 2003; Fenistein et al., 2004) illustrated in Figure 10.13a. A typical split-bottom cell consists primarily of a cylindrical container oriented with its axis vertical, and one end, the "bottom" of the cylindrical cell (of radius $r = R_o$) is split at radius $r < R_o$. Typically, the inner part of the bottom is rotated and the outer part is fixed to the "outer" vertical cylinder walls. The cell is filled with granular materials to a height H. At sufficiently low fill levels, a cylindrically symmetric—primarily vertical—shear band is produced in the bulk of the granular materials (Fenistein et al., 2006). Near the base, the shear band is narrow and centered at the split; with increasing distance from the bottom, the band moves slightly inward, and its width increases (Figure 10.13b). As Fenistein and vanHecke demonstrated (2003), the split-bottom cell allows one to create wide and tunable shear zones away from any sidewalls or free surface. For the purposes of studying segregation, this allows one to isolate shear-rate gradients and their associated dynamics in a direction distinct from that of gravity.

While one might expect this to result in horizontal segregation patterns associated with the horizontal shear-rate gradients, the dominant segregation effect still appears to be vertical. Within the shear band, less dense particles rise and denser particles sink (Figure 10.13e through g). The temporal details indicate that particles differing only in density behave as one might expect from an emulsion of immiscible fluids of different densities in the fluid-like shear zone. Within a single rotation of the base, the segregated zone of less dense particles extends at least

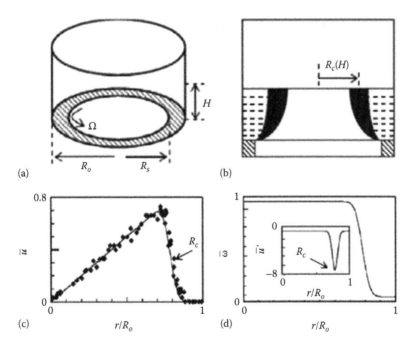

Figure 10.13 Pictures illustrating segregation results in the sheared band within a split-bottom cell. (a) Sketch of a split-bottom cell; Ω is the rotation speed of the base. (b) Sketch of bulk shear band (black area). (c) Measured surface velocity as a function of radial position (symbols) for 1 and 3 mm glass particle mixtures described in Hill and Fan (2008). The solid line is the shear band velocity described by Fenistein et al. (2004) (d) The radial dependence of $\omega = u/r$ and $u' = rd\omega/dr$.   (*Continued*)

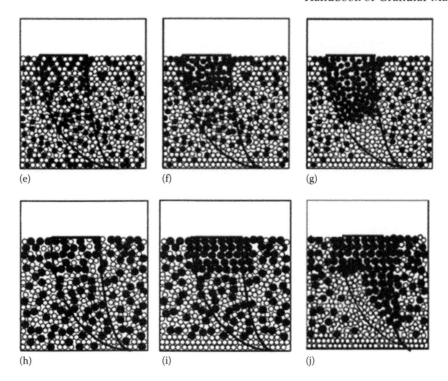

Figure 10.13 (*Continued*)   Pictures illustrating segregation results in the sheared band within a split-bottom cell. (e–g) Sketch showing the progression of segregation associated with particles of different densities with the shear band outlined. (h–j) Sketch showing the progression of segregation associated with particles of different sizes, never evolving to a fully vertically segregated state.

1–2 layers and a small segregated region of denser particles grows near the base of the cell. With further rotations, the two segregated zones grow toward each other until they meet in the middle. The segregation is primarily limited to the bulk shear zone; there is no compelling evidence of horizontal segregation in the bulk, and so one concludes that the vertical segregation (in the direction of gravity) is much faster than any horizontal segregation (in the direction of the shear-rate gradient). Vertical segregation by density of equal-sized particles appears to be much quicker and complete than that by size (e.g., Figure 10.13h through j). The steady-state segregation pattern in the latter case involves only a thick layer of large/small particles on the top/bottom.

If the bottom of the cell is rotated for a sufficiently long period of time, a horizontal segregation pattern develops in the majority of the sheared region, with the large/small particles at the outside/inside. Computational simulations (Fan and Hill, 2010) indicate that this horizontal pattern is not so much due to a shear-driven effect but, instead, a combination of vertical segregation and advection of the large particles outward associated with a slow convection roll.

10.5.2 Shear-Driven Segregation in a Couette Cell: With and Without Gravity

A more complete way to isolate the effect of a shear-rate gradient from gravity is by shearing a granular mixture in microgravity. Louge et al. (2001) and Xu et al. (2001,2003) performed such a study using a racetrack-shaped shear cell and, additionally, in a canonical Couette cell. Some results from their Couette cell are shown in Figure 10.14. The shear rate and the velocity fluctuations, or granular temperatures, varied smoothly from one cell wall to the other. In all cases, the segregation patterns involved the more massive particles—denser or larger—segregated to regions of lower granular temperature and lower shear rates.

These segregation patterns could be modeled in the context of kinetic theory (originally developed for agitated molecules in a dense gas) applied to colliding particles in rapid granular flows (Xu et al., 2003). The derivations from kinetic theory are quite involved and discussed further in this Chapter. The essence is that it considers the statistics of energy and momentum exchange between particles during collisions and the subsequent evolution of species energies, momenta, and concentrations. The mechanisms for segregation may be expressed as a competition between a pressure diffusion force dependent on the gradient of the particle pressure, a thermal diffusion force dependent on the gradient of the granular temperature, and the ordinary diffusion force dependent on the gradient of the particle concentration or number density (e.g., Jenkins and Mancini, 1989; Arnarson and Willits, 1998). Xu et al. (2003) showed that kinetic theory works well to predict qualitatively segregation trends, but in the most common form to this point, where assumptions of binary collisions and molecular chaos are applied, the theory overpredicts the degree of segregation, particularly for dense flows.

Segregation simulations (Figure 10.15a) and ground-based experiments (Figure 10.15b) in a Couette cell performed by Conway et al. (2006) exhibit similar behavior for low solid fractions. More massive particles segregate toward regions of high density and lower shear rates (Figure 10.15c and d). However, for higher solid fractions (above ~0.3), the large particles segregated toward regions of higher solid fractions (Figure 10.15e and f). The segregation patterns were modeled reasonably well by kinetic theory for all cases *except* those for the high solid fraction mixtures. They reasoned that, at higher solid fraction, the clustering of particles away from the cell walls was associated with an additional mechanism (e.g., kinetic sieving) that was omitted from their kinetic theory (e.g., Willits and Arnarson, 1999).

10.5.3 Shear-Driven Segregation in a Vertical Chute

A vertical chute with roughened walls is a third geometry recently used to investigate the effect of a shear rate and its gradient on segregation. If the vertical walls are roughened (as in Figure 10.16a), the geometry ensures that the flow is nonuniform (Figure 10.16b through d) with high shear rates near the walls and a plug-like flow in the middle. Fan and Hill (2011a) performed soft-sphere discrete element method (DEM) simulations on mixtures of different-sized particles and found that,

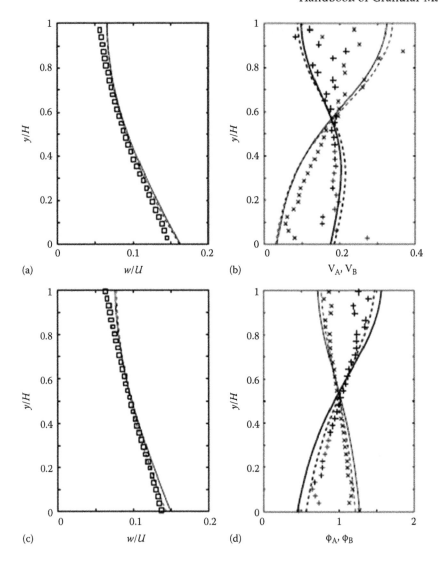

Figure 10.14   Data from granular mixtures sheared in a Couette cell under microgravity from Xu et al. (2003). The two boundaries are bumpy and move in opposite directions at velocities of $U/2$ and $-U/2$. All data are plotted as profiles; y is the distance from one wall to the other, where H is the width of the gap. (a and b) Data from a mixture of particles differing only in density; $\rho_A/\rho_B = 0.5842$. (c and d) Data from a mixture of particles differing only in size; $d_A/d_B = 1.25$ (a) Average normalized velocity fluctuations w. (b) Particle solid fractions of the less dense (+) and denser (×) particles. (c) Average normalized velocity fluctuations w and (d) Solid fractions of the larger (+) and smaller (×) particles. Lines indicate predictions from kinetic theory as outlined in Xu et al. (2003).

while the local solid fraction and other kinematics are nonuniform (Figure 10.16b through d), the segregation trend varies with a global measure of the internal structure that loosely follows the global solid fraction. For systems of low solid fraction, the large particles accumulate within regions of low shear rate, low granular temperature, and high solid fraction regions (Figure 10.16e and f). Similar to the Couette studies of Conway et al. (2006) shown in Figure 10.15, the segregation fluxes for these cases where the system solid fractions are low agree quantitatively with predictions from kinetic theory. The segregation remained qualitatively the same for slightly higher solid fractions (up to ~47%, Fan and Hill, 2011a), the large particles segregating to lower shear rates and granular temperatures. However, for these intermediate solid fractions, the flux decreased, and kinetic theory overpredicted the segregation "strength." At the highest solid fractions (above ~50%) the segregation reversed in the simulations, and large particles segregated to higher shear rates and granular temperatures (Figure 10.16g and h). Kinetic theory as classically applied did not capture this reversal.

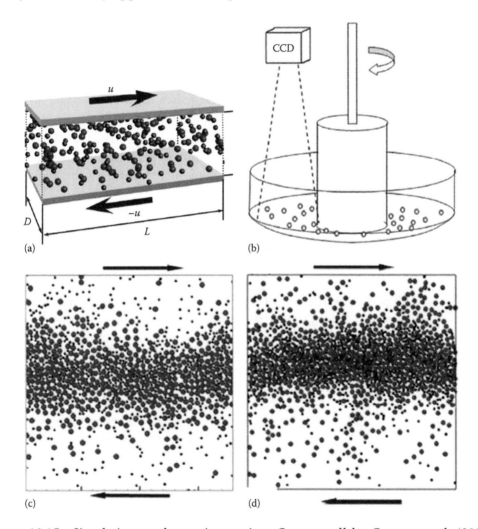

Figure 10.15 Simulations and experiments in a Couette cell by Conway et al. (2006). (a) Sketch of simulations in a Couette or a shear cell and (b) sketches of experiments in the Couette cell. The tilted base allows gravity to compensate for centrifugal effects in the rotated system. (c and d) Snapshots from simulations of lower and higher solids fractions, respectively. (the larger particles are twice the diameter of the smaller particles and of equal density). *(Continued)*

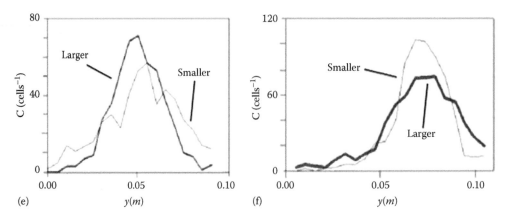

Figure 10.15 (*Continued*) Simulations and experiments in a Couette cell by Conway et al. (2006). (e and f) Concentration profiles of the constituents from the simulations shown in (c and d), respectively. The labels in (e and f) indicate the corresponding particle size (larger or smaller). The average concentration of the system depicted in (c) and (e) is 0.25, and the average concentration of the system depicted in (d) and (f) is 0.30.

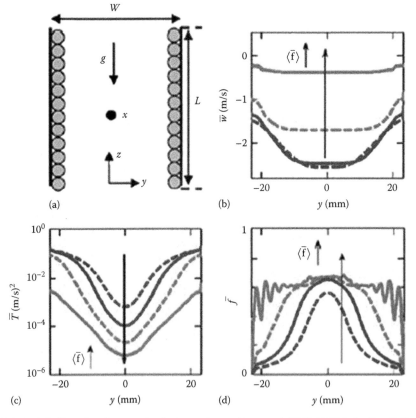

Figure 10.16 (a) Sketch of a vertical chute from Fan and Hill (2011a). (b–d) Profiles of kinematic quantities for four mixtures (average system solid fraction = 0.21, 0.34, 0.47 and 0.60). (b) Average streamwise velocity; (c) average granular temperature; and (d) average local solid fraction of the mixture. (*Continued*)

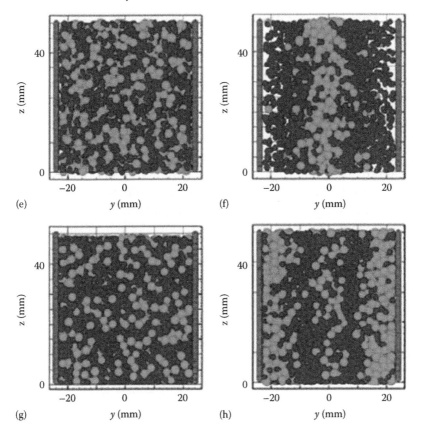

Figure 10.16 (*Continued*)   (e–h) Snapshots of two mixtures from (b–d) at the point of system initiation ($t \sim 0$) and at steady state. (e–f) snapshots of the system with solid fraction ~0.21 (e) at $t \sim 0$ s, (f) at steady state ($t \sim 50$ s), (g and h) snapshots of the system with solid fraction ~0.60 (e) at $t \sim 0$ s, (f) at steady state ($t \sim 400$ s). Each system is a 50/50 mixture of 2 mm particles, dark; 3 mm particles, light.

Fan and Hill demonstrated that this segregation reversal is correlated with the system structure: as the solid fraction increases, the system structure changes from one dominated by binary collisions to one dominated by a single cluster that can involve >60% of the particles and span the system. The segregation reversal correlates to the point at which the biggest cluster contains more than 34% of the particles in the system, which occurs somewhere between a total solid fraction of 0.5 and 0.6. Fan and Hill suggested using the cluster size as an order parameter for the phase change, of which the segregation behavior is a signature. As the cluster size increased, geometric filtering (e.g., sieving) might change the dynamics, which could be modeled by considering both the kinetic stress gradient (that scales similarly to the granular temperature gradient) and the manner in which the contact stress and kinetic stress are partitioned among the mixture constituents (e.g., Fan and Hill, 2011b).

10.6 Complex Segregation in Simple Systems

There are many systems in which gravity, shear rate, granular temperature, and other segregation drivers work together to determine the important details for segregation flux and subsequent segregation patterns. The outcome can be relatively simple segregation patterns, as in the radial segregation patterns in the drum, or they may be much more complex and difficult to predict. Examples in relatively simple systems include longer rotated cylindrical and rectangular drums. Figure 10.17a shows the well-known axial segregation pattern that can appear in long rotated drums (e.g., Oyama, 1939; Oyama and Ayaki, 1956). While some data suggest that axial banding is driven by surface effects (e.g., Hill and Kakalios, 1994), other data suggest that the banding pattern is driven by subsurface mechanisms that may be related to other dynamics such as shear-rate gradients

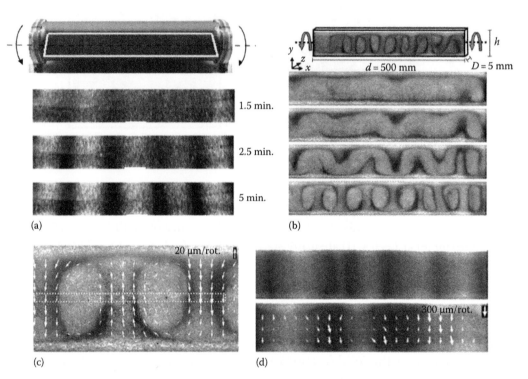

Figure 10.17 Pictures of experiments of axially rotated containers with mixtures of different-sized particles illustrating relatively simple systems where effects of gravity and shear-rate gradients combined may give rise to complicated segregation patterns. (a) Axially segregation of different-sized particles in a cylindrical partially filled drum. (From Hill, K.M. et al., *Phys. Rev. E*, 56, 4386, 1997.) (b) Segregation of different-sized particles in a thin rotated box of a solid fraction of ~0.65. (From Rietz and Stannarius, *Phys. Rev. Lett.* 100, 078002, 2008). (c) Pattern from experiments in (b) with velocities superposed. (d) Analogous results from those shown in (b and c), obtained for a fill level of ~0.58 (less than the critical amount required for the complex circulation and segregation patterns of (b and c)).

(e.g., Hill et al., 1997; Nakagawa et al., 1997). In a rectangular drum, even more complex patterns can arise including elliptical swirls (Figure 10.17b and c) and simple banding patterns (Figure 10.17d) (Rietz and Stannarius, 2008). Simultaneous concentration and velocity data such as those shown in Figure 10.17 (c–d) suggest a link between shear-rate gradients and segregation, though a predictive model for these effects is not yet available.

It is apparent from the wide range of segregation patterns observed that, even for relatively simple systems, segregation behavior is complex and even the most basic trends are sensitive to local and global boundary conditions. It seems unlikely that a single model framework could fully predict segregation behaviors of such a wide range of systems. By focusing on one regime of agitated systems, however, one may obtain deeper insights on some of the segregation behavior. In the next section, we focus on a few such modeling frameworks.

10.7 Modeling Frameworks for Segregation in Dense Sheared Mixtures

One of the challenges of modeling segregation is capturing the wide variation of segregation behavior with boundary condition and method of agitation. Modeling efforts vary widely, a reflection of how the dominant segregation mechanisms can change from one set of boundary conditions to the next. Some efforts have focused on capturing particle-scale dynamics and the subsequent segregation trends (e.g., BNP vs. RBNP). Other efforts have focused on continuum-style models that incorporate segregation trends into macroscopic system dynamics to predict the subsequent segregation patterns that arise.

This section is intended as a sampling of efforts used to capture different aspects of segregation in gravity-driven sheared granular mixtures. The first two sections illustrate how the physics of segregation by particle density (in Section 10.7.1) and by particle size (in Section 10.7.2) may be modeled at the particle scale. Section 10.7.3 illustrates a model framework in which particle-scale segregation by particle density is combined with a continuum-style model to predict the evolution of segregation in a complex sheared system. Section 10.7.4 illustrates another model framework for different-sized particles based on conservation laws and an adaptation of particle-scale segregation laws into a continuum framework.

10.7.1 A Particle-Scale Dynamics Model for the Brazil Nut Problem: Kinetic Sieving and Squeeze Expulsion

This section focuses a model designed to capture the size-dependent segregation process that drives large particles to the top of a sheared gravity-driven flow in a relatively high solid fraction mixture, that is, the dynamics that lead to the "BNP." The focus of the model is on the segregation dynamics within the flow itself, rather than in the geometric configuration—for example, a rotated drum or a chute flow—in which it might appear.

Savage and Lun (1988) proposed that two primary mechanisms drive the segregation process. The first mechanism they referred to as a "random fluctuating sieve," in other words, "a gravity-induced, size-dependent, void-filling mechanism." Essentially, in a sheared flow, as layers of particles—and vacancies—shear past one another, particles are pulled in the streamwise direction and also "feel" the pull of gravity, normal to the flow, toward the layer below. The voids within each layer are randomly generated, but there is a higher probability of one large enough to accommodate a small particle, so these have a statistically higher chance to move downward. The focus of modeling efforts is on this "kinetic sieving" mechanism, the statistical distribution of void sizes, and the dependence of the void availability on shear rate.

Savage and Lun (1988) hypothesized that a second complimentary mechanism is associated with imbalances in contact forces on individual particles, which squeeze it out of its own layer into an adjacent one. They referred to this mechanism as the "squeeze expulsion" mechanism, which balances the preferential downward movement of small particles from kinetic sieving with a non size preferring motion upward. Since the small particles tend to drift downward relative to the mixture, all particles are more likely to experience imbalances in vertical contact forces. This squeeze expulsion mechanism is not size preferential and, other than the underlying driving force to maintain mass balance, there is no inherent preferential direction for the layer transfer. Thus, somewhat less attention has been given to the underlying physical dynamics of this second mechanism.

The theory associated with these two complimentary mechanisms predicts N_i, the rate of a particle of type i, small or large ($i = S$ or L), falling into a layer below (N_i) based on the statistically determined size distribution of void spaces in the layer below and local shear rate:

$$N_i = \int_{E_i}^{\infty} n_v \frac{n_i}{N} n_c p(E) \, dE \tag{10.5}$$

where the integral is over all voids in a region of interest greater in size than the diameter of particles of type i; n_c represents the number of particles of either size that have the potential of being within a particular "capture area" per time, dependent on the local shear rate, the "capture diameter" of the region (see Figure 10.18), and the number of particles per unit area in a layer; n_i/N is the number fraction of particles of type i in the region of interest, so this multiplied by the bracketed term gives the potential of particles of type i for being within a particular "capture area" per unit time. n_v is the number of voids per unit area within a layer. $E = D_V/\bar{D}$, the ratio of the diameter of a particular void to the average particle diameter; $p(E)dE$ is the probability that a void has a particular size or void ratio between E and $E + dE$. Savage and Lund showed that using Stirling's approximation for a random distribution of states and maximizing information entropy (Jaynes 1963; Brown 1978)

$$p(E)dE = \frac{1}{E - E_m} \exp\left\{-\frac{E - E_m}{\bar{E} - E_m}\right\} dE \tag{10.6}$$

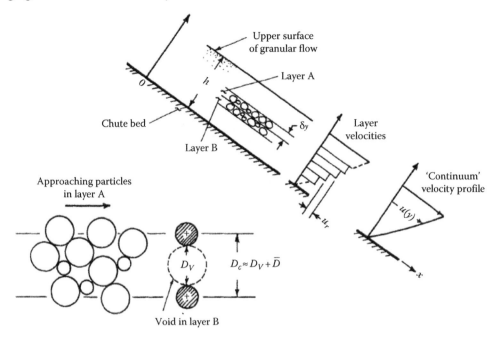

Figure 10.18 Figures from Savage and Lun (1988) illustrating the framework of their model—a systems of layers of particles sheared passed one another with random voids of different sizes (upper right) and the process of capture of particles from one layer into the layer below (lower left).; $D_c \approx D_V + \bar{D}$ is the capture diameter of the region, where D_V is the associated void diameter, and $\bar{D}$ is the mean particle diameter.

where
  $\bar{E}$ is the ratio of the mean void diameter $\bar{D}_V$ to $\bar{D}$ (see Figure 10.18)
  E_m is the minimum possible value for $\bar{E}$ for the system investigated ($E_m = 0.1547$, for close-packed monosized spheres)

Solving the integral in Equation 10.5 gives the rate of capture of particles of type i into the layer in the following:

$$N_i = n_v \frac{n_i}{N} n_p u_r \bar{D} \left(E_i + \bar{E} - E_m + 1 \right) \exp\left\{ -\frac{E_i - E_m}{\bar{E} - E_m} \right\}. \tag{10.7}$$

For continuum or averaged percolation velocities, N_i can be expressed as a mass flux normal to the direction of flow relative to the mixture

$$\rho_i q_i = -m_i N_i \tag{10.8}$$

where
  q_i is the average volumetric velocity of the particles of type i normal to the flow (a volumetric segregation velocity)
  m_i is the mass of an individual particle of type i
  ρ_i is the density of particles of type i

$$\rho_i = f_i \rho_m \tag{10.9}$$

where
> f_i is the fraction of volume filled by particles of type i
> ρ_m is the material density of the particles

According to this formulation, both types of particles in a mixture should have a downward segregation flux. The distinction between the two constituents is that the small particles move downward at a higher rate. For mass conservation, the downward "segregation flux" of the two constituents must be balanced by an upward flux. Savage and Lun (1988) hypothesized that this was achieved through the squeeze expulsion mechanism, where all particles were equally likely to respond to an imbalance of forces to move into an available space in *any* direction. The net flux associated with this squeeze expulsion mechanism could be expressed by ρq_{SE}, where $\rho = \rho_S + \rho_L = \rho_m (f_S + f_L)$, and mass balance requires

$$\rho_S q_S + \rho_L q_L + \rho q_{SE} = 0. \tag{10.10}$$

The net volumetric flux of each species can be expressed in terms of the volumetric segregation flux and the segregation flux

$$q_{i,NET} = q_i + q_{SE}. \tag{10.11}$$

Combining Equations 10.10 and 10.11 it is clear that the net mass flux of one species normal to the flow is equal and opposite that of the other species:

$$q_{L,NET} = -q_{S,NET}. \tag{10.12}$$

The net fluxes may be written in terms of the probabilistic expressions derived for the rate of transfer of each particle into the layer in the following:

$$\rho_L q_{L,NET} = \rho_L (q_L + q_{SE}) = \frac{\rho_S \rho_L}{\rho} (q_L - q_S) = -\frac{\rho_S}{\rho} m_L N_L + \frac{\rho_L}{\rho} m_S N_S. \tag{10.13}$$

This equation provides a quantity—the total flux of each component—that can be measured in experiments. Savage and Lun (1988) compared their predictions with published experimental data for the special case of a small fraction of small particles. They calculated parameters for some theoretical configurations, such as the number density of voids in a region of interest, the average void sizes, and a normalized mean voids area for close-packed and simple cubic configurations, and used these to compare the dependence of net flux on particle size ratio with experiments in gravity-driven chute flow (Bridgwater et al., 1978; Cooke and Bridgwater, 1979). Results, shown in Figure 10.19, indicate that the theory does predict the experimental fluxes assuming parameters associated with a solid fraction of 0.546, particularly for particle mixtures whose sizes are not too different. For particle diameter ratios smaller than approximately 0.155, small particle can spontaneously percolate through spaces among the large particles. This is a dynamic not accounted for in the theory, so it is reasonable that the theory would fail for particle diameter ratios approaching this value.

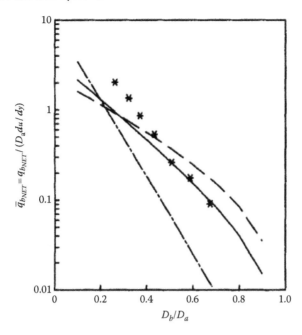

Figure 10.19   Normalized net flux of small particles vs. particle size ratio. * denotes experimental data from Bridgwater et al. (1978) and Cooke and Bridgwater (1979). The lines represent theoretical predictions using specific solid fractions, equal to 0.524, 0.546, and 0.571, providing for the number density of voids and the average void sizes. Parameters associated with a number density of 0.546 (solid line) appear to match with the most data. Plot from Savage and Lun (1988).

Strengths of this model include the fundamental mathematical and physical approach to the problem with minimal need for fitting parameters. The exception is that gravity does not appear explicitly, indicating that the direction of segregation must be determined by the user, so that there is no capability in the model to reflect a possible reversal of segregation reported in experimental observations of equal-density different-sized particles by Thomas (2000). The segregation rate is gravity independent, and one might imagine that gravity should play a role in the rate at which a particle falls into a void. Changing gravity could, however, simultaneously change shear rate and availability of voids, and the two effects may offset.

10.7.2   Segregation by Density Influenced by the Effective Temperature

Although the essential qualitative details of density-driven segregation often resemble those of size-driven segregation, the most common mechanism attributed to segregation by density is fundamentally different, more continuum in nature. Segregation according to density is most commonly attributed to the Archimedean buoyancy force, essentially, the force on an object due to the difference between the weight of the object and that of the displaced particles.

Sarkar and Khakhar proposed a model that combined this effective buoyancy force with an effective temperature. For a single particle of density ρ_1 and effective volume $\forall$ among a sea of particles of effective average density ρ_m, one can write this force buoyancy force as

$$F_B = \forall (\rho_1 - \rho_m) g \tag{10.14}$$

(e.g., Alam et al., 2006), where g is the gravitational acceleration. For a sheared mixture of heavy and light particles where the sheared layer is inclined from the horizontal by an angle θ (as in the rotated drum in Figure 10.4), one may write

$$F_B = -(m_D - \rho_m \forall) g_\perp \tag{10.15}$$

(e.g., Sarkar and Khakhar, 2008). Here,
   m_D is the mass of a particle denser than the surrounding medium
   ρ_m is the effective density of the medium immediately surrounding the particle
      in question
   $g_\perp = g \cos \theta$ is the component of gravity normal to the average flow

One benefit of a relatively simple force law like that in Equation 10.15 is that it provides an easy-to-use building block for a predictive model for large-scale pattern evolution in segregating system. Sarkar and Khakhar suggested that the buoyancy force be related to the velocity of each constituent normal to the flow by

$$F_B = v_D \xi_D \tag{10.16}$$

where ξ_D is the friction coefficient of the heavier particles. The fluctuation–dissipation relation relates the friction coefficient to an effective temperature T_E

$$\xi_D = \xi_{DL} = \frac{T_E}{D}. \tag{10.17}$$

The segregation flux is defined as

$$J_D^s = (v_D - v_m) n_D \tag{10.18}$$

where
   v_D is the velocity of a dense particle normal to the flow direction
   v_m is the number average velocity
   n_D is the number density of dense particles

The average number velocity is $v_m = f v_D + (1 - f) v_{DL}$, where f is the number fraction of heavy particles and v_i is the average velocity normal to the flow for species i (D or LD for dense or less dense particles, respectively). Using Equations 10.17 and 10.18, the segregation flux J_D^s may be rewritten as

$$J_D^s = -nD (m_D - m_{LD}) g_\perp f (1 - f) / T_E. \tag{10.19}$$

This is nearly a predictive form for segregation, with the exception of the uncertainty of the effective temperature and diffusion, for which different forms have been proposed. Sarkar and Khakhar pointed out that the effective temperature may be related to an apparent viscosity and diffusivity analogous to the drag coefficient with Stokes flow of a sphere,

$$\xi_D = \xi_{DL} = 6\pi\eta dD \tag{10.20}$$

where
   η is the viscosity of the granular fluid
   d is the particle size

This yields a relationship between the effective temperature and the viscosity and diffusivity

$$T_E = 6\pi\eta dD. \tag{10.21}$$

Balancing the segregation flux (from Equation 10.19) with a diffusion flux for dense particles ($J_D = -nD\partial f/\partial y$), Sarkar and Khakhar derived a predictive form for the concentration of dense particles and its gradient that agreed well with their experimental measurements of the sheared boundary layer in the drum.

10.7.3 Macro-Scale Model: Interplay between Segregation, Advection, and Diffusion

In contrast with the previous two sections, where the models focused on the segregation dynamics at the particle-scale level, in this section, we focus on a model developed for the evolution of macro-scale segregation pattern.

We discuss a model first proposed by Khakhar et al. (1997b) for mixtures differing only in density. This approach can be very fruitful in understanding the origin of pattern formation in certain systems (e.g., stratigraphic patterns in nature) and also in diagnosing segregation problems from the segregation pattern outcome. While the particular model and solutions described by Khakhar et al. (1997a,b) are derived for a particular set of boundary conditions, the approach can be applied to a wide range of systems. The model we consider here has been used to describe pattern evolution in a rotated drum based on segregation and diffusion dynamics in the flowing layer (See "B" in Figure 10.5a) and advection throughout the drum proposed by Khakhar et al. (1997a,b).

The starting point of the model is a description of the mixing of non cohesive identical particles as they are sheared in the flowing layer of a circular drum (Figure 10.20 contains an illustration of some of the geometric variables used in this framework). The model approximates the flowing layer as steady, thin, and nearly flat, with a shape of a parabola. Assuming the particle volume fraction in the bed and the layer are nearly uniform and L is the half-length of the flowing

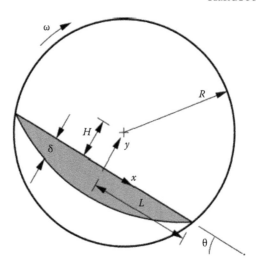

Figure 10.20   Sketch of flow in a slowly rotated drum from Khakhar et al. (1997b) illustrating parameters relevant to the model proposed by Khakhar et al. (1997a,b). The origin
$(x = 0, y = 0)$ is placed at the top of the center of the flowing layer, the center of a half-filled
drum.

layer (the radius of a half-filled circular drum), the spatial dependence of the
thickness of the boundary layer may be expressed by

$$\delta(x) = \delta_{x=0} \times \left(1 - \frac{x^2}{L^2}\right). \tag{10.22}$$

The shear rate is approximated as constant, that is, the velocity profile in the
x-direction is approximated as linearly dependent on the local depth of the flowing layer

$$v_x(x,y) = 2u(x) \times \left(1 + \frac{y}{\delta(x)}\right) \tag{10.23}$$

where $u(x)$ is the depth-averaged velocity in the x-direction, assumed to be uniform based in part on experimental data (Khakhar et al., 1997a); $u(x) = Q/\delta(x)$,
where Q is the volumetric flow rate of particles per unit width in the layer. In a
half-filled drum, this may be expressed as

$$Q = \frac{\omega L^2}{2} \times \left(1 - \frac{x^2}{L^2}\right) \tag{10.24}$$

(e.g., Rajchenbach, 1990). Thus, $u(x) = \omega L^2/2$. Particles enter the flowing layer at
the upstream end with a speed of ωx (and leave at the downstream end with speed
of ωy). Conservation of mass then dictates that the velocity in the y-direction
$v_y(x,y) \sim -\omega x \left(\frac{y}{\delta(x)}\right)^2$.

This model, coupled with solid body rotation in the bed, is sufficient to compute a Poincaré section for the circular mixer, shown in Figure 10.21a. As is

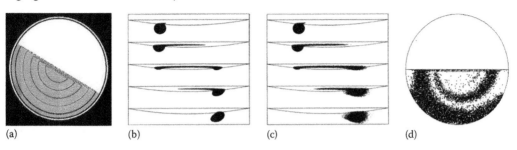

(a)　　　　　　　　(b)　　　　　　　　(c)　　　　　　　　(d)

Figure 10.21　Results from the model proposed by Khakhar et al. (1997a,b) as illustrated in Hill et al., (2001b) (a) A Poincaré section corresponding to a half-filled circular drum. The positions of the particles are marked after every half revolution. Different initial positions are marked with different shaded lines. (b and c) Evolution of a blob of marked particles within the flowing layer without (b) and with (c) diffusion term. (d) Result of a blob of marked particles evolving in a drum with advection and diffusion.

standard in fluid mechanics, the flow is interpreted in a continuum sense and collisional diffusion, the random-like motion of individual particles as they collide with other particles is ignored in the particle trajectory calculations. The plot is trivial for the circular mixer; the system is regular; particles can cross streamlines only by collisional diffusion.

The effect of collisional mixing/diffusion is incorporated in terms of a model, developed by Savage (1993), in which the diffusion coefficient is written as

$$D_{coll} = g(\eta)d^2\frac{\partial v_x}{\partial y} \tag{10.25}$$

where

　　$g(\eta)$ is a coefficient based on the solid volume fraction η
　　d is a particle diameter

$$\frac{\partial v_x}{\partial y} = \frac{-2u(x)}{\delta(x)} = \frac{-\omega L^2}{\delta(x)} \tag{10.26}$$

D_{coll} can be treated as a constant (c.f. Hill et al., 1999a) where $g(\eta)$ is approximated from experimental results as $g(\eta) \approx 0.025$, $\delta(x)$ is approximated by $\delta_{x=0}$, and so $D_{coll} \approx 0.025d^2\omega L^2/\delta_{x=0}$.

To account for diffusional mixing, a white noise term S is added to the particle advection in the y-direction. The particle paths may then be described in terms of the rate change of the position of a particle

$$\frac{dx}{dt} = v_x \tag{10.27}$$

$$\frac{dy}{dt} = v_y + S \tag{10.28}$$

where S is the white noise term which, upon integration over a time interval Δt, gives a Gaussian random number with variance $2D_{coll}\Delta t$. Diffusion along the layer (x-direction) is neglected, since it is masked by convection (i.e., particles advected relative to their neighbor much faster than they diffuse).

Figure 10.21b and c show simulation results of the time evolution of a blob of tracer particles during a typical mixing simulation. Figure 10.21b shows the evolution without collisional diffusion, whereas Figure 10.21c includes diffusion. In a typical experiment, the blob is deformed into a filament by the shear flow and blurred by collisional diffusion until particles exit the layer. This is the only form of mixing in mixers with circular cross sections. Particles then execute a solid body rotation in the bed, reenter the layer, and the process repeats. The evolution of a blob in the entire mixer is shown in Figure 10.21d.

The part of the model pertaining specifically to segregation is based on an effective buoyancy force (Khakhar et al., 1997b) that gives rise to drift/segregation velocities relative to the mean mass velocity. As for the case for diffusion, the effects are present only in the direction normal to the flow, as the segregation rate is slow compared to the time it takes for particles to traverse the flowing layer. To determine the effect of density in segregating particles normal to the flow, one can consider the effect of particle weight in combination with the effective buoyant force. First, there is an assumption of an effective medium density of ρ_m

$$\rho_m = (\phi_D \rho_D + \phi_{LD} \rho_{LD}). \tag{10.29}$$

Here
   ρ_i is the density of particles of type i (dense, D, or less dense, LD)
   ϕ_i is the volumetric concentration of particles of type i (where $\phi_1 + \phi_2 = 1$)

The magnitude of the force is written (as in Section 10.7.2) as

$$F_{B,i} = (\rho_m \forall - m_i) g \cos \theta. \tag{10.30}$$

where
   m_i is the mass of a particle of type i
   $\forall$ is the particle volume
   θ is the angle of inclination of the inclined, gravity-driven flow

The model incorporates a linear drag law $F_{D,i} = c_D v_{seg,i}$ with drag coefficient c_D and near-constant segregation velocity $v_{seg,i}$, so force balance trivially gives $v_{seg,i} = F_{B,i}/c_D$. Since c_D represents the resistance to motion, it likely scales inversely with the collisional diffusion coefficient D_{coll}; it is represented as $c_D = T/D_{coll}$. (For now, T is taken as constant, but see discussion in Section 10.7.2 and Sarkar and Khakhar (2008) for discussion of c_D and T in the context of the energy dissipation theorem where T is a temperature.) Putting these expressions together

$$v_{seg,i} = D_{coll} ((\phi_D \rho_D + \phi_{LD} \rho_{LD}) - \rho_i) \forall g \cos \theta / T \tag{10.31}$$

or, arbitrarily replacing D and LD with i and j,

$$v_{seg,i} = D_{coll}(1 - \phi_i)(\rho_j - \rho_i)\mathbb{V}g\cos\theta/T \tag{10.32}$$

We can assign a dimensionless segregation velocity $\beta \sim dg\rho_D\mathbb{V}\cos\theta/2T$ so that

$$v_{seg,D} = 2\beta D_{coll}(1 - \phi_D)(\bar{\rho} - 1)\mathbb{V}g\cos\theta/d \tag{10.33}$$

and

$$v_{seg,LD} = 2\beta D_{coll}(1 - \phi_{LD})(1 - \bar{\rho})\mathbb{V}g\cos\theta/d \tag{10.34}$$

where $\bar{\rho} = \rho_{LD}/\rho_D$. As for the diffusion term, the segregation velocity $v_{seg,i}$ is simply added to the velocity in the direction normal to the flow

$$\frac{dx}{dt} = v_x \tag{10.35}$$

$$\left(\frac{dy}{dt}\right)_i = v_y + S + v_{seg,i}. \tag{10.36}$$

Some model results are compared with experimental results in Figure 10.22 (from Khakhar et al., 1997b) and 0.16 (from Hill et al., 1999a). One of the significant strengths of this modeling framework lies in its flexibility. With a few simplifications in the development of a theoretical expression for the movement of the different particles, a concise set of expressions was obtained, capable of predicting segregation evolution in a relatively wide variety of settings. In particular, one simply needs to replace L with the length of the flowing layer in the actual system (e.g., Figure 10.20). If the drum is noncircular, the constant value of L is replaced with a nonconstant value $L(t)$ representative of the effect of the geometry on the periodically changing length of the flowing layer. Another strength of this type of model is that it lends insight into a rich interplay between two competing effects in granular mixtures: chaotic advection or mixing that arises in the

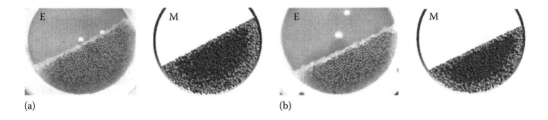

(a) (b)

Figure 10.22 Steady-state segregated patterns from experiments with steel and glass beads ("E") and from theoretical predictions with $\bar{\rho} = 0.263$ ("M") from Khakhar et al. (1997b). (a) Results from rotation in a half-filled cylindrical drum. (b) Results from rotation in an approximately quarter-filled cylindrical drum. The denser beads in the results appear darker but in the experiments they also appear shiny, and the less dense beads appear white in both.

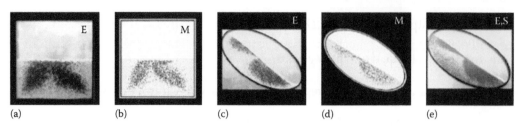

(a)   (b)   (c)   (d)   (e)

Figure 10.23   Steady-state segregated patterns from experiments with steel (dark) and glass (white) beads ("E"); from theoretical predictions with $\bar{\rho} = 0.263$ ("M"), with dense particles marked blue and lighter particles yellow, and with glass beads of 1 mm (blue), 2 mm (white), and 3 mm (red) diameter ("E,S") from Hill et al. (1999a). (a and b) Results from a half-filled square drum. (c–e) Results from a half-filled elliptical drum.

noncircular mixers, a dynamic common in fluids, and flow-induced segregation, a phenomenon without parallel in fluids (as discussed in more detail in Hill et al. 1999a,b) (Figure 10.23).

One weakness, perhaps, is that the model is based on a number of assumptions and empirical relations regarding the form of the flow, the magnitude of the diffusion coefficient, and the scale of the segregation velocity, so it is not quite directly adaptable to other systems without analogous empirical relations for the flow, diffusion, and segregation in these systems. Yet, in many systems, these measurements are relatively easy to make and the insights gained even from approximations of these factors can be instructive.

In the last section, we describe a somewhat different modeling approach for segregation of gravity-driven dense sheared granular mixtures, where the average flow is solved from a continuum approach, and rules for segregating mixtures of different-sized particles are imposed based in part on the form of the kinetic sieving model of Savage and Lun (1988) described in Section 10.7.1.

10.7.4   Binary Mixture Theory for Particle Size Segregation in Shallow Granular Free-Surface Flows

Gray and Thornton (2005) proposed a continuum approach to modeling kinetic sieving in dense sheared flows in which individual constituent momentum balances are used to model the percolation of the large and small particles. The model predictions are in many ways similar to those of Savage and Lun (1988), but the segregation velocity is dependent on the normal component of gravity, which automatically defines the segregation direction and ensures that the kinetic sieving process does not act in the absence of gravity. Similar to the effective buoyancy model described in Section 10.7.2, the Gray and Thornton model may be used with other details of the system dynamics to predict both the temporal and spatial evolution of the particle size distribution in a wide range of sheared flows.

The approach to dealing with the mixtures is addressed directly, similar to mixture theory (e.g., Truesdell, 1984; Morland, 1992), which models a system in such a way that every point in the material may be simultaneously occupied by both phases. As such, the relationship between partial variables—the contributions to each quantity (density, pressure, velocity) from each constituent—and the intrinsic or physical values for these variables must be determined. Many of these are simply additive, based on the concentration of each component. For example, the large and small particles occupy volume fractions f_L and f_S, of the total volume in space, and the volume fraction of the mixture is $f = f_L + f_S$. The model concerns a binary mixture of large and small spherical particles with the same material density ρ_m; the local bulk mass density of species i is $\rho_i = \rho_m f_i$, and the local mass density of the mixture is $\rho = \sum \rho_i$. The local pressure born by the mixture is the sum of the pressures born by each of the components: $p = \sum p_i$. On the other hand, classic mixture theory assumes that the velocity of the two components is the same at each point in space, in segregating granular materials that is not the case, and the local average velocity is $u = \sum \phi_i u_i$.

Another deviation from mixture theory occurs in the manner with which the local pressure is born by each of the components. Standard mixture theory states that all fields other than the velocity field are related linearly by their concentrations in the mixture. Gray and Thornton suggested, however, that pressure was rather divided in such a way as to reflect the likely manifestation of kinetic sieving in a continuum formulation. They suggested that, while the small particles move downward, or percolate, through the matrix, they bear less of the overburden pressure than their concentration would suggest, and the large particles carry more of the load. In other words, the partial pressures may not be proportional to their concentrations, that is, if the partial pressure coefficients are defined as

$$p_S = \psi_S p \quad \text{and} \quad p_L = \psi_L p \tag{10.37}$$

then $\psi_S + \psi_L = 1$, but $\psi_S < \phi_S$, and $\psi_L > \phi_L$. To see how this may used to predict the evolution of segregation, it is helpful first to consider the mass and momentum balance for constituent i in the mixture:

$$\frac{\partial}{\partial t} \rho_i + \bar{\nabla} \cdot \rho_i \bar{u}_i = 0 \tag{10.38}$$

$$\frac{\partial}{\partial t} (\rho_i \bar{u}_i) + \bar{\nabla} \cdot (\rho_i \bar{u}_i \otimes \bar{u}_i) = \bar{\nabla} \cdot \sigma_i + \bar{F}_i + \bar{f}_{ij} \tag{10.39}$$

Here, the stress tensor is denoted by σ_i, body forces (per unit volume) by $\bar{F}$, and forces on constituent i (per unit volume) solely due to interactions with constituent j by $\bar{f}$. $\otimes$ represents the dyadic cross product. Summing the momentum balance equation over the components in the direction normal to the flow, the y-direction as in Figure 10.20, assuming normal accelerations are negligible and gradients are zero in the x- and z-directions and that the only body force is gravity, yields

$$0 = dp/dy - \rho g \cos \theta \tag{10.40}$$

or, integrating for a mixture of depth h,

$$p = \rho g (h - y) \cos \theta \tag{10.41}$$

The normal component of the momentum balance equation for constituent i therefore may be written as

$$0 = \frac{dp_i}{dy} - \rho_i g \cos \theta - f_{ij,y} = \frac{d(\psi_i p)}{dy} - \phi_i \rho g \cos \theta - f_{ij,y}. \tag{10.42}$$

For the interaction force in the normal direction, Gray and Thornton (2005) took some insight from Darcy's law relating the percolation of fluids through porous solids (e.g., Morland, 1992), assuming a linear velocity-dependent drag force $-\rho_i c_D (v^i - v)$ and a grain–grain surface interaction force $p d \psi_i / dy$. Gray and Chugunov (2006) added a remixing force $\rho d \phi_i / dy$ (e.g., Bear, 1972; Morland, 1992) similar to Fickian diffusion. The interaction force was then written as

$$f_{ij,y} = p \frac{d \psi_i}{dy} - \rho_i c_D (v^i - v) - \rho_i d \frac{d \phi_i}{dy}. \tag{10.43}$$

Using this in Equation 10.42 (with a little rearranging) yields a segregation flux

$$\phi_i c_D (v^i - v) = (\psi_i - \phi_i) g \cos \theta + \phi_i d \frac{d \phi_i}{dy} \tag{10.44}$$

With Equation 10.44 and a form for the partial pressure coefficients ψ_i, this provides a predictive form for the evolution of the relative concentrations ϕ_i, that is, the segregation concentrations, within a sheared layer. To obtain such a predictive form, Gray and Thornton (2005) suggested a relatively simple form for ψ_S and ψ_L based on particle concentrations:

$$\psi_S = \phi_S - b \phi_S \phi_L, \quad \text{and} \quad \psi_L = \phi_L + b \phi_S \phi_L \tag{10.45}$$

where b is a segregation coefficient to be determined. These satisfy the basic requirements of the partial pressures, that is, $\psi_S + \psi_L = 1$; $\psi_S < \phi_S$, and $\psi_L > \phi_L$. Substituting Equation 10.45 into Equation 10.44, and the equation of mass balance for each species, Equation 10.38, one can obtain a concentration evolution equation for each species:

$$\frac{\partial}{\partial t} \phi_L + \bar{\nabla} \cdot \phi_L \bar{u}_L + q \frac{d}{dy} (\phi_S \phi_L) + \frac{d}{dy} D \frac{d \phi_i}{dy} = 0 \tag{10.46}$$

where $q = bg \cos \theta / c_D$ may be thought of as a segregation coefficient and $D = d / c_D$ as a diffusion coefficient.

Similar to the buoyancy model proposed by Khakhar et al. (1997a,b), a key strength of this model is its adaptability to a wide range of systems. A hyperbolic theory without diffusion (Gray and Thornton, 2005) has been used to construct

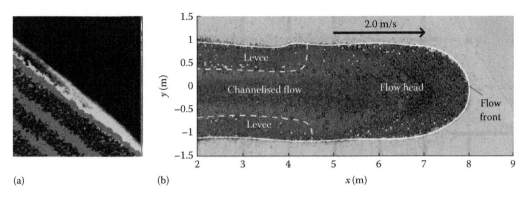

(a)   (b) x (m)

Figure 10.24   Two pictures of segregation in avalanching granular flows where the model framework of Gray and colleagues (e.g., Gray and Thornton, 2005; Gray and Chugunov, 2006) has been successfully applied. (a) Picture of a section of stratification in heap flow with active flowing layer highlighted formed using a composite image from the system at different times. (From Gray, J.M.N.T. and Ancey, C., J. *Fluid Mech.*, 629, 387, 2009.) (b) Picture illustrating levee formation and the flow front of a large-scale experimental debris flow where both the flow head and levee were more highly concentrated with large particles (colored in the photograph). (From Johnson, C.G. et al., *J. Geophys. Res.*, 117, F01032/1–23, 2012.)

the 2D steady-state and time-dependent exact solutions in uniform shear flows (e.g., Gray et al., 2006; Shearer et al., 2008). In these systems, the evolution of concentration shocks and the formation of sharply segregated inversely graded layers are particularly revealing. The version described in this section includes a diffusion term (Gray and Chugunov, 2006) which smears out the sharp concentration shocks. The outcome is a parabolic theory that is more difficult to solve for exact solutions. The hyperbolic theory has been used, particularly in a depth-averaged formulation, to provide a framework for some of the more complex segregation patterns such as stratification in heap flows (e.g., Gray and Ancey, 2009) and coarsening of avalanching granular flow fronts and levees (e.g., Johnson et al., 2012) (Figure 10.24a and b).

It should also be noted that, within the predictive expression of the model, there are several assumptions that require further investigation. Among these are the form of the partial pressures as well as the linear drag and diffusion assumptions and their corresponding coefficients. Recent variations (e.g., those of May et al., 2010a,b; Marks and Einav, 2011; Marks et al., 2012; Thornton et al., 2012; Weinhart et al., 2013; Hill and Tan, 2014) have shown how incorporating nonlinear drag and diffusion, nonconstant segregation rates, and the effects of shear-rate gradients can improve the predictive power of the model. Kinetic theory models and simulations such as discrete element models (c.f. Chapter **COREY CHAPTER**) provide techniques for deriving these terms in conjunction with ongoing experiments in drums and chute flows.

10.8 Other Mitigating Factors: Effects of Interstitial Fluids

We conclude this chapter by briefly considering the importance of fluids in some segregating mixtures. Indeed, in many industrial and natural segregating systems, interstitial fluid (e.g., air in industrial powder or grains or water and mud in debris flow) is unavoidable and may alter or even drive (e.g., water over a riverbed) the system dynamics. Still, many of the dynamics discussed in this chapter are relevant to systems strongly influenced by the presence of fluid.

Burtally et al. (2002) demonstrated that the presence of an interstitial fluid can have a strong influence on a segregation pattern in a granular system, particularly for smaller particles. They vibrated mixtures of micron-sized glass and brass spheres and found that, in the presence of air at atmospheric pressures, the segregation was reversed compared to that associated with buoyancy (i.e., the denser (brass) particles rose and the lighter (glass) particles sunk). When the air pressure was significantly reduced, the segregation effect vanished completely.

The influence of a fluid is somewhat different but at least as important in the dynamics of a riverbed. In this case, the moving fluid drives all of the particle movement; the slope of the free surface is low enough that no particle rearrangement (segregation, mixing, or otherwise) would take place without the moving fluid. Still, in this dramatically different free-surface sheared granular system, there is often a segregation pattern with the large particles concentrated on top. The dynamics leading to this effect are often viewed as significantly different, typically described to the relative ability of the fluid to carry larger and smaller particles. Specifically, the flows generate stresses large enough to move the smaller, but not the larger, particles in the bed. Over time, the smaller particles are removed from the surface of the bed and the larger particles left behind on the surface. This has the effect of sheltering the smaller particles below, and therefore, the top surface is often referred to as an "armor layer" (e.g., Parker et al., 1982; Parker and Sutherland, 1990).

On the other hand, even with the dramatically different riverbed systems, there are situations where the relatively simple segregation dynamics associated with dry flows are relevant. Examples have been reported in bedload sheets, a term used describe the mode of transport of sediment along riverbeds where at higher flows much of a thin surface of a riverbed may move together. There is evidence that under these transport conditions, kinetic sieving similar to that seen in gravity-driven flows is responsible for segregating particles of different sizes in a riverbed (e.g., Recking et al., 2009; Frey and Church, 2011). As Frey and Church have suggested, there is considerable potential for the application of models of "simple" granular physics such as the segregation discussed here to a wide range of apparently more complex multiphase systems.

10.9 Conclusion

As described in this chapter, segregation dynamics are considerably more complex than they appear at first glance. As discussed in Sections 10.2 through 10.5, the sensitivity of the most basic behaviors to boundary conditions, excitation method, and other details makes apparently simple "up-or-down" segregation a multifaceted problem to understand. Then, as described in Section 10.6, complexities of the final segregation pattern, most evident to the casual observer, are even more confounding because of the influence of details such as system advection that can quite literally turn the segregation behavior on its side and worse to create a segregation pattern quite different than the original segregation behavior that underlies the unmixing of the material. Different complimentary modeling efforts such as the few presented in Section 10.7 in combination with increasingly faster computational speeds and sophisticated experimental capabilities have helped detangle some of the underlying problems in segregation behaviors and emergent patterns. Ultimately, these efforts will help provide the necessary tools to model more fundamental as well as applied problems in industry and nature, where the effects of fluids may also become important. Not only do the details of segregation impede efficient particle processing (in industry) and preferentially distribute resources important for local ecology (in nature). Additionally, the local particle property distribution—altered by segregation—can significantly influence the behavior of granular materials locally and globally (e.g., Rognon et al., 2007; Yohannes and Hill, 2010; Hill and Yohannes, 2011). Thus, the interplay between segregation and particle flow is also important for maximizing the production of particle-based materials in industry, and for environmental problems from river restoration to hazard mitigation (e.g., debris and lahar flows).

References

Alam, M., Trujillo, L., and Herrmann, H.J. (2006). Hydrodynamic theory for reverse Brazil Nut segregation and the non-monotonic ascension dynamics. *J. Stat. Phys.*, 124, 587–623.

Alonso, M., Satoh, M., and Mihanami, K. (1991). Optimum combination of size ratio, density ratio and concentration to minimize free surface segregation. *Powder Technol.*, 68, 145–152.

Arnarson, B.O., and Willits, J.T. (1998). Thermal diffusion in binary mixtures of smooth, nearly elastic spheres with and without gravity. *Phys. Fluids*, 10, 1324–1328.

Barentin, C., Azanza, E., and Pouligny, B. (2004). Flow and segregation in sheared granular slurries. *Europhys. Lett.*, 66, 139–145.

Bear, J. (1972). *Dynamics of Fluids in Porous Media*. Elsevier.

Breu, A.P.J., Ensner, H.-M., Kruelle, C.A., and Rehberg, I. (2003). Reversing the Brazil-nut effect: Competition between percolation and condensation. *Phys. Rev. Lett.*, 90, 14302/1–4.

Bridgwater, J.R., Cooke, M. H., and Scotta, M. (1978). Inter-particle percolation: Equipment development and mean percolation velocities. *Trans. Instn Chem. Eng.* 56, 157–167.

Brown, C.B. (1978). The use of maximum entropy in the characterization of granular media. In *Proceedings of the US Japan Seminar on Continuum Mechanical and Statistical Approaches in the Mechanics of Granular Materials*, Cowin, S.C. and Satake, M. (eds.), pp. 98–108. Tokyo, Japan: Gakujutsu Bunken Fukyukai.

Burtally, N., King, P.J., and Swift, M.R. (2002). Spontaneous air-driven separation in vertically vibrated fine granular mixtures. *Science*, 295 1877–1879.

Cantelaube, F. and Bideau, D. (1995). Radial segregation in a 2d drum: An experimental analysis. *Europhys. Lett.*, 30, 133–137.

Clément, E., Rajchenbach, J., and Duran, J. (1995). Mixing of granular material in a bidimensional rotating drum. *Europhys. Lett.*, 30, 7–12.

Conway, S. L., Liu, X., and Glasser, B.J. (2006). Instabitily-induced clustering and segregation in high-shear Couette flows of model granular materials. *Chem. Eng. Sci.*, 61, 6404–6423.

Cooke, M.H. and Bridgwater, J.R. (1979). Interparticle percolation: A statistical mechanical interpretation. *Ind. Eng. Chem. Fundam.*, 18, 25–27.

Drahun, J.A. and Bridgwater, J. (1983). The mechanisms of free surface segregation. *Powder Technol.*, 36, 39–53.

Fan, Y. and Hill, K.M. (2010). Shear-driven segregation of dense granular mixtures in a split-bottom cell. *Phys. Rev. E*, 81, 041303/1–12.

Fan, Y. and Hill, K.M. (2011a). Phase transitions in shear-induced segregation of granular materials. *Phys. Rev. Lett.*, 106, 218301/1–4.

Fan, Y. and Hill, K.M. (2011b). Theory for shear-induced segregation of dense granular mixtures. *New J. Phys.*, 13, 095009.

Félix, G. and Thomas, N. (2004). Evidence of two effects in the size segregation process in dry granular media. *Phys. Rev. E*, 70, 051307/1–16.

Fenistein, D. and van Hecke, M. (2003). Wide shear zones in granular bulk flow. *Nature* (Lond.) 425, 256.

Fenistein, D., van de Meent, J. W., and van Hecke, M. (2004). Universal and wide shear zones in granular bulk flow. *Phys. Rev. Lett.*, 92, 094301/1–4.

Fenistein, D., van de Meent, J. W., and van Hecke, M. (2006). Core precession and global modes in granular bulk flow. *Phys. Rev. Lett.*, 96, 118001/1–4.

Frey, P. and Church, M. (2009). How river beds move. *Science* 325, 1509–1510.

Frey, P. and Church, M. (2011). Bedload: A granular phenomenon. *Earth Surf. Process. Landforms*, 36, 58–69.

Gray, J.M.N.T. and Ancey, C. (2009). Segregation, recirculation and deposition of coarse particles near two-dimensional avalanche fronts. *J. Fluid Mech.*, 629, 387–423.

Gray, J.M.N.T. and Chugunov, V.A. (2006). Particle-size segregation and diffusive remixing in shallow granular avalanches. *J. Fluid Mech.*, 569, 365–398.

Gray, J.M.N.T., Shearer, M., and Thornton, A.R. (2006). Time-dependent solutions for particle-size segregation in shallow granular avalanches. *Proc. R. Soc. A*, 462, 947–972.

Gray, J.M.N.T. and Thornton, A.R. (2005). A theory for particle size segregation in shallow granular free-surface flows. *Proc. R. Soc. A*, 461, 1447–1473.

Hill, K.M., Caprihan, A., and Kakalios, J. (1997). Axial segregation of granular media rotated in a drum mixer: Pattern evolution. *Phys. Rev. E*, 56, 4386–4393.

Hill, K.M. and Fan, Y. (2008). Isolating segregation mechanisms in a split-bottom cell. *Phys. Rev. Lett.*, 101, 088001/1-4.

Hill, K.M., Fan, Y., Zhang, J., Van Niekerk, C., Zastrow, E., Hagness, S.C., and Bernhard, J.T. (2010). Granular segregation studies for the development of a radar-based three-dimensional sensing system. *Granular Matter*, 12, 201–207.

Hill, K.M., Gioia, G., and Amaravadi, D. (2004). Radial segregation patterns in rotating granular mixtures: Waviness selection. *Phys. Rev. Lett.*, 93, 224301.

Hill, K.M., Gioia, G., and Tota, V.V. (2003). Structure and kinematics in dense free-surface granular flow. *Phys. Rev. Lett.*, 91, 064302.

Hill, K.M. and Kakalios, J. (1994). Reversible axial segregation of binary mixtures of granular materials. *Phys. Rev. E*, 49(5), pt. A, R3610–R3613.

Hill, K.M. and Tan, D. (2014). Segregation in dense sheared flows: Gravity, temperature gradients, and stress partitioning. *Journal of Fluid Mechanics*, 756, 54–88.

Hill, K.M. and Yohannes, B. (2011). Rheology of dense granular mixtures: Boundary pressures. *Phys. Rev. Lett.*, 106, 058302.

Hong, D.C. (1999). Condensation of hard spheres under gravity. *Physica (Amsterdam)*, 271A, 192–199.

Hong, D.C., Quinn, P.V., and Luding S. (2001). Reverse Brazil nut problem: Competition between percolation and condensation. *Phys. Rev. Lett.*, 86:3423–3426.

Jain, N., Ottino, J., and Lueptow, R. (2005a). Regimes of segregation and mixing in combined size and density granular systems: An experimental study. *Granular Matter*, 7, 69–81.

Jain, N., Ottino, J., and Lueptow, R. (2005b). Combined size and density segregation and mixing in noncircular tumblers. *Phys. Rev. E*, 71, 051301/1–10.

Jaynes, E.T. (1963). Information theory and statistical mechanics. In *Statistical Physics* III., Ford, K.W. (ed.), Lectures from Brandeis Summer Institute 1962. New York: W.A. Benjamin, Inc., 1963., p.181.

Jenkins J.T. and Mancini, F. (1989). Kinetic theory for binary mixtures of smooth, nearly elastic spheres. *Phys. Fluids A*, 1, 2050–2057.

Jenkins, J. and Yoon, D.K. (2002). Segregation in binary mixtures under gravity. *Phys. Rev. Lett.*, 88, 194301/1–4.

Johnson, C.G., Kokelaar, B.P., Iverson, R.M., Logan, M., LaHusen, R.G., and Gray, J.M.N.T. (2012). Grain-size segregation and levee formation in geophysical mass flows. *J. Geophys. Res.*, 117, 01032/1–23.

Khakhar, D.V., McCarthy, J.J., and Ottino, J.M. (1997B). Radial segregation of granular mixtures in rotating cylinders. *Phys. Fluids*, 9, 3600–3614.

Khakhar, D.V., McCarthy, J. J., Shinbrot, T., and Ottino, J.M. (1997A). Transverse flow and mixing of granular materials in a rotating cylinder. *Phys. Fluids*, 9, 31–43.

Krishnan, G.P., Beimfohr, S., and Leighton, D.T. (1996). Shear-induced radial segregation in bidisperse suspensions. *J. Fluid Mech.* 321, 371–393.

Louge, M.Y., Jenkins, J.T., Xu, H., and Arnarson, B.O. (2001). Granular segregation in collisional shearing flows. In *Mechanics for a New Millennium*, Aref, A. and Phillips, J. (eds.), pp. 239–252. Kluver Academic Publishers.

Marks, B. and Einav, I. (2011). A cellular automaton for segregation during granular avalanches. *Granular Matter*, 13, 211–214.

Marks, B., Rognon, P., and Einav, I. (2012). Grainsize dynamics of polydisperse granular segregation down inclined planes. *J. Fluid Mech.*, 690, 499–511.

May, L.B.H., Golick, L.A., Phillips, K.C., Shearer, M., and Daniels, K.E. (2010a). Shear-driven size segregation of granular materials: Modeling and experiment. *Phys. Rev. E*, 81, 051301.

May, L.B.H., Shearer, M., and Daniels, K.E. (2010b). Scalar conservation laws with nonconstant coefficients with application to particle size segregation in granular flows. *J. Nonlinear Sci.*, 20, 689–707.

Metcalfe, G. and Shattuck, M. (1996). Pattern formation during mixing and segregation of flowing granular materials. *Phys. A*, 233, 709–717.

Morland, L.W. 1992 Flow of viscous fluids through a porous deformable matrix. *Surv. Geophys.*, 13, 209–268.

Nakagawa, M., Altobelli, S.A., Caprihan, A., and Fukushima, E. (1997). NMRI study: Axial migration of radially segregated core of granular mixtures in a horizontal rotating cylinder. *Chem. Eng. Sci.*, 52, 4423–4428.

Oyama, Y. (1939). *Bull. Inst. Phys. Chem. Res.* (in Japanese) *Tokyo. Rep.* 5, 600.

Oyama, Y. and Ayaki, K. (1956) Studies on the mixing of particulate solids. *Kagaku Kikai* 20, 6.

Parker, G., Klingeman, P.C., and McLean, D.G. (1982). Bedload and size distribution in paved gravel-bed streams. *ASCE J. Hydraulics Div.*, 108, 544–571.

Parker, G. and Sutherland, A.J. (1990). Fluvial armor. *ASCE J. Hydraulic Eng.*, 28, 529–544.

Pouliquen, O. and Vallance, J.W. (1999). Segregation induced instabilities of granular fronts. *Chaos*, 9, 621–630.

Quinn, P.V. and Hong, D.C. (2000). Liquid-solid transition of hard spheres under gravity. *Phys. Rev. E.*, 62, 8295–8298.

Rajchenbach, J. (1990). Flow in powders: From discrete avalanches to continuous regime. *Phys. Rev. Lett.*, 65, 2221–2224.

Recking, A., Frey, P., Paquier, A., and Belleudy, P. (2009). An experimental investigation of mechanisms responsible for bedload sheet production and migration. *J. Geophys. Res. Earth Surface*, 114, F03010.

Ristow, G.H. (1994). Particle mass segregation in a two-dimensional rotating drum. *Europhys. Lett.*, 28, 97–101.

Rognon, P.G., Roux, J.-N., Naam, M., and Chevoir, F. (2007). Dense flows of bidisperse assemblies of disks down an inclined plane. *Phys. Fluids*, 19, 058101.

Rosato, A.D., Prinz, F.B., Strandburg, K.J., and Swendsen, R.H. (1986). Monte Carlo simulation of particulate matter segregation. *Powder Technol.*, 49, 59–69.

Rosato, A.D., Strandburg, K.J., Prinz, F.B., and Swendsen, R.H. (1987). Why the Brazil nuts are on top: Size segregation of particulate matter by shaking. *Phys. Rev. Lett.*, 58, 1038–1041.

Sarkar, S. and Khakhar, D.V. (2008). Experimental evidence for a description of granular segregation in terms of the effective temperature. *EPL*, 83, 54004/1–6.

Savage, S.B. (1993). Disorder, diffusion and structure formation in granular flow. In *Disorder and Granular Media*, Bideau, D. and Hansen, A. (eds.), pp. 255–285. Amsterdam, the Netherlands: Elsevier Science.

Savage, S.B. and Lun. C.K.K (1988). Particle size segregation in inclined chute flow of dry cohesionless granular solids. *J. Fluid. Mech.*, 189, 311–335.

Shearer, M., Gray, J.M.N.T., and Thornton, A.R. (2008). Stable solutions of a scalar conservation law for particle-size segregation in dense granular avalanches. *Eur. J. Appl. Math.*, 19, 61–86.

Shi, Q., Sun, G., Hou, M., and Lu, K. (2007). Density-driven segregation in vertically vibrated binary granular mixtures. *Phys. Rev. E*, 75, 061302/1–4.

Shinbrot, T. and Muzzio, F.J. (1998). Reverse buoyancy in shaken granular beds. *Phys Rev. Lett.*, 81, 4365–4368.

Thomas, N. (2000). Reverse and intermediate segregation of large beads in dry granular media. *Phys. Rev. E*, 62, 961–975.

Thornton, A., Weinhart, T., Luding, S., and Bokhove, O. (2012). Modeling of particle size segregation: Calibration using the discrete particle method. *Int. J. Mod. Phys. C*, 23(8).

Truesdell, C. (1984). *Rational Thermodynamics*. Berlin, Germany: Springer.

Trujillo, L., Alam, M., and Herrmann, H.J. (2003). Segregation in a fluidized binary granular mixture: Competition between bouyancy and geometric forces. *Europhys. Lett.*, 64, 190–196.

Weinhart, T., Luding, S., and Thornton, A.R. (2013). From discrete particles to continuum fields in mixtures. In *Powders and Grains* 2011, vol. 1542, pp. 1202–1205. American Institute of Physics.

Williams, J.C. (1963). The segregation of powders and granular materials. *Fuel Soc. J.*, 14, 29–34.

Williams, J. C. (1968). The mixing of dry powders. *Powder Technol.*, 2, 13–20.

Willits, J.T. and Arnarson, B.O. (1999). Kinetic theory of a binary mixture of nearly elastic disks. *Phys. Fluids*, 11(10), 3116–3122.

Woodhouse, M.J., Thornton, A.R., Johnson, C.G., Kokelaar, B.P., and Gray, J.M.N.T. (2012). Segregation-induced fingering instabilities in granular free-surface flows. *J. Fluid Mech.*, 709, 543–580.

Xu, H., Louge, M.Y., and Jenkins, J.T. (2001). Flow development of a collisional granular flow. In *Powders and Grains*, Kishino, Y. (ed.), pp. 359–362. Kluver Academic Publishers.

Xu, H., Louge, M.Y., and Reeves, A. (2003). Solutions of the kinetic theory for bounded collisional granular flows. *Continuum Mech. Thermodyn.*, 15, 321–349.

Section III

Extensions of Granular Research

CHAPTER 11

Suspension Mechanics and Its Relation to Granular Flow

Jeffrey F. Morris

CONTENTS

11.1 Introduction

This chapter addresses suspensions and considers their relationship to granular flows. By a suspension, we typically mean a mixture composed of solid particles within a fluid, under conditions where the particles remain suspended as opposed to settling. For example, under normal gravity environments, a fine sand that settles slowly in water may be readily redispersed by the water flowing over it, and this mixture may be termed a suspension. The same fine sand in air would rapidly settle to form a layer of contacting particles that might still flow, but would be considered a granular flow. A third form is a slurry, defined as a material needing stirring or other agitation to maintain the particles in suspension but whose properties may be otherwise described roughly as suspensions.

It is illuminating to observe that it is standard to say that one studies "suspensions," while it would not convey the right meaning to say that one studies "grains": the motion is critical in the terminology granular flow. Of course, we are primarily interested in suspensions under flow as well, but a difference in the common experience with suspensions and granular materials is encountered in this distinction: suspensions are expected to be fluid, whereas for granular materials the expectation depends upon conditions. It takes little consideration to realize that the reason for this difference in expectation regarding the flowability of the material is related to the larger role of the intervening fluid in suspensions relative to its role in granular materials and also because grains tend to settle to a dense packing without flow.

While we may consider dust suspended in air or another gas as a suspension, the overwhelming majority of interest in suspensions is in cases where the particles are suspended in a liquid and so, for this work, a liquid suspending fluid will be assumed. Although many points made will not be dependent on this choice, two points that are critical in practice do rely upon the suspending fluid being a liquid. The first is that solids in a liquid form a system in which the two phases have similar densities, so that gravity need not be a dominating factor. For many cases of practical relevance these differ by no more than a factor of two, for example, for coal and soil in water. By tuning the densities, the influence of gravity can be effectively eliminated. The second point is that the dynamic viscosity of the liquid is orders of magnitude larger than that of a typical gas. This means that the interactions of particles are affected by viscous interactions, and direct contact is not generally the dominant interaction mode.

The assumption in much of suspension mechanics has been that particle surfaces do not touch. The basis for this assumption is mathematically sound for spherical particles under viscous flow conditions, that is, when the Reynolds number $Re = \rho U a / \eta_0 \ll 1$. Considering a shear flow $u_x = \dot{\gamma} y$ and then $U = \dot{\gamma} a$, the lubrication interaction associated with particles approaching one another (squeezing flow) causes a divergent resistance to relative motion as the gap between surfaces goes to zero, that is, $R \sim h^{-1}$ where h is the point of the closest approach.

The basic model of a suspension is hard spheres in viscous liquid, leading to overdamped dynamics with strong hydrodynamic influence; other forces may be included in the model that leads to differences in behavior, but this basic model is at the heart of our understanding of suspension behavior. By contrast, the basic model of a flowing granular material is that of elastic (or perhaps inelastic) spheres in a gas, interacting primarily through dissipative collisions between particles. More recently, a significant body of work on "wet granular materials," in which the fluid is a liquid, has appeared (Cassar et al. 2005; Boyer et al. 2011a). The slow flow regime of granular dynamics where inertial collisions are of minimal influence is most closely related to suspension dynamics. However, even in this regime, gravitational effects are strong in granular flows in terrestrial environments and need not be so in suspensions.

This chapter does not serve as a thorough review of the behavior of suspensions. For this, reviews by Stickel and Powell (2005), Wagner and Brady (2009), or Morris (2009) may be consulted; the book by Guazzelli and Morris (2012) provides an introduction to many of the main topics in suspension mechanics, while that by Mewis and Wagner (2012) describes the rheology of suspensions in detail with a focus on colloidal dispersions. The goal of this chapter is to provide guidance to what is known about suspensions and its relation to granular matter, and to this end the behavior of suspensions in its broad outlines is presented with reference to certain ways this may be illuminating to granular flow.

To make such comparisons, the differences and similarities are worthwhile to consider, and we begin with what appear to be the key differences. The first very readily noted difference is the flexibility in solid fraction, ϕ, for suspension, while granular materials are found in a rather narrow range of solid fractions close to ϕ_{max}. This is evidence of the strong and essential influence of gravity forces in granular matter, owing to the large $\Delta\rho = \rho_p - \rho_g \approx \rho_p$ between particles and surrounding gas. By contrast, $\Delta\rho$ can be essentially zero in a carefully designed suspension; recipes exist in which not only density but also refractive index of particles and fluid can be matched (Bailey and Yoda 2003; Dijksman et al. 2012). The density matching allows ϕ to be varied such that flow properties may be studied over its full range $0 < \phi < \phi_{max}$. Rheology is thus a highly developed subject for suspensions and dispersions and shows incredibly the diverse influence of solid interactions: dilute gels to close-packed glasses, shear thinning and thickening, normal stresses, and history dependence (thixotropy). Suspension study thus may offer some information elucidating granular matter behavior, particularly in terms of rheology and the structural effects caused by flow, and perhaps even in terms of how particle phase dilation may be related to the flow.

In addressing the similarities in their behavior, it is useful to consider that a static dense-packed mass of particles could become either a granular flow or a suspension. And it is near this limit, as $\phi \to \phi_{max}$, that the behaviors of granular materials and suspensions appear to be quite similar; considering the so-called hard-sphere state, flow arrest and yielding would seem to be cases of great similarity in the two, and understanding their difference is a valuable avenue for distinguishing key physics. Here, granular matter appears to offer elucidation to suspensions, as the crucial role of contact forces including friction upon the rheology are well developed for granular flows (Cundall and Strack 1979; Rao and Nott 2008), whereas contact forces have been recognized as relevant (Davis et al. 2003) but not thoroughly studied for suspensions.

We first provide a guide for discussion of suspensions, in terms of dimensional analysis. This will provide a framework for the discussion that follows as well as allow certain basic expectations to be discussed—it is the violation of these expectations that makes the behavior of suspensions rich and interesting, but it is essential to have the expectations of the basic model of the material in order to be prepared to examine such phenomena. The rheology of suspensions will then be discussed, followed by a section devoted to the description of the particle arrangement, or microstructure, under flow. A section describing bulk flow phenomena in suspensions is presented, followed by a concluding section in which we gather the key points and reiterate points about their relation to granular matter.

11.2   Relevant Variables and Dimensional Analysis

Before describing the behavior of suspensions, it is useful to consider the parameters that influence their properties and dynamics. These will be described and the dimensionless parameters that arise from them described.

The suspensions of interest are solid particles in a Newtonian suspending fluid of density ρ and viscosity η_0. We will primarily be concerned with the spherical particles of radius denoted as a; although relaxation of the geometric constraint to consider rough particles is a key consideration in relation to granular flows, we will not address extended particles such as fibers or rods here, and for the interested reader a good starting point on the shear flow properties of such materials is given by Petrie (1999), while interesting sedimentation dynamics are described in the review by Guazzelli and Hinch (2011).

The flow of primary interest here is a shear flow of the form $u_x = \dot{\gamma}y$, with $\dot{\gamma}$ the shear rate of the flow. The flow may vary spatially and retain its form, as occurs, for example, in a pressure-driven flow in a planar channel, where $\dot{\gamma} = (dP(x)/dx)y$ with y the cross-channel coordinate measured from the centerline. Similarly, for pressure-driven flow in a pipe with axis in the z-direction, we have radial variation of the shear rate, $\dot{\gamma} = [dP(z)/dz](r/2)$. In rheometric apparatus, care is generally taken to minimize variation of the shear rate. A translational velocity may be of interest, for example, for particles in sedimentation at speed U.

Dimensionless numbers characterizing the motion of the suspension typically begin from the Reynolds number on the particle scale $Re = \rho U a/\eta_0$, which for shear flow is $Re = \rho \dot{\gamma} a^2/\eta_0$. The bulk of work on suspensions has considered $Re \ll 1$, assuming $Re = 0$ to yield linear Stokes flow on the particle scale. This has been instrumental in the development of analysis (Happel and Brenner 1965; Kim and Karrila 1989) and in the development of simulation tools such as Stokesian dynamics (Brady and Bossis 1988; Phung et al. 1996; Sierou and Brady 2001) and related methods (Ladd 1988; Melrose and Ball 2004; Meng and Higdon 2008). When the particle density, ρ_p, differs from the fluid density, $\Delta\rho$, the Stokes number characterizing particle inertia to fluid viscosity is given by $St = (\rho_p/\rho)Re$. When the suspending fluid is a gas, $\rho_p/\rho \gg 1$ and a finite St and $Re = 0$ description may be valid. This is a case where a suspension mechanical approach applying Stokes regime fluid mechanics, coupled with collisional mechanics of the particles, has been applied to describe granular materials (Sangani et al. 1996), as well as to address more moderate St where the collisions are avoided but interesting dynamics arise due to the particle inertia alone (Subramanian and Brady 2006). Recent research has addressed the influence of finite fluid mechanical inertia on the properties of suspensions, and this will be noted in the appropriate places later. This work has been greatly facilitated by the application of the lattice Boltzmann method for suspensions developed by Ladd (1994a,b), Aidun and Lu (1995), and Aidun et al. (1998).

In suspensions, it is common to consider particles whose sizes range down to the submicron scale, where Brownian motion and colloidal forces have pronounced effects (Russel et al. 1991). The influence of Brownian motion relative to shear flow is captured through a Péclet number given by $Pe = (\dot{\gamma}a^2)/D_0 = (6\pi\eta_0\dot{\gamma}a^2)/kT$, where $D_0 = kT/(6\pi\eta_0 a)$ is the Stokes–Einstein diffusion coefficient, k is Boltzmann's constant, and T is the temperature. The first form shows that Pe may be interpreted as a ratio of the hydrodynamic diffusion scaling with the shear rate and particle size $(\dot{\gamma}a^2)$ as well as a dimensionless function of the volume fraction not shown. It is more common, however, to interpret Pe as the ratio of a diffusive timescale a^2/D_0, relative to the flow timescale given by $\dot{\gamma}^{-1}$. When $Pe = 0$, a Brownian suspension will approach a true equilibrium state through its thermal motions. Interparticle forces of many sorts are possible in a liquid medium.

Coulombic forces from net electrostatic charge and polarization-induced forces (e.g., van der Waals forces) play a prominent role even for particles larger than the micron scale, below which Brownian motion dominates. A dimensionless form for all interparticle forces is not possible but, assuming pairwise forces, it is common to use a ratio of a characteristic value of the interparticle force near contact of a pair (F_0) with the representative shearing force at particle scale $\eta\dot{\gamma}a^2$ to form the dimensionless interparticle force $F^* = F_0/(\eta\dot{\gamma}a^2)$.

Finally, the particle volume fraction, ϕ, is a critical dimensionless parameter. For monodisperse suspensions of spheres of radius a, $\phi = n(4\pi/3)a^3$ with n the number of particles per volume. As noted in the introduction, with gravitational effects eliminated when the difference in phase densities vanishes, $\Delta\rho = 0$, a suspension with any volume fraction $0 < \phi \leq \phi_{max}$ may be formed and studied, a feature unique to suspensions. When interest is in the properties approaching the maximum packing limit, ϕ_{max}, it is common to describe the properties of the suspension in terms of a deviation from the limiting value i.e., in terms of $1 - \phi/\phi_{max}$.

11.3 Rheology

The rheological response of suspensions has always been an essential reason for their study. The well-known prediction of Einstein (1906) for the first effects of rigid particles on the viscosity of suspensions, $\eta_E = \eta_0(1 + 5\phi/2)$, was determined based on an energy dissipation argument. Early work measuring the viscosity of suspensions (Mooney 1951; Thomas 1965; Lewis and Nielsen 1968) focused on the influence of the particle volume fraction and showed that, following the initial linear increase with ϕ predicted by the Einstein viscosity, the increase becomes progressively more rapid and tends to diverge at a maximum packing fraction ϕ_{max} (e.g., in the Krieger (1972) form of the effective viscosity $\eta_s = \eta(1-\phi/\phi_{max})^{-\alpha}$ with $\alpha \approx 2$). Here, ϕ_{max} is an empirical quantity whose value varies from about 0.58 to 0.68, with the random close packing value of $\phi_{rcp} \doteq 0.64$ often taken in the absence of other information.

As suspensions involve particles in liquids, the physicochemical environment allows for a diverse range of interparticle forces. When particle interactions are appropriate, aggregation can occur leading to gelation (Weitz et al. 1985; Brinker and Scherer 1990) and time-dependent rheology with thixotropy (viscosity breakdown under shear and recovery at rest; Mewis and Wagner 2009) among the most applicationally relevant behaviors. These behaviors have little relation to granular materials and are left aside.

The increase in viscosity with the volume fraction of solids in the suspension does not in itself imply any non-Newtonian behavior, as the stress can remain strictly linear in the shear rate (Guazzelli and Morris 2012). In fact, when we consider the hard-sphere suspension in a Newtonian liquid under Stokes flow conditions, the expectation is that the rheology should be Newtonian. This is a basic result of dimensional analysis; since there is no intrinsic rate associated with either the fluid or the particles, the only rate is that set by the flow shear rate.

An alternative perspective is that there is no intrinsic force (or stress) scale to balance the shear force defined by the viscous stress $\eta_0\dot{\gamma}$ and the particle surface area, scaling as a^2. As a result, the shear stress should scale as $\Sigma_{xy} \sim \eta_0\dot{\gamma}$ and a dimensionless function of the solid fraction $\eta_r(\phi)$. The subscript here implies a "reduced" viscosity as $\eta_r\eta_0$ represents the measured effective viscosity.

For spherical particles in the limit of $Pe \to \infty$, we might expect, based on the reversibility of Stokes flow, that a noncolloidal suspension shows only a ϕ-dependence effective viscosity: neither shear rate dependent nor normal stresses are expected. In fact, even suspensions of particles carefully tailored to adhere to the hard-sphere description exhibit shear thinning and thickening in Newtonian suspending liquids. Note that if we consider Brownian colloidal suspensions, rate dependence is expected and most aspects of this behavior are well understood. Shear thinning occurs due to saturation of the Brownian contribution to the stress, Σ_{xy}^{Br}, and the overall response of both shear and normal stresses is strongly rate-dependent: the shear rate "competes" with the thermal motion, so that the dimensionless shear rate, or a Péclet number $Pe = (6\pi\eta\dot{\gamma}a^2)/kT$, is the appropriate independent variable to describe both the influence of particle size and shear rate, as well as the much smaller variability in temperature. The variation of the viscosity and normal stresses with shear rate is illustrated in schematic fashion in Figure 11.1. The thinning ends and shear thickening of a mild form occur above a critical shear rate that depends on solid fraction. This shear thickening arises due to the accumulation of particles asymptotically in contact with a preferred direction roughly along the compressional axis of the shear flow; the accumulation in this preferred direction gives rise to extended zones of correlation of clusters. This thickening associated with the balance of hydrodynamics and Brownian motion is much less abrupt than what has come to be known as "discontinuous shear thickening" (DST). The phenomenon of DST appears to be related to frictional contacts (Seto et al. 2013), a point which was developed in work by Brown and Jaeger (2012). Noncolloidal suspensions often also exhibit shear thinning, a behavior that appears likely to also involve contact networks in some fashion. Examples are found in the work of Zarraga et al. (2000) and Dai et al. (2013). It appears likely that surface effects play a role in this thinning, but conclusive study is lacking.

The normal stress response of suspensions has received rather less attention than the effective viscosity. However, since the first measurement of normal stress differences in suspensions, with the work of Gadala-Maria (1979), the normal stresses and their relation to the microstructure of the material on one hand (see Section 11.4) and to the bulk flow response on the other (see Section 11.5) have garnered increasing attention. Restricting attention to a simple shear flow, we first define the flow, gradient, and vorticity directions as the 1, 2, and 3, directions, respectively ($1 = x$, $2 = y$, $3 = z$ in the shear flow $u_x = \dot{\gamma}y$). The first and second normal stress differences are given, respectively, by

$$N_1 = \Sigma_{11} - \Sigma_{22} \quad \text{and} \quad N_2 = \Sigma_{22} - \Sigma_{33}. \tag{11.1}$$

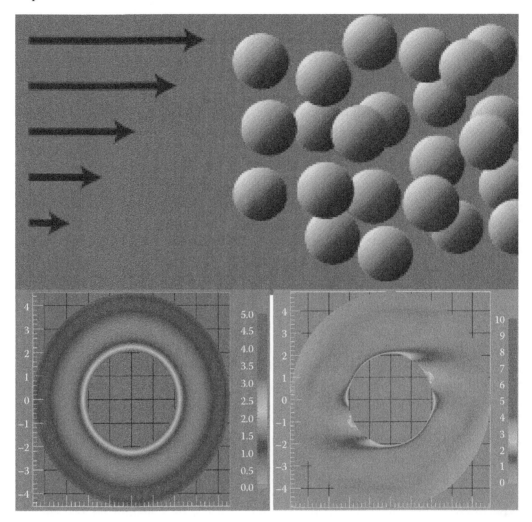

Figure 11.1 At top is a schematic showing spherical particles, typically relatively monodisperse with radius a in (Newtonian) liquid, with the mixture subjected to shear flow. A key issue in suspensions is the structure developed, and in the lower panels is the pair distribution function $g(r)$ for a suspension of solid fraction $\phi = 0.45$ shown in a plane for the equilibrium at lower left, and for the plane of shear at dimensionless shear rate of $Pe = (6\pi\eta_0\dot\gamma a^2)/kT = 25$ at lower right. The strong anisotropy induced by flow at elevated shear rate is evident.

These quantities have been well studied in polymeric fluids (Bird et al. 1987), where $N_1 > 0$ and $|N_1/N_2| \gg 1$, and the anisotropic elongation of polymer chains provides the basis for these properties. The situation is rather different because the microscopic basis for the stresses is different. What has most commonly been observed in noncolloidal suspensions is that $N_1 < 0$ and $N_2 < 0$, with $|N_2/N_1| \approx 3\text{–}4$ in experiments (Zarraga et al. 2000; Dai et al. 2013) while simulations tend to find $|N_2/N_1|$ closer to unity for noncolloidal or strongly sheared Brownian suspensions (Foss and Brady 2000; Sierou and Brady 2002); see Figure 11.1. Yeo and Maxey

(2010) used a simulation method different from Stokesian dynamics (namely, the force-coupling method) to examine noncolloidal suspensions and found similar magnitudes of the normal stresses. This disagreement has been argued to be rather noteworthy as it suggests some possibility of missing physics in the simulations; Denn and Morris (2014) have discussed this point and the fact that other work (Dbouk et al. 2013) has found a positive N_1 for similar conditions. The possible role of near-contact or true-contact forces playing a role that must be accounted for on a case-by-case basis should be kept in mind in consideration of the normal stress differences in suspensions. Recent studies of the normal stress response in noncolloidal suspension using an inclined trough approach (Couturier et al. 2011) provide N_2 and find a lower measurable limit of $\phi \approx 0.2$, above which N_2 was approximately linear in ϕ.

In addition to the normal stress differences, the particle pressure, defined

$$\Pi = -\frac{\Sigma_{11} + \Sigma_{22} + \Sigma_{33}}{3}, \tag{11.2}$$

is of interest. The concept of the particle pressure has been discussed for decades, appearing in the book by Wallis (1969) and, in a context more directly in line with interests here, was related to suspended particle migration by Jenkins and McTigue (1990) and Nott and Brady (1994). Following earlier examinations at infinite Pe by Sierou and Brady (2002), Yurkovetsky and Morris (2008) showed that Π is, in fact, the nonequilibrium continuation of the osmotic pressure; this work found $\Pi > 0$ (a dispersive pressure) and $\Pi/|N_2| > 5$ for $Pe = 1000$ and $\phi \geq 0.45$ based on the particle configurations and motions from accelerated Stokesian dynamics simulation. Yeo and Maxey (2010; 2013) have provided Π data using the force-coupling method for noncolloidal suspensions, finding similar values. Experimental determination of the normal stresses by considering the suction pressure induced in the liquid directly, following the basic notion of an osmotic pressure, has been used to estimate Π for noncolloidal suspensions by Deboeuf et al. (2009), who found values in line with simulation and modeling (e.g., Morris and Boulay 1999). As this method of measurement is related to the tendency of the particle phase to dilate—requiring liquid to be sucked into the expanded interstitial space—there is a relation to the granular phenomenon of Reynolds dilatancy.

Our focus has been on the behavior of suspensions under Stokes flow conditions. When inertia at the particle scale plays a role, implying an appreciable value of $Re = \rho \dot{\gamma} a^2 / \eta_0$, the rheology of the suspension takes on an additional rate dependence. The role of inertia in the rheology of suspensions has been addressed at single particle (Mikulencak and Morris 2004; Vivek Raja et al. 2010) and bulk scale (Shakib-Manish et al. 2002; Kromkamp et al. 2005; Kulkarni and Morris 2008; Haddadi and Morris 2014). At sufficiently large inertia, the expectation of a Bagnold scaling with stresses scaling as $\rho U^2 \sim (\dot{\gamma} a)^2$ is expected but this area is not well studied. Experiments (Koos et al. 2012) for suspensions under large inertia conditions have found rather limited rate dependence, and further work to understand this surprising result is warranted.

Rheological relations to granular matter: Several issues noted earlier suggest important links between suspensions and granular flows. Specifically, we note the role of contact forces, inertia, and the concept of a dispersive pressure. Recent work in suspensions has found that particle contact can have a pronounced effect on the rheology of suspensions. This contact influence is suggested (Boyer, Pouliquen and Guazzelli 2011b) to begin at solid fractions as low as $\phi \approx 0.2$, and thus, suspensions provide a system where the relative influence of contact and hydrodynamic forces can be examined. This could provide valuable insight about how the material behavior approaches that of granular media, where contact is a dominant interaction. Inertial effects with solid body collisions between particles are associated with granular flows, and development of a clear understanding of the conditions for the onset of collisions (and the contribution of collisions to momentum transfer) in shear flows of suspensions could bring the study of these two materials closer. Finally, the basic idea that particles under shear may dilate their structure, that is, expand their volume when considered as a continuum phase, is one which is seen in the extremes of osmotically driven diffusion (which is driven strictly by kT for hard spheres) and slowly sheared granular materials (where contact forces are the dominant stress mechanism). Intermediate between these conditions, suspensions exhibit a particle pressure in which various stress mechanisms (hydrodynamic, Brownian, and interparticle force including possibly contact) are at play. Understanding this general tendency of the material to spread connects to thermodynamic concepts and appears to be a useful organizing principle that links suspensions to granular flow.

11.4   Microstructure

By microstructure, we here imply the average arrangement of particles relative to one another. In particular, the pair distribution function is of interest, but we will also note some other results that bear relevance to force chain concepts in granular materials.

The pair distribution function is defined as

$$g(r) = \frac{P_{1/1}(r|0)}{n} \tag{11.3}$$

where
  $P_{1/1}(r|0)$ is the probability of finding a particle centered at r given that a particle is centered at the origin
  r is the vector between centers of the pair of interest
  $g(r)$ represents the relative likelihood of finding particles at a given separation

This quantity is discussed here only for monodisperse spheres, where data are well developed and the relation to the structure in hard-sphere gases or liquids is clear. It is known that $P_{1/1}$ and hence g may be mathematically related to many-body distributions (McQuarrie 2000). Here, we are interested in the information provided by the pair distribution in terms of the anisotropy of the sheared suspension (or granular material).

While there is not a large body of experimental study of the pair distribution function in sheared suspensions, work by Parsi and Gadala-Maria (1987) showed that the near-contact structure for a noncolloidal suspension was strongly anisotropic, with buildup of pair correlation in the compressional portion of the contact surface[*] and a depletion in the extensional zone. More recent experiments examining the microstructure by direct imaging of colloidal dispersions at large Pe includes a comparison with Stokesian dynamics simulation by Gao et al. (2010), while anisotropic structure has been considered in scattering studies by Newstein et al. (1999). The anisotropy observed by Parsi and Gadala-Maria (1987) was shown to reverse after a reversal of the direction of shearing. This work motivated theoretical analysis at the pair level (Brady and Morris 1997; Bergenholtz et al. 2002), which showed that the balance of advection and weak diffusion at large Pe could explain the development of this boundary layer/wake (buildup/depletion) structure, illustrated for the equilibrium (unsheared) and a moderate Pe condition in Figure 11.2 from the work of Morris and Katyal (2002). In the limit of infinite Pe, Stokes reversibility leads to an apparent paradox, as experiments and simulation find greater and greater anisotropy while the reversibility suggests a return to isotropy. The presence of any small non-hydrodynamic influence breaks this symmetry, implying the "pure-hydrodynamic limit" to be singular (Brady and Morris 1997), and thus,

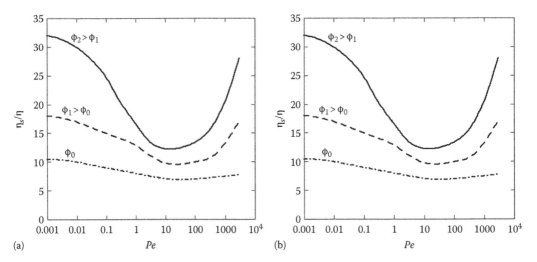

Figure 11.2 Plots illustrating the typical form of the rheology in a suspension of Brownian particles, plotted as a function of the Péclet number, $Pe = (6\pi\eta\dot{\gamma}a^2)/kT$: (a) shear viscosity with increasing ϕ and (b) first and second normal stress differences. (Adapted from Frank, M. et al., $J.$ $Fluid$ $Mech.$, 493, 363, 2003, following forms observed by Foss, D.R. and Brady, J.F., $J.$ $Fluid$ $Mech.$, 407, 167, 2000 and references therein.)

---

[*] For a shear flow of $u_x = \dot{\gamma}y$, the quadrants where $xy < 0$ are compressional and the quadrants where $xy > 0$ are extensional.

jamming tendencies seen by numerical study in this limit (Ball and Melrose 1995) are perhaps pathological as they may be relieved by a number of mechanisms (short-range repulsion or weak Brownian motion, for example).

The results shown in Figure 11.2 are from Stokesian dynamics simulation (Morris and Katyal 2002), similar to earlier observations by Phung et al. (1996) and Foss and Brady (2000) for a strongly sheared colloidal dispersion. Carried to higher Pe as seen in Figure 11.3, the structure at large Pe agrees qualitatively

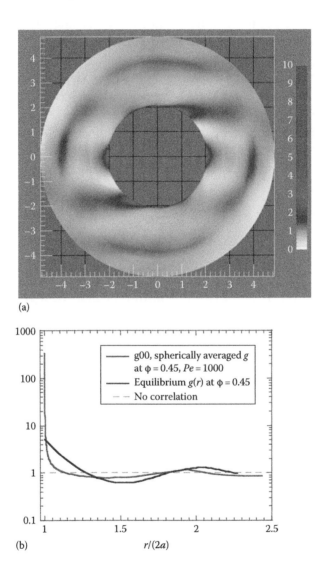

(a)

(b) $r/(2a)$

Figure 11.3   The pair distribution function determined by Stokesian dynamics simulation for hard spheres at $\phi = 0.45$ and $Pe = 1000$ shown (a) in the full shear plane and (b) in spherical average as a function of pair separation. In (b), a comparison is made to the radial dependence of the isotropic equilibrium structure, which was shown in a planar view in Figure 11.2. Shear flow is as in Figure 11.2, left to right and increasing velocity up the page.

well with the findings for experiments (Parsi and Gadala-Maria 1987) and simulation (Drazer et al. 2004; Sierou and Brady 2004) for noncolloidal suspensions. In
these simulations, a small interparticle repulsion is typically used. The finding of a
large accumulation of particle pairs at contact, particularly in compression owing
to the extreme anisotropy, explains the development of the normal stress response
in suspensions (Foss and Brady 2000) and with it the tendency for particles to
migrate in nonuniform shear flows (Morris 2009). In Figure 11.3a, note that the
high-correlation "boundary layer" on the upstream side is compressed so close to
contact that it is difficult to visualize, and here g becomes very large, leading to
the large spherical average seen in part (b) of the same figure, while in the downstream "wake," $g = O(1)$.

The pair structure is the most readily defined and best-studied structural measure in suspensions. However, longer-range structures such as clusters are also
of interest, although it may be difficult to define these objects precisely if contact is not made. These larger-scale structures are able to make larger contributions to the stress, as the stress associated with a body is given by the force
or traction moment, well known as an xF stress in molecular systems but of
the form

$$S = \frac{1}{2} \int (x\boldsymbol{\sigma}\cdot\boldsymbol{n} + \boldsymbol{\sigma}\cdot\boldsymbol{n}x)\,dS \tag{11.4}$$

with the integral taken over the surface of the body (or cluster) where $\boldsymbol{\sigma}$ is the
continuum stress and $\boldsymbol{n}$ the outward pointing normal ($\boldsymbol{\sigma}\cdot\boldsymbol{n}$ represents the local
surface traction). In a simulational study of bidisperse suspensions, where the vis-
cosity of the mixture is generally lower than an equal solid fraction of either the
small or large particles alone (Shapiro and Probstein 1992), Chang and Powell
(1993) found that the smaller particles tend to break up the clusters formed by
larger particles, leading to a substantial drop in the stress for a given shear rate.
In a study of frictional interactions on shear thickening, Seto et al. (2013) found
that the development of a frictional contact network was directly associated with
the onset of strong shear thickening, and it was the development of a connected
network in all directions that led to discontinuous shear thickening.

Microstructure relations to granular flow: The microstructural anisotropy of suspensions has been closely associated with the non-Newtonian rheology of suspensions. Similar ideas are seen in the work of Alam and Luding (2003), where the
first normal stress difference of a 2D granular material was probed; a change of
sign in N_1 with increasing solid fraction was found to be related to the anisotropy
of structure and the resulting influence on collisional stresses. Force (or stress)
chain concepts are better developed in granular materials (Mueth et al. 1998; Majmudar and Behringer 2005) than in suspensions. Recent work showing contact
to play a prominent role in dense suspensions may bring added interest in this
concept to suspensions. The idea of shear-driven jamming leading to fragile cluster structures (Cates et al. 1998) appears to be applicable to either suspensions or
granular materials.

11.5   Bulk Flow Phenomena

One of the features which has resulted in significant interest in suspension flows over the last three decades has been their behavior in bulk flows. Early evidence of plug flow (Karnis et al. 1966) in a pressure-driven situation suggested nonuniform particle concentration, but the experimental data did not allow a conclusion in this regard. Leighton and Acrivos (1987) were the first to clearly describe a shear-induced migration of particles, with the system studied showing migration from a high-shear Couette viscometer gap to the low-shear well beneath the gap. The phenomenology of particle flux satisfying $j \propto -\nabla\dot{\gamma}$ proves successful in a number of situations, although it is in conflict with certain experiments in curvilinear flow as described by Morris and Boulay (1999), where a description of the flux is given in the form $j \propto \nabla \cdot \Sigma^P$ with Σ^P the particle contribution to the bulk stress, following earlier work by Nott and Brady (1994).

The bulk flow of suspensions is of practical interest, as it is related to such applications as mold flows of ceramic precursors, coating of paper, and coal slurry flow. It also raises basic fundamental questions regarding the coupling of mass and momentum conservation in these flows. This coupling is especially interesting because of the strong dependence of the rheology on the solid fraction, which because of the relatively large size of the suspended particles (in comparison to the microscopic objects in other complex fluids, e.g., polymeric molecules) leads to rapid migration.

As developed by Morris (2009), a problem that illustrates how the bulk rheology and migration are related to the observed concentration field is what is termed "viscous resuspension." Here, a suspension of heavy particles with gravity acting along the direction y is subjected to a possibly nonuniform shear flow $u_x(y)$. In the absence of the shearing flow, the particles will settle to form a densely packed bed. When the shear flow is imposed above a certain stress level, however, the particle bed will flow. It is necessary to introduce an additional dimensionless parameter to characterize this behavior, and this is conveniently given by a dimensionless gravitational force or buoyancy number $B = \Delta\rho g a/\eta_0\dot{\gamma}$ (here $\dot{\gamma}$ represents the average of the shear rate across the flow). A uniform shear stress is obtained if the flow is driven by relative motion of two large parallel plates. However, the particle momentum balance in the direction of gravity is

$$\frac{\partial \Sigma_{yy}^P}{\partial y} = -\phi\Delta\rho g, \tag{11.5}$$

a balance of "hydrostatic" form in which the excess weight of the particles must be supported by a variation of the particle stress Σ^P. There are many ways to satisfy this balance: For a large density difference ($\Delta\rho$) or small shear rate (large B), the flow will segregate to a dilute rapidly sheared upper layer and a dense slowly shearing lower layer. By contrast, for small density difference or at a large shear rate, the particles will be relatively uniform and the shearing will take place throughout. In the densely packed condition, this excess weight could be borne by static contact. However, when the particles are well dispersed but denser than the fluid, this balance makes clear that it is a non-Newtonian normal stress that balances the excess weight.

This normal stress is present in the absence of the density difference, but has little dynamical consequence unless there is a stress imbalance, in which case it is associated with migration as noted earlier. Migration of the particles relative to suspending fluid is seen in a number of geometries in suspensions. The particle migration behaviors seen for neutrally buoyant particles include

- Outward migration in a wide-gap Couette flow

- Migration toward the centerline of a pressure-driven flow in a pipe or channel

- Little or no migration in parallel-plate torsion

We emphasize here that these are the observations for particles density matched with the suspending fluid. With the benefit of being able to eliminate density differences, these migration phenomena can all be related to the stresses developed by the material under flow, that is, to the bulk rheology. Migration is of significant practical interest as it can lead to local accumulations in process flows, as suggested by the accumulation predicted (Miller et al. 2009) at the upstream corner in a 4:1 channel contraction for a concentrated suspension in Figure 11.4 (along with an upstream migration leading to elevated concentration at the center of the channel); qualitatively similar observations appear in experimental studies using NMR imaging (Altobelli et al. 1997; Moraczewski et al. 2005).

When migration interacts with buoyancy, some rather unexpected results may arise. Morris and Brady (1998) showed that when a suspension of heavy particles is in pressure-driven flow, a weak density difference between solid and fluid leads to a higher concentration of particles riding over a less-concentrated (and therefore lighter) layer. This heavy-over-light configuration is understood as resulting from

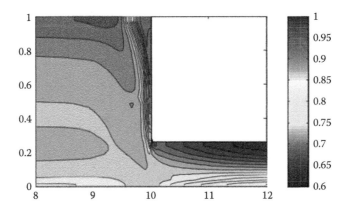

Figure 11.4 Predicted ϕ/ϕ_{max} contours of particle fraction in a 4:1 channel contraction of a suspension of bulk concentration $\phi/\phi_{max} = 0.74$ injected as a uniform mixture of 10 channel widths upstream, showing accumulation at the corner. The large channel width is 18 particle diameters.

migration toward the centerline coupled with relatively weak settling—so long as migration due to shear-induced normal stress dominates over settling, such a scenario is expected. This observation was also made in magnetic resonance imaging of suspension flow in a pipe by Altobelli et al. (1991).

Lenoble et al. (2005) studied a suspension of heavy particles at an initial solid fraction of $\phi = 0.59$, a mixture they in fact termed a "wet granular material," in a simple shear flow configuration (between parallel plates) and found a sheared layer over a static layer. Granular avalanche flows under liquid have also been studied to determine the normal stress response, specifically through the pore pressure at the base of the flow (Cassar et al. 2005), and a model for the development of the steady state in such flows was presented by Pailha and Pouliquen (2009).

In pressure-driven flows in conduits of noncircular cross section, secondary flows are predicted to arise due to the presence of the second normal stress difference (Giesekus 1965). In Figure 11.5, adapted from Ramachandran and Leighton (2008), the form of the secondary flow seen in a conduit of elliptical cross section is seen; as detailed by these authors, the driving force for the circulation (as indicated by the arrows) is proportional to the second normal stress difference and the local particle pressure. The importance of such flows in establishing the particle concentration is an interesting but little-studied issue, with the work by Ramachandran and Leighton pointing out that the rheological response of a suspension of noncolloidal particles is largely insensitive to the size of the particles, whereas the migration phenomena associated with rheological driving forces scale roughly as the particle size squared (Morris and Boulay 1999; Guazzelli and Morris 2012). Thus, the secondary flows may lead to stronger effects than migration in certain geometries, especially for small particles. Zrehen and Ramachandran (2013) have experimentally demonstrated the role of secondary flows in concentrated suspensions. Secondary flows associated with normal stress differences may also be of interest in granular materials.

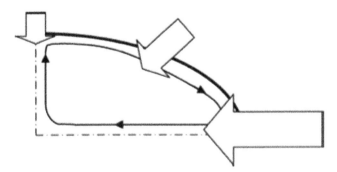

Figure 11.5   Secondary flow in an elliptical conduit of a material exhibiting second normal stress as does a concentrated suspension, with arrows depicting the nonuniform normal stress driving force (not to scale). (Adapted from Ramachandran, A. and Leighton, D.T., *J. Fluid Mech.*, 603, 207, 2008.)

Important unresolved issues in the bulk flow of suspensions are found in consideration of the boundary conditions on the flow. When the boundary of the flow is a solid wall, the finite size of the particles introduces an important limitation. A no-particle-flux boundary condition must be applied at a distance r away from the boundary (assuming monodisperse particles—the problem is more complex for polydisperse suspensions), but particle slip is appropriate, as particles may roll along the boundary and even, assuming a lubricating film, slip with only a very weak singularity logarithmic in the gap from particle surface to wall. This suggests that the particles and liquid must have separate boundary conditions, although such conditions have typically not been applied. This neglect of particle-scale phenomena is questionable particularly when flows may be only tens of particles in depth, or when the flow may encounter a corner (as in a bifurcating channel) and separation may occur. When particles flow adjacent to a liquid–fluid interface, as at the upper surface of an inclined plane flow (Timberlake and Morris 2005) or at the boundary of a torsional cone-and-plate or parallel-plate rheometric flow, the particles may protrude into the interface, and this particle-scale evidence of the normal stress response of the mixture is critical to the measurement obtained as described by Brown and Jaeger (2012); Garland et al. (2013) show through particle pressure measurements by the suction technique outlined in Deboeuf et al. (2009) that the measured normal stress is limited by the Laplace pressure associated with a particle deforming the interface at a level comparable to its radius a, namely, $p_\gamma \sim \gamma/a$ where γ is the interfacial tension.

11.6 Concluding Remarks and Connection with Granular Flows

Recent work has shown that in the very dense state, there are similiarities in granular and suspension flows. Since at the jamming limit it is no longer possible to distinguish suspensions and granular material, there is an essential similarity between the two for very concentrated flows. In this limit, contacts between particles are a dominant stress transmission mechanism. Contacts have historically received little attention in suspension mechanics, whereas in the study of granular materials they are essential.

The importance of contacts in granular materials is accompanied by the strong influence of gravity in most granular flows. This highlights an essential difference in perspective and approach to study of suspensions relative to granular material. A granular material in inclined plane flow will only flow in such a situation once it is tilted beyond a certain angle (Pouliquen 1999), and the bed will dilate as the rate increases—it is thus natural to consider the granular system behavior as a function of applied stress, allowing the volume to vary. By elimination of the density effects, suspensions can be made to flow at any solid loading up to the jamming condition and any shear rate, and thus, the traditional approach in suspensions has been to fix the solid fraction and vary the rate to determine the stress response. However, the use of normal stress control, allowing ϕ to vary, in suspensions (Prasad and Kytomaa 1995; Boyer et al. 2011a) has recently provided excellent understanding of the way in which suspension and granular rheology may be unified. One may thus expect that granular flow understanding, for example, of how contact networks behave in shear may illuminate dense suspension flow studies.

Away from jamming, where fluid gaps remain between most, if not all, near-neighbor particles, suspensions are closer to classical fluids. However, the presence of the solids imparts important differences, in particular leading to a nonlinear rheology, including not only rate dependence of the viscosity but also normal stresses. The special value of the suspension as a model system is that these rheological characteristics, which become very pronounced at maximum packing as all of the rheological responses diverge for $\phi \to \phi_{max}$, can be studied over a wide range of conditions. Thus their basis in the microstructure of the particles has been appreciated, connecting the bulk behavior to the microscale in a way that mirrors developments in liquid-state theory (Nazockdast and Morris 2013). While this approach is an active area of study, it is understood at least in its broad outlines and may be expected to provide guidance to similar approaches in other complex fluids. As the structural issues are similar in their influence for granular materials, even if the stresses may not be transmitted in the same fashion for the two, suspension structure-property understanding may also guide efforts in granular materials.

Bulk flow behavior is an area where suspensions and granular materials appear the most different, especially to a practicing engineer. A suspension flows due to gravitational head, whereas a granular material may form arches in a hopper (the Janssen effect) and its flow out will be independent of the level in the hopper. The several points noted at the outset of the chapter come together in this example: the solid fraction approaches maximum packing with small changes in ϕ effecting large changes in flowability, with friction playing a pronounced role. Similarly, a suspension can be transported in a pipeline by pumping using understanding of simple fluids as a firm basis, whereas a granular material must be fluidized and carried along with a significant "slip velocity" between the mean fluid and solid velocities. Bulk flows of granular materials will thus often require different approaches than can be successfully applied for suspensions; that said, remarkable control can be exerted over this process as shown, for example, by the success of certain industrial processes involving conveyance of powders or sand-like granular media.*

It appears fair to say that suspensions lie intermediate between pure fluids and granular materials, defining the latter to depend critically upon contact interactions and to be strongly influenced by gravity in typical cases. The suspension can be freely varied from a liquid carrying dilute solids to an almost-jammed material nearly indistinguishable from a granular material. Recent research has tended to seek to establish the similarities between granular and suspension rheologies, although the differences in the two materials may teach us more. Regardless, there is a continuous connection between fundamentally interesting and practically

---

* An example is seen (at) http://www.neu-process.com.

important materials, from the fluid to suspension to granular material, and the consideration of all of these materials in conjunction offers valuable insight and opportunities.

Acknowledgments

The author was partially supported in the preparation of this work by the NSF grants DMR PREM 0934206 and CBET 0847271.

References

Aidun, C. K. and Lu, Y. 1995. Lattice Boltzmann simulation of solid particles suspended in fluid. *J. Stat. Phys.* **81**, 49–61.

Aidun, C. K., Lu, Y., and Ding, E. 1998. Direct analysis of particulate suspensions with inertia using the discrete Boltzmann equation. *J. Fluid Mech.* **373**, 287–311.

Alam, M. and Luding, S. 2003. First normal stress difference and crystallization in a dense sheared granular fluid. *Phys. Fluids* **15**, 2298.

Altobelli, S. A., Fukushima, E., and Mondy, L. A. 1997. Nuclear magnetic resonance imaging of particle migration in suspensions undergoing extrusion. *J. Rheol.* **41**, 1105–1115.

Altobelli, S. A., Givler, R. C., and Fukushima, E. 1991. Velocity and concentration measurements of suspensions by nuclear magnetic resonance imaging. *J. Rheol.*, **35**, 721.

Bird, R. B., Armstrong, R. C., and Hassager, O., 1987. Dynamics of Polymeric Liquids, 2nd ed. Vol. 1. Wiley & Sons. New York.

Bailey, B. C. and Yoda, M. 2003. An aqueous low-viscosity density- and refractive index-matched suspension system. *Exp. Fluids* **35**, 1–3.

Ball, R. C. and Melrose, J. R. 1995. Lubrication breakdown in hydrodynamic simulations of concentrated colloids. *Adv. Colloid Interface Sci.* **59**, 19.

Bergenholtz, J., Brady, J.F., and Vicic, M. 2002. The non-Newtonian rheology of dilute colloidal suspensions. *J. Fluid Mech.* **456**, 239–275.

Boyer, F., Guazzelli, É., and Pouliquen, O. 2011a. Unifying suspension and granular rheology. *Phys. Rev. Lett.* **107**, 188301.

Boyer, F., Pouliquen, O., and Guazzelli, É. 2011b. Dense suspensions in rotating-rod flows: normal stresses and particle migration. *J. Fluid Mech.* **686**, 5–25.

Brady, J. F. and Bossis, G. 1988. Stokesian dynamics. *Annu. Rev. Fluid. Mech.* **20**, 111–157.

Brady, J. F. and Morris, J. F. 1997. Microstructure of strongly sheared suspensions and its impact on rheology and diffusion. *J. Fluid Mech.* **348**, 103–139.

Brinker C. J. and Scherer G. W. 1990. *Sol-gel Science: The Physics and Chemistry of Sol-gel Processing*, Academic Press, San Diego.

Brown, E. and Jaeger, H. M. 2012. The role of dilation and confining stresses in shear thickening of dense suspensions. *J. Rheol.* **56**, 875.

Cassar, C., Nicolas, M., and Pouliquen, O. 2005 Submarine granular flows down inclined planes. *Phys. Fluids* **17**, 103301.

Cates, M. E., Wittmer, J. P., Bouchaud, J. P., and Claudin, P. 1998. Jamming, force chains, and fragile matter. *Phys. Rev. Lett.* **81**, 1841.

Chang, C. Y. and Powell, R. L. 1993. Dynamic simulation of bimodal suspensions of hydrodynamically interacting spherical particles. *J. Fluid Mech.* **53**, 1.

Couturier, É., Boyer, F., Pouliquen, O., and Guazzelli, É. 2011. Suspensions in a tilted trough: Second normal stress difference. *J. Fluid mech.* **686**, 5–25.

Cundall, P. A. and Strack, O. D. L. 1979. A discrete numerical model for granular assemblies. *Géotechnique*, **29**, 47–65.

Dai, S.-C., Bertevas, E., Qi, F., and Tanner, R. L. 2013. Viscometric functions for non-colloidal sphere suspensions with Newtonian matrices. *J. Rheol.* **57**, 493510

Davis, R. H., Zhao, Y., Galvin, K. P., and Wilson, H. J. 2003. Solid-solid contacts due to surface roughness and their effects on suspension behaviour. *Phil. Trans. R. Soc Lond. A* **361**, 871.

Dbouk, T., Lobry, L., and Lemaire, E. 2013. Normal stresses in concentrated non-Brownian suspensions. *J. Fluid Mech.* **715**, 239–272.

Deboeuf, A., Gauthier, G., Martin, J., Yurkovetsky, Y., and Morris, J. F. 2009. Particle pressure: A bridge from osmotic pressure to granular dilation. *Phys. Rev. Lett.* **102**, 108301.

Denn, M. M. and Morris, J. F. 2014. Rheology of non-Brownian suspensions. *Ann. Rev. Chem Biomol. Eng.* **5**, 203–328.

Dijksman, J. A., Rietz, F., Lorincz, K. A., van Hecke, M., and Losert, W. 2012. Refractive index matched scanning of dense granular materials. *Rev. Sci. Instrum.* **83**, 011301.

Drazer, G., Koplik, J., Khusid, B., and Acrivos, A. 2004. Microstructure and velocity fluctuations in sheared suspensions. *J. Fluid Mech.* **511**, 237–263.

Einstein, A. 1906. Eine neue Bestimmung der Molek uldimensionen. *Ann. Phys.* **19**, 289–305.

Foss, D. R. and Brady, J. F. 2000. Structure, diffusion and rheology of Brownian suspensions by stokesian dynamics simulation. *J. Fluid Mech.* **407**, 167.

Frank, M., Anderson, D. Weeks, E. R., and Morris, J. F. 2003. Particle migration in pressure-driven flow of a Brownian suspension *J. Fluid Mech.* **493**, 363.

Gadala-Maria, F. 1979. The rheology of concentrated suspensions. PhD dissertation, Stanford University, Stanford, CA.

Gao, C., Kulkarni, S. D., Gilchrist, J. F., and Morris, J. F. 2010. Direct investigation of anisotropic suspension structure in pressure-driven ow. *Phys. Rev. E.* **81**, 041403.

Garland, S., Gauthier, G., Martin, J., and Morris, J. F. 2013. Normal stresses measurements in sheared non-Brownian suspensions. *J. Rheol.* **57**, 71.

Giesekus, H. 1965. Sekundarstromungen in viskoelastischen Flussigkeiten bei stationarer und periodischer Bewegung. *Rheol. Acta* **4**, 85–101.

Guazzelli, É. and Hinch, E. J. 2011. Fluctuations and instability in sedimentation. *Ann. Rev. Fluid Mech.* **43**, 87–116.

Guazzelli, É. and Morris, J. F. 2012. *A Physical Introduction to Suspension Dynamics.* Cambridge University Press, Cambridge, U.K.

Haddadi, H. and Morris, J. F. 2014. Microstructure and rheology of finite inertia neutrally buoyant suspensions. *J. Fluid Mech.* **749**, 431–459.

Happel, J. and Brenner, H. 1965. *Low Reynolds Number Hydrodynamics.* Prentice-Hall: Englewood Cliffs, NJ; Martinus Nijhoff: Boston.

Jenkins, J. T. and McTigue, D. F. 1990. Transport process in concentrated suspensions: the role of particles fluctuations. In *Two Phase Flows and Waves.* ed. D. D. Joseph and D. G. Schaeffer) Springer, Berlin.

Karnis, A., Goldsmith, H. L., and Mason, S. G. 1966. The kinetics of flowing dispersions: Concentrated suspensions of rigid particles. *J. Colloid Interface Sci.* **22**, 531–553.

Kim, S. and Karrila, S. 1989. *Microhydrodynamics: Principles and Selected Applications.* Butterworth-Heinemann: Stoneham, MA.

Koos, E., Linares-Guerrero, E., Hunt, M. L., and Brennen, C. E. 2012. Rheological measurement of large particles in high shear rate flows. *Phys Fluids* **24**, 013302.

Krieger, I. M. 1972. Rheology of monodisperse latices. *Adv. Colloid Interface Sci.* **3**, 111.

Kromkamp, J., van den Ende, D. T. M., Kandhai, D., van der Sman, R. G. M., and Boom, R. M. 2005. Shear-induced self-diffusion and microstructure in non-Brownian suspensions at non-zero Reynolds numbers. *J. Fluid Mech.* **529**, 253–278.

Kulkarni, P. M. and Morris, J. F. 2008. Suspension properties at finite Reynolds number from simulated shear flow. *Phys. Fluids* **20**, 040602.

Ladd, A. J. C. 1994a. Numerical simulations of particulate suspensions via a discretized Boltzmann equation. Part 1. Theoretical foundation. *J. Fluid Mech.* **271**, 285–309.

Ladd, A. J. C. 1994b. Numerical simulations of particulate suspensions via a discretized Boltzmann equation. Part 2. Numerical results. *J. Fluid Mech.* **271**, 311–339.

Ladd, A. J. C. 1988. Hydrodynamic interactions in a suspension of spherical particles *J. Chem. Phys.* **88**, 5051.

Leighton, D. T. and Acrivos, A. 1987. The shear-induced migration of particles in concentrated suspensions. *J. Fluid Mech.* **181**, 415.

Lenoble, M., Snabre, P., and Pouligny, B. 2005. The flow of a very concentrated slurry in a parallel-plate device: Influence of gravity. *Phys. Fluids* **17**, 073303.

Lewis, T. B. and Nielsen, L. E. 1968. Viscosity of dispersed and aggregated suspensions of spheres. *Trans. Soc. Rheol.* **12**, 421–433.

Majmudar, T. S. and Behringer, R. P. 2005. Contact force measurements and stress-induced anisotropy in granular materials. *Nature* **435**, 1079–1082.

McQuarrie, D. A. 2000. *Statistical Mechanics*, University Science Books: Sausalito, CA.

Melrose, J. R. and Ball, R. C. 2004. Continuous shear thickening transitions in model concentrated colloids. The role of interparticle forces. *J. Rheol.* **48**, 937–960.

Meng, Q. and Higdon, J. J. L. 2008. Large scale dynamic simulation of plate-like particle suspensions. Part I: Non-Brownian simulation. *J. Rheol.* **52**, 1.

Mewis, J. and Wagner N. J. 2009. Thixotropy. *Adv. Colloid Interface Sci.* **147**, 214–227.

Mewis, J. and Wagner, N. J. 2012. *Colloidal Suspension Rheology*. Cambridge University Press.

Mooney, M. 1951. The viscosity of a concentrated suspension of spherical particles. *J. Colloid Sci.* **6**, 162–170.

Moraczewski, T., Tang, T. H., and Shapley, N. C. 2005. Flow of a concentrated suspension through an abrupt axisymmetric expansion measured by nuclear magnetic resonance imaging. *J. Rheol.* **49**, 1409–1428.

Morris, J. F. 2009. A review of microstructure in concentrated suspensions and its implications for rheology and bulk flow. *Rheol. Acta.* **48**, 909.

Morris, J. F. and Brady, J. F. 1998. Pressure-driven flow of a suspension: Buoyancy effects. *Int. J. Multiph. Flow* **24**, 105.

Morris J. F. and Boulay, F. 1999. Curvilinear flows of noncolloidal suspensions: The role of normal stresses. *J. Rheol.* **43**, 1213.

Morris, J. F. and Katyal, B. 2002. Microstructure from simulated Brownian suspension flow at large shear rate. *Phys .Fluids* **14**, 1920.

Mikulencak, D. R. and Morris, J. F. 2004. Stationary shear flow around fixed and free bodies at finite Reynolds number. *J. Fluid Mech.* **520**, 215–242.

Miller, R. M., Singh, J. P. B., and Morris, J. F. 2009. Suspension flow modeling for general geometries. *Chem. Eng. Sci.* **64**, 4597–4610.

Mueth, D. M., Jaeger, H. M., and Nagel, S. R. 1998. Force distribution in a granular medium. *Phys. Rev. E* **57**, 3164.

Nazockdast, E. and Morris, J. F. 2013. Microstructural theory and the rheology of concentrated colloidal suspensions. *J. Fluid Mech.* **713**, 420–452.

Newstein, M. C., Wang, H., Balsara, N. P. et al. 1999. Microstructural changes in a colloidal liquid in the shear thinning and shear thickening regimes. *J. Chem. Phys.* **111**, 4827.

Nott, P. R. and Brady, J. F. 1994. Pressure-driven flow of suspensions: Simulation and theory. *J. Fluid Mech.* **275**, 157.

Pailha, M. and Pouliquen, O. 2009. A two-phase flow description of the initiation of underwater granular avalanches. *J. Fluid Mech.* **633**, 115–135.

Parsi, F. and Gadala-Maria, F. 1987. Fore-and-aft asymmetry in a concentrated suspension of solid spheres. *J. Rheol.* **31**, 725.

Petrie, C. S. 1999. The rheology of fibre suspensions. *J. Non-Newtonian Fluid Mech.* **87**, 369–402.

Phung, T. N., Brady, J. F., and Bossis, G. 1996. Stokesian dynamics simulation of Brownian suspensions. *J. Fluid Mech.* **313**, 181.

Pouliquen, O. 1999. Scaling laws in granular flows down rough inclined planes. *Phys. Fluids* **11**, 542–548.

Prasad, D. and Kytomaa, H. K. 1995. Particle stress and viscous compaction during shear of dense suspensions. *Int. J. Multiph. Flow*, **21**, 775.

Ramachandran, A. and Leighton, D. T. 2008. The influence of normal stress difference induced secondary flows on the shear-induced migration of particles in concentrated suspensions. *J. Fluid Mech.* **603**, 207–243.

Rao, K. K. and Nott, P. R. 2008. *An Introduction to Granular Flow*. Cambridge University Press, Cambridge, U.K.

Russel, W. B., Saville D. A., and Schowalter, W. R. 1989. *Colloidal Dispersions*. Cambridge University Press, Cambridge, U.K.

Sakib-Manesh, A., Raiskinmaki, P., Koponen, A., Kataja, M., and Timonen, J. 2002. Shear stress in a couette flow of liquid-particle suspensions. *J. Stat. Phys.* **107**, 67–84.

Sangani, A. S., Mo, G., Tsao, H., and Koch, D. L. 1996. Simple shear flows of dense gas-solid suspensions at finite Stokes numbers. *J. Fluid Mech.* **313**, 309–341.

Shapiro, A. P. and Probstein, R. F. 1992. Random packings of spheres and fluidity limits of monodisperse and bidisperse suspensions. *Phys. Rev. Lett.* **68**, 1422.

Seto, R. Mari, R., Morris, J. F., and Denn, M. M. 2013. Discontinuous shear thickening of frictional hard-sphere suspensions. *Phys. Rev. Lett.* **111**, 218301.

Sierou, A. and Brady, J. F. 2001. Accelerated Stokesian dynamics simulations. *J. Fluid Mech.* **448**, 115–146.

Sierou, A. and Brady, J. F. 2002. Rheology and microstructure in concentrated noncolloidal suspensions. *J. Rheol.* **46**, 1031–1056.

Sierou, A. and Brady, J. F. 2004. Rheology and microstructure in concentrated non-colloidal suspensions. *J. Rheol.* **46**, 1031–1056.

Stickel, J. J. and Powell, R. L. 2005. Fluid mechanics and rheology of dense suspensions. *Ann. Rev. Fluid Mech.* **37**, 129.

Subramanian, G. and Brady, J. F. 2006. Trajectory analysis for non-Brownian inertial suspensions in simple shear flow. *J. Fluid Mech.* **559**, 151–203.

Subramanian, G. and Koch, D. L. 2006. Centrifugal forces alter streamline topology and greatly enhance the rate of heat and mass transfer from neutrally buoyant particles to a shear flow. *Phys. Rev. Lett.* **96**, 134503.

Thomas, D. G. 1965. Transport characteristics of suspension: VIII. A note on the viscosity of Newtonian suspensions of uniform spherical particles. *J. Colloid Sci.*, **20**, 267–277.

Timberlake, B. D. and Morris, J. F. 2005. Particle migration and free-surface topography in inclined plane flow of a suspension. *J. Fluid Mech.* **538**, 309–341.

Vivek Raja, R., Subramanian, G., and Koch, D. L. 2010. Inertial effects on the rheology of a dilute emulsion. *J. Fluid Mech.* **646**, 255–296.

Wagner, N. J. and Brady, J. F. 2009. Shear thickening in colloidal dispersions. *Phys. Today* **62**, 27–32.

Wallis, G. B. 1969. *One-Dimensional Two-Phase Flow*. McGraw-Hill, New York.

Weitz, D. A., Huang, J. S., Lin, M. Y., and Sung, J. 1985. Limits of the fractal dimension for irreversible kinetic aggregation of gold colloids. *Phys. Rev. Lett.* **54**, 1416.

Yeo, K. and Maxey, M. R. 2010. Dynamics of concentrated suspensions of noncolloidal particles in couette flow. *J. Fluid Mech.* **649**, 205–231.

Yeo, K. and Maxey, M. R. 2013. Dynamics and rheology of concentrated, finite-Reynolds-number suspensions in a homogeneous shear flow. *Phys. Fluids.* **25**, 053303.

Yurkovetsky, Y. and Morris, J. F. 2008. Particle pressure in a sheared Brownian suspension. *J. Rheol.* **52**, 141.

Zarraga, I. E., Hill, D. A., and Leighton, D. T. 2000. The characterization of the total stress of concentrated suspensions of noncolloidal spheres in Newtonian fluids. *J. Rheol.* **44**, 185.

Zrehen, A. and Ramachandran, A. 2013. Demonstration of secondary currents in the pressure-driven flow of concentrated suspensions through non-axisymmetric conduits. *Phys. Rev. Lett.* **110**, 018306.

CHAPTER 12

Wet Foams, Slippery Grains

Brian P. Tighe

CONTENTS

The preceding chapters dealt with disordered collections of solid particles in the form of either granular materials or suspensions. Here instead we consider dispersions of a fluid—either a gas or a liquid—inside a liquid. When the dispersed phase is a gas, the resulting material is called a liquid foam; when both phases are liquids, the material is an emulsion. On the mesoscale, foams and emulsions can be viewed as packings of, respectively, gas bubbles or liquid droplets. Despite being composed entirely of fluids, they resist shear deformation elastically and are therefore solids. Our main goal here is to explore the structural and mechanical properties of this solid state, both on the microscopic and on the macroscopic scale.

The intrinsic stress scales of foams and emulsions are much softer than those in granular materials. Their elasticity is generated by the surface tension of the interfaces separating the dispersed and continuous phases, which in dishwashing detergent is on the order of 10 mN/m. Interfacial tensions depend on the properties of the two phases and the *surfactants* (*surface active agents*) that are added to increase the lifetime of interfaces. In addition to this intrinsic softness, there is a second important difference from granular matter. Namely, bubbles and droplets do *not* support static (Coulomb) friction, but they *do* experience rate-dependent dissipative forces, sometimes called viscous friction. In this sense, one could say that foams and emulsions are composed of "slippery grains." Along with the molecular and the macroscopic scale, the physics of foams and emulsions occurs in thin liquid films (~100 nm) and between the bubbles and droplets they form (10 μm–1 cm).

Our focus will be primarily on those aspects of foam and emulsion physics that bear the closest analogy to granular matter. These emerge at and above the mesoscale, that is, the bubble/droplet/grain scale. An important common feature of foams, emulsions, and granular media is the *jamming transition* (discussed in Chapter 7); they each undergo a nonequilibrium phase transition in which rigidity is developed or lost as the packing fraction ϕ (i.e., the volume fraction of the dispersed phase) is varied [2,21]. This is illustrated in a foam in Figure 12.1.

In the context of foams, jamming occurs in the so-called wet foam limit. In wet foams, bubbles are nearly spherical, as opposed to the polygonal shape characteristic of dry foams. The terms "wet" and "dry" refer to the liquid fraction $\epsilon = 1 - \phi$ in the foam. Variations in liquid fraction are clearly visible in Figure 12.2, which depicts a foam undergoing drainage, as one might see in the head of a beer. Close to the critical packing fraction ϕ_c, frictionless sphere packings are an excellent model for the structure and mechanical response of wet foams, and the distance to the critical packing fraction $\Delta\phi = \phi - \phi_c$ represents an important macroscopic parameter that determines a foam's response. Ironically, packings of frictionless spherical particles are often studied as simplified models of granular materials. In their simplicity, these models are actually more accurate representations of wet foams than granular media.

Of course, understanding granular media by analogy is neither the only nor the principal reason to study foams and emulsions. Surfactant chemistry, thin film

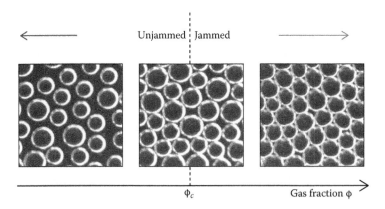

Figure 12.1 Bubble monolayer of varying gas fraction. The bubbles are nearly circular in cross section and form permanent contacts at a critical gas fraction ϕ_c. Above ϕ_c, the system responds elastically to small deformations: it is a solid. (Image courtesy M. van Hecke.)

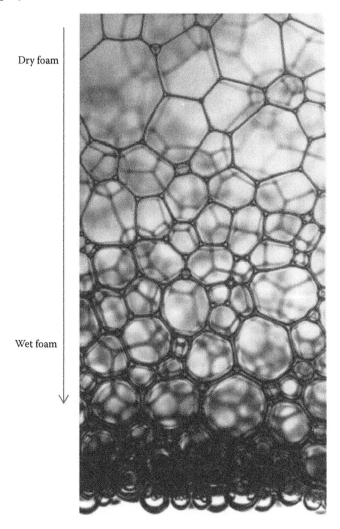

Dry foam

Wet foam

Figure 12.2   Foam undergoing drainage; wetness increases from top to bottom. Bubbles in a wet foam approach a spherical shape, while those in a dry foam are more polygonal. (Image courtesy D. Durian.)

flow, and the physics of drainage, coarsening, and rupture—the processes driving the topological evolution of material structure—are all important topics that will not be discussed here in any detail. The dry foam limit is also a rich subject in its own right, with important connections to the mathematics of minimal surfaces and spatial tessellations. The interested reader will find greater detail on many of the topics discussed here—and an introduction to many that are omitted—in "The Physics of Foams" by Weaire and Hutzler [28], along with review articles by Mason [18], Höhler and Cohen-Addad [13], and Hilgenfeldt et al. [12]. Introductions to jamming can be found in Chapter 7 and in the review article of Van Hecke [27].

12.1   Structure

A liquid foam is a dispersion comprising a gas phase and a liquid phase; shaving cream and the froth in the head of a beer are two familiar examples. The gas phase forms bubbles that are separated by thin liquid films. The gas is thus called the dispersed phase, while the liquid is the continuous phase. There is an energetic cost associated with any interface between phases and an associated interfacial tension. This tension is, ultimately, the reason a material made up of two fluids can display solid-like responses; it allows the thin liquid films to resist deformation.

An emulsion is also a dispersion of one fluid in another, but in this case the two fluids are both liquids. Mayonnaise is a familiar oil-in-water emulsion. The dispersed phase in an emulsion forms droplets, rather than bubbles. For most of the physics discussed in this chapter, the difference between a gaseous and a liquid dispersed phase is negligible, and we shall speak freely of foams, emulsions, bubbles, and droplets without distinction.

12.1.1   Surfactants

Anyone who has made soap bubbles from dishwashing detergent or left a bottle of salad dressing standing overnight knows that foams and emulsions are not thermodynamically stable. Because of the energy cost associated with interfaces, the dispersed and continuous phases seek to minimize their surface area and eventually phase separate. Often this is undesirable—think of your salad dressing. It can be postponed by introducing surfactant molecules, which consist of a hydrophilic head group and a hydrophobic tail. When dissolved in water, they spontaneously adsorb at an interface with the head group pointing toward the water (see Figure 12.3a).

Surfactants slow down the processes of coarsening and rupture by which phase separation takes place. Due to a mechanism discussed in the following section, regions of high interfacial curvature set up a large pressure difference between the liquid and gas phases. This sets up a pressure gradient within the liquid phase that drives liquid out of the thin films (to the right and left in Figure 12.3a), causing them to thin further—this is drainage. As this process continues, fluctuations in film thickness eventually become comparable to the thickness itself and the film ruptures. Surfactants slow drainage and rupture by establishing a disjoining pressure between the two interfaces of a thin film. The disjoining pressure is a consequence of Coulomb repulsion (as in Figure 12.3a), steric exclusion, or other molecular interactions. The disjoining pressure keeps the films from becoming thin enough to rupture.

Surfactants can be divided into two distinct classes—mobile and immobile—that have a qualitative impact on the flow of foams and emulsions. The distinction hinges on how easily surfactant molecules diffuse, both on the interface and from the bulk to the interface. Mobile surfactants diffuse easily and permit slip when the continuous phase flows. Immobile surfactants tend to be larger molecules that lock together, leading to a no-slip boundary condition at the interface.

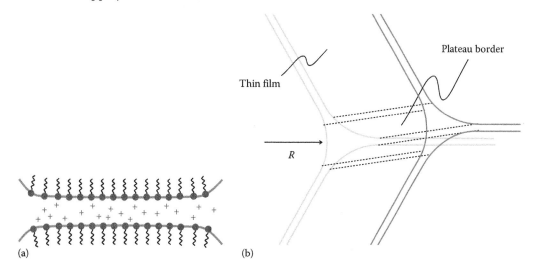

Figure 12.3 (a) Thin film stabilized by surfactants with hydrophilic head groups and hydrophobic tails. Repulsion between the surfactants establishes a disjoining pressure that obstructs drainage and rupture. (b) The intersection of thin films forms a Plateau border. The radius of curvature R at the border is smaller in dry foams than in wet foams.

12.1.2 Plateau Borders and Vertices

Figure 12.3b depicts three thin films. The juncture of multiple thin films is called a Plateau border, after the Belgian physicist who first described the equilibrium structure of dry foams [24]. In three dimensions thin films define a surface, and Plateau borders are quasi-1D objects. The point of intersection of several Plateau borders is, in turn, known as a vertex or node. Increasing the liquid content causes the bubbles to become more rounded, ultimately attaining a circular/spherical shape. In the dry foam limit, the thin film thickness goes to zero and the radius of curvature R at each vertex vanishes. The liquid content in Figure 12.3 is therefore seen to be low—the foam is dry rather than wet.

The analogy between foams and emulsions and granular media is strongest in the wet limit; nevertheless, dry foams are fundamentally significant in their own right, and Plateau's rules for the structure of an ideal dry foam ($\phi = 1$) can be regarded as a tessellation of space. These rules (for dry 3D foam) [24] are as follows:

1. All intersections of thin films, that is, edges or Plateau borders, are three-fold and symmetric with an angle of 120°.

2. All intersections of edges, that is, vertices or nodes, are four-fold and symmetric with an angle of $\cos^{-1}(1/3) \approx 109.5°$.

Note that thin films in polydisperse foam structures have nonzero mean curvature; bubbles in a dry foam are not simply polyhedra. Their curvature is intimately related to capillary pressure, discussed later. Curving thin films are apparent in

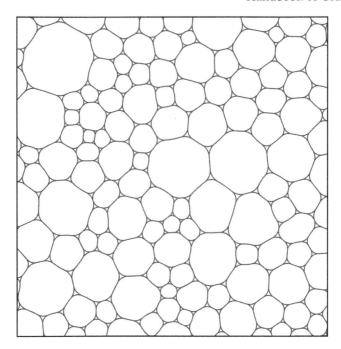

Figure 12.4   Numerically generated 2D dry foam with liquid fraction $\epsilon = 0.03$. (Image courtesy of S. Cox).

Figure 12.4, which depicts a 2D computer-generated foam with gas fraction $\phi = 0.97$. The vertices are not truly point-like because of the finite wetness of the foam.

12.1.3   Coordination

Foam topology changes with wetness. One simple measure of topology is how many adjacent bubbles, on average, each bubble has. Consider a dry 2D foam; for example, each bubble must average $Z = 6$ neighboring bubbles with which it shares a thin film. This is a consequence of Euler's theorem relating the number of bubbles N_b, the number of thin films or edges N_e, and the number of vertices N_v in a spatial tessellation

$$N_b - N_e + N_v = \chi. \tag{12.1}$$

The quantity χ, called the Euler characteristic, is equal to 1 for a non periodic tessellation. Plateau's rules require $N_e = (3/2)N_v$ in two dimensions (each edge is shared by two vertices); it follows that there are $Z = 2N_e/N_v = 6$ neighbors per bubble in the limit of large system size.

In two dimensions, it is possible to construct a slightly wet foam without changing the coordination by "decorating" the vertices of an ideal dry foam with Plateau borders containing liquid [3]. Substantial changes to the wetness, however, must also change Z. Bubbles become increasingly spherical in shape, as evident in

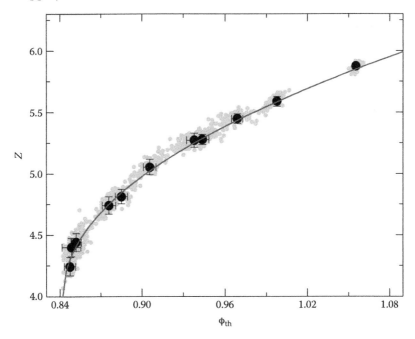

Figure 12.5 Contact number Z as a function of gas fraction in a disordered foam mono-layer. The contact number jumps to the isostatic value $Z_c = 4$ at a gas fraction of $\phi_c \approx 0.84$ and grows as a square root for higher values of ϕ. (Image from Katgert, G. and van Hecke, M., *EPL*, 92, 34002, 2010.)

the monolayer of bubbles depicted in Figure 12.1. At the critical packing frac-tion ϕ_c, the bubbles are perfectly spherical and just touch; this is the (un)jamming transition. A system above ϕ_c can elastically support a shear stress on long time scales, and is therefore a solid, while systems below ϕ_c flow under arbitrarily weak shear. The critical packing fraction is $\phi_c \approx 0.64$ in three dimensions and 0.84 in two, but the precise value depends on the polydispersity of the system.

Later in this chapter, we will investigate the macroscopic mechanical response of materials near ϕ_c. It will emerge that many properties near unjamming exhibit critical scaling, that is, they have power-law dependence on $\Delta\phi = \phi - \phi_c$. One structural quantity that displays scaling is the mean coordination. As discussed in Chapter 6, frictionless spheres at the jamming point have $Z_c = 2d$ contacts on average. Figure 12.5 shows that Z in a disordered monolayer grows above ϕ_c[*] as [15]

$$Z - 4 \propto (\phi - \phi_c)^{1/2}. \tag{12.2}$$

---

[*] The x-axis of the figure is labeled ϕ_{th} because, as is convention in simulations, the packing fraction is calculated by treating each bubble as a perfect disk. This double counts area where the disks overlap, the error is negligible near ϕ_c but permits ϕ to exceed unity in drier foams.

Equation 12.2 shows that compressing a foam or emulsion near ϕ_c creates anomalously many new contacts. If the bubbles were all to move together "smoothly" or affinely—a concept defined more precisely later—the change in coordination would grow linearly with $\phi - \phi_c$.

12.1.4 Coarsening

In addition to drainage and rupture, foams evolve due to coarsening, the process whereby gas is exchanged between neighboring bubbles. The gas flux is driven by the pressure difference between bubbles, and we show in the following section that higher pressures are found in smaller bubbles. One thus anticipates that gas will be driven from small bubbles to large, and in a coarsening foam big bubbles grow and small bubbles shrink. Von Neumann provided an elegant expression for the rate of growth of the area A_n of a bubble with n neighbors in a 2D dry foam:

$$\frac{dA_n}{dt} = \frac{2\pi}{3}\sigma\kappa(n-6),$$

(12.3)

where κ is the permeability constant of the thin films. Surprisingly, the right-hand side of Equation 12.3 depends only on the coordination of the bubble, a quantity describing topology and not geometry. There is no explicit size dependence, though as one would expect, n is positively correlated with bubble size. The area of a bubble with six sides—which is the average—will not change, while those with fewer than six neighbors shrink and those with more than six neighbors grow.

12.2 Mesoscopic Mechanics

Foam bubbles or emulsion droplets in contact experience both elastic and viscous interactions. Suppose, an interface between two fluid phases experiences a tension σ, where $\sigma\,dA$ is the energy cost of increasing the surface area of the interface by an infinitesimal amount dA. The dispersed phase can be treated, to very good approximation, as an incompressible fluid; hence, the bubbles and droplets assume the surface area–minimizing shape consistent with their boundary conditions and the constraint of fixed volume. This is why an isolated bubble is spherical. There is another related consequence of interfacial tension: an interface can only support curvature in mechanical equilibrium if there is a pressure difference Δp between the fluids on either side of the interface. Consider the surface in Figure 12.6. If the two solid arcs have radii R_1 and R_2 ("radii of curvature") and subtend angles $d\theta_1$ and $d\theta_2$, the net downward force on the surface due to tension is $\sum F_z = \sigma(R_1 + R_2)d\theta_1\,d\theta_2$. Balancing this force with $\Delta p\,dA = \Delta p\,R_1 R_2\,d\theta_1\,d\theta_2$ gives

$$\Delta p = \sigma\left(\frac{1}{R_1} + \frac{1}{R_2}\right).$$

(12.4)

This is the law of Young and Laplace. The quantity in parentheses is known as the mean curvature of the surface. The Young–Laplace law says that the pressure is

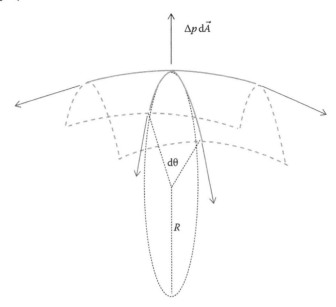

Figure 12.6   In a curved interface, surface tension generates a net force normal to the surface. This must be balanced by a pressure difference across the interface (black arrow).

set by the mean curvature and is higher on the side of the interface toward which the interface curves. The quantity Δp is referred to as the Laplace pressure or the capillary pressure.

In the context of foams and emulsions, the Young–Laplace law is often applied across the thin film separating two bubbles or droplets. Such a film has two interfaces; the continuous phase participates in both. Because of the thinness of the film, we can approximate the two interfaces as locally parallel. Labeling the bubbles/droplets i and j, their pressure difference Δp_{ij} is

$$\Delta p_{ij} = 2\sigma\left(\frac{1}{R_1} + \frac{1}{R_2}\right). \tag{12.5}$$

By considering nearly spherical bubbles ($R_1 \approx R_2$), one sees that pressure is higher in smaller bubbles. The thin film separating two bubbles will curve inward with respect to the large bubble and outward with respect to the small bubble. Also note that since the mean curvature is set uniquely by the pressure difference between the two bubbles, the mean curvature of the film separating them is every-where constant. It immediately follows that in two dimensions, where a surface has just one radius of curvature, the edges connecting Plateau borders are arcs of a circle.

12.2.1   Elastic Forces between Bubbles

In a wet foam, nearly spherical bubbles press against each other, creating small circular facets separated by thin liquid films, with both elastic and viscous inter-actions. Elastic forces in a wet foam are associated with the potential energy stored when a bubble is distorted from its initially spherical shape. Because the bubbles are incompressible, the deformed bubble assumes the surface area–minimizing shape compatible with its boundary conditions [16]. This is a nonlocal criterion, and generally, the force a bubble feels from any one of its neighbors is not independent of its other neighbors. Nevertheless, we will determine an approximate, pairwise expression for the elastic forces through an ansatz for the shape that the bubble assumes after deformation.

Consider two bubbles pressed into contact by body forces F. The force F represents the sum of body forces, such as gravity, and the net force imposed by other contacting bubbles. We assume that the deviation from a spherical shape is small and that the bubble's deformed shape is a truncated sphere [16]; see Figure 12.7a. Deformed bubbles do *not*, in fact, assume this shape; the sharp corner at the edge of the thin film, in particular, is unrealistic. Nevertheless, the anszatz suffices to give an approximate idea of the elastic interaction between bubbles of similar size.

Given the truncated sphere of Figure 12.7, the Young–Laplace law can be used to obtain an approximate relation between the imposed force and the resulting deformation. For each bubble, the body force F must be balanced by a force with magnitude $\Delta p\left(\pi r_f^2\right)$ applied by the fluid in the thin film, where $r_f \approx \sqrt{2R\delta}$ is the radius of the facet and $\delta = \delta(F)$ is the difference between the sphere radius and the distance from the bubble center to the center of the facet. The radius of the

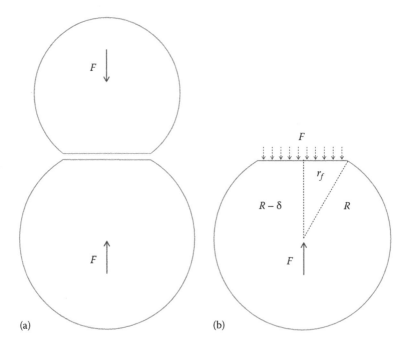

(a)   (b)

Figure 12.7 (a) A contact formed between two bubbles in a wet foam subject to body forces F. The bubbles' shapes are approximated as truncated spheres. A thin film of liquid separates two circular facets. (b) Each facet has a radius r_f and an effective overlap δ. A disjoining pressure in the thin film applies a force on the facet.

truncated sphere R is fixed by requiring that the bubble's volume remain equal to $(4/3)\pi R_0^3$, where R_0 is the undeformed radius; it thus depends on F through δ. For small deformations, the pressure difference is approximately that in the absence of the contact, $\Delta p \approx 2\sigma/R_0$. Hence,

$$F \approx \frac{4\pi\sigma R}{R_0}\delta = 4\pi\sigma\delta + O(\delta^2). \tag{12.6}$$

Equation 12.6 resembles the force law of a spring with a spring constant $k = 4\pi\sigma$ set by the interfacial tension. Therefore, the contact of Figure 12.7, which involves two bubbles, acts like two such springs connected in series. Writing the total overlap $\delta = \delta_0 + \delta_1$ as the sum of the overlaps in each bubble, the effective force law of the contact can be written as

$$F = k_{\text{eff}}\delta, \tag{12.7}$$

with $1/k_{\text{eff}} = 1/4\pi\sigma + 1/4\pi\sigma$, or

$$k_{\text{eff}} = 2\pi\sigma. \tag{12.8}$$

Equation 12.7 resembles the force law of a simple linear spring. The result is only approximate due to the truncated sphere assumption. More accurate calculations that do not presume the deformed shape *a priori* are available in the literature [16,19]. These show that Equation 12.8 overestimates the magnitude of k_{eff} and misses logarithmic corrections that cause the stiffness to vanish as $F \to 0$. Nevertheless, Equation 12.8 establishes useful intuition.

Note an important difference between bubble–bubble contacts and the grain–grain contacts of granular media: while surface tension gives rise to a linear force law (k_{eff} independent of δ), granular contacts are Hertzian, with k_{eff} proportional to $\delta^{1/2}$. Perhaps surprisingly, in the limit $\delta \to 0$ the stiffness of granular contacts is always *smaller* than the stiffness of bubble–bubble or droplet–droplet contacts, even though the relevant material parameters (Young's modulus of a grain and the Laplace pressure of a droplet or bubble) are well separated.

12.2.2 Viscous Forces between Bubbles

Dissipation mechanisms in foams and emulsions are complex, and exact expressions are available only for simple geometries such as bubble trains in a tube [4]. When adjacent bubbles are in relative motion, there is fluid flow in the thin film that separates them; see Figure 12.8. To determine the dissipation rate in the film, one must know both its shape and the form of the velocity profile. Neither is trivial. Moreover, the shape of the film is free to change in response to the flow, hence the two are coupled and the shape must be determined self-consistently with the velocity profile. The velocity profile is also sensitive to boundary conditions at the interface between dispersed and continuous phases. As noted earlier, different surfactants yield different boundary conditions.

For mobile surfactants, the interface provides little resistance to flow, there is slip at the interface, and the velocity profile resembles plug flow; see Figure 12.8a.

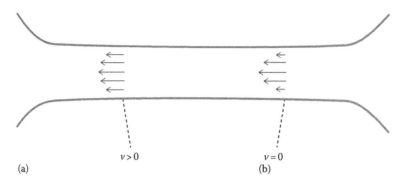

Figure 12.8   Flow of liquid in the thin film between two droplets with (a) mobile and (b) immobile surfactants. The form of the velocity profile depends on the boundary conditions at the interface.

Immobile surfactants do not permit slip, which leads to a parabolic Poiseuille profile. Velocity gradients are thus substantially larger in the presence of immobile surfactants, and there is considerable dissipation along the length of the film. In foams with mobile surfactants, by contrast, dissipation is concentrated in locations where geometry requires the continuous phase to change directions, such as edges and vertices.

A canonical problem in the physics of foams and emulsions is the drag force F_v experienced by a bubble moving with velocity v past a wall. It is customary to express F_v in terms of the dimensionless capillary number $Ca = \eta v/\sigma$, where η is the viscosity of the continuous phase. Bretherton showed [4] that for immobile surfactants the force is

$$F_v \propto (Ca)^n, \tag{12.9}$$

with $n = 2/3$. More recently, Denkov and coworkers showed $n = 1/2$ for mobile surfactants [7].

To describe dissipation in the bulk, one needs to know the viscous force law between two bubbles or droplets. Theoretical and numerical work suggests that Equation 12.9 describes drag between two adjacent bubbles (with v now representing the *relative* velocity) in periodic structures; it is likely, but not yet certain, that it also holds in disordered structures.

A significant complication, however, is that *macroscopic* response in disordered foams and emulsions does not trivially reflect the *microscopic* viscous force law between bubbles. While it is true that over some range of strain rates $\dot{\gamma}$ the shear stress is proportional to $\dot{\gamma}^\beta$, in general $\beta \neq n$. This effect is dramatically illustrated in the data of Figure 12.9a through c [14]. The first plot shows the force on a layer of bubbles with immobile surfactant under a moving wall; it displays the expected 2/3 exponent. The force between the two rows of monodisperse bubbles sheared past one another, plotted in Figure 12.9b, also displays a 2/3 exponent; this is a strong evidence that the viscous force law between two bubbles has the same exponent as that between a bubble and a wall. However, when a disordered collection

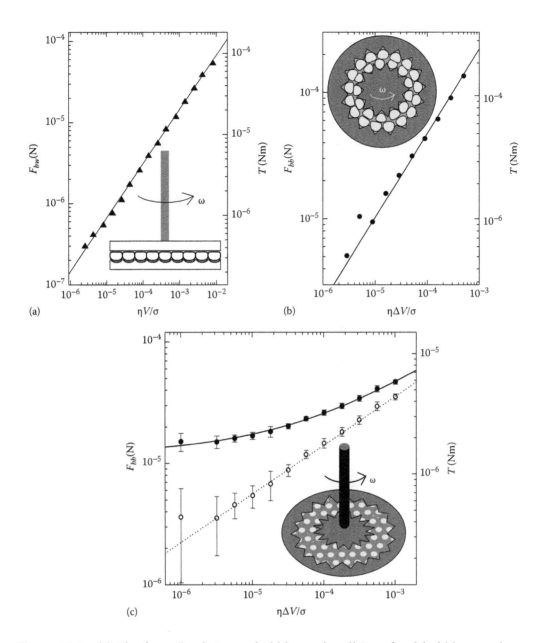

Figure 12.9 (a) The force F_{bw} between bubbles and wall in a fixed bubble monolayer sheared under a smooth glass plate. The solid line has slope 2/3. (b) The force F_{bb} between two ordered lanes of bubbles. The solid line has slope 2/3. (c) The force F_{bb} between adjacent bubbles in a disordered monolayer. The dashed line has slope 0.4. All data are plotted versus capillary number. (Adapted from Katgert, G. Et al., *Phys. Rev. Lett.*, 101, 058301, 2008.)

of the same bubbles is sheared (Figure 12.9c), the force displays a plateau at low capillary number and an exponent $\beta \approx 0.4$ at higher Ca, anomalous behavior that can be associated with the jamming transition.

12.3 Macroscopic Response

Above a critical packing fraction ϕ_c, foams and emulsions "jam" into a solid phase that can sustain finite shear stresses under shear strain. We will model the macroscopic mechanical response of foams and emulsions above, but still near, this rigidity transition within the context of Durian's "bubble model" for wet foams [8].

12.3.1 The Bubble Model

The bubble model is defined on the mesoscale; its constituent elements are perfectly spherical bubbles. Bubbles in contact exchange pairwise elastic and viscous forces. Two bubbles are said to be in contact whenever they overlap, that is, whenever

$$\delta_{ij} = R_i + R_j - |\vec{r}_i - \vec{r}_j| > 0; \tag{12.10}$$

see Figure 12.10. When they are in contact, they experience an elastic force linear in the overlap,

$$\vec{f}_{ij}^{\text{el}} = -k_{ij}\,\delta_{ij}\,\hat{r}_{ij}. \tag{12.11}$$

The spherical cap model described earlier gives $k = 2\pi\sigma$. As noted earlier, this approximation overestimates the spring constant between two bubbles of similar

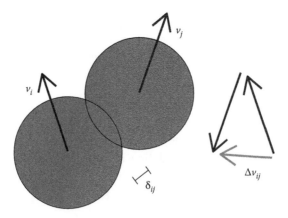

Figure 12.10 The bubble model for wet foams. Bubbles are treated as disks or spheres that exchange elastic forces proportional to their overlap δ_{ij}. Viscous forces oppose their relative velocity $\Delta\vec{v}_{ij}$.

size; it also misses the fact that bubbles become softer with increasing size. Thus, a more common choice is [8]

$$k_{ij} \propto \frac{\sigma d}{R_i + R_j},\tag{12.12}$$

where d is the mean bubble size. In the following discussion, the difference between these expressions will not play any role. Largely for reasons of convenience, in the bubble model the nonlinear viscous force law of Equation 12.9 is replaced with a linear drag,

$$\vec{f}_{ij}^{\text{visc}} = -b\,\Delta\vec{v}_{ij},\tag{12.13}$$

where

  b is a viscosity
  Δv_{ij} is the relative velocity between bubbles i and j, evaluated at the contact

For slowly driven systems, the effect of inertia can be neglected; equivalently, the bubbles', masses can be set to zero. The dynamics are then termed overdamped, and the system is described by a coupled set of first-order differential equations,

$$\sum_j \left[\vec{f}_{ij}^{\text{el}} + \vec{f}_{ij}^{\text{visc}}\right] = \vec{0}.\tag{12.14}$$

This amounts to a statement of force balance. Torque balance can also be incorporated by allowing the bubbles to rotate.

12.3.2 Quasistatic Deformation

A fundamental characterization of any solid is its response to shear. When a volume V is slowly sheared through a strain γ (see Figure 12.11), dissipation can be

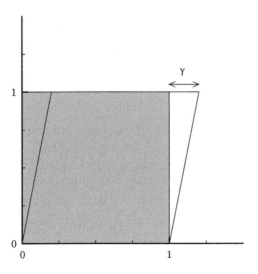

Figure 12.11   Simple shear of a unit square.

neglected and the material resides at every moment at a local minimum of the elastic potential energy U. This is quasistatic deformation. The work done on the material is simply the change in potential energy $\Delta U = U' - U$ during the shear, which defines the shear modulus G_0

$$\frac{\Delta U}{V} = \frac{1}{2}G_0\gamma^2. \tag{12.15}$$

The shear stress σ, not to be confused with interfacial tension, is given by the first derivative of the energy density $\Delta U/V$ with respect to strain

$$\sigma = \frac{\partial}{\partial\gamma}\left(\frac{\Delta U}{V}\right) = G_0\gamma. \tag{12.16}$$

Note the similarity of this relation to the more familiar Hooke's law, which relates the force on a spring $f = k_s x$ to its extension x. Just as a spring constant describes the stiffness of a spring, the shear modulus G_0 describes the stiffness of a material with respect to shear deformations.

The bubble model treats foams and emulsions as disordered collections of soft, frictionless spheres. Chapter 6 discusses the structure and mechanical response of frictionless sphere packings, which correspond to the bubble model (1) in the absence of driving or (2) subject to oscillatory driving in the limit where the driving frequency is vanishingly small. This latter case is referred to as quasistatic driving. Near the critical packing fraction ϕ_c, the quasistatic shear modulus of the bubble model is

$$G_0 \propto k\,\Delta\phi^{1/2}, \tag{12.17}$$

where

   $\Delta\phi = \phi - \phi_c$ is the distance to the critical packing fraction
   k is the (typical) effective spring constant of a contact, proportional to the Laplace pressure

Foams and emulsions soften as their packing fraction decreases—their shear modulus gets smaller and ultimately vanishes continuously at the critical packing fraction ϕ_c. Hence, the *collective* response of a foam or emulsion can be much softer than its individual contacts.

Our goal in the present section is to explore the energetics of quasistatic deformations. We will relate the energy of deformation, and hence the shear modulus, to the highly inhomogeneous spatial character of the bubbles' displacement field (Figure 12.12). Later, we will build on these observations to describe *viscoelastic* response in the bubble model, which includes driving at finite frequency.

The shear modulus relates stress and strain. These are macroscopic quantities used in coarse-grained or continuum descriptions of materials. The bubble model takes a mesoscopic perspective, viewing the material as a collection of individual bubbles; the shear modulus is related to the bubbles' displacements via Equation 12.15. To develop this relationship, the elastic energy stored by the deformation,

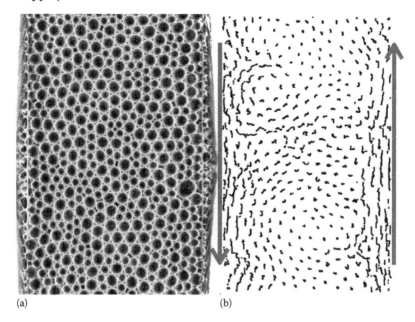

(a)   (b)

Figure 12.12   (a) A monolayer of bubbles undergoing steady shear. (b) Tracks of the the same bubbles over a fixed time interval. Arrows indicate the shearing direction. (Image courtesy M. Möbius.)

ΔU, is first expressed in terms of, and then expanded in powers of, the displacements $\{\vec{u}_i\}_{i=1...N}$ of the N bubbles in the system. This harmonic approximation is familiar from condensed matter physics, where it is regularly applied to periodic structures [5]. However, periodicity is not a requirement; the method can be fruitfully applied to foams, emulsions, grains, glasses, and other disordered materials.[*]

In the bubble model, pairs of bubbles in contact act like springs. The energy stored in the spring is

$$\epsilon_{ij} = \frac{1}{2} k_{ij} \delta_{ij}. \tag{12.18}$$

All contacts are assumed to have the same spring constant k. The effective overlap between bubbles δ_{ij} is zero when the bubbles just touch, at which point the spring is at its rest length and stores no energy. The elastic potential energy of a packing before it is sheared is thus

$$U = \sum_{\langle ij \rangle} \epsilon_{ij}^0 = \frac{1}{2} k \sum_{\langle ij \rangle} (\delta_{ij}^0)^2, \tag{12.19}$$

where the sum runs over all the pairs of particles in contact and the superscript 0 indicates quantities in the packing's initial configuration. When a packing is at

---

[*] The work of Alexander [1] eloquently advocates this approach.

ϕ_c and the bubbles just touch, the initial compression $\delta_{ij}^0 = 0$ at every contact. For denser packings $\delta_{ij}^0 > 0$ is finite and, intuitively, its mean or typical value δ^0 is proportional to $\Delta\phi$:

$$\frac{\delta^0}{d} \propto \Delta\phi. \tag{12.20}$$

Consider a shear strain γ. In a periodic system, shear can be imposed by "tilting" the packing's unit cell, as illustrated in Figure 12.11 for a 2D system. If the lattice vectors of the unit cell are $\vec{L}_1 = (L, 0)$ and $\vec{L}_2 = (0, L)$ prior to the shear, then after a "simple shear" they are $\vec{L}_\alpha' = (\mathbf{I} + \hat{\mathbf{D}})\vec{L}$, where

$$\hat{D} = \begin{pmatrix} 0 & \gamma \\ 0 & 0 \end{pmatrix}. \tag{12.21}$$

The term "simple shear" distinguishes Equation 12.21 from "pure shear," where the deformation gradient $\hat{D}$ is symmetric and therefore has no rotational component. In response to the shear, each bubble i will displace some amount $\vec{u}_i$. Examples of the displacement field for two different distances to unjamming are shown in Figure 12.13b and d. A striking feature of the displacements close to unjamming is their spatially disordered character; they do not smoothly follow the imposed deformation, and the similarity to the experimental system of Figure 12.12 is apparent. For comparison, we call a particle displacement *affine* if it follows the deformation gradient, that is, if a particle initially at $\vec{r}_i$ displaces by $\vec{u}_i^{\text{aff}} \equiv \hat{D}\vec{r}_i$. Hence, the displacements in Figure 12.13d have a large non affine component, while the displacements in the denser packing of Figure 12.13b have a smaller non affine contribution. We will now show that this disordered, non affine motion goes hand-in-hand with packings' increasing softness as they approach unjamming.

Assume that the imposed shear strain γ is small enough that the bubbles initially in contact remain in contact and also that no new contacts are formed.[*] The displacements $\vec{u}_i$ and $\vec{u}_j$ therefore change only the magnitude of the contact between bubbles i and j. In the initial pre-shear state, the bubbles exchange a force $\vec{f}_{ij}^0 = -\vec{f}_{ji}^0 = -k\delta_{ij}^0 \hat{n}_{ij}$, where $\hat{n}_{ij}$ is the normal vector pointing from the center of j to i. The displacements of the bubbles can change both the overlap δ and the orientation of the contact $\hat{n}$. The change in energy of the contact is

$$\delta\epsilon_{ij} = \frac{1}{2}k\left[\delta_{ij}^2 - \left(\delta_{ij}^0\right)^2\right]. \tag{12.22}$$

---

[*] One might call this "strong linear response," linearizing the equations of motion around a static packing of bubbles in mechanical equilibrium, which is convenient for calculations. A weaker, experimentally verifiable requirement would be linear superposition of the response.

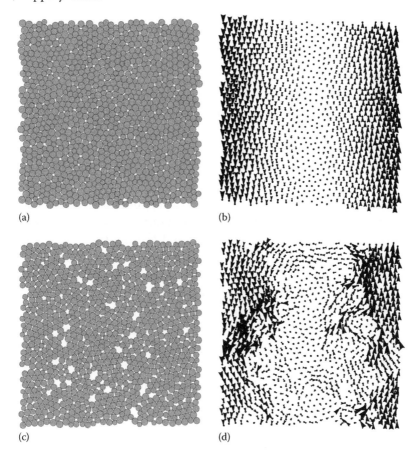

Figure 12.13 (a,c) Two numerically generated foams in the bubble model; that in (c) is closer to unjamming. (b,d) Bubble displacements in the foams of (a,c) under shear. The displacements of (d) are clearly more strongly non affine than those in (b).

By Taylor expanding and keeping only leading order terms in the displacements, we find

$$\delta\epsilon_{ij} = \frac{1}{2}k_{ij}^{\parallel}\left(\Delta u_{ij}^{\parallel}\right)^2 - \frac{1}{2}k_{ij}^{\perp}\left(\Delta u_{ij}^{\perp}\right)^2, \tag{12.23}$$

where for convenience we have introduced the effective spring constants

$$\begin{aligned} k_{ij}^{\parallel} &= k \\ k_{ij}^{\perp} &= \frac{\delta_{ij}^0}{\Delta r_{ij}^0}k, \end{aligned} \tag{12.24}$$

the relative normal motion $\Delta\vec{u}_{ij}^{\parallel}$, and the relative tangential motion $\Delta\vec{u}_{ij}^{\perp}$,

$$\begin{aligned} \Delta\vec{u}_{ij}^{\parallel} &= (\vec{u}_j - \vec{u}_i)\cdot\hat{n}_{ij} \\ \Delta\vec{u}_{ij}^{\perp} &= \vec{u}_j - \vec{u}_i - \Delta\vec{u}_{ij}^{\parallel}. \end{aligned} \tag{12.25}$$

This naming convention is standard but can be confusing; note that $\Delta u^{\parallel}$ is parallel to the vector $\hat{n}$ that points from one bubble to another, while $\Delta u^{\perp}$ is perpendicular to $\hat{n}$, hence parallel to the interface at the contact. Under the convention of Equation 12.23, normal motion ($\Delta u^{\parallel}$) stretches or compresses a spring, while tangential motion ($\Delta u^{\perp}$) occurs when two bubbles slide past each other.

Considering separately the two terms in the energy expansion of Equation 12.23 explains qualitatively why the bubble displacements in Figure 12.13d are so disordered [1,9]. The first term is the usual expression for the energy required to stretch or compress a spring by a distance $\Delta u^{\parallel}$. The second term, because it is negative, is destabilizing: tangential motion can *lower* the total energy. Moreover, the effective tangential stiffness $k^{\perp}$ is proportional to the overlap in the pre-sheared state. Bubbles have very small overlaps close to the critical packing fraction, so the tangential stiffness is much smaller than the bare spring constant, $k^{\perp}/k \ll 1$. Thus, the system can easily lower its elastic energy by having particles slide past each other. Because the stiffness $k^{\perp}$ is small, the tangential motion $\Delta u^{\perp}$ will tend to be large compared to the normal motions. And as the contact normals are randomly oriented, bubbles performing sliding motion will not necessarily be following the deformation gradient. Therefore, non affinity reflects a competition between the globally imposed deformation $\hat{D}$ and a local preference for sliding motion.

Let us now quantify the earlier observation. The energy of a quasistatic shear deformation satisfies

$$\frac{1}{2}G\gamma^2 = \frac{1}{2V}\sum_{\langle ij \rangle}\left[\frac{1}{2}k_{ij}^{\parallel}\left(\Delta u_{ij}^{\parallel}\right)^2 - \frac{1}{2}k_{ij}^{\perp}\left(\Delta u_{ij}^{\perp}\right)^2\right]. \tag{12.26}$$

This expression relates the macroscopic shear modulus to the microscopic (relative) motion of the bubbles. We now drop subscripts and prefactors and speak only of the *typical* normal and tangential motions, $\Delta u^{\parallel}$ and $\Delta u^{\perp}$. When the packing is sheared, it responds so as to minimize the elastic energy ΔU. Assuming the two terms in the energy expansion are of similar order [9], it then follows that (in three dimensions)

$$G\gamma^2 \sim \frac{k}{d}\left(\frac{\Delta u^{\parallel}}{d}\right)^2 \sim \frac{k\delta^0}{d^2}\left(\frac{\Delta u^{\perp}}{d}\right)^2. \tag{12.27}$$

Using the fact that $G_0 \propto k\Delta\phi^{1/2}$ and $\delta^0 \propto d\Delta\phi$, we have [9]

$$\Delta u^{\parallel} \sim \Delta\phi^{1/4}d\gamma$$

$$\Delta u^{\perp} \sim \frac{1}{\Delta\phi^{1/4}}d\gamma. \tag{12.28}$$

These relations confirm the qualitative picture intuited earlier. As the foam or emulsion gets closer to the critical packing fraction ϕ_c, the typical relative normal motion vanishes, while the typical tangential or sliding motion actually diverges. This explains the strongly non affine motion evident in Figures 12.12 and 12.13d: it is a direct consequence of the material's softening on approach to the critical packing fraction.

12.3.2.1   Storage and Loss

The static shear modulus G_0 describes a material's stiffness against shear. As foams and emulsions approach the unjamming point, that is, the critical packing fraction ϕ_c or, equivalently, zero confining pressure p, their shear modulus becomes increasingly smaller. Thus, foams and emulsions near (un)jamming are soft for two reasons: first, because the Laplace pressure is much smaller than the elastic moduli of conventional solids, and second, because of the collectively non affine motion of the bubbles in response to shear.

We now consider shear at finite frequency ω. In response to such shear, the bubbles' velocities will also be finite, and therefore, viscous forces will cause the system to dissipate energy. We will consider two sorts of driving at finite rate: oscillatory driving at frequency ω and steady shear flow at strain rate $\dot{\gamma}$. One of our goals will be to explain the appearance of *shear thinning*, a phenomenon in many complex fluids whereby increasing the driving rate decreases the material's resistance to flow. It is familiar to anyone who has tried to increase the flow rate of ketchup from a bottle: tapping the bottle can speed up the flow, but tapping a little bit harder produces a dramatic increase in the rate of flow and soggy French fries.

To characterize the response of foams and emulsions to driving at finite frequency, one can extend the concept of a shear modulus to incorporate frequency dependence. Consider a system subjected to an imposed shear strain with amplitude γ

$$\gamma(t) = \gamma \cos \omega t. \tag{12.29}$$

In response to this strain, the material will develop a shear stress $\sigma(t)$. In linear response, the stress will also be sinusoidal with frequency ω, but it need not be in phase with $\gamma(t)$

$$\begin{aligned}\sigma(t) &= \sigma(\omega)\cos(\omega t + \delta(\omega)) \\ &= [\sigma(\omega)\cos\delta(\omega)]\cos\omega t - [\sigma(\omega)\sin\delta(\omega)]\sin\omega t.\end{aligned} \tag{12.30}$$

Here,
  $\delta(\omega)$ is a phase shift, not to be confused with the overlap between bubbles δ_{ij}
  $\sigma(\omega)$ is the (as-yet-undetermined) frequency-dependent stress amplitude

The phase shift between the stress and strain is directly analogous to that found in a driven, damped harmonic oscillator. In the absence of damping, a Hookean spring oscillates in phase with the driving. (That is, imposing a sinusoidal driving force $f(t) = f_0 \cos \omega t$ results in a displacement $x(t) = x_0 \cos \omega t$ in phase with the driving.) If the spring is damped (e.g., by immersing it within a Newtonian fluid with viscosity, a mechanical element known as a "dashpot"), the force is now out of phase with the displacement. In the quasistatic limit the force is proportional to the velocity, $f = b_f \dot{x}$, which is $\pi/2$ out of phase with the displacement. $f(t) = b_f \omega x_0 \cos(\omega t - \pi/2)$.

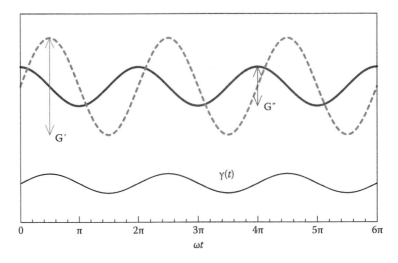

Figure 12.14 Oscillatory driving at frequency ω. When a material is driven by a sinusoidal strain $\gamma(t)$ (thin curve), the stress is also sinusoidal and has in-phase and out-of-phase contributions (thick curves). Here, the stress is vertically offset for clarity. The amplitude of the in-phase contribution (thick dashed curve) is the storage modulus $G'(\omega)$, while that of the out-of-phase contribution (thick solid curve) is the loss modulus $G''(\omega)$.

Real materials, which can store energy elastically *and* dissipate energy viscously, lie somewhere between a spring and a dashpot.[*] For sufficiently small strain amplitude, we expect to find a regime of linear response in which the stress amplitude is linearly proportional to the strain amplitude, $\sigma(\omega) \propto \gamma$. In this linear regime, the macroscopic response can be characterized by the storage modulus $G'(\omega)$ and loss modulus $G''(\omega)$,

$$G'(\omega) = \frac{\sigma(\omega)\cos\delta(\omega)}{\gamma} \text{ (Figure 12.14)},\tag{12.31}$$

and

$$G''(\omega) = \frac{\sigma(\omega)\sin\delta(\omega)}{\gamma},\tag{12.32}$$

which describe the in-phase and out-of-phase components of the stress response, respectively.

Using complex notation, it is conventional to write the strain as

$$\gamma(t) = \gamma e^{\iota\omega t},\tag{12.33}$$

---

[*] In fact, it is common to model viscoelastic materials with "circuit diagrams" composed of springs and dashpots. The simplest materials combining elastic storage and viscous loss are (1) a spring and a dashpot in series and (2) a spring and a dashpot in parallel. The former (the Maxwell fluid) is a fluid because for low frequencies it can be deformed without storing energy. The latter (the Kelvin–Voigt solid) is a solid because the spring continues to be compressed even as $\omega \to 0$.

and the stress as

$$\sigma(t) = G^*(\omega)\gamma(t), \tag{12.34}$$

where the complex shear modulus $G^*(\omega)$ is

$$G^*(\omega) = \frac{\sigma(\omega)}{\gamma} = G'(\omega) + \imath G''(\omega). \tag{12.35}$$

Complex notation makes it apparent that $G^*(\omega)$ is the generalization of the shear modulus to finite frequency response. G' and G'' describe the elastic and viscous contributions to the stress, respectively. Note that the strain rate $\dot{\gamma}$ is, again in complex notation,

$$\dot{\gamma}(t) = \imath \omega \gamma(t). \tag{12.36}$$

One may also define a complex viscosity $\eta^*(\omega)$ such that

$$\sigma(t) = \eta^*(\omega)\dot{\gamma}(t), \tag{12.37}$$

or

$$\eta^*(\omega) = \frac{G^*(\omega)}{\imath \omega}. \tag{12.38}$$

To illustrate the complex shear modulus in foams, Figure 12.15a shows experimental data for G^* in a coarsening foam for different sample ages [6]. The complex shear modulus evolves as the material coarsens. The low-frequency plateau in the loss modulus reflects the foam's structural evolution due to coarsening; absent such non steady state effects, the loss modulus must be odd in ω, while the storage modulus must be even. Figure 12.15b shows G^* in the bubble model in a broader frequency window for various values of the excess contact number Δz (recall that $\Delta z \propto \Delta\phi^{1/2}$) [25]. Coarsening is not included in the bubble model, so there is no evolution with sample age. There is, however, clear dependence on proximity to the unjamming transition—as the material approaches unjamming G' shifts downward while G'' shifts up. In solids, the storage modulus is larger than the loss modulus at low frequencies; hence, the shift in G^* presages the loss of rigidity at unjamming. Both the experimental and numerical system display a characteristic $\omega^{1/2}$ frequency dependence with increasing ω [10,17]. We will show that this behavior is also associated with the jamming transition.

Oscillatory Driving in the Bubble Model

Consider the bubble model under oscillatory driving (at finite frequency). When foams and emulsions are driven at a sufficiently high rate, the quasistatic limit breaks down. In qualitative terms, this happens when viscous forces are large enough that bubbles' displacements no longer minimize the elastic potential energy ΔU. (Figure 12.15).

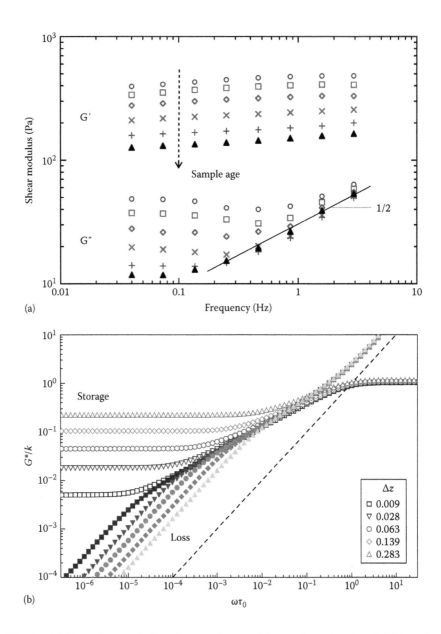

Figure 12.15 (a) Experimental data from a foam with gas fraction $\phi = 0.926$ undergoing oscillatory shear. Both the storage and loss modulus decrease due to coarsening as the sample ages. (From Cohen-Addad, S. et al., *Phys. Rev. E*, 57, 6897, 1988.) (b) Storage and loss modulus in the bubble model for varying excess coordination Δz. The system does not coarsen. The dashed line has slope 1. (Adapted from Tighe, B.P., *Phys. Rev. Lett.*, 107, 158303, 2011.)   (*Continued*)

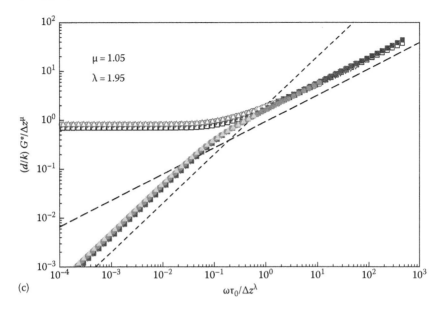

(c)

Figure 12.15 (*Continued*)   (c) Critical scaling collapse of the data in Figure 12.15b. Note that the rescaling is expressed here in terms of excess coordination $\Delta z \propto \Delta\phi^{1/2}$, hence the expected critical exponents are $\mu = 1$ and $\lambda = 2$. The dashed lines have slopes of 1 and μ/λ. High-frequency data ($\omega\tau_0 > 0.3$) are outside the critical scaling regime and not plotted. (Adapted from Tighe, B.P., *Phys. Rev. Lett.*, 107, 158303, 2011.)

We wish to characterize the dissipation due to the viscous forces in a packing, which are modeled as a force linear in relative velocity. Macroscopically, the rate P at which energy is dissipated is related to the loss modulus via

$$\frac{P}{V} = \frac{1}{2}[\eta'(\omega)]^2\dot{\gamma}^2. \tag{12.39}$$

In oscillatory shear $\dot{\gamma} = i\omega\gamma$ and taking the magnitude of both the loss modulus η' and $\dot{\gamma}$, we have

$$\frac{P}{V} = \frac{1}{2}\left[\frac{G''(\omega)}{\omega}\right](\omega\gamma)^2. \tag{12.40}$$

G'' can be related to the motions of individual bubbles, which dissipate energy as they slide around each other. A pair of bubbles in contact dissipates energy at a rate $\vec{f}_{ij}^{\text{visc}} \cdot \Delta\vec{v}_{ij}$; the total energy dissipated is the sum over all bubbles in contact. Equating this sum with the expression given earlier yields

$$G''(\omega) = \frac{b}{\omega V}\sum_{\langle ij\rangle}\left\{\left[\frac{\Delta u_{ij}^{\parallel}(\omega)}{\gamma}\right]^2 + \left[\frac{\Delta u_{ij}^{\perp}(\omega)}{\gamma}\right]^2\right\}. \tag{12.41}$$

In the low-frequency limit, the bubble displacements are, to very good approximation, the same as under quasistatic driving, with the magnitude of the typical sliding motion much larger than that of the typical normal motion. The normal motion can, therefore, be neglected and, defining the zero-frequency limit of the loss modulus to be

$$\eta_0 \omega = \lim_{\omega \to 0} G''(\omega), \tag{12.42}$$

where η_0 is called the **dynamic viscosity**, we have

$$\eta_0 \sim \frac{b}{d} \left(\frac{\lim_{\omega \to 0} \Delta u^\perp(\omega)}{\gamma d} \right)^2 \sim \frac{b}{d} \frac{1}{\Delta \phi^{1/2}}, \tag{12.43}$$

Thus, the diverging amplitude of relative sliding motion sets the dynamic viscosity near unjamming [9,25,29].

To summarize, in the low-frequency limit $\omega \to 0$, the storage modulus of the bubble model approaches its quasistatic value,

$$G'(\omega) \simeq G_0 \sim \frac{k}{d} \Delta \phi^{1/2}, \tag{12.44}$$

while the loss modulus is

$$G''(\omega) \simeq \eta_0 \omega \sim \frac{b}{d} \frac{\omega}{\Delta \phi^{1/2}}. \tag{12.45}$$

Both of the expressions given earlier hold when driving is sufficiently slow that the bubbles follow their quasistatic trajectories. For this to be true, the influence of dissipation must be weak. This observation allows us to infer a self-consistent criterion for the breakdown of the quasistatic regime: quasistatic rheology cannot hold above the frequency scale ω^* such that the energy scales associated with dissipation and storage are comparable. These energy scales can themselves be estimated from the quasistatic shear modulus and dynamic viscosity. Recalling that $G^*(\omega)$ has units of stress or, equivalently, energy per unit volume, ω^* must be

$$\omega^* = \frac{G_0}{\eta_0} \sim \frac{\Delta \phi}{\tau_0}, \tag{12.46}$$

where $\tau_0 = b/k$ is a microscopic timescale; it describes, for example, the exponential relaxation of two overlapping bubbles.

Equations 12.44 and 12.45 must break down for frequencies $\omega \gtrsim \omega^*$, with the storage and loss moduli crossover from their quasistatic form to some other, as-yet-undetermined form. ω^* is referred to as the *crossover frequency*.[*]

Critical Finite Frequency Response

We will determine the form of the response above ω^* by means of a *critical scaling ansatz*, making certain well-motivated assumptions about the forms of the storage

---

[*] The crossover frequency ω^* is related to, but distinct from, the crossover frequency found in undamped sphere packings near jamming.

and loss moduli and deducing the necessary consequences of those assumptions. The result is a set of predictions for the form of $G^*(\omega)$ near unjamming that correctly reproduces more detailed calculations [25]. Critical scaling can, in practice, feel like handwaving and trickery, but it provides a deeply insightful perspective on the system's response.

According to the arguments outlined earlier, we have

$$\frac{d}{k} G'(\omega) \sim \Delta\phi^{1/2} \quad \omega \lesssim \omega^*. \tag{12.47}$$

Below ω^*, the storage modulus $G'(\omega)$ is (1) independent of frequency (to leading order) and (2) vanishes as the square root of the distance to jamming (Equation 12.44), while the dynamic viscosity is (1) independent of frequency and (2) diverges as the square root of $\Delta\phi$ (Equation 12.45). The question is: how do these functions depend on ω and $\Delta\phi$ above the crossover frequency?

We assume a power-law dependence on ω and $\Delta\phi$ for both

$$\frac{d}{k} G'(\omega) \sim \Delta\phi^q (\omega\tau_0)^\Delta \quad \omega \gtrsim \omega^*, \tag{12.48}$$

$$\frac{d}{k} G''(\omega) \sim \Delta\phi^{q'} (\omega\tau_0)^{\Delta'} \quad \omega \gtrsim \omega^*. \tag{12.49}$$

While we do not yet know the exponents q, q', Δ, and Δ', we note an important property of the crossover frequency ω^*: it vanishes at the unjamming transition. Since the quasistatic regime resides at frequencies *lower* than ω^*, this means that, at the unjamming transition, there is no quasistatic response! It also means that Equations 12.48 and 12.49 describe the response down to zero frequency *even at unjamming* ($\Delta\phi \to 0$). For this reason, these relations are sometimes said to describe the "critical regime" of response.

We can now determine the exponents from plausibility arguments. At unjamming the finite-frequency response must not dissipate infinite energy, nor should deformation be possible without storing energy. Put another way, the loss modulus must remain finite and the storage modulus cannot vanish as $\Delta\phi \to 0$, and so

$$q = q' = 0, \tag{12.50}$$

that is, G' and G'' are independent of $\Delta\phi$ in the critical regime.

The exponents Δ and Δ' are determined by requiring that, for $\Delta\phi > 0$, the critical regime scaling (Equations 12.48 and 12.49) smoothly crosses over to the quasistatic regime scaling (Equations 12.44 and 12.45) as ω is lowered through ω^*. These requirements can be expressed as

$$(G' \text{ in the QS regime at } \omega^*) \quad \Delta\phi^{1/2} \sim (\omega^*\tau_0)^\Delta \quad (G' \text{ in the critical regime at } \omega^*), \tag{12.51}$$

and

$$(G'' \text{ in the QS regime at } \omega^*) \quad \frac{\omega}{\Delta\phi^{1/2}} \sim (\omega^*\tau_0)^{\Delta'} \quad (G'' \text{ in the critical regime at } \omega^*). \tag{12.52}$$

Recalling that $\omega^* \sim \Delta\phi/\tau_0$, we find

$$\Delta = \frac{1}{2} = \Delta'. \tag{12.53}$$

Thus, the critical scaling ansatz implies that

$$G'(\omega) \sim \frac{k}{d}\sqrt{\omega\tau_0}, \tag{12.54}$$

and

$$G''(\omega) \sim \frac{k}{d}\sqrt{\omega\tau_0} \tag{12.55}$$

in the critical regime.

We can now test our critical scaling ansatz. Note that, according to the earlier discussion, the form of the complex shear modulus can be written compactly as

$$\frac{G^*(\omega,\Delta\phi)}{k\,\Delta\phi^\mu/d} = \mathcal{G}^*\left(\frac{\omega\tau_0}{\Delta\phi^\lambda}\right), \tag{12.56}$$

with $\mu = 1/2$ and $\lambda = 1$. $\mathcal{G}(x) = \mathcal{G}'(x) + \iota\mathcal{G}''(x)$ is a dimensionless function with two distinct scaling regimes

$$\mathcal{G}'(x) \sim \begin{cases} \text{const} & x \ll 1 \\ x^\Delta & x \gg 1. \end{cases} \tag{12.57}$$

and

$$\mathcal{G}''(x) \sim \begin{cases} x & x \ll 1 \\ x^\Delta & x \gg 1. \end{cases} \tag{12.58}$$

While the storage modulus is a function of both ω and $\Delta\phi$, the quantity $G/\Delta\phi^\mu$ depends only on the ratio $\omega/\Delta\phi^\lambda$. The physically important quantity is not the absolute frequency, but its magnitude compared to ω^*, which scales as $\Delta\phi^\lambda$; this determines the response regime.

An important consequence of Equation 12.56 is that the quantities on the left-hand side of each relation can be plotted versus $\omega\tau_0/\Delta\phi^\lambda$, and all of the data—not only for different frequencies, but also for different packing fractions—will all lie on a single master curve given by $\mathcal{G}'(x)$ or $\mathcal{G}''(x)$. This *scaling collapse* is demonstrated in Figure 12.15(c). In the present context, where we have already determined the values of the exponents μ, μ', Δ, Δ', and λ, scaling collapse serves to validate our predictions. However, it often arises that we expect a set of experimental or numerical data to be related via critical scaling, but do not know the critical exponents. In this situation, scaling collapse can estimate their values. To do this, a scaling function such as $\mathcal{G}'$ or $\mathcal{G}''$ is assumed, and the data are plotted accordingly. A numerical fit for best collapse determines the optimal values for the critical exponents. Note: the value of the critical point ϕ_c may also be a parameter to be determined. For example, one could plot $G'd/k(\phi - \phi_c)^\mu$ versus $\omega\tau_0/(\phi - \phi_c)^\lambda$ and vary μ, λ, and ϕ_c until the data collapse is optimized. Note that Δ is a property of the scaling function and must be equal to μ/λ and can be omitted from the parameters to be varied.

There are two important consequences of scaling near unjamming, in particular, the $\omega^{1/2}$ scaling of the shear modulus in the critical regime Equations 12.54 and 12.55. First, even though the viscous forces in the bubble model are linear, the macroscopic stresses have a nonlinear frequency dependence. This is a collective effect that cannot be explained by considering the dynamics of a small number of bubbles. Conversely, the rate dependence of macroscopic response measurements do not necessarily reflect the rate dependence of the microscopic interactions between bubbles. Second, the frequency-dependent storage and loss moduli of Equations 12.54 and 12.55 also show that wet foams are shear thinning. In the critical regime, the amplitude of the complex viscosity $|\eta^*(\omega)| = |G^*(\omega)|/\omega$ is

$$|\eta^*(\omega)| \sim \frac{b}{d}\frac{1}{\sqrt{\omega\tau_0}}. \tag{12.59}$$

This decreases with increasing frequency—we can quadruple the driving frequency ω, and the stress amplitude will only double. This sublinear growth of stress with driving rate is shear thinning[*] and it emerges in the bubble model as a critical effect associated with proximity to the unjamming transition [25].

12.3.2.2 Steady Flow

We now turn to steady flow at finite strain rate $\dot{\gamma}$, for which the strain amplitude grows linearly in time and thus is not small. In steady flow, the bubble model is not only viscous and elastic but also *plastic*, because the network of contacts between bubbles is continuously evolving. The assumptions of quasistaticity and infinitesimal deformations (linear response) of the previous sections no longer apply. Our discussion will be limited to the introduction of the empirical Herschel–Bulkley "flow law," which gives shear stress as a function of strain rate in steady state. Such a functional relationship is known as a constitutive relation or rheological curve and provides a surprisingly reasonable description of the steady flow of a number of complex fluids, including foams, emulsions, and soft colloidal suspensions. We conclude the chapter by revisiting the concept of critical scaling in the context of steady flow.

The Herschel–Bulkley law is

$$\sigma = \sigma_y[1 + (\dot{\gamma}\tau_x)^\beta]. \tag{12.60}$$

---

[*] Technically, the term "shear thinning" is reserved for steady flow, where the inverse timescale ω is replaced by the strain rate $\dot{\gamma}$. A system is then shear thinning whenever η decreases under increasing $\dot{\gamma}$.

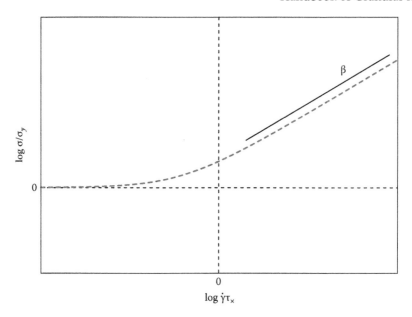

Figure 12.16   The Herschel–Bulkley flow law features a plateau at low strain rates giving way to a power law $\dot{\gamma}^{\beta}$ for $\dot{\gamma}\tau_{\times} > 1$.

The constitutive relation is plotted on a log–log scale in Figure 12.16. There are three fitting parameters: σ_y, $\tau_{\times}$, and the exponent β. The yield stress σ_y dominates the stress at low strain rates, that is, $\sigma \simeq \sigma_y$. It represents the minimum stress needed to sustain steady flow; a nonzero value reflects the fact that the material deforms elastically, instead of flowing, for small stresses. We anticipate that the yield stress will depend on packing fraction ϕ and will vanish at ϕ_c, where the material no longer displays elastic response. The parameter $\tau_{\times}$ is a crossover time scale, reminiscent of the crossover frequency ω^* in oscillatory response. For strain rates that are fast compared to $1/\tau_{\times}$, the yield stress no longer dominates the stress. Instead, the stress grows nonlinearly with the strain rate, $\sigma \sim \dot{\gamma}^{\beta}$. The exponent β is called the Herschel–Bulkley exponent and is less than one in many materials.

For several specific parameter values, the Herschel–Bulkley law reduces to other named constitutive relations. The case $\beta = 1$ is known as the Bingham law. The case $\sigma_y = 0$ is called a power-law fluid. The case where both $\beta = 1$ and $\sigma_y = 0$ is, of course, a Newtonian fluid.

Critical Scaling in Steady Flow

Because oscillatory response in the bubble model displays dynamic critical scaling, one might expect to find the same in steady flow. As noted earlier, the yield stress must vanish at ϕ_c, and so we assume a critical scaling in $\Delta\phi$ of the form

$$\frac{d}{k}\sigma_y \sim \Delta\phi^y. \tag{12.61}$$

The prefactor d/k renders the expression dimensionless, and the yield stress exponent y is undetermined. Similarly, the crossover time scale τ_x should diverge on approach to ϕ_c,

$$\frac{\tau_x}{\tau_0} = A\Delta\phi^\Gamma, \tag{12.62}$$

for some constant A and some exponent Γ. (Recall that the crossover frequency ω^* in oscillatory response vanishes at ϕ_c.)

Inserting these expressions into the Herschel–Bulkley law gives

$$\frac{d}{k}\sigma = \Delta\phi^y\left[\mathrm{const} + \left(A\frac{\dot{\gamma}\tau_0}{\Delta\phi^\Gamma}\right)^\beta\right] \phi \geq \phi_c. \tag{12.63}$$

As in oscillatory flow, the energy dissipated should remain finite as $\phi = \phi_c$, hence the prefactor of $\dot{\gamma}^\beta$ must be independent of $\Delta\phi$, and

$$y = \Gamma\beta. \tag{12.64}$$

For packing fractions $\phi > \phi_c$, we divide by $\Delta\phi^y$ in order to arrive at an expression such that σ and $\dot{\gamma}$ always appear in combination with a power of $\Delta\phi$:

$$\frac{d}{k}\left[\frac{\sigma}{\Delta\phi^y}\right] = \mathrm{const} + \left[A\tau_0\frac{\dot{\gamma}}{\Delta\phi^\Gamma}\right]^{y/\Gamma} \tag{12.65}$$

This allows us to seek a scaling collapse and numerically determine the best exponents as follows. If the critical scaling ansatz is correct (and if the rheology is indeed described by a Herschel–Bulkley law), it should be possible to plot the quantities in square braces, $\sigma/\Delta\phi^y$ versus $\dot{\gamma}/\Delta\phi^\Gamma$, and, by varying y and Γ, find values of the exponents for which the rheological data collapse to the shape in Figure 12.16. The exponent $\beta = y/\Gamma$ can then be read off from the form of the scaling function.

While such scaling collapse is not (yet) available for experimental data from foams or emulsions, it has been investigated for suspensions of soft colloidal particles made from N-isopropylacrylamide (NIPA) gel [20]. This is a useful model system because the particles (de)swell in response to a change in temperature, permitting experimental control of the packing fraction ϕ. Rheological data are plotted in Figure 12.17, which shows collapse for a yield stress exponent $y \approx 2$ and a Herschel–Bulkley exponent $\beta \approx 0.5$. (Note that data for $\phi < \phi_c$, which do not display a yield stress, are also collapsed by the same rescaling.) The data in Figure 12.17 provide experimentally determined scaling exponents for at least one material near the jamming transition. These exponents are correctly predicted by the scaling model of Tighe et al. [26]. However, simulations of the bubble model and related models provide differing results [11,22,23,26]. This may be due at least in part to a sensitive dependence on the form of the elastic and viscous interactions between particles, including the type of surfactant [26].

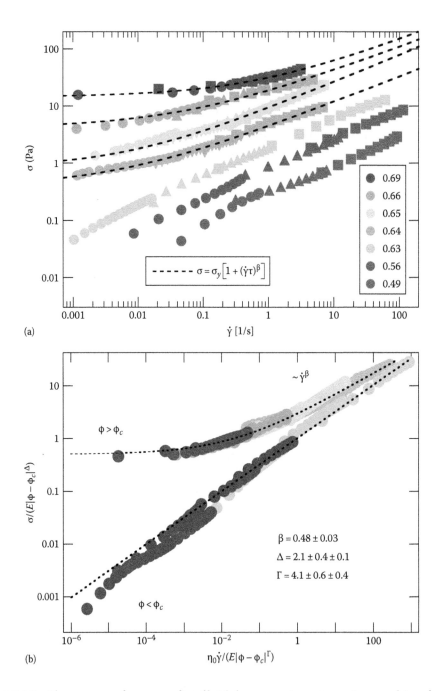

Figure 12.17   Flow curves from a soft colloidal suspension at varying packing fraction φ. The data collapse in the scaled coordinates $\dot{\gamma}|\Delta\phi|^{\Gamma}$ and $\sigma/|\Delta\phi|^{\Delta}$. Here, the exponent Δ is equivalent to y in the text, and E and η_0 are material parameters. (Adapted from Nordstrom, K.N. et al., *Phys. Rev. Lett.*, 105, 175701, 2010.)

Note that, as for oscillatory driving, steady flow in the critical regime is shear thinning. In this regime, the flow curve is

$$\sigma \sim \dot{\gamma}^{\beta}, \tag{12.66}$$

corresponding to a steady viscosity

$$\eta = \frac{\sigma}{\dot{\gamma}} \sim \frac{1}{\dot{\gamma}^{1-\beta}}. \tag{12.67}$$

Because $\beta < 1$, the viscosity decreases with increasing shear rate, and the material is shear thinning. Hence, shear thinning is again a consequence of critical scaling near the jamming transition.

References

1. S. Alexander. Amorphous solids: Their structure, lattice dynamics and elasticity. *Phys. Rep.*, 296:65–236, 1998.

2. F. Bolton and D. Weaire. Rigidity loss transition in a disordered 2d froth. *Phys. Rev. Lett.*, 65:3449–3451, December 1990.

3. F. Bolton and D. Weaire. The effects of plateau borders in the two-dimensional soap froth i. decoration lemma and diffusion theorem. *Phil. Mag. B*, 4:795–809, 1991.

4. F. P. Bretherton. The motion of long bubbles in tubes. *J. Fluid Mech.*, 10:166, 1961.

5. P. M. Chaikin and T. C. Lubensky. *Principles of Condensed Matter Physics*. Cambridge University Press, Cambridge, U.K., 2000.

6. S. Cohen-Addad, H. Hoballah, and R. Höhler. Viscoelastic response of a coarsening foam. *Phys. Rev. E*, 57:6897–6901, June 1998.

7. N. D. Denkov, S. Tcholakova, K. Golemanov, K. P. Ananthpadmanabhan, and A. Lips. The role of surfactant type and bubble surface mobility in foam rheology. *Soft Matter*, 5:3389–3408, 2009.

8. D. J. Durian. Foam mechanics at the bubble scale. *Phys. Rev. Lett.*, 75:4780–4783, December 1995.

9. W. G. Ellenbroek, E. Somfai, M. van Hecke, and W. van Saarloos. Critical scaling in linear response of frictionless granular packings near jamming. *Phys. Rev. Lett.*, 97:258001, 2006.

10. A. D. Gopal and D. J. Durian. Relaxing in foam. *Phys. Rev. Lett.*, 91:188303, 2003.

11. T. Hatano. Scaling properties of granular rheology near the jamming transition. *J. Phys. Soc. Jpn.*, 77:123002, 2008.

12. S. Hilgenfeldt, S. Arif, and J.-C. Tsai. Foam: A multiphase system with many facets. *Phil. Trans. R. Soc. A*, 366:2145–2159, 2008.

13. R. Höhler and S. Cohen-Addad. Rheology of liquid foam. *J. Phys. Cond. Matt.*, 17(41):R1041, 2005.

14. G. Katgert, M. E. Möbius, and M. van Hecke. Rate dependence and role of disorder in linearly sheared two-dimensional foams. *Phys. Rev. Lett.*, 101:058301, 2008.

15. G. Katgert and M. van Hecke. Jamming and geometry of two-dimensional foams. *EPL*, 92(3):34002, 2010.

16. M.-D. Lacasse, G. S. Grest, and D. Levine. Deformation of small compressed droplets. *Phys. Rev. E*, 54:5436–5446, November 1996.

17. A. J. Liu and S. R. Nagel. Nonlinear dynamics: Jamming is not just cool any more. *Nature*, 396:21–22, 1998.

18. T. G. Mason. New fundamental concepts in emulsion rheology. *Curr. Opin. Coll. Interfaces Sci.*, 4:231–238, 1999.

19. D. C. Morse and T. A. Witten. Droplet elasticity in weakly compressed emulsions. *EPL (Europhysics Letters)*, 22(7):549, 1993.

20. K. N. Nordstrom, E. Verneuil, P. E. Arratia, A. Basu, Z. Zhang, A. G. Yodh, J. P. Gollub, and D. J. Durian. Microfluidic rheology of soft colloids above and below jamming. *Phys. Rev. Lett.*, 105(17):175701, October 2010.

21. C. S. O'Hern, L. E. Silbert, A. J. Liu, and S. R. Nagel. Jamming at zero temperature and zero applied stress: The epitome of disorder. *Phys. Rev. E*, 68:011306, 2003.

22. P. Olsson and S. Teitel. Critical scaling of shear viscosity at the jamming transition. *Phys. Rev. Lett.*, 99:178001, 2007.

23. M. Otsuki and H. Hayakawa. Critical behaviors of sheared frictionless granular materials near the jamming transition. *Phys. Rev. E*, 80:011308, July 2009.

24. J. A. F. Plateau. *Statique Expérementale et Théorique des Liquides soumis aux seules Forces Moléculaires*. Gauthier-Villars, Paris, France, 1873.

25. B. P. Tighe. Relaxations and rheology near jamming. *Phys. Rev. Lett.*, 107:158303, October 2011.

26. B. P. Tighe, E. Woldhuis, J. J. C. Remmers, W. van Saarloos, and M. van Hecke. Model for the scaling of stresses and fluctuations in flows near jamming. *Phys. Rev. Lett.*, 105(8):088303, August 2010.

27. M. van Hecke. Jamming of soft particles: Geometry, mechanics, scaling and isostaticity. *J. Phys. Cond. Matt.*, 22:033101, 2010.

28. D. Weaire and S. Hutzler. *The Physics of Foams*. Oxford University Press, Oxford, U.K., 1999.

29. M. Wyart, H. Liang, A. Kabla, and L. Mahadevan. Elasticity of floppy and stiff random networks. *Phys. Rev. Lett.*, 101(21):215501, November 2008.

CHAPTER 13

Introduction to Colloidal Suspensions

Piotr Habdas

CONTENTS

13.1 Basic Definitions and Some History

Colloidal suspensions consist of particles (1 nm to 10 μm) suspended in a solvent (Hamley 2000, Jones 2002). The size and shape of the particles are variable, and their interactions are tunable. Colloidal particles are found in a wide range of substances, including paints, foods, and biological fluids. Because of their mesoscopic size, colloidal suspension time and length scales are more accessible in experimental study than those in atomic systems, and colloidal suspensions are often studied in analogy to atomic systems. The size of the colloids allows one to treat the background liquid as approximately continuum, providing a thermal bath (Figure 13.1).

Early studies of pollen grains suspended in water (Brongniart 1827, Brown 1828) showed that these small objects exhibited irregular motion, initially attributed to external forces such as temperature differences between the liquid and the surroundings or surface tension effects or external vibrations. It was not until 1880s that Leon Gouy (1888) carried out a series of experiments that demonstrated convincingly that the irregular motion was indeed a fundamental physical property (Haw 2002), with a much deeper meaning about the "continuous" description of the world.

One issue that hindered the explanation of Brownian motion was that experimentalists did not know exactly which physical quantity to measure. It was not

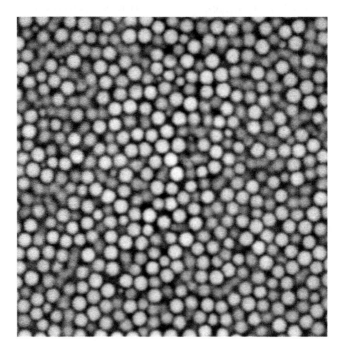

Figure 13.1 A snapshot of colloidal suspension of polymethyl metacrylate colloidal particles suspended in mixture of decalin and cyclohexyl bromide that matches both the density and the index of refraction of colloidal particles. Image was obtained with a confocal microscope, 100× objective and the particles are 2.1 μm in diameter.

until 1905 when Albert Einstein (1905) (and later Marian von Smoluchowski 1906, Paul Langevin 1908) provided a theoretical framework for Brownian motion and showed that mean displacement of particles in a liquid was the relevant parameter and that it should increase with the square root of time. Quantitative measurements by Jean Baptiste Perrin (1908) in 1908 showed that colloidal particles exhibit thermal motion. Not only were his studies in agreement with theoretical predictions, but they also provided definitive evidence for the "discontinuous" (atomistic) view of matter and thus Maxwell's kinetic theory of matter (Haw 2002, Weeks 2007).

13.1.1 Brownian Motion and the Stokes–Einstein Equation

Brownian motion is a manifestation of macroscopic particles under constant bombardment by the molecules of the liquid in which they are suspended. Since at any instant of time there are more collisions of liquid molecules with a macroscopic particle on one side of the particle than another, the result is a constantly fluctuating net force and irregular motion.

In such random motion, the mean of the total displacement is zero, but the mean of the square of the displacement of the particle grows linearly with time:

$$\left\langle (\vec{r}(t))^2 \right\rangle = At \tag{13.1}$$

where
 $\vec{r}$ is the displacement vector
 t is time
 A is a constant related to the diffusion constant D

This results directly from considering the equation of motion for a particle:

$$m\frac{d^2\vec{r}}{dt^2} + C\frac{d\vec{r}}{dt} = F_{random} \tag{13.2}$$

where
 m is the particle's mass
 C is the coefficient for a linear drag
 F_{random} is the random force resulting from collisions of liquid molecules with the colloidal particle

The drag is assumed to obey Stokes's law:

$$C = 6\pi\eta a \tag{13.3}$$

where
 η is the liquid viscosity
 a is the particles radius

Since the forces are random in all directions, x, y, and z, $\langle x \rangle^2 = \langle y \rangle^2 = \langle z \rangle^2$ and $\langle \vec{r}^2 \rangle = 3\langle x^2 \rangle = 3\langle y^2 \rangle = 3\langle z^2 \rangle$. For any specific dimension (say, x) $d(x^2)/dt = 2x(dx/dt)$ and Equation 13.2 becomes

$$\frac{C}{2} \frac{d(x^2)}{dt} = xF_{random} - mx\frac{d^2x}{dt^2} \tag{13.4}$$

Using the identity

$$x\frac{d^2x}{dt^2} = \frac{d}{dt}\left(x\frac{dx}{dt}\right) - \left(\frac{dx}{dt}\right)^2$$

and taking the average of each term, we obtain

$$\frac{C}{2} \frac{d\langle x^2 \rangle}{dt} = \langle xF_{random} \rangle - m\frac{d}{dt}\left\langle x\frac{dx}{dt}\right\rangle - m\left\langle \left(\frac{dx}{dt}\right)^2 \right\rangle \tag{13.5}$$

Since the force is random, the first term on the right-hand side of Equation 13.5 vanishes. The second term on the right-hand side of Equation 13.5 is also zero since there is no correlation between the particle's position and its velocity.

Using the theorem of equipartition of energy, $mv_x^2/2 = k_BT/2$, we can rewrite the third term on the right-hand side of Equation 13.5 as

$$\frac{d(x^2)}{dt} = 2\frac{k_BT}{C} \tag{13.6}$$

and, taking into account all three dimensions, finally conclude that

$$\langle \vec{r}^2 \rangle = 6\frac{k_BT}{C}t = 6Dt \tag{13.7}$$

where k_B is the Boltzmann's constant and

$$D = \frac{k_BT}{C} = \frac{k_BT}{6\pi\eta a} \tag{13.8}$$

is the diffusion coefficient D for a sphere diffusing in a liquid (Einstein 1905, Jones 2002). This is known as the Stokes–Einstein equation.

Einstein expressed the diffusion coefficient D in terms of the particle radius a, fluid dynamic viscosity η, absolute temperature T, gas constant R, and Avogadro's Number N_A (Einstein 1905):

$$D = \frac{RT}{N_A} \frac{1}{6\pi\eta a} \tag{13.9}$$

The formula is historically important since it was used to make the first absolute measurement of N_A, consequently confirming molecular nature of matter.

The Stokes–Einstein equation allows one to determine Boltzmann's constant experimentally (Nakroshis et al. 2003) if the liquid viscosity and particle size is known or, alternatively, the diffusion constant D and particle size or the liquid viscosity (Weeks 2007).

Measuring changes in the diffusion constant D shows that it decreases as the volume fraction (fraction of space occupied by particles) increases since there is less space for the particles to diffuse through. When the particles have essentially no free space, the diffusion constant goes to zero and this is often considered to be an indication of the transition to a colloidal glass state. The diffusion constant goes also to zero as the particle size increases, the thermal motion vanishes, and one approaches a granular limit. Analogous Brownian-like motion can be observed in granular matter that is rapidly vibrated, thus maintaining a gaseous state. In this state, one observes deviations from the Maxwellian velocity distribution (Brilliantov and Pöschel 2004) and breakdown of the equipartition principle (Feitosa and Menon 2002). A qualitative change in the motion from normal Brownian to nearly ballistic motion of a large particle (Santos and Dufty 2001) can also be seen. Both the equipartition principle and fluctuation–dissipation relations break down (Bodrova et al. 2009) in granular gases composed of viscoelastic particles.

13.1.2 Interactions between Colloids

Charged particles in an electrolyte solution experience both a van der Waals attraction and a repulsion originating from electric double layer, explained by the DVLO theory (Derjaguin, Landau, Verwey and Overbeek). The competition between these forces results in a stable equilibrium (Derjaguin and Landau 1941, Verwey and Overbeek 1948). Adding polymers to the suspension also strongly influences interactions between colloidal particles as they can either attach to the particle surface causing steric stabilization or disperse in the solution causing depletion attraction (Hamley 2000, Jones 2002). Here we consider each of the individual forces in turn.

13.1.2.1 van der Waals Forces

The van der Waals force between any pair of atoms or molecules is due to fluctuating dipoles in the atoms (Jones 2002). Attractive van der Waals forces between colloidal particles can be considered as a result of dispersion interactions between the molecules on each particle. Neglecting many-body interactions and assuming the potential is pairwise additive, the resulting pairwise summation can be performed analytically by integrating the pair potential for molecules in microscopic volumes on both particles. The resulting potential depends on the colloidal particle shape and their separation (Hamley 2000).

Specifically, for two spherical particles of radius R, in the regime where the interparticle distance r is small with respect to the particle size $r \ll R$, the Derjaguin approximation (Derjaguin 1934) can be used to relate the potential between two curved surfaces to that between two flat surfaces (Russel et al. 1989, Hamley 2000). Then,

$$V = -\frac{HR}{12r} \tag{13.10}$$

where H is the Hamaker constant. This is true in vacuum, but if there is a liquid present between the colloidal particles, the van der Waals potential is significantly reduced (Hamley 2000). In such a case, Hamaker constant is replaced by effective Hamaker constant, which is the sum of particle–particle plus medium–medium contributions.

13.1.2.2 Electric Double-Layer Forces

A particle that is charged at the surface attracts counterions and an ionic environment is formed around it. Thus, a layer of counterions will be attracted to the surface by the electrostatic field. Close to the colloidal particle, counterions are predominant due to the strong electrostatic forces. Therefore, an electric double layer is created (Hamley 2000, Jones 2002).

Most of the counterions form a diffuse concentration profile away from the particle surface. There are two models that describe this diffusive layer. In the diffusive double-layer model, the concentration of this diffuse region of counterions decreases gradually away from the surface (Figure 13.2a) (Hamley 2000). In the Stern model, the interface between the inner region of the counterion environment is a sharp plane (Stern plane) and the inner region consists of a single layer of counterions called the Stern layer (Figure 13.2b) (Hamley 2000).

The diffusive double layer can be described by the Gouy–Chapman equation, which is a solution of the Poisson–Boltzmann equation for a planar diffusive double layer.

The distribution of charges in the electrolyte solution is described by Boltzmann equation:

$$n(z) = n_0 e^{\frac{-zeV(x)}{k_B T}} \tag{13.11}$$

where ze is the charge of ions.

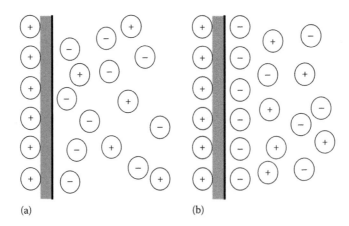

(a)   (b)

Figure 13.2 Models for the electric double layer around a charged colloidal particle: (a) diffuse double layer model; (b) Stern model.

The potential V is determined by the distribution of net charge $\rho(z)$ through the Poisson equation

$$\rho(z) = -\varepsilon\varepsilon_0 \left(\frac{d^2V}{dx^2} \right) \tag{13.12}$$

where
  ε is dielectric permittivity (dielectric constant)
  ε_0 is dielectric permittivity of free space (Griffiths 1999)

The Poisson–Boltzmann equation relates the electric potential to the distribution and concentration of charged matter. If only counterions are present, these two equations can be combined to give the Poisson–Boltzmann equation.

$$\frac{d^2V}{dx^2} = -\left(\frac{zen_0}{\varepsilon\varepsilon_0} \right) e^{\left(-\frac{zeV}{k_BT} \right)} \tag{13.13}$$

Often, at a point where the electrical potential is V, the concentrations of positive and negative ions are given by

$$c_\pm = c_0 e^{\frac{\pm zeV}{k_Bt}} \tag{13.14}$$

where c_0 is the number density of each ionic species of valence z. The net charge density is then

$$\rho = ze(c_+ - c_-) \tag{13.15}$$

By inserting this into the Poisson's equation, an expression for the potential as a function of distance from the charged plane can be obtained:

$$\frac{d^2V}{dx^2} = -\frac{\rho}{\varepsilon} \tag{13.16}$$

where ε is the permittivity of the solution. A general solution is quite complex, but if it is assumed that the surface potential V_0 is much smaller than k_BT, then $zeV_0/(k_BT) \ll 1$ and the potential reduces to

$$V = V_0 e^{(-\kappa x)} \tag{13.17}$$

decreasing exponentially with distance and

$$\kappa = \left(\frac{e^2 \sum_i c_i z_i}{\varepsilon k_B T} \right)^{1/2} \tag{13.18}$$

The quantity $1/\kappa$ has dimensions of length and is called the Debye screening length (Hamley 2000, Jones 2002). It represents the distance over which the induced electrostatic forces are appreciable.

13.1.2.3 Steric Stabilization of Colloidal Particles

If long-chain molecules are attached to the colloidal particles, the limited inter-particle penetration of the polymer chains leads to an effective repulsion that stabilizes the suspension against flocculation. The colloids are said to be "sterically stabilized"; the most effective steric stabilizers are block or graft copolymers where one type of block is soluble in the dispersion medium and the other is insoluable so that it attaches to the colloid particle (Hamley 2000). The range of the interaction is controlled by the distance from the surface that the polymer chains occupy. This is controlled by the length of the polymer chain, the density at which chains are grafted, and the strength of the interaction between polymer segments and the solvent.

Stabilized polymeric particles are typically used to form a nearly monodispersed colloidal suspension. Sterically stabilized colloidal suspensions can behave as nearly hard spheres; the lack of total hardness is due to the small but finite compressibility of the polymer stabilizing hairs (Pusey and van Megen 1986, Weeks 2010).

13.2 Experimental Methods

13.2.1 Microscopy

There are several types of optical microscopy that can be used to study the dynamics and the structure of colloidal suspensions. Bright-field microscopy is perhaps the simplest, with light focused onto the sample and an objective lens on the other side collecting the light. In bright-field microscopy, the image quality can be determined by the sample itself. Variations in refractive index of the components of the colloidal suspension can also influence the image quality of the sample.

Another method, fluorescence microscopy, requires particles to be labeled with a fluorescent dye. Fluorescent molecules absorb short-wavelength light and radiate slightly a longer wavelength (lower energy). A dichroic mirror is responsible for distinguishing between the excitation and emitted light (Weeks 2010). An advantage of fluorescence microscopy is that only a fraction of the colloidal particles may be dyed, making it easier to distinguish individual particles. Figure 13.3 shows an image of about 20% of colloidal particles dye.

A disadvantage of fluorescence microscopy is photobleaching, a phenomenon in which after dyed molecules absorb the excitation light, but before they emit light, they chemically react with oxygen to form a nonfluorescent molecule. Photobleaching manifests itself as the image becoming gradually darker (Weeks 2010).

Confocal microscopy is another fluorescence microscopy technique, using laser light to excite fluorescence. Typically, the laser beam is reflected by scanning mirrors that sweep the beam in the x and y directions on the sample. Fluorescent light is sent back through the microscope and becomes descanned by the same mirrors. The defining feature of confocal microscopy is that, before reaching the detector, the fluorescent light is focused onto a screen with a pinhole. All of the light originating from the focal point of the microscope passes through the pinhole, while

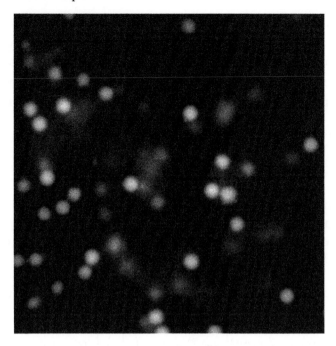

Figure 13.3   A video snapshot of polymethyl metacrylate particles, 2.1 mm in diameter obtained with fluorescent setup and 100× objective. This is a dense sample in the glass regime. Since only 20% of colloidal particles are dyed, there are voids in between dyed fluorescent colloidal particles.

any out-of-focus light is blocked by the screen. Thus, the pinhole filters out the out-of-focus light, allowing a strong and clean signal to come from the in-focus region. The pinhole is conjugate to the focal point of the lens, meaning that a point in focus at the focal point of the objective lens is reimaged onto the pinhole, hence termed "confocal." By rejecting out-of-focus light, a crisp 2D image can be obtained, as shown in Figure 13.1. By collecting a sequence of these 2D images, a 3D image can be created.

13.2.2   Light Scattering

The microscopic Brownian motion underlying the colloidal glass transition can also be studied using light scattering techniques. Dynamic light scattering (DLS) is based on the random Brownian motion of colloidal particles (van Megen and Underwood 1993). The intensity of spatially coherent light scattered off colloidal particles fluctuates on the microsecond to second timescales, and these can be traced to movements of colloidal particles within the scattering volume. These movements change the particles' spatial arrangement and, since different arrangements produce different scattered intensities, the light scattered also fluctuates. The size of the particles can be determined since the intensity from the light scattered by small particles fluctuates more rapidly than that scattered off larger particles.

The scattered intensity at two close times is correlated and useful information about particle size and size distribution can be determined using the intensity–intensity temporal autocorrelation function. The incident beam has a wave vector $\vec{k}_{in}$ and the beam scattered off the colloidal particles has a wave vector $\vec{k}_{out}$. The momentum transfer (scattering vector) $\vec{q} = \vec{k}_{out} - \vec{k}_{in}$ has a magnitude

$$\vec{q} = \frac{4\pi n_0 \sin(\theta/2)}{\lambda_0} \tag{13.19}$$

where
λ_0 is the incident laser wavelength
n_0 is the refractive index of the sample
θ is the angle at which the detector is located with respect to the sample cell (Alsayed 2006)

The intensity autocorrelation function measured from scattered light is

$$G^{(2)}(\tau) = \langle I(0)I(\tau)\rangle = \left\langle |E(0)|^2|E(\tau)|^2 \right\rangle = \lim_{T\to\infty} \left(\frac{1}{2T}\right) \int\limits_{-T}^{+T} I(t)I(t+\tau)dt \tag{13.20}$$

where
G is the unnormalized measured quantity from the DLS apparatus
The superscript 2 denotes that it is related to the light intensity I rather than to the electric field vector $\vec{E}$.

This second-order autocorrelation function is related to the normalized electric field correlation function $g^{(1)}(\tau) = \left\langle |E|^2 \right\rangle$ by the Siegert relation

$$G^{(2)}(\tau) = B \cdot \left(1 + f^2 \left|g^{(1)}(\tau)\right|^2\right) \tag{13.21}$$

where
τ is the delay time
B is the baseline measured when $\tau \to \infty$
f is an instrumental parameter related to the coherent properties of the laser beam

For a colloidal suspensions where the particles are monodispersed

$$\left|g^{(1)}(\tau)\right| = e^{-\Gamma \cdot \tau} \tag{13.22}$$

where Γ is the decay rate. For a dilute suspension, Γ depends on the random diffusion of particles and

$$\Gamma = D \cdot q^2 \tag{13.23}$$

where

$\vec{q}$ is the scattering vector defined in Equation 13.19

D is the diffusion constant that is given by the Stokes–Einstein relation (Equation 13.8)

In a colloidal suspension with polydisperse particles, other methods of analysis must be employed (Koppel 1972, Poon et al. 1997, Alsayed 2006).

13.3 Phase Behavior of Colloidal Suspensions

13.3.1 Crystallization of Nearly Hard-Sphere Colloids

Hard-sphere colloids are defined as particles that repel only when in contact. When the volume fraction ϕ (a ratio between the volume of all colloidal particles and the total volume of the sample) of hard-sphere colloidal suspensions increases to $\phi = 0.494$, the system starts to form crystals and completely crystallizes at $\phi = 0.545$ (Figure 13.4). Between $\phi = 0.494$ and $\phi = 0.545$, the system is in the coexistence phase. Thus, at $\phi = 0.545$, there is a transition from a disordered arrangement of particles, similar to a liquid, to a crystalline ordered packing. Colloidal crystals possess long-range order that results in beautiful opalescence.

This transition is driven by entropy, not temperature. This seems illogical as it suggests that a regular crystalline phase has higher entropy than a disordered liquid-like arrangement. The explanation is that, while the crystalline packing does lose some entropy due to the long-range order, it gains more entropy because its packing is more efficient and each particle has more local space to explore. (Alternately, a random packing jams at a lower volume fraction, keeping the random arrangement but losing all freedom of motion.) This can be quantified through a connection with the free volume (Jones 2002). Since two spheres cannot overlap, there is an effective repulsive force between the spheres that has an entropic origin. In the van der Waals theory of nonideal gases, the entropy per atom (for a perfect gas) is

$$S_{ideal} = k_B \ln\left(f\frac{V}{N}\right) \tag{13.24}$$

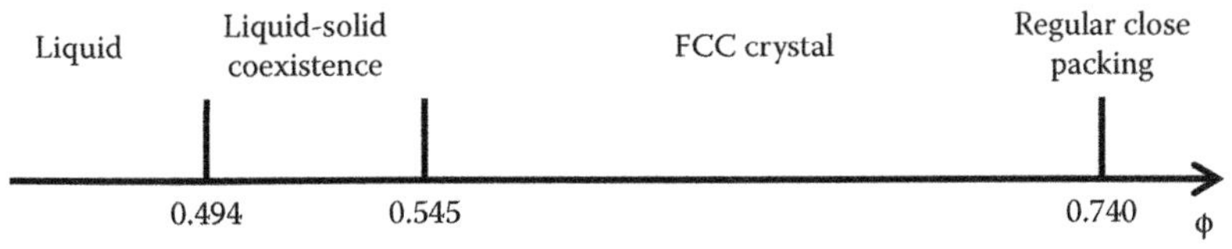

Figure 13.4 Equilibrium phase diagram for uniformly sized hard spheres. The liquid-crystal coexistence region is $0.494 < \phi < 0.545$. Face-centered cubic structure for volume fraction $\phi > 0.545$ and regular close packing 0.740. (From Mau, S.C. and Huse, D.A., *Phys. Rev. E*, 59, 4396, 1999.)

where

    N is the number of atoms

    V is the ideal volume

    f is a constant

    k_B is the Boltzmann's constant

If b is a finite volume of each gas particle, then the volume accessible to any given atom decreases from V to $V - Nb$, and the expression for the entropy must be modified to

$$S = k_B \ln\left(f\, \frac{V - Nb}{N}\right) \tag{13.25}$$

or

$$S = S_{ideal} + k_B \ln\left(1 - \frac{fN}{V}\right) \tag{13.26}$$

Finally, for low-volume fraction, the logarithm can be expanded to give

$$S = S_{ideal} - k_B f\left(\frac{N}{V}\right). \tag{13.27}$$

The corresponding free energy is then

$$F = F_{ideal} - k_B T\left(\frac{N}{V}\right) f. \tag{13.28}$$

The negative contribution to free energy implies an effective repulsion that causes the particles to arrange on a crystalline lattice. Even though this theory is only for a pair of atoms, and at volume fractions at which colloids crystallize, there are many-body interactions to be considered; this development provides a basis for modern statistical mechanical theories of the hard-sphere fluid (Jones 2002).

Crystalline packing of particles can take three forms: hexagonal close packing, face-centered cubic, or random hexagonal close packing. In the hexagonal close packing, every other layer is the same, whereas in the face-centered cubic packing every third layer is the same. In random hexagonal close-packed arrangement, layers are hexagonal closely packed, but with a very large number of stacking faults.

As the volume fraction of the colloidal suspension increases beyond $\phi = 0.545$, the colloidal particles arranged in a regular structure are increasingly closer and the system reaches the regular closed-packed structure, which is the maximum packing for hard spheres.

Granular materials can also crystallize. Granular materials are athermal and energy has to be input continuously to rearrange granular particles. Typically, this is done by shaking or shearing a macroscopic cell filled with granular particles. Shear-induced crystal formation can occur in dense granular fluids. Mueth (2000) used MRI and X-ray tomography to study the internal velocity of a Couette flow. Profiles of mean density and mass flow rate were obtained and a "shear band"

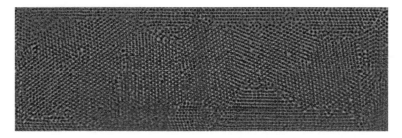

Figure 13.5 Transversal view of the shear cell 10 mm from the top of the system after $5 \cdot 10^5$ shear cycles. (Reprinted with permission from Panaitescu, A., Reddy, K. A., and Kudrolli, A., Nucleation and crystal growth in sheared granular sphere packings, *Phys. Rev. Lett*, 108, 108001–1008005, 2012. Copyright 2012 by the American Physical Society.)

a narrow region of a few grain diameters that is induced by boundary driving, was observed (Tsai et al. 2003).

A number of computer simulations have demonstrated crystalline ordering of hard spheres under steady shearing (Erpenbeck 1984, Lutsko 1996, Polashenski et al. 2002, Sierou and Brady 2002). In simulations at constant granular temperature, an ordering transition occurred when the shear rate exceeded a critical value set by the mean collision time (Erpenbeck 1984, Lutsko, 1996). Simulations that include the interaction between grains and interstitial fluid (Sierou and Brady 2002) showed ordering that depends on the filled volume fraction.

Tsai et al. (2003) experimentally investigated both the internal spatial structure and the time evolution of a deep granular flow in a circular channel. They identified a crystallization transition by observing a steep compaction step and simultaneous rise in the amplitude of a spectral peak. They also noticed that the disordered and ordered states have distinct flow rheology: the ordered state consisted of sliding hexagonal planes with nearly uniform motion parallel to the shearing velocity, while the disordered flow was nearly Poiseuille-like. The ordering occurred with and without interstitial fluid, and was destroyed by sufficient polydispersity.

More recently, the development of ordered phases in disordered frictional granular sphere packings was studied with experiments using a cyclic shear apparatus that allowed 3D visualization with a refractive index matching technique (Panaitescu et al. 2012). Figure 13.5 shows granular particles organized in crystalline domains after half a million of shear cycles. Crystallization of granular particles is obvious and its development is similar to those reported in colloidal systems (Gasser et al. 2001).

13.3.2 Colloidal Glass Transition

If a slight polydispersity, measured by the standard deviation of the particle size distribution, is introduced, a metastable phase diagram can be achieved (Figure 13.6). As the volume fraction approaches $\phi = 0.49$, the system goes into a supercooled state in which particles can diffuse, but only over very short distances

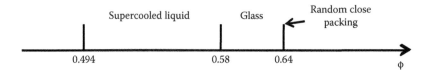

Figure 13.6 Metastable state diagram for hard spheres. Note that polydispersity of at least ~5% is required to achieve various metastable states. Supercooled liquid region is $0.494 < \phi < 0.58$, colloidal glass transition occurs at $\phi \approx 0.58$, and the maximum random packing is at $\phi = 0.64$.

before they collide with other particles. Because of the diffusion, a system in the supercooled state will eventually crystallize. Increasing the volume fraction still further puts the system into a glassy state at around $\phi = 0.58$. In this state, the system does not crystallize, at least not on experimental timescales. The volume fraction of this arrested glass state can be increased further, up to $\phi = 0.64$ (random close packing), the highest volume fraction possible for a random distribution of spheres.

This lack of diffusive motion of particles is one of the defining features of the glass state. Another is *aging* (described more fully in Section 13.3.2.4), which refers to the slow evolution of system properties over time.

The slight polydispersity in particle size allows the system to avoid the crystalline phase and reach the metastable glass state. Above $\phi = 0.58$, the system is metastable; with polydispersity, the random close-packing volume fraction shifts to higher values.

13.3.2.1 Colloidal and Molecular Glasses

As a molecular glass is cooled, its viscosity rises by many orders of magnitude. The molecular glassy state is conventionally defined as the state in which the material cannot reach equilibrium within a time period of 100 s (Casalini et al. 2001) or when its viscosity reaches 10^{13} poise (Angell 1995). Increasing the volume fraction in colloidal suspensions causes their viscosity to increase by many orders of magnitude (Weeks 2010), although not by quite as large an amount as in molecular glasses (Cheng et al. 2002). In colloidal suspensions, it is often difficult to measure precisely the viscosity (Larson 1999) or determine the precise volume fraction (Cheng et al. 2002, Hunter and Weeks 2012). Nonlinear phenomena such as shear thinning and shear thickening (Figure 13.7), exhibited at high-volume fractions (Larson 1999), further complicate the issue.

Nevertheless, the viscosity increase is well fit by standard models, for example, the Doolittle equation (Doolittle 1951) modified to depend on ϕ (Marshall and Zukoski 1990, Hunter and Weeks 2012):

$$\frac{\eta}{\eta_0} = Ce^{\frac{D\phi}{\phi_m - \phi}} \tag{13.29}$$

where $C = 1.20$, $D = 1.65$, and $\phi_m = 0.638$. Interestingly, they found ϕ_m, which corresponds to maximum packing, to be 0.638, very close to the random close

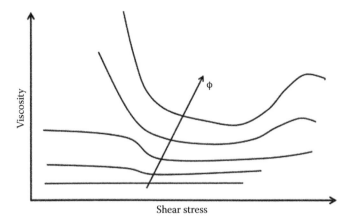

Figure 13.7   Plot of viscosity versus shear stress showing shear thinning and shear thickening for colloidal suspensions of various volume fractions. (Redrawn from Larson 1999.)

packing at $\phi_{rcp} = 0.64$. Moreover, the viscosity diverges at ϕ_m, where the particle dynamics are completely arrested at random closed packing (Cheng et al. 2002).

In more recent work, Cheng et al. (2002) reported measurements of the low-shear viscosity for dispersions of colloidal hard spheres up to $\phi = 0.56$. Nonequilibrium theories based on solutions to the two-particle Smoluchowski equation or ideal mode coupling approximations did not capture observed viscosity divergence (Cheng et al. 2002), although the Doolittle and Adam–Gibbs equations still appeared to hold.

The origin of the viscosity increase with volume fraction is still a matter of debate. Unfortunately, viscosity measurements are trustworthy only up to volume fractions below the colloidal glass transition ($\phi = 0.58$). Thus, most conclusions and equations rely on extrapolation, and some models even show a lack of viscosity divergence (Cheng et al. 2002).

13.3.2.2   Relaxation Time

Another way of probing the dynamics of repulsive colloidal suspensions is by determining the relaxation times. Relaxation times can be determined from light scattering studies, that is, from the intermediate scattering function (ISF) or from mean square displacement (MSD) versus lag time relationships obtained either from scattering experiments or from microscopy studies. For dilute colloidal suspensions, ISF is well fitted by a single exponential function. With the increase of volume fraction, ISF develops a two-step relaxation (Götze and Sjögren 1992). The initial decay corresponds to the motion of a particle in the cage formed by its neighbors; the final decay corresponds to the relaxation of the cage (Figure 13.8).

The relaxation time τ_α, which corresponds to the time associated with the cage relaxations, is obtained by fitting the ISF to a stretched exponential function. It has

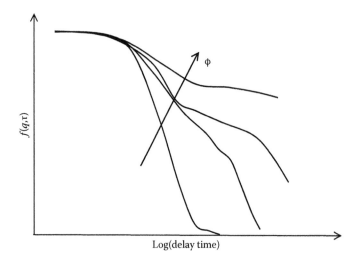

Figure 13.8   Intermediate scattering function versus log of a delay time for a range of volume fractions. (Redrawn from van Megen, W. and Underwood, S.M., *Phys. Rev. E*, 47, 248, 1993.)

been found that structural relaxation times diverge as colloidal glass transition is approached (van Megen et al. 1998).

Recently, Brambilla and colleagues studied hard-sphere samples with $\phi > 0.58$ using DLS (Brambilla et al. 2009, El Masri et al. 2009). Some of their results agreed quite well with the earlier work of van Megen and colleagues (1998). However, above a critical volume fraction ϕ_c where mode coupling theory (MCT) predicts a divergence of τ_α, they found finite values of τ_α, suggesting that their samples were not yet glasses and that ϕ_c of MCT is not equivalent to ϕ_g for their samples. The discrepancy could be due to various experimental difficulties, including the precise volume fractions and/or particle polydispersity (Hunter and Weeks 2012, Poon et al. 2012).

13.3.2.3   Dynamical Heterogeneities

Dynamical heterogeneities refer to dynamic characteristics such as mobility that are spatially heterogeneous. Such behaviors show, for example, that various regions of a dense colloidal suspension relax at different rates. As the glass transition is approached and the system becomes increasingly jammed and congested, particles move more cooperatively; one thus can identify localized regions that exhibit cooperative motion.

As the glass transition is approached, the size of the cooperative groups of particles and the timescale of their motion increase (Kegel and van Blaaderen 2000, Weeks et al. 2000). Rearranging particle displacements are small, especially in glassy colloidal suspensions, and careful analysis is needed in order to distinguish these motions from the Brownian noise (Hunter and Weeks 2012).

Rearranging regions can be obtained directly from microscopy images. Figure 13.9a shows a raw confocal microscope image, a 2D slice in the bulk of the

(a)   (b)

Figure 13.9   (a) Confocal microscope image of a sample with $\phi = 0.60$. (b) Difference between the left image and one taken 150 s later. Grey image indicates that nothing has moved. Black indicates a particle that existed at the earlier time and white a particle existed at the later time.

sample at $\phi = 060$. Figure 13.9b shows the difference between particle positions in the left image and an image taken 150 s later. Grey regions indicate areas where particles move relatively little during this period of time; black and white show particles moving cooperatively. A rearranging region is clearly visible in the center of the difference image. In general, neighboring particles that are rearranging tend to move in similar directions. As the glass transition is approached, the regions of rearranging particles and the timescale of the rearrangements grow larger, but since volume fraction increases, particle displacements during a given rearrangement are smaller (Weeks and Weitz 2002). While it is believed that dynamical heterogeneities play a fundamental role in the glass transition, experimental confirmation is not definitive.

Confocal microscopy has been particularly successful in studying dense hard-sphere colloidal suspensions. The advantage of confocal microscopy over other microscopy techniques is the possibility of imaging fully 3D samples. String-like regions of highly mobile particles were observed in dense repulsive colloidal suspensions (Kegel and van Blaaderen 2000, Weeks et al. 2000), and the sizes of the mobile regions increased dramatically as the colloidal glass transition was approached. Figure 13.10 shows an image of a region of fast-moving particles obtained in a 3D colloidal suspension near the glass transition (Dinsmore et al. 2001).

Dynamical heterogeneities have also been observed in granular media (Dauchot et al. 2005). The dynamical behavior in granular media near jamming resembles that of liquids close to the glass transition, slowing down dramatically (Knight et al. 1995, D'Anna and Gremaud 2001). Typical grain trajectories show a cage effect similar to that observed in colloidal suspensions (Weeks et al. 2000, Weeks and Weitz 2002) and in molecular dynamics simulations of molecular

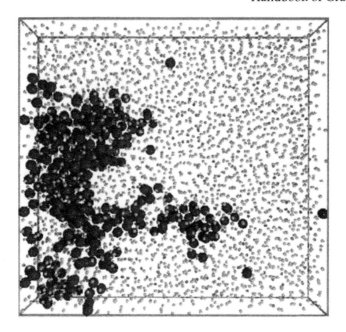

Figure 13.10 Snapshot of positions of the fast particles (large spheres that are drawn to scale) and the other particles (shown as smaller spheres that are not to scale) in a supercooled liquid at $\phi = 0.56$. (Reprinted from Dinsmore, A.D. et al., *Appl. Opt.*, 40, 4152, 2001.)

glasses (Kob and Andersen 1995a,b). The dynamics are also strongly correlated near the jamming transition (Dauchot et al. 2005).

13.3.2.4 Aging

Glasses age; since they are out of equilibrium, their properties change with time. If a colloidal suspension is quenched (i.e., centrifuged) into a volume fraction in the colloidal glass regime, particles can initially move relatively quickly, but over time the particles' motion slows down (van Megen et al. 1998). Cages formed by a particle's neighbors are nearly permanent, and consequently particles spend almost all of their time stuck in these cages.

Dynamical heterogeneities have been observed in aging colloidal glasses (Lynch et al. 2008, Yunker et al. 2009), with particles occasionally undergoing irreversible rearrangements. The average size of the rearranging regions remains approximately constant during aging, although the domain size of particles surrounding irreversible rearrangements does increase. Particle motion is related to the local structure (Cianci et al. 2006). The majority of particles pack into tetrahedra, which become more regular with aging. Colloidal observations are in reasonable agreement with simulations, which found that subtle structural changes during aging (Kob et al. 2000a,b).

Aging effects have also been observed in granular materials. Memory effects have been seen in vibration-induced compaction of granular materials, with an abrupt change in shaking intensity bringing about a granular analog of aging (Josserand et al. 2000). Aging has also been induced in granular systems using ultrasound in creep experiments (Espíndola 2012).

13.3.3 Depletion

If another species of colloidal particles is added to the original suspension, an attractive interaction due to osmotic pressure, termed a *depletion force*, arises. The size of the second particle species should be intermediate between that of the colloidal particles and of the solvent molecules. Typically, a polymer that does not adsorb to the colloidal particles surface is used.

Figure 13.11 shows colloidal particles depicted as large spheres and the depletant particles as small spheres. The region around each colloidal particle is called the depletion zone since the center of a smaller particle must lie outside. As colloidal particles, driven by Brownian or other motion, approach each other, the depletion zones overlap and the depletant particles are excluded from the volume in between. A net osmotic pressure is then developed between the bulk solution and the depletion zone that leads to an attractive force between the colloidal particles (Jones 2002).

For a hard-sphere dilute solution of colloidal particles, the osmotic pressure p_{osm} is given by the ideal gas expression

$$p_{osm} = \frac{N}{V} k_B T \qquad (13.30)$$

where N is the number of colloidal particles in a volume V of the solution.

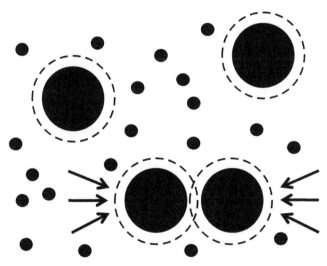

Figure 13.11 A schematic illustration of depletion interaction. Small colloidal particles are excluded from coming closer than a certain distance, approximately their size, to the surface of big colloidal particles because of high entropic cost of configurational distortion. If two big colloidal particles come to each other close enough to exclude small colloidal particles from the region between them, a net osmotic force presses the particles together, resulting in the depletion attraction.

The net interaction potential between the particles F_{dep} is then

$$F_{dep} = -p_{osm} V_{dep} \qquad\qquad (13.31)$$

where V_{dep} is the total volume between the particles from which the polymers are excluded. For two spheres of radius R, at a center-to-center separation r, from geometry

$$V_{dep} = \frac{4\pi}{3}\left(R + L^3\right)\left(1 - \frac{3r}{4(R+L)} + \frac{r^3}{16(R+L)^3}\right) \qquad\qquad (13.32)$$

where L is the thickness of the depletion region (dashed circle around colloidal particles in Figure 13.11) (Jones 2002). The concentration of the depletant particles controls the depth of the potential well while the depletant particle size determines the spatial extent (Eckert and Bartsch 2002, Pham et al. 2002).

13.3.4   Reentrant Glass Transition

The depletion interaction gives rise to a new glass state in colloids, the *attractive glass* (Eckert and Bartsch 2002, Pham et al. 2002). The attractive glass, in contrast to the hard-sphere glass, possesses both a repulsive and attractive potential and is therefore a more accurate model of molecular glasses.

13.3.4.1   Dynamics and Structure of Colloidal Suspensions with Short-Range Attraction

Experimental light scattering studies (Eckert and Bartsch 2002, Pham et al. 2002) found that the reentrance transition from ergodic liquid to bonded repulsive glass is due to the existence of two qualitatively distinct types of glasses, dominated by (respectively) repulsion and attraction. For low attraction strengths, a repulsive hard-sphere glass is formed. As the attraction strength increases, particles cluster, opening up holes that allow particles to escape from cages. The repulsive glass melts when the characteristic time of the attraction-dominated particle dynamics becomes comparable to that of cage opening. The density autocorrelation function decays to zero, what is indicative of a liquid region (curve 3 in Figure 13.12). Increasing the attraction still further leads to an arrest where the strong attraction between particles creates long-lived bonds and prevent structural rearrangement, giving rise to an attraction-dominated glass (curve 6 in Figure 13.12) with a distinctly different dynamics.

The presence of the attractive glass gives rise to a new phase diagram (Figure 13.13) that accounts for both volume fraction and potential.

Computer simulations have found four distinct states in attractive hard-sphere systems at high concentrations: nonbonded repulsive glass (hard-sphere or repulsive glass) and bonded repulsive glasses (attractive glass), dense gels, and ergodic fluids. Moreover, bonding, typically associated as the main reason for formation of the attractive glasses, is found to be a fluctuating characteristic. Bonds constantly break and reform, although a particle is still trapped topologically by its

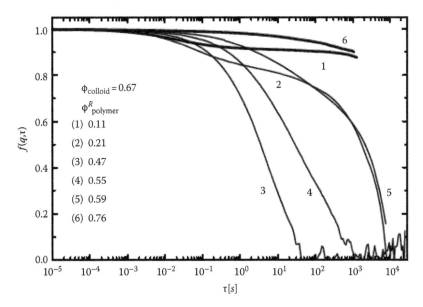

Figure 13.12 Density autocorrelation functions $f(q,t)$ (probed at the same scattering vector). The dynamics is probed at a scattering vector corresponding to the peak maximum of the structure factor of the pure colloid suspension at its glass-transition volume fraction. The polymer reservoir volume fractions ϕ^R are indicated by the numbers at the curves. The thick solid lines correspond to nonergodic (glassy) samples. (Reprinted with permission from Eckert, T. and Bartsch, E., Re-entrant glass transition in a colloid-polymer mixture with depletion attractions, *Phys. Rev. Lett*, 89, 125701, 2002. Copyright 2002 by the American Physical Society.)

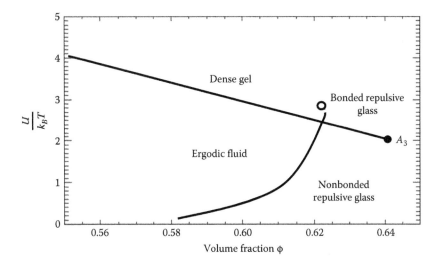

Figure 13.13 A sketch of state diagram of a colloid–polymer mixture redrawn from Zaccarelli and Poon (2009). Their simulations show that four distinct states exist: ergodic fluid, nonbonded repulsive glass (hard-sphere glass), bonded repulsive glass (attractive glass), and dense gel.

cage of neighbors. In nonbonded repulsive glasses, the localization of the particles results from crowding at high particle concentrations. Each particle is surrounded by a cage of neighbors, unable to escape on short timescales. In bonded repulsive glasses, attractions between the particles cause bonding. Particles can rattle by stretching bonds, but with strong enough attraction, the bonds are effectively permanent. It is unclear what mechanisms of particle arrest dominate in the ergodic fluid state.

Computer simulations showed that mobile particles can escape from the cages (Vollmayr-Lee 2004) in two ways: reversible jumps occur in which a particle jumps back and forth between several average positions, while irreversible jumps are where a particle does not return to one of its former average positions. The ratio of irreversible to reversible jumps increased with increasing temperature or, in case of colloidal suspensions, decreasing volume fraction.

Other simulations found that short-ranged attractive glasses are unstable in time (Zaccarelli et al. 2003). The absence of a clear short-ranged attractive plateau of density–density correlation functions in attractive glasses seems to suggest that the attractive cages are not as stable as those in hard-sphere glasses. In the bonded repulsive glass, particles are more likely to break the bond confinement, and so a sharp transition from nonbonded repulsive glass to bonded repulsive glass is obscured.

Finally, light scattering studies (Eckert and Bartsch 2002, Pham et al. 2002) revealed that attractive and repulsive glasses show qualitatively different aging behavior, although some form of aging is present in all glasses (Kob and Barrat 1997). It is theorized that the activated processes seen in simulations (Zaccarelli et al. 2003) are responsible for the final decays of the dynamics in the attractive glasses and their aging behavior.

Microscopy allows a direct observation of the particles and their dynamics during the reentrant glass transition (Kaufman and Weitz 2006, Simeonova et al. 2006, Latka 2009). Simeonova et al. (2006) observed the particle displacement distribution and its moments to broaden with increasing attractive potential. Displacement distributions in the glass states were non-Gaussian, but did not differ significantly from each other or from that in the attractive liquid state. There were also no significant differences in local structure and density distributions in real space between the glass phases.

Kaufman and Weitz (2006) did find qualitatively different dynamics in repulsive and attractive glasses. Repulsive glasses are well described by cage rattling and escape models. Attractive glasses, however, show very little particle rattling. Instead, there are abrupt cooperative motions enabling large jumps in a short time.

Latka (2009) focused on what they termed "motional events" of particles more mobile than the average. A transition from a hard-sphere arrested phase to a liquid-like phase is characterized by an increase in event speed and the event rate of moving particles. Increasing the solvent polymer concentration does not cause motional particles to move longer distances, but they did move faster. The transition to the liquid region was also characterized by a growing number of particles participating in motional events. Particles experiencing motional events were increasingly spatially correlated with increasing attractive potential; the particles moved in clusters, and the distribution of the cluster size was broader and shifted to larger average values with increasing attraction.

Recently, Zhang et al. (2011) studied domains of particles that rearrange in a correlated manner in hard-sphere and attractive glasses. The size and shape of these spatially clustered cooperative rearrangement regions (CRRs) are closely related to the macroscopic properties of glasses. Cooperative rearrangements and heterogeneous dynamics were observed in both types of glasses, but, compared to repulsive glasses, attractive glass dynamics were heterogeneous over a wider range of time and length scales, and their rearrangements involved more particles. Additionally, clusters of rearranging particles were observed to form string-like structures in repulsive glasses and compact structures in attractive glasses (Figure 13.14).

Molecular dynamics simulations have found that the change in dynamics induced by short-ranged attractions in a dense model liquid is also seen on the microscopic scale (Geissler and Reichman 2005). When attraction is weak, mobilized particles make discrete jumps between cage structures. For a range of intermediate attraction strengths, fluid and uniform motion is observed, leading to the speculation that interparticle attraction is responsible for binding small transient clusters that move on a timescale comparable to their lifetimes. Strong attractions restore some discreteness of particle displacements, presumably because transient clusters are too large to move appreciably on relevant timescales.

Experimental studies of dense colloidal suspensions are difficult due to the high viscosity of colloidal suspensions at high-volume fractions. Regions such as the A_3 point—the point at which hard-sphere and attractive glasses become indistinguishable—and beyond remain inaccessible to experimental studies. Before the A_3 point, there is a *glass-to-glass transition*, raising the possibility of creating a material (by tuning the interaction between particles) in which the transition between the attractive and repulsive glass takes place without any

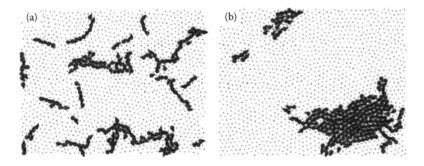

Figure 13.14 Snapshots of cooperatively rearranging particles for repulsive glass (a) and attractive glass (b). The large spheres are drawn to scale and represent the 10% fastest particles. The rest of the particles are shown as small dots, reduced in size for clarity. Arrows indicate the direction of motion. (Reprinted with permission from Zhang, Z. et al., 2012. Copyright 2012 by the American Physical Society.)

intermediate liquid phase. A small change in attraction strength between colloidal particles could produce a significant change in macroscopic properties (like elasticity) of the material without significant structural change (Dawson 2000, Sciortino 2002).

It has also been suggested that the glass transition in molecular glasses has a universal dependence on temperature and pressure and that colloidal systems, because of the fine control over the interaction potential, can shed insight into this relationship (Voigtmann and Poon 2006).

13.3.4.2 Vibrational Properties of Dense Colloidal Suspensions

Information about structure can be obtained from particles that do not exhibit any motional events. Vibrational properties of repulsive glassy colloidal suspensions have been studied recently (Chen et al. 2010b, Ghosh 2010) and show similarities with those predicted for zero-temperature sphere packings and molecular glasses. The boson peak, at which there is an excess of modes with respect to what is expected for sound, shifts from low to high frequency as the system becomes more concentrated (Figure 13.15) (Chen et al. 2010b).

As the system is compressed beyond the jamming transition (O'Hern et al. 2003), the boson peak decreases in height and shifts toward higher frequencies (this is also seen in polymer glasses) (Hong et al. 2008). Soft low-frequency modes have been found to be very different in nature from the usual acoustic vibrations of regular solids. Particles in areas exhibiting soft modes are more likely to rearrange.

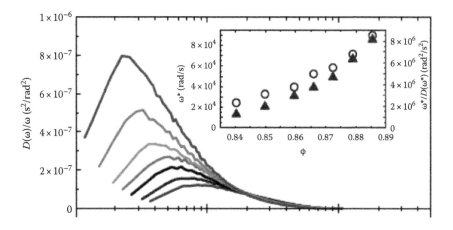

Figure 13.15 Density of states (DOS) normalized by frequency versus frequency as a function of packing fraction. The boson peak decreases with increasing volume fraction. The boson peak frequency (circles, left) and inverse peak height of DOS (triangles, right) versus ϕ. (Reprinted with permission from Chen, K., Ellenbroek, W. G., Zhang, Z. et al., Low-frequency vibrations of soft colloidal glasses, *Phys. Rev. Lett*, 105, 025501, 2010. Copyright 2010 by the American Physical Society.)

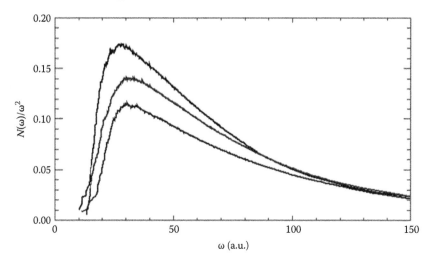

Figure 13.16   Normalized accumulated fractional number $N(\omega)/\omega^2$ versus ω for a repulsive glass (top line), and two polymer concentrations: $c_p = 0.2$ mg/mL (middle line) and $c_p = 0.8$ mg/mL (bottom line), all at $\phi = 0.66$. Thus, peak decreases with increasing c_p. (Unpublished data by Habdas 2012)

Thus, there is the possibility of using vibrational properties to identify regions susceptible to failure (Chen et al. 2011). There is some evidence that this is also true in attractive glasses (Habdas 2012), with preliminary results suggesting that increasing interparticle attractions causes the DOS to decrease (Figure 13.16).

Such results seem to indicate that the "springs" between colloidal particles stiffen with increasing interparticle attraction; however, final experimental confirmation is needed.

Vibrational behavior has also been explored in granular packings (Brito et al. 2010).

Principal component analysis of the covariance matrix of the position of individual grains finds that spectral properties of the covariance matrix show large, collective fluctuation modes. Eigenmode analysis reveals large-scale dynamic structures that appear as the system approaches the jamming transition.

13.3.4.3   Aging in Dense Colloidal Suspensions with Short-Range Attraction

Aging in dense colloidal suspensions with short-range attraction has been studied primarily through computer simulations (Foffi et al. 2004). While the aging dynamics of repulsive glasses are similar to those observed in atomic and molecular systems, the aging dynamics of attractive glasses do show some novel features (Foffi et al. 2004). In systems with short-ranged attractive potentials, an efficient competition between attraction and excluded volume generates a highly nontrivial equilibrium dynamics and two kinetically distinct glasses. In particular, the first peak in the partial static structure factor decreases with the waiting time t_w, whereas it grows in repulsive and molecular glasses (Figure 13.17).

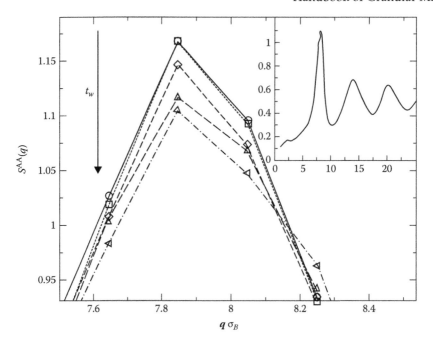

Figure 13.17 First peak of the static structure factor at different waiting time $t_w = 10^1$ (circles), 10^2 (squares), 10^3 (diamonds), 10^4 (upward triangles), and 10^5 (leftward triangles) for an attractive system. The inset shows structure factors for the same t_w. (Reprinted with permission from Foffi, G., Zaccarelli, E., Buldyrev, S., Sciortino, F., and Tartaglia, P., Aging in short-ranged attractive colloids: A numerical study, *J. Chem. Phys.*, 120, 8824–8830, 2004. Copyright 2004, AIP Publishing LLC.)

The decay of correlation functions becomes slower and slower on increasing t_w in both repulsive and attractive glasses, the strength of the relaxation is much more pronounced in attractive glasses. Repulsive case shows the emergence of a plateau of MSD, whose time duration increases significantly with t_w, which does not occur in attractive glasses. It appears that, in short-ranged attractive colloids, activated processes can be associated with thermal fluctuations of order of the inter particle potential depth that allow particles to escape from the bonds. These processes generate a finite bond lifetime and, at the same time, destabilize the attractive glass.

13.4 Jamming

In Section 13.3.4, we described two distinctive arrested states: nonbonded repulsive glass and bonded repulsive glass. The theory of *jamming* attempts to unite features of these two states (and other materials like molecular glasses, colloidal gels, foams, and granular materials) (Figure 13.18). Like granular materials, colloidal suspensions can jam, transitioning from a freely flowing to a disordered jammed state (Siemens and van Hecke 2010). A phase diagram for jammed materials

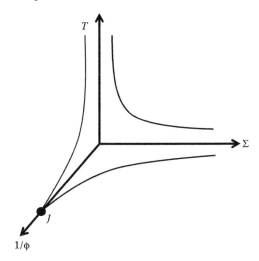

Figure 13.18 Jamming diagram proposed by Liu and Nagel (1998), revised by O'Hern et al. (2003), and experimentally determined by Trappe et al. (2001). The diagram illustrates that many disordered materials are in a jammed state for low temperature T, low load Σ, and large density ϕ, but can become unjammed when these parameters are varied. For frictionless soft spheres, there is a well-defined jamming transition indicated by point J on the inverse density axis, which exhibits similarities to a critical phase transition.

(Liu and Nagel 1998, O'Hern et al. 2003, Trappe et al. 2001) posits a universal jamming transition that describes how a material changes from unjammed to jammed, including the glass transition. Colloidal glasses certainly satisfy the main criterion of a jammed system, with a microscopically disordered structure that can support a finite stress without plastic deformation or flow (Weeks 2007).

The volume fraction or density, applied stress, and temperature are typically the variables that control the thermodynamics of the jamming transition, with unjamming occurring with decreasing density (Pusey and van Megen 1986, Mason and Weitz 1995, Agarwal et al. 2011) or increasing load (Joshi and Reddy 2008, Lee et al. 2009, Shahin and Joshi 2011). The curvature in Figure 13.18 reflects the idea that a material can be kept in a jammed state with sufficiently high density even if the temperature is raised dramatically; and likewise at sufficiently low temperature and high density a material remains jammed even with a very high applied stress (Weeks 2007). The point J reflects that below a certain volume fraction particles no longer touch each other and thus the system cannot be jammed at all. This is particularly clear in the case of granular materials, which are considered to be at $T = 0$ and which unjam below random loose packing (Weeks 2007).

The jammed surfaces depicted in the diagram are typically not sharp and are governed by when the system relaxation time exceeds experimental constraints. However, Liu and Nagel (1998) have shown that point J is well defined. This point exists for systems with repulsive, finite-range potentials (systems with frictionless particles that do not interact beyond a certain distance, which defines their diameter); at point J, particles just come into contact, and at further compression particles interact and the pressure and zero-frequency shear modulus are nonzero (Liu and Nagel 2010).

Point J has very special properties. It occurs at random close-packing density and provides a definition of it. Point J is an isostatic point, where the number of particle contacts equals the number of force balance equations to describe them. As a result, it is a purely geometrical point: the properties of the state at point J are independent of potential. Also, point J appears to be a zero-temperature mixed phase transition, with a discontinuity in the number of contacts, characteristic of first-order phase transitions, and diverging length scales, characteristic of second-order phase transitions.

Some models proposed for the glass transition have scaling exponents that, within mean-field theory, are the same as those for point J (Liu and Nagel 2010). Many of these models exhibit dynamical slowing down and kinetic heterogeneities that are trademark features of the glass transition. Hence jamming may be a universal phenomenon related to glassy behavior.

For thermal system such as colloidal suspensions, it has been shown (Zhang et al. 2009), using both colloidal experiments and computer simulations, that the overlap distance between neighboring particles vanishes at the $T = 0$ jamming transition. The maximum in the height of the first peak of the pair-correlation function shifts to higher packing fractions as the temperature is increased from zero. This maximum is a vestige of one of the most important length scales that define the zero-temperature jamming transition at point J (i.e., the overlap length between neighbors). At point J, this length scale vanishes because the system is isostatic and on the brink of mechanical failure.

Recently, Agarwal et al. (2011) used self-suspended nanoparticles densely grafted with polymer chains as a model soft glassy material to study jamming. They showed that, contrary to expectations, increasing temperature enhances jamming, and that subsequent lowering of the temperature unjams the system. It is possible that there are several different jamming transitions; some materials may share a common transition, but perhaps eventually distinct categories of these transitions, with sharp differences between categories will be found (Weeks 2007).

13.5   Rheology of Colloidal Suspensions

Rheology studies allow one to determine rheological behavior of colloidal suspensions, specifically if colloidal glasses are truly solid and how they yield and flow under application of stress. The existence of a well-defined boundary between fluid-like and solid-like states is most clearly shown by the rheological properties of the samples.

One of the most direct manifestations of the onset of a glassy state is shown by the mechanical response of the suspension, which can be measured with a rheometer (Mason and Weitz 1995, Weitz 2011). Mason and Weitz used a controlled-strain rheometer and applied a known strain at a given frequency and measured the resultant stress at the same frequency. Frequency-dependent measurements of the real (elastic) and imaginary (viscous) moduli found that the viscous modulus dominates at lower volume fractions ϕ, and both moduli increase with frequency. However, for higher ϕ's, the elastic modulus begins to dominate over an extended range of frequencies and it develops a plateau, varying only very slowly with frequency, while viscous modulus exhibited a definite minimum. At higher frequencies, both moduli began to increase, with viscous modulus rising more sharply. This was direct evidence that the colloidal suspension behaves like a glass and that since there was no interparticle interaction energy, the solid-like behavior of the particles results from purely entropic origins (Mason and Weitz 1995, Weitz 2011).

For weakly attractive particles the authors found that rheological data for every sample could be scaled onto a single master curve (Trappe and Weitz 2000). They also found that for a fixed interaction energy and changing volume fraction, samples with a higher ϕ had a larger elastic modulus and occupied the lower frequency side of the master curve. As ϕ decreased, the elastic modulus decreased, and the data fell to the higher frequency side of the master curve. Thus, the dominant response at low frequencies is the elastic response, whereas at high frequencies it is the viscous response (Trappe and Weitz 2000, Weitz 2011).

Interestingly, Trappe et al. (2001) found that behavior of the weakly attractive colloidal particles exhibited exactly the same sort of behavior as that predicted for a jamming transition for granular particles. In the colloids, the attractive energy holds the system together; for granular systems it is excluded volume effects. In this case the temperature axis in the jammed diagram (Figure 13.18) is replaced by the $k_B T / U$ where U is the measure of the attractive energy of the interparticle potential (Weitz 2011).

Sheared systems also exhibit dynamical heterogeneities similar to the thermally activated rearrangements in supercooled colloidal liquids. As the glass transition is approached, molecules collaborate or cooperate on larger length scales in order to rearrange (Adam and Gibbs 1965). Particles that undergo plastic rearrangements in a sheared sample are distributed nonuniformly in space. One of the main questions studied in rheological investigations of colloidal suspensions is that of the orientation of the rearranging regions with respect to the direction of the shear (Hunter and Weeks 2012). Recent experiments confirm that rearrangements are anisotropic, with long-range correlations due to elastic effects (Chikkadi 2011). The sizes of the dynamically heterogeneous regions appear smaller in sheared samples (Yamamoto and Onuki 1998, Zausch et al. 2008), although there is some ambiguity in how one defines the extent of a plastic rearrangement in a sheared sample (Chen et al. 2010a).

Spatial heterogeneities are also noticeable on a macroscopic scale. When complex fluids are sheared, the majority of the strain can be localized within a narrow band, with the rest of the sample remaining relatively unstrained. This phenomenon, called *shear banding* (Dhont 1999, Dhont et al. 2003, Dhont and Briels 2008, Manneville 2008, Ovarlez et al. 2009, Schall and van Hecke 2010), has been observed in dense colloidal suspensions (Ballesta et al. 2008, Moller et al. 2008, Rogers et al. 2008, Fielding et al. 2009, Besseling et al. 2010, Chen et al. 2010b, Chikkadi et al. 2011) and attractive colloidal glasses (Pham et al. 2008).

Since shearing complex fluids has an extreme industrial importance, studies of shear in a variety of complex fluids, including colloidal suspensions, are an active research area (Hunter and Weeks 2012).

Sheared granular materials also exhibit a range of interesting phenomena, including shear banding (Hu and Molinari 2004), particle mixing and segregation (Ottino and Khakhar 2000, Golick and Daniels 2009), force networks (Hartley and Behringer 2003), and crystallization. One of the most striking consequences of shearing is the "Brazil nut effect" where large granular particles rise to the top of sheared or shaken container of mixed small and big granular particles (Rosato et al. 1987, Jullien and Meakin 1990, Jullien and Meakin 1992, Golick and Daniels 2009).

13.6 Nonspherical Dense Colloidal Suspensions

In this section, we will briefly discuss aspherical colloidal particles. Experimental and computational studies of soft materials find significant effects of the particle geometry on the properties of amorphous structures. These include the fractal dimension of colloidal aggregates in diffusion-limited cluster aggregation (Mohraz et al. 2004), percolation threshold of anisometric particles (Garboczi et al. 1995, Yi and Sastry 2004), and maximum packing density of particulate materials (Donev et al. 2004, Mohraz et al. 2004).

Molecular MCT (Franosch et al. 1997, Schilling and Scheidsteger 1997) suggests that, in three dimensions, hard ellipsoids with an aspect ratio greater than 2:5 can form an orientational glass in which the rotational degrees of freedom become glassy while the center-of-mass motion remains ergodic (Letz et al. 2000, Zheng et al. 2011). Such behavior is similar to that seen in liquid crystals. Anisotropic particles also enable exploration of the dynamic heterogeneity in the rotational degrees of freedom. Since monodispersed spheres quench to a glass in three dimensions but not two dimensions, typically bidispersed or highly polydispersed spheres are used to study the glass in two dimensions, whether in experiments (König et al. 2005, Mazoyer et al. 2009, Zhang et al. 2009), simulations (Speedy 1999), or theory (Bayer et al. 2007, Hajnal et al. 2009, Hajnal et al. 2011). Zheng et al. (2011) found that monodispersed ellipsoids of intermediate aspect ratio are excellent glass formers in two dimensions because their shape effectively frustrates crystallization and nematic order. Interestingly, the sample undergoes not one, but two distinct glass transitions: one where the rotational motion drastically slows down, and a second corresponding to a slow down in translational motion. Large translational motions typically occur for particles within aligned domains, while large rotation motions are more common between the domains (Zheng et al. 2011, Weeks 2011).

Brownian motion enables ellipsoids to either rotate or translate (most easily along their long axis). At higher area fractions, diffusion becomes difficult because particles become increasingly constrained by their neighbors. In order for particles to rearrange, they must move cooperatively in groups. Similar behavior, where particles rearrange in groups, has been seen near the glass transition in spherical colloidal suspensions (Kegel and van Blaaderen 2000, Weeks et al. 2000). In both spherical and ellipsoidal colloidal suspensions, the size of such regions grows sharply as the glass transition was approached.

Granular materials composed of nonspherical particles have been also studied. Jamming in 3D piles of large aspect ratio particles showed that it depends not only on the particle aspect ratio, but also on the container in which they are confined (Desmond and Franklin 2006). The particle aspect ratio also has a strong influence on the mechanical and structural properties of granular packings (Hidalgo et al. 2009).

13.7 Summary and Outlook

Colloidal suspensions are ideal models for studying a wide assortment of fundamental problems in materials science and soft condensed matter physics. The glass transition in molecular liquids, packing density of granular materials, and self-assembly in interacting, Brownian systems are all well studied by analogy with colloidal systems. Colloidal suspensions are also technologically important in the manufacturing of advanced materials such as photonic crystals or chemical sensors (Mohraz and Solomon 2005).

Suspension of nearly hard-sphere colloids with added nonadsorbing, random-coil, nearly ideal polymers has proven to be extremely fruitful for modeling the glass transition. Detailed studies of this simple experimental system over more than a decade have uncovered a range of interesting behaviors, especially in the nonequilibrium regime. The simplicity of the experimental system has greatly facilitated comparison with theory and simulation, which in turn has led to a detailed understanding of many of the experimental observations (Weeks 2007).

It is also plausible that conventional molecular glasses and colloidal glasses have in common the physical interpretation of the way their glass transition changes with pressure and temperature. Thus colloidal systems can be used to gain insight into some open questions being asked in the field of molecular glasses (Voigtmann and Poon 2006).

References

Adam, G. and Gibbs, J. H. 1965. On the temperature dependence of cooperative relaxation properties in glass-forming liquids. *J. Chem. Phys.* 43: 139–146.

Agarwal, P., Srivastava, S., and Archer, L. A. 2011. Thermal jamming of a colloidal glass. *Phys. Rev. Lett.* 107: 268302.

Alsayed, A. M. 2006. Melting in temperature sensitive suspensions. PhD thesis.

Angell, C. A. 1995. Formation of glasses from liquids and biopolymers. *Science* 267: 1924–1935.

Ballesta, P. et al. 2008. Slip and flow of hard-sphere colloidal glasses. *Phys. Rev. Lett.* 101: 258301.

Bayer, M. et al. 2007. Dynamic glass transition in two dimensions. *Phys. Rev. E* 76: 011508.

Besseling, R et al. 2010. Shear banding and flow-concentration coupling in colloidal glasses. *Phys. Rev. Lett.* 105: 268301.

Bodrova, A. S., Brilliantov, N. V., and Loskutov, A. Y. 2009. Brownian motion in granular gases of viscoelastic particles. *J Exp Theor. Phys.* 109: 946–953.

Brambilla, G., Masri, D. E., Pierno, M., Berthier, L., Cipelletti, L., Petekidis, G., and Schofield, A. B. 2009. Probing the equilibrium dynamics of colloidal hard spheres above the modecoupling glass transition. *Phys. Rev. Lett.* 102: 085703.

Brilliantov, B. V. and Pöschel, T. 2004. *Kinetic Theory of Granular Gases*. Oxford University Press, Oxford, U.K.

Brito, C., Dauchot, O., Biroli, G., and Bouchaud, J.-P. 2010. Elementary excitation modes in a granular glass above jamming. *Soft Matter* 6: 3013–3022.

Brongniart, A. K. 1827. Memoire sur la génération et le développment de l'embryon dans les végétaux phanérogames. *Annales. Sci. Naturelles* 12: 41.

Brown, R. 1828. A brief account of microscopical observations made on the particles contained in the pollen of plants. *Phil. Trans* 4: 171–173.

Casalini, R., Capaccioli, S., Lucchesi, M., Rolla, P. A., Paluch, M., Corezzi, S., and Fioretto, D. 2001. Effect of pressure on the dynamics of glass formers. *Phys. Rev. E.* 64: 041504.

Chen, D. et al. 2010a. Microscopic structural relaxation in a sheared supercooled colloidal liquid *Phys. Rev. E* 81: 011403.

Chen, K. et al. 2010b. Low-frequency vibrations of soft colloidal glasses. *Phys. Rev. Lett.* 105: 025501.

Chen, K. et al. 2011. Measurements of correlations between low-frequency vibrational modes and particle rearrangements in quasi-two-dimensional colloidal glasses. *Phys. Rev. Lett.* 107: 108301.

Cheng, Z., Zhu, J., Chaikin, P. M., Phan, S. E., and Russel, W. B. 2002. Nature of the divergence in low shear viscosity of colloidal hard-sphere dispersions. *Phys. Rev. E* 65: 041405.

Chikkadi, V et al. 2011. Long-range strain correlations in sheared colloidal glasses. *Phys. Rev. Lett.* 107: 198303.

Cianci, G. C., Courtland, R. E., and Weeks, E. R. 2006. Correlations of structure and dynamics in an aging colloidal glass. *Solid State Comm.* 139: 599–604.

D'Anna, G. and Gremaud, G. 2001. The jamming route to the glass state in weakly perturbed granular media. *Nature* 413: 407.

Dauchot, O., Marty, G., and Biroli, G. 2005. Dynamical heterogeneity close to the jamming transition in a sheared granular material. *Phys. Rev. Lett.* 95: 265701.

Dawson, K. et al. 2000. Higher-order glass-transition singularities in colloidal systems with attractive interactions. *Phys. Rev. E* 63: 011401-1.

Derjaguin, B. V. 1934. Untersuchungen über die Reibung und Adhäsion, IV Theorie des Anhaftens kleiner Teilchen. *Kolloid Zeits* 69: 155–164.

Derjaguin, B. V. and Landau, L. 1941. Theory of the stability of strongly charged lyophobic sols and of the adhesion of strongly charged particles in solutions of electrolytes. *Acta. Phys. Chem. URSS* 14: 633.

Desmond, K. and Franklin, S. V. 2006. Jamming of three-dimensional prolate granular materials. *Phys. Rev. E* 73: 031306.

Dhont, J. K. G. 1999. A constitutive relation describing the shear-banding transition. *Phys. Rev. E* 60: 4534–4544.

Dhont, J. K. G et al. 2003. Shear-banding and microstructure of colloids in shear flow. *Faraday Discuss.* 123: 157–172.

Dhont, J. K. G. and Briels, W. 2008. Gradient and vorticity banding. *Rheol. Acta* 47: 257–281.

Dinsmore, A. D., Weeks, E. R., Prasad, V., Levitt, A. C., and Weitz, D. A. 2001. Three-dimensional confocal microscopy of colloids. *Appl. Opt.* 40: 4152–4159.

Donev, A. et al. 2004. Improving the density of jammed disordered packings using ellipsoids. *Science* 303: 990.

Doolittle, A. K. 1951. Studies in Newtonian flow. II. The dependence of the viscosity of liquids on free-space. *J. App. Phys.* 22: 1471–1475.

Eckert, T. and Bartsch, E. 2002. Re-entrant glass transition in a colloid-polymer mixture with depletion attractions. *Phys. Rev. Lett.* 89: 125701.

Einstein, A. 1905. Über die von der molekularkinetischen Theorie der Wärme geforderte Bewegung von in ruhenden Flüssigkeiten suspendierten Teilchen. *Ann. Phys.* 17: 549–560.

El Masri, D., Brambilla, G., Pierno, M., Petekidis, G., Schofield, A. B., Berthier, L., and Cipelletti, L. 2009. Dynamic light scattering measurements in the activated regime of dense colloidal hard spheres. *J. Stat. Mech.* 2009: P07015.

Erpenbeck J. 1984. Shear viscosity of the hardsphere fluid via nonequilibrium molecular dynamics. *Phys. Rev. Lett.* 52: 1333–1335.

Espíndola, D., Galaz, B., and Melo, F. 2012. Ultrasound induces aging in granular materials. *Phys. Rev. Lett.* 109: 158301.

Feitosa, K. and Menon, N. 2002. Breakdown of energy equipartition in a 2D binary vibrated granular gas. *Phys. Rev. Lett.* 88: 198301.

Fielding, S. M., Cates, M. E., and Sollich, P. 2009. Shear banding, aging and noise dynamics in soft glassy materials. *Soft Matter* 5: 2378–2382.

Foffi, G., Zaccarelli, E., Buldyrev, S., Sciortino, F., and Tartaglia, P. 2004. Aging in short-ranged attractive colloids: A numerical study. *J. Chem. Phys.* 120: 8824–8830.

Franosch, T. et al. 1997. Theory for the reorientational dynamics in glass-forming liquids. *Phys. Rev. E* 56: 5659–5674.

Garboczi, E. J. et al. 1995. Geometrical percolation threshold of overlapping ellipsoids. *Phys. Rev. E* 52: 819–828.

Gasser, U., Weeks, E. R., Schofield, A., Pusey, P. N., and Weitz, D. A. 2001. Real space imaging of nucleation and growth in colloidal crystallization. *Science* 292: 258.

Geissler, P. L. and Reichman, D. R. 2005. Short-ranged attractions in jammed liquids: How cooling can melt a glass. *Phys. Rev. E* 71: 031206.

Ghosh, A. et al. 2010. Density of states of colloidal glasses. *Phys. Rev. Lett.* 104: 248305.

Golick, L. A. and Daniels K. E. 2009. Mixing and segregation rates in sheared granular materials. *Phys. Rev. E* 80: 042301.

Gouy, L. 1888. Note sur le mouvment brownien. *J. Phys. Serie II* 7: 561.

Götze, W. and Sjögren, L. 1992. Relaxation processes in supercooled liquids. *Rep. Progr. Phys.* 55: 241.

Griffiths, D. J. 1999. *Introduction to Electrodynamics*. Prentice-Hall, Inc, Upper Saddle River, NJ, p. 07458.

Habdas, P. 2012. Unpublished data.

Hajnal, D., Brader, J., and Schilling, R. 2009. Effect of mixing and spatial dimension on the glass transition. *Phys. Rev. E* 80: 021503.

Hajnal, D., Oettel, M., and Schilling, R. 2011. Glass transition of binary mixtures of dipolar particles in two dimensions. *J. Non-Cryst. Solids* 357: 302.

Hamley, I. W. 2000. *Introduction to Soft Matter*. John Wiley & Sons, Ltd.

Hartley, R. R. and Behringer, R. P. 2003. Logarithmic rate dependence of force networks in sheared granular materials. *Nature* 421: 928–931.

Haw, M. D. 2002. Colloidal suspensions, Brownian motion, molecular reality: A short history. *J. Phys.: Condens. Matter* 14: 7769–7779.

Hidalgo, R. C., Zuriguel, I., Maza, D., and Pagonabarraga, I. 2009. Role of particle shape on the stress propagation in granular packings. *Phys. Rev. Lett.* 103: 118001.

Hong, L. et al. 2008. Pressure and density dependence of the boson peak in polymers. *Phys. Rev. B* 78: 13201.

Hu, N. and Molinari, J. F. 2004. Shear bands in dense metallic granular materials. *J Mech. Phys Solids* 52: 499–531.

Hunter, G. L. and Weeks, E. R. 2012. The physics of the colloidal glass transition. *Rep. Prog. Phys.* 75: 066501.

Jones, R. A. L. 2002. *Soft Condensed Matter*. Oxford University Press, Oxford, U.K.

Joshi, Y. M. and Reddy, G. R. K. 2008. Aging in a colloidal glass in creep flow: Time-stress superposition. *Phys. Rev. E* 77: 021501.

Josserand, C., Tkachenko, A. V., Mueth, D. M., and Jaeger, H. M. 2000. Memory effects in granular materials. *Phys. Rev. Lett.* 85: 3632–3635.

Jullien, R. and Meakin, P. 1990. A mechanism for particle size segregation in three dimensions. *Nature* 344: 425–427.

Jullien, R. and Meakin, P. 1992. Three-dimensional model for particlesize segregation by shaking. *Phys. Rev. Lett.* 69: 640–643.

Kaufman, L. J. and Weitz, D. A. 2006. Direct imaging of repulsive and attractive colloidal glasses. *J. Chem. Phys.* 125: 074716–1.

Kegel, W. K. and van Blaaderen, A. 2000. Direct observation of dynamical heterogeneities in colloidal hard-sphere suspensions. *Science* 287: 290–293.

Knight, J. B., Fandrich, C. G., Lau, C. N., Jaeger, H. M., and Nagel, S. R. 1995. Density relaxation in a vibrated granular material. *Phys. Rev. E* 51: 3957–3963.

Kob, W. and Andersen H. C. 1995a. Testing mode-coupling theory for a supercooled binary Lennard-Jones mixture I: The van Hove correlation function. *Phys. Rev. E* 51: 4626–4641.

Kob, W. and Andersen H. C. 1995b. Testing mode-coupling theory for a supercooled binary Lennard-Jones mixture. II. Intermediate scattering function and dynamic susceptibility. *Phys. Rev. E* 52: 4134–4153.

Kob, W. and Barrat, J.-L. 1997. Aging effects in a Lennard-Jones glass. *Phys. Rev. Lett.* 78: 4581.

Kob, W., Barrat, J. L., Sciortino, F., and Tartaglia, P. 2000a. Aging in a simple glass former. *J. Phys.: Condens. Matter* 12: 6385–6394.

Kob, W., Sciortino, F., and Tartaglia, P. 2000b. Aging as dynamics in configuration space. *Europhys. Lett.* 49: 590–5966.

Koppel, D. E. 1972. Analysis of macromolecular polydispersity in intensity correlation spectroscopy: The method of cumulants. *J. Chem. Phys.* 57: 4814.

König, H. et al. 2005. Experimental realization of a model glass former in 2D. *Eur. Phys. J. E* 18: 287–293.

Langevin, P. 1908. Sur la théorie du mouvement brownien. *C. R. Acad. Sci., Paris* 146: 530–533.

Larson, G. L. 1999. *The Structure and Rheology of Complex Fluids*. Oxford University Press, Oxford, U.K.

Latka, A. et al. 2009. Particle dynamics in colloidal suspensions above and below the glass-liquid re-entrance transition. *Europhys. Lett.* 86: 58001.

Lee, H. et al. 2009. Direct measurement of molecular mobility in actively deformed polymer glasses. *Science* 323: 231–234.

Letz, M., Schilling, R., and Latz, A. 2000. Ideal glass transitions for hard ellipsoids. *Phys. Rev. E* 62: 5173–5178.

Liu, A. J. and Nagel, S. R. 1998. Nonlinear dynamics: Jamming is not just cool any more. *Nature* 396: 21–22.

Liu, A. J. and Nagel, S. R. 2010. The jamming transition and the marginally jammed solid. *Annu. Rev. Condens. Matter Phys.* 1: 347–369.

Lutsko, J. 1996. Molecular chaos, pair correlations, and shearinduced ordering of hard spheres. *Phys. Rev. Lett.* 77: 2225.

Lynch, J. M., Cianci, G. C., and Weeks, E. R. 2008. Dynamics and structure of an aging binary colloidal glass. *Phys. Rev. E* 78: 031410.

Manneville, S. 2008. Recent experimental probes of shear banding. *Rheol. Acta* 47: 301–318.

Marshall, L. and Zukoski, C. F. 1990. Experimental studies on the rheology of hard sphere suspensions near the glass transition. *J. Phys. Chem.* 94:1164–1171.

Mason, T. G. and Weitz, D. A. 1995. Linear viscoelasticity of colloidal hard sphere suspensions near the glass transition. *Phys. Rev. Lett.* 75: 2770–2773.

Mau, S. C. and Huse, D. A. 1999. Stacking entropy of hard-sphere crystals. *Phys. Rev. E* 59: 4396–4401.

Mazoyer, S. et al. 2009. Dynamics of particles and cages in an experimental 2D glass former. *Europhys. Lett.* 88: 66004.

van Megen, W., Mortensen, T. C., Williams, S. R., and Muller, J. 1998. Measurement of the self-intermediate scattering function of suspensions of hard spherical particles near the glass transition. *Phys. Rev. E* 58: 6073–6085.

van Megen, W. and Underwood, S. M. 1993a. Dynamic-light-scattering study of glasses of hard colloidal spheres. *Phys. Rev. E* 47, 248–261.

van Megen, W. and Underwood, S. M. 1993b. Glass transition in colloidal hard sphere: Mode-coupling theory analysis. *Phys. Rev. Lett.* 70: 2766–2769.

Mohraz, A. et al. 2004. Effect of monomer geometry on the fractal structure of colloidal rod aggregates. *Phys. Rev. Lett.* 92: 155503.

Mohraz, A. and Solomon, M. J. 2005. Direct visualization of colloidal rod assembly by confocal microscopy. *Langmuir* 21: 5298–5306.

Moller, P. C. F. et al. 2008. Shear banding and yield stress in soft glassy materials. *Phys. Rev. E* 77: 041507.

Mueth, D. et al. 2000. Signatures of granular microstructure in dense shear flows. *Nature* 406: 385–389.

Nakroshis, P., Amoroso, M., Legere, J., and Smith, C. 2003. Measuring Boltzmann's constant using video microscopy of Brownian motion. *Am. J Phys.* 71: 568.

O'Hern, C. S., Silbert, L. E., and Liu, A. J. 2003. Jamming at zero temperature and zero applied stress: The epitome of disorder. *Phys. Rev. E* 68: 011306.

Ottino, J. M. and Khakhar, D. V. 2000. Mixing and segregation of granular materials. *Annu. Rev. Fluid Mech.* 32: 55–91.

Ovarlez, G. et al 2009. Phenomenology and physical origin of shear-localization and shear-banding in complex fluids. *Rheol. Acta* 48: 831–844.

Panaitescu, A., Reddy, K. A., and Kudrolli, A. 2012. Nucleation and crystal growth in sheared granular sphere packings. *Phys. Rev. Lett.* 108: 108001–1008005. http://prl.aps.org/abstract/PRL/v108/i10/e108001

Perrin, J. 1908. La loi de Stokes et le mouvement brownien. *Comptes Rendus Acad. d. Sci, Paris* 147: 475.

Pham, K. N. et al. 2002. Multiple glassy states in a simple model system. *Science* 296: 104–106.

Pham, K. N. et al. 2008. Yielding behavior of repulsion-and attraction-dominated colloidal glasses. *J. Rheol.* 52: 649–676.

Polashenski, W., Zamankhan, P., Makiharju, S., and Zamankhan, P. 2002. Fine structures in sheared granular flows. *Phys. Rev. E* 66: 021303.

Poon, W. C. K., Pirie, A. D., Haw, M. D., and Pusey, P. N. 1997. Non-equilibrium behaviour of colloid-polymer mixtures. *Phys. A* 235: 110–119.

Poon, W. C. K., Weeks, E. R., and Royall, C. P. 2012. On measuring colloidal volume fractions. *Soft Matter* 8: 21–30.

Pusey, P. N. and van Megen, W. 1986. Phase behaviour of concentrated suspensions of nearly hard colloidal spheres. *Nature* 320: 340–342.

Rogers, R. B. and Lagerlöf, K. P. D. 2008. Crystallography of ordered colloids using optical microscopy. 1. Parallel-beam technique. *Appl. Opt.* 47: 284–295.

Rosato, A., Strandburg, K. J., Prinz, F., and Swendsen, R. H. 1987. Why the Brazil nuts are on top: Size segregation of particulate matter by shaking. *Phys. Rev. Lett.* 58: 1038–1040.

Russel, W. B., Saville, D. A., and Schowalter W. R. 1989. *Colloidal Dispersions.* Cambridge University Press, Cambridge, U.K.

Santos, A. and Dufty, J. W. 2001. Nonequilibrium phase transition for a heavy particle in a granular fluid. *Phys. Rev. E* 64: 051305.

Schall, P. and van Hecke, M. 2010. Shear bands in matter with granularity. *Annu. Rev. Fluid Mech.* 42: 67–88.

Schilling, R. and Scheidsteger, T. 1997. Mode coupling approach to the ideal glass transition of molecular liquids: Linear molecules. *Phys. Rev. E* 56: 2932–2949.

Sciortino, F. 2002. One liquid, two glasses. *Nat. Mater.* 1: 145–146.

Shahin, A. and Joshi, Y. M. 2011. Prediction of long and short time rheological Behavior in soft glassy materials. *Phys. Rev. Lett.* 106: 038302.

Siemens, A. O. N. and van Hecke, M. 2010. Jamming: A simple introduction. *Phys. A* 389: 4255–4264.

Sierou, A. and Brady, J. 2002. Rheology and microstructure in concentrated non-colloidal suspensions, *J. Rheol.* 46: 1031.

Simeonova, N. B. et al. 2006. Devitrification of colloidal glasses in real space. *Phys. Rev. E* 73: 041401.

von Smoluchowski, M. 1906. Zur kinetischen Theorie der Brownschen Molekular-bewegung und der Suspensionen. *Ann. Phys.* 21: 756.

Speedy, R. J. 1999. Glass transition in hard disc mixtures. *Chem. Phys.* 110: 4559–4565.

Trappe, V. and Weitz, D. A. 2000. Scaling of the viscoelasticity of weakly attractive particles. *Phys. Rev. Lett.* 85: 449–452.

Trappe V. et al. 2001. Jamming phase diagram for attractive particles. *Nature* 411: 772–775.

Tsai, J.-C., Voth, G. A., and Gollub, J. P. 2003. Internal granular dynamics, shearinduced crystallization, and compaction steps. *Phys. Rev. Lett.* 91: 064301.

Verwey, E. J. W. and Overbeek, J. T. G. 1948. *Theory of the Stability of Lyophobic Colloids*. Elsevier, Amsterdam, the Netherlands.

Voigtmann, T. and Poon, W. C. K. 2006. Glasses under high pressure: A link to colloidal science? *J. Phys.: Condens. Matter* 18: L465–L469.

Vollmayr-Lee, K. 2004. Single particle jumps in a binary Lennard-Jones system below the glass transition. *J. Chem. Phys.* 121: 4781.

Weeks, E. R. 2007. *Soft Jammed Materials: Statistical Physics of Complex Fluids*, pp. 2-1–2-87, Tohoku University Press, Sendai, Japan.

Weeks, E. R. 2010. Microscopy of soft materials. Book chapter in *Experimental and Computational Methods in Soft Condensed Matter*, ed. J. S. Olafsen, Cambridge University Press, Cambridge, U.K.

Weeks, E. R. 2011. Two for one in a colloidal glass. *Physics* 4: 61.

Weeks, E. R. and Weitz, D. A. 2002. Properties of cage rearrangements observed near the colloidal glass transition. *Phys. Rev. Lett.* 89: 095704.

Weeks, E. R. et al. 2000. Three-dimensional direct imaging of structural relaxation near the colloidal glass transition. *Science* 287: 627–631.

Weitz, D. A. 2011. *Glasses and Grains*, pp. 25–39. Springer Basel AG Poincaré Seminar.

Yamamoto, R. and Onuki, A. 1998. Dynamics of highly supercooled liquids: Heterogeneity, rheology, and diffusion. *Phys. Rev. E* 58: 3515–3529.

Yi, Y. B. and Sastry, A. M. 2004. Analytical approximation of the percolation threshold for overlapping ellipsoids of revolution. *Proc. R. Soc. Lond., Ser. A* 460: 2353–2380.

Yunker, P. et al. 2009. Irreversible rearrangements, correlated domains, and local structure in aging glasses. *Phys. Rev. Lett.* 103: 115701–115704.

Zaccarelli, E. et al. 2003. Activated bond-breaking processes preempt the observation of a sharp glass-glass transition in dense short-ranged attractive colloids. *Phys. Rev. Lett.* 91: 108301.

Zaccarelli, E. and Poon, W. C. K. 2009. Colloidal glasses and gels: The interplay of bonding and caging. *PNAS* 106: 15203–15208.

Zausch, J. et al 2008. From equilibrium to steady state: The transient dynamics of colloidal liquids under shear. *J. Phys.: Condens. Matter* 20: 404210.

Zhang, Z. et al. 2009. Thermal vestige of the zero-temperature jamming transition, *Nature* 459: 230.

Zhang, Z. et al. 2011. Cooperative rearrangement regions and dynamical heterogeneities in colloidal glasses with attractive versus repulsive interactions. *Phys. Rev. Lett.* 107, 208303.

Zheng, Z., Wang, F., and Han Y. 2011. Glass transitions in quasi-two-dimensional suspensions of colloidal ellipsoids. *Phys. Rev. Lett.* 107: 065702.

Index

Printed and bound by CPI Group (UK) Ltd, Croydon, CR0 4YY

22/10/2024

01777614-0015